MULTIPLE PARTON INTERACTIONS AT THE LHC

ADVANCED SERIES ON DIRECTIONS IN HIGH ENERGY PHYSICS

ISSN: 1793-1339

This is the best review series in high energy physics today. It comprehensively reviews the most important developments in each sector of high energy physics and is of lasting use to all researchers. All volumes are edited by eminent physicists — researchers who have themselves made substantial contributions to their respective fields of research.

Published

Vol. 29 – *Multiple Parton Interactions at the LHC* (eds. Paolo Bartalini and Jonathan Richard Gaunt)

Vol. 28 – *The State of the Art of Neutrino Physics: A Tutorial for Graduate Students and Young Researchers* (ed. A. Ereditato)

Vol. 27 – *Technology Meets Research 60 Years of CERN Technology: Selected Highlights* (eds. C. Fabjan, T. Taylor, D. Treille and H. Wenninger)

Vol. 26 – *The Standard Theory of Particle Physics: Essays to Celebrate CERN's 60th Anniversary* (eds. L. Maiani and L. Rolandi)

Vol. 25 – *Massive Neutrinos: Flavor Mixing of Leptons and Neutrino Oscillations* (ed. H. Fritzsch)

Vol. 24 – *The High Luminosity Large Hadron Collider: The New Machine for Illuminating the Mysteries of Universe* (eds. O. Brüning and L. Rossi)

Vol. 23 – *60 Years of CERN Experiments and Discoveries* (eds. H. Schopper and L. Di Lella)

Vol. 22 – *Perspectives on String Phenomenology* (eds. B. Acharya, G. L. Kane and P. Kumar)

Vol. 21 – *Perspectives on Supersymmetry II* (ed. G. L. Kane)

Vol. 20 – *Lepton Dipole Moments* (eds. B. Lee Roberts and William J. Marciano)

Vol. 19 – *Linear Collider Physics in the New Millennium* (eds. K. Fujii, D. J. Miller and A. Soni)

Vol. 18 – *Perspectives on Supersymmetry* (ed. G. L. Kane)

Vol. 17 – *Perspectives on Higgs Physics II* (ed. G. L. Kane)

Vol. 16 – *Electroweak Symmetry Breaking and New Physics at the TeV Scale* (eds. T. L. Barklow, S. Dawson, H. E. Haber and J. L. Siegrist)

Vol. 15 – *Heavy Flavours II* (eds. A. J. Buras and M. Lindner)

Vol. 14 – *Precision Tests of the Standard Electroweak Model* (ed. P. Langacker)

Vol. 13 – *Perspectives on Higgs Physics* (ed. G. L. Kane)

Vol. 12 – *Advances of Accelerator Physics and Technologies* (ed. H. Schopper)

Vol. 11 – *Quantum Fields on the Computer* (ed. M. Creutz)

Vol. 10 – *Heavy Flavours* (eds. A. J. Buras and M. Lindner)

Vol. 9 – *Instrumentation in High Energy Physics* (ed. F. Sauli)

The complete list of titles in the series can be found at
http://www.worldscientific.com/series/asdhep

Advanced Series on
Directions in High Energy Physics — Vol. 29

MULTIPLE PARTON INTERACTIONS AT THE LHC

Editors

Paolo Bartalini
Central China Normal University, China

Jonathan Richard Gaunt
CERN, Switzerland

NEW JERSEY • LONDON • SINGAPORE • BEIJING • SHANGHAI • HONG KONG • TAIPEI • CHENNAI • TOKYO

Published by

World Scientific Publishing Co. Pte. Ltd.
5 Toh Tuck Link, Singapore 596224
USA office: 27 Warren Street, Suite 401-402, Hackensack, NJ 07601
UK office: 57 Shelton Street, Covent Garden, London WC2H 9HE

Library of Congress Cataloging-in-Publication Data
Names: Bartalini, Paolo (Professor of particle physics), editor. | Gaunt, Jonathan Richard, editor.
Title: Multiple parton interactions at the LHC / editors,
Paolo Bartalini (Central China Normal University, China),
Jonathan Richard Gaunt (CERN, Switzerland).
Other titles: Advanced series on directions in high energy physics ; v. 29.
Description: Singapore ; Hackensack, NJ : World Scientific, [2018] |
Series: Advanced series on directions in high energy physics, ISSN 1793-1339 ; vol. 29 |
Includes bibliographical references and index.
Identifiers: LCCN 2018002458| ISBN 9789813227750 (hardcover ; alk. paper) |
ISBN 9813227753 (hardcover ; alk. paper)
Subjects: LCSH: Hadron interactions. | Partons. | Scattering (Physics)
Classification: LCC QC793.5.H328 M85 2018 | DDC 539.7/216--dc23
LC record available at https://lccn.loc.gov/2018002458

British Library Cataloguing-in-Publication Data
A catalogue record for this book is available from the British Library.

For any available supplementary material, please visit
https://www.worldscientific.com/worldscibooks/10.1142/10646#t=suppl

Desk Editor: Ng Kah Fee

Typeset by Stallion Press
Email: enquiries@stallionpress.com

Printed in Singapore

Contents

Chapter 1

Introduction

P. Bartalini* and J. R. Gaunt†

**Central China Normal University (CCNU)*
Luoyu Road, 152, Wuhan, Hubei 430079, China
†CERN Theory Division
CH-1211 Geneva 23, Switzerland

This book represents a first attempt to provide a comprehensive reference on multiple parton interactions (MPI), a concept that has a deep impact on a wide range of research lines in high energy particle and nuclear physics, in particular at hadron and ion colliders.

Hadrons, and ions that are assembled from them, are not fundamental objects but are composed from multiple quarks and gluons (collectively, "partons") held together via the strong force. The composite nature of hadrons and ions means that in a generic hadron–hadron collision (also known as "minimum bias" or MB collision), multiple parton–parton collisions are an inevitability, where such scatters will typically be soft in nature. This will of course remain true when we select events with a hard interaction, as is done in many physics analyses that seek to test the Standard Model to high precision, or seek new physics beyond this model — the hard interaction will be accompanied by an "underlying event" (UE) of additional soft interactions. Understanding and accurately modeling the additional soft scatters is vital to ensure proper interpretation of such hard events at hadronic colliders. Due to the predominantly non-perturbative nature of the MPI, their typically large number and the potentially complex interplay between them, these additional scatters may at present only be modeled in an approximate way in the context of Monte Carlo (MC) event generators, where these generators provide fully-exclusive final states (events) that may be compared directly with the experimental results. The MC event generators have various model parameters (some of which are related to the MPI models, and others being related to other physics aspects), and these have to be fitted to the experimental data in a complex exercise known as tuning.

In principle, a much more clean environment to study MPI physics is in events with two (or more) separate hard processes in an individual hadron–hadron collision — this is known as double parton scattering [DPS] (or triple parton scattering [TPS], etc.). Here, the hard nature of the scatters (and their limited number) ensures that this process is much more amenable to first-principle theoretical approaches to its description, with the potential to develop and prove factorization formulae as have already been extensively used for single parton scattering processes. Such events with multiple hard scatters are rare, but are becoming more common as the energy of the hadronic colliders we use increases, and can form backgrounds to certain single scattering processes that are suppressed by small or multiple coupling constants. Insight gained during the study of hard MPI can potentially be used to improve the modeling of the soft MPI in MB/UE — this principle is already at work in several of the existing MC models, where the soft MPI model is an extension of the one used for harder MPI.

MPI is also of interest from the perspective of understanding hadron/ion dynamics and structure. In particular, it reveals information on the correlation between partons in the proton, information inaccessible in other processes. One topic that is of particular recent interest is the final state in high multiplicity (HM) proton–proton and proton–nucleus collisions, which appears to exhibit collective-like patterns typical of larger systems in the experimental measurements. This indicates that the dynamics of multi-particle production in such collisions may be different than that in lower multiplicity collisions.

In this book we have collected together a set of review chapters summarizing the state of our knowledge on the physics and modeling of MPI, and the latest experimental measurements. These have been contributed by scientists who are leading the effort in their research line. Given the deep interplay between the various MPI-related research areas, there were various possibilities for how we could have organized the chapters of this book. We chose to adopt a strategy in which the monograph is subdivided into two parts: the first one is dedicated to the theory and experimental measurements of the hard MPI (i.e., DPS, TPS,...) and the second one focuses on the modeling of soft MPI and their measurement in the context of the underlying event as well as in minimum bias and high multiplicity collisions. As emphasized above, however, these are not distinct research areas and important knowledge about one can be gained by studying the other. Across all the books, the emphasis is on the physics at the current Large Hadron Collider (LHC) experiment.

Chapter 2 opens the first part of this book with a state-of-the-art review of the DPS theory, discussing the progress towards establishing a rigorous factorization formula for this process, as well as the interplay between DPS and other reaction mechanisms. Chapter 3 focuses on the link between inclusive and exclusive cross-sections of general hard MPI processes and on the prescriptions for the corresponding experimental tests. A systematic review of the variety of possible correlations between partons that can affect DPS cross-sections is presented in Chapter 4, along with a review of progress in modeling these correlations, and a discussion of prospects of measuring and/or constraining such correlations during the high-luminosity phase of the LHC. In Chapter 5 the theory of DPS is discussed further, and a model is developed for the DPS parton densities at low scale by linking DPS with elastic and inelastic diffraction. The discussion in this chapter also extends to the interplay with the UE, anticipating concepts that are addressed in further details in the second part of the book. Chapters 6–8 provide a state-of-the-art review on experimental analyses and measurements of DPS, focusing on the present and future measurements at the LHC and considering the final states with jets, heavy vector bosons and heavy flavors, respectively. The discussion of the feasibility of DPS measurements is extended to proton–nucleus (pA) and nucleus–nucleus (AA) collisions in Chapter 9. This chapter also discusses triple and n-parton scatterings in these contexts.

The second part of the book opens with Chapter 10, an account of the very first Monte Carlo MPI model developed in response to data coming from the S$p\bar{p}$S collider in the eighties, and its development all the way up to its modern incarnation in the widely-used Pythia event generator. After this discussion of MPI MC modeling, the rich phenomenology of soft MPI is discussed in different steps: Chapter 11 reviews the UE measurements, and Chapter 12 addresses the Minimum Bias measurements. Chapter 13 provides a bottom line to the first three chapters of the second part, discussing the modern techniques of MC parameter optimization against experimental data, or tuning, and reviewing the state-of-the-art tunes. The relation between soft MPI, diffractive processes and saturation effects is discussed in Chapter 14, where the relevant phenomenology is also presented. Chapter 15 reports the striking experimental results on the high multiplicity final states of pp and pA interactions, indicating collective-like patterns typical of larger systems. The subsequent chapters, in different respects, address the interpretation of such phenomenology: in Chapter 16, reviewing the event shapes measurements at the LHC, the usage of these observables to track events with significant jet contributions in the high multiplicity events is investigated. Chapter 17 presents DIPSY, a Monte Carlo based on a BFKL-inspired initial-state dipole evolution model with a specific treatment of high string densities occurring in events with several MPI. In Chapter 18 the high multiplicity events in pp collisions are discussed in the framework of a model combining a Color Glass Condensate (CGC) effective theory of multiple particle production with a state-of-the-art hadronization model based on the Lund-string fragmentation. The last chapter of the book presents EPOS, a Monte Carlo relying on the Parton Based Gribov Regge Theory which adopts a collective evolution of matter in the secondary scattering stage of all reactions, from pp to AA.

Part I

Hard MPI: The Double Parton Scattering (DPS)

Chapter 2

Double Parton Scattering Theory Overview

Markus Diehl* and Jonathan R. Gaunt†

**Deutsches Elektronen-Synchroton DESY,*
Notkestraße 85, 22607 Hamburg, Germany
†CERN Theory Division,
1211 Geneva 23, Switzerland

The dynamics of double hard scattering in proton–proton collisions is quite involved compared with the familiar case of single hard scattering. In this contribution, we review our theoretical understanding of double hard scattering and of its interplay with other reaction mechanisms.

1. Introduction

The most familiar mechanism for hard processes in proton–proton collisions is single parton scattering (SPS): two partons, one from each proton, undergo a hard scattering that produces heavy particles or particles with high transverse momenta. For the cross-section one then has a factorization formula, containing a parton distribution function (PDF) for each proton and a parton-level cross-section for the hard subprocess. Double parton scattering (DPS) occurs if in the same proton–proton collision two partons in each proton initiate two separate hard scattering processes. The corresponding factorization formula contains two parton-level cross-sections and a double parton distribution (DPD) for each proton. The two hard scatters are separated by a finite distance $\boldsymbol{y}$ in the plane transverse to the colliding proton momenta, so that a DPD depends not only on the momentum fractions x_1 and x_2 of two partons, but also on the transverse distance $\boldsymbol{y}$ between them. Very roughly, DPDs should grow like the square of two ordinary PDFs when x_1 and x_2 become small. The importance of DPS compared with SPS is hence increased in this small x region, which for a given final state becomes more and more important with growing collision energy.

The single and double parton distributions just described are integrated over transverse parton momenta; they are often called "collinear" distributions, and the associated formalism is called "collinear factorization". The information on transverse parton momenta is retained in so-called TMDs (transverse momentum-dependent distributions). The corresponding TMD factorization formulas allow one to compute cross-sections differential in the transverse momentum $\boldsymbol{q}$ of a heavy particle (e.g., a Z or a Higgs boson) in the region where $\boldsymbol{q}$ is much smaller than the boson mass. The TMD concept can be extended to DPS processes, for instance to describe the region of low transverse boson momenta $\boldsymbol{q}_1$ and $\boldsymbol{q}_2$ in W^+W^+ or HZ production. This is especially valuable because the importance of DPS compared with SPS is much higher in the cross-section for measured small $\boldsymbol{q}_1$ and $\boldsymbol{q}_2$ than it is in the integrated cross-section.

Factorization for SPS processes has been derived within QCD to a high level of rigor, as reviewed for instance in Ref. 1. It is an ongoing effort to bring factorization for DPS to a comparable standard. In the present contribution, we review the status of this effort. Note that we are discussing so-called *hard scattering factorization* here, which is based on separating dynamics at different distance scales. This is distinct from "high-energy" or "small x factorization", where the separation criterion is rapidity. Some discussion of this concept in the context of DPS is given in Chapter 17.

2. Cross-Section Formula

Let us start with a main theory result: the cross-section formula for DPS. Consider the production of two particles with invariant masses Q_1, Q_2 and transverse momenta $\boldsymbol{q}_1$, $\boldsymbol{q}_2$. We require that Q_1 and Q_2 be large and generically denote their size by Q. Instead of a heavy particle, one may also have a system of particles with large invariant mass, for instance a dijet.

Collinear factorization allows us to compute the cross-section integrated over $\boldsymbol{q}_1$ and $\boldsymbol{q}_2$:

$$\frac{d\sigma_{\text{DPS}}}{dx_1\, dx_2\, d\bar{x}_1\, d\bar{x}_2} = \frac{1}{C} \sum_{a_1 a_2 b_1 b_2} \int_{x_1}^{1-x_2} \frac{dx_1'}{x_1'} \int_{x_2}^{1-x_1'} \frac{dx_2'}{x_2'} \int_{\bar{x}_1}^{1-\bar{x}_2} \frac{d\bar{x}_1'}{\bar{x}_1'} \int_{\bar{x}_2}^{1-\bar{x}_1'} \frac{d\bar{x}_2'}{\bar{x}_2'}$$
$$\times \sum_R {}^{R}\hat{\sigma}_{a_1 b_1}(x_1' \bar{x}_1' s, \mu_1^2)\, {}^{R}\hat{\sigma}_{a_2 b_2}(x_2' \bar{x}_2' s, \mu_2^2)$$
$$\times \int d^2\boldsymbol{y}\, \Phi^2(y\nu)\, {}^{R}F_{b_1 b_2}(\bar{x}_i', \boldsymbol{y}; \mu_i, \bar{\zeta})\, {}^{R}F_{a_1 a_2}(x_i', \boldsymbol{y}; \mu_i, \zeta). \qquad (1)$$

Note that we use boldface for any vector $\boldsymbol{w}$ in the transverse plane and denote its length by $w = |\boldsymbol{w}|$. There are strong indications[2] that TMD factorization in SPS works only for the production of colorless particles, so that we make the same restriction for DPS. The differential cross-section for transverse momenta

$|\boldsymbol{q}_1|, |\boldsymbol{q}_2| \sim q_T$ much smaller than Q reads

$$\frac{d\sigma_{\text{DPS}}}{dx_1\, dx_2\, d\bar{x}_1\, d\bar{x}_2\, d^2\boldsymbol{q}_1\, d^2\boldsymbol{q}_2} = \frac{1}{C} \sum_{a_1 a_2 b_1 b_2} \hat{\sigma}_{a_1 b_1}(Q_1^2, \mu_1^2)\, \hat{\sigma}_{a_2 b_2}(Q_2^2, \mu_2^2)$$
$$\times \int d^2\boldsymbol{y}\, \frac{d^2\boldsymbol{z}_1}{(2\pi)^2}\, \frac{d^2\boldsymbol{z}_2}{(2\pi)^2}\, e^{-i(\boldsymbol{q}_1\boldsymbol{z}_1 + \boldsymbol{q}_2\boldsymbol{z}_2)}\, \Phi(y_+\nu)\, \Phi(y_-\nu)$$
$$\times \sum_R {}^R F_{b_1 b_2}(\bar{x}_i, \boldsymbol{z}_i, \boldsymbol{y}; \mu_i, \bar{\zeta})\, {}^R F_{a_1 a_2}(x_i, \boldsymbol{z}_i, \boldsymbol{y}; \mu_i, \zeta). \tag{2}$$

These formulas are quite complex. In the following we briefly explain their different ingredients, and the physics behind them.

We begin with the simplest ones. The variables x_i and $\bar{x}_i$ are given by

$$x_i = Q_i\, e^{Y_i}/\sqrt{s}, \quad \bar{x}_i = Q_i\, e^{-Y_i}/\sqrt{s} \quad (i = 1, 2), \tag{3}$$

where Y_i is the centre-of-mass rapidity of the system i and $\sqrt{s}$ the overall collision energy. C is a combinatorial factor, equal to 2 if the systems 1 and 2 are identical, and equal to 1 otherwise.

The parton-level cross-sections $\hat{\sigma}$ are precisely the same as the ones in the corresponding SPS cross-sections, except for the superscript R in Eq. (1), which will be explained below. They include the effects of hard QCD radiation in the process. In TMD factorization, $\hat{\sigma}$ receives only virtual corrections, since hard real radiation tends to knock $\boldsymbol{q}_i$ out of the region $q_T \ll Q_i$. As a consequence, the momentum fractions of the partons entering the hard subprocesses are fixed to x_i and $\bar{x}_i$ by external kinematics. In collinear factorization, $\hat{\sigma}$ includes real emission, which allows for momentum fractions $x_i' \geq x_i$ and $\bar{x}_i' \geq \bar{x}_i$.

The joint distribution of two partons in a proton is quantified by the DPDs F, which have two labels a_i for the parton type, two momentum fraction arguments x_i, and two factorization scales μ_i (they can be chosen separately, which is useful if Q_1 and Q_2 are of different size). In the TMD case there are two transverse position arguments $\boldsymbol{z}_i$, which are Fourier conjugate to the transverse parton momenta $\boldsymbol{k}_i$. The structure $\int d^2\boldsymbol{z}_i\; e^{-i\boldsymbol{q}_i\boldsymbol{z}_i}\, F(\bar{x}_i, \boldsymbol{z}_i, \ldots)\, F(x_i, \boldsymbol{z}_i, \ldots)$ in Eq. (2) is the same as in the corresponding factorization formula for SPS — in momentum space it corresponds to a convolution product $\int d^2\boldsymbol{k}_i\; F(\bar{x}_i, \boldsymbol{q}_i - \boldsymbol{k}_i, \ldots)\, F(x_i, \boldsymbol{k}_i, \ldots)$.

As already mentioned, a DPD also depends on the distance $\boldsymbol{y}$, which in collinear factorization literally corresponds to the transverse distance between the two active partons in the proton, and thus to the distance between the two hard-scattering processes. In the TMD case, $\boldsymbol{y}$ corresponds to the average distance between the partons in the scattering amplitude and its conjugate, as can be seen in Eqs. (4) and (5) below. Notice that in the cross-section, $\boldsymbol{y}$ is not Fourier conjugate to any observable momentum, unlike $\boldsymbol{z}_i$.

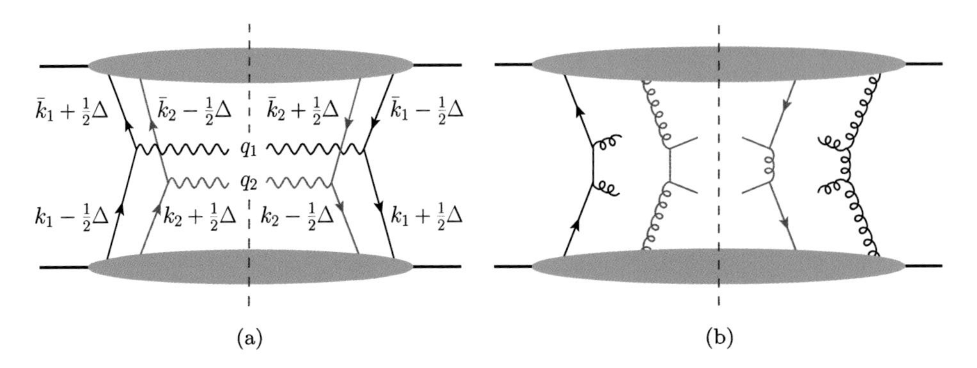

Fig. 1 (a) Tree-level graph for the production of two electroweak gauge bosons by DPS (often called double Drell–Yan). The blobs represent DPDs. The graph is for the cross-section, with the vertical line indicating the final state cut. (b) Graph for double dijet production with DPDs for quark–gluon interference.

It is instructive to see how the distance $\boldsymbol{y}$ emerges from the analysis of Feynman graphs in momentum space. The longitudinal momentum fractions of each parton are fixed by the final state kinematics and thus must be equal in the scattering amplitude and its conjugate. By contrast, the transverse parton momenta can differ by an amount $\boldsymbol{\Delta}$ or $-\boldsymbol{\Delta}$ as shown in Fig. 1(a). The momentum mismatch for the first and the second parton is opposite in sign, so that the transverse momentum of the spectator partons is the same in the amplitude and its conjugate. Since this momentum mismatch is not observable, one has an integral of the form $\int d^2\boldsymbol{\Delta}\; F(\bar{x}_i, -\boldsymbol{\Delta}, \ldots)\, F(x_i, \boldsymbol{\Delta}, \ldots)$ in the cross-section. A Fourier transform from $\boldsymbol{\Delta}$ to $\boldsymbol{y}$ gives the form shown in Eqs. (1) and (2). More detail on the tree-level derivation of the factorized structure in collinear factorization is given in Chapter 3.

Let us now turn to the quantum numbers of the partons. Even in an unpolarized proton, two extracted partons can have correlations between their polarizations. The labels a_i, b_i in the cross-section formulas refer not only to the type of the parton but also to its polarization, and one must sum over all allowed combinations. An example for a polarization-dependent DPD is $F_{\Delta q \Delta q}$, which corresponds to the difference of distributions for two quarks with equal helicities and for two quarks with opposite helicities.

Not only the transverse momentum of a parton can differ between the amplitude and its conjugate, but also its color. The different possible color combinations in DPDs and the parton-level cross-sections are specified by the label R. For the production of colorless particles there is only one possible color structure for $\hat{\sigma}$, which hence requires no index R in the TMD formula (2). Let us explain the meaning of R for the simplified setting of the tree graph in Fig. 1(a). In each DPD, one can couple the two parton lines with momentum fraction x_1 ($\bar{x}_1$) to be in the color representation $R = 1, 8, \ldots$ (the two other lines then are in the conjugate representation, because all four lines must couple to an overall singlet). In color singlet distributions ${}^1F$, partons with equal momentum fractions thus have equal color — this is the only possible combination for single parton distributions. Color non-singlet DPDs describe color correlations. At the end of Section 6 we will see that they are suppressed by Sudakov logarithms if the scale of the hard process is large.

Finally, there also exist DPDs describing the interference between different parton types in the amplitude and its conjugate, be it between different quark flavors, between quarks and antiquarks, or between quarks and gluons. For ease of notation, they are not included in the cross-section formulas (1) and (2). An example for quark–gluon interference in double dijet production is given in Fig. 1(b). Parton-type interference distributions do not have any dynamical cross talk with gluon DPDs, which have the strongest enhancement at small x_i. In many situations, one can therefore expect them to play only a minor role. A detailed discussion of correlations in DPDs can be found in Chapter 4.

DPDs can be defined via operator matrix elements, which provides a solid field theoretical basis for their investigation. For a double quark TMD one writes

$$\begin{aligned} {}^{R}F_{a_1 a_2}(x_i, \boldsymbol{z}_i, \boldsymbol{y}; \mu_i, \zeta) &= 2p^+ \int dy^- \, \frac{dz_1^-}{2\pi} \, \frac{dz_2^-}{2\pi} \, e^{i(x_1 z_1^- + x_2 z_2^-)p^+} \\ &\quad \times \langle p | \, \mathcal{O}_{a_2}(0, z_2) \, \mathcal{O}_{a_1}(y, z_1) \, | p \rangle \times \{\text{soft factor}\}, \end{aligned} \tag{4}$$

where we use light-cone coordinates $w^{\pm} = (w^0 \pm w^3)/\sqrt{2}$ for any four-vector w^{μ}. It is understood that $\boldsymbol{p} = \boldsymbol{0}$ and that the proton spin is averaged over. The bilinear operators $\mathcal{O}$ are the same as in the definition of a single parton TMD. They are given by

$$\mathcal{O}_a(y, z) = \bar{q}\left(y - \frac{1}{2}z\right) W^{\dagger}\left(y - \frac{1}{2}z\right) \Gamma_a \, W\left(y + \frac{1}{2}z\right) q\left(y + \frac{1}{2}z\right)\bigg|_{z^+ = y^+ = 0} \tag{5}$$

with a past-pointing light-like Wilson line

$$W(\xi) = \mathrm{P}\exp\left[igt^a \int_0^{\infty} ds \; nA^a(\xi - sn) \right], \tag{6}$$

where P denotes path-ordering and n is a light-like vector ($n^- = 1$, $n^+ = 0$, $\boldsymbol{n} = \boldsymbol{0}$). The dynamical origin of this Wilson line is explained in Section 6. Γ_a is a Dirac matrix and determines the quark polarization. In particular, unpolarized quarks correspond to $\Gamma_q = \frac{1}{2}\gamma^+$, and longitudinal quark polarization is described by $\Gamma_{\Delta q} = \frac{1}{2}\gamma^+\gamma_5$.

The "soft factor" in Eq. (4) originates from soft gluon exchange in the physical scattering process and gives rise to the dependence on a parameter ζ, as explained in Section 6. Such a dependence is already present in single parton TMDs. Moreover, the operator (5) and the soft factor contain ultraviolet divergences, which require renormalization. This brings in the dependence on the renormalization scales μ_i.

Finally, the dependence of the DPD on R arises from the color indices of the operators $(\bar{q}W^{\dagger})_{i'}$ and $(Wq)_i$ in Eq. (5) and from the soft factor. Again, more detail is given in Section 6.

The preceding discussion can be repeated for antiquarks or gluons, with different operators $\mathcal{O}_a$. The definition of collinear DPDs $F_{a_1 a_2}(x_i, \boldsymbol{y}; \mu_i, \zeta)$ reads as in Eq. (4) but with $\boldsymbol{z}_i = \boldsymbol{0}$. Note that in the color non-singlet case, the soft factor and the dependence on ζ do not drop out in the collinear case. Putting $\boldsymbol{z}_i$ to zero introduces additional ultraviolet divergences, so that the renormalization and hence the μ_i dependence are quite different between TMDs and collinear distributions, as we will see later.

The role of the function Φ in Eqs. (1) and (2) will be explained in Section 4. It is closely related to the fact that the cross-section of a physical process receives not only contributions from DPS, but also from SPS and possibly other mechanisms. In Section 3, we give an overview of these.

3. Power Behavior

The factorization of cross-sections into perturbative hard-scattering subprocesses and non-perturbative quantities like parton distributions is based on an expansion in the small parameter Λ/Q. Here Q denotes the scale of the hard scattering and Λ a typical hadronic scale. For simplicity, we treat the size of the transverse momenta $\boldsymbol{q}_1$ and $\boldsymbol{q}_2$ in TMD factorization as order Λ here. The case where they are much larger than a hadronic scale (but still much smaller than Q) is discussed in Section 8.

Dimensional analysis of the TMD factorization formulas for SPS and DPS reveals that the two mechanisms have the same power behavior:

$$\frac{d\sigma_{\text{SPS}}}{d^2\boldsymbol{q}_1\, d^2\boldsymbol{q}_2} \sim \frac{d\sigma_{\text{DPS}}}{d^2\boldsymbol{q}_1\, d^2\boldsymbol{q}_2} \sim \frac{1}{\Lambda^2 Q^4}. \tag{7}$$

The situation changes if one integrates over $\boldsymbol{q}_1$ and $\boldsymbol{q}_2$. In DPS both are of order Λ since they originate from the transverse momenta of partons inside the colliding protons. In SPS this holds only for the sum $\boldsymbol{q}_1 + \boldsymbol{q}_2$, whilst the individual momenta $\boldsymbol{q}_1$ and $\boldsymbol{q}_2$ (and thus their difference) are only limited by the available phase space and can hence be of order Q. One thus obtains for the integrated cross-sections

$$\sigma_{\text{SPS}} \sim 1/Q^2, \quad \sigma_{\text{DPS}} \sim \Lambda^2/Q^4, \tag{8}$$

where DPS has become power suppressed because it populates a smaller phase space. However, DPS can still be important in this case, for instance if SPS is suppressed by coupling constants (the production of W^+W^+ or W^-W^- is a prominent example). Generically, DPS is enhanced if the momentum fractions x in the hard scattering subprocesses become small, as already noted in the introduction.

There are further mechanisms that contribute at the same power to the cross-section as the terms in Eq. (7) or Eq. (8), as shown in Ref. 3. In TMD factorization,

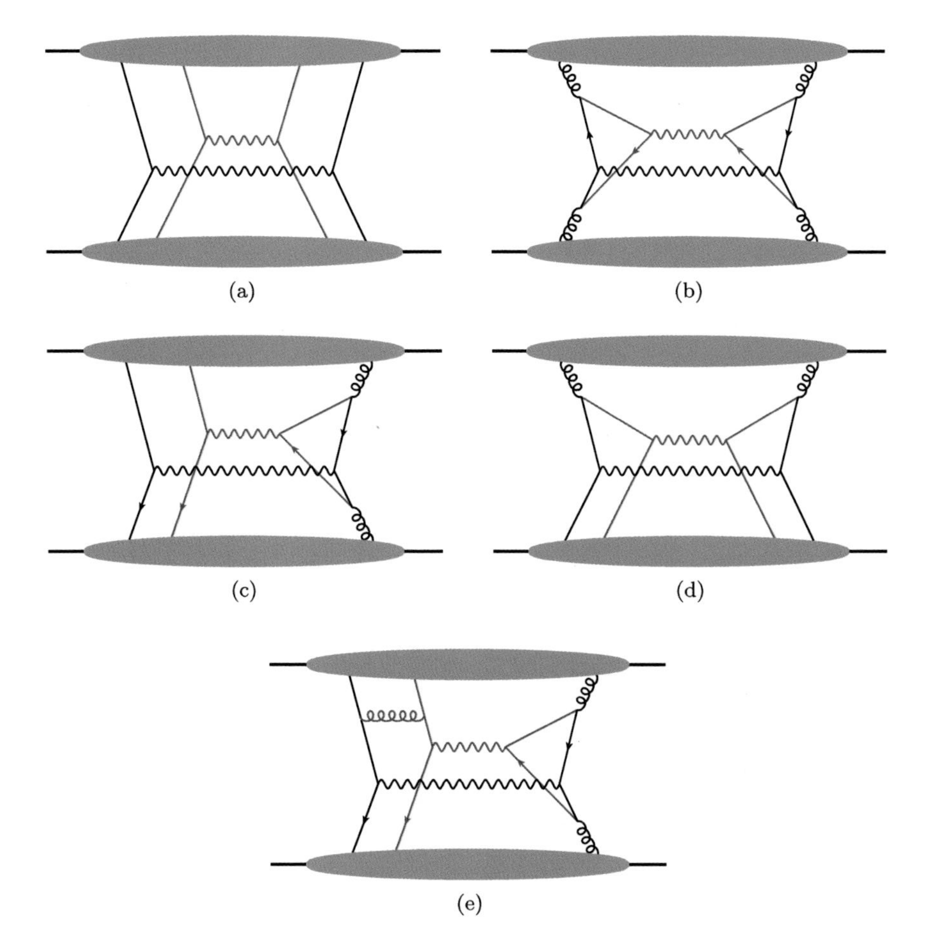

Fig. 2 Different contributions to the production of two electroweak gauge bosons: DPS (a), SPS (b) and their interference (c). Graphs (d) and (e) involve higher-twist distributions. Internal lines in the hard scattering are off shell by order Q. A vertical line for the final state cut is not shown for simplicity.

the leading power contributions are from SPS, from DPS and from the interference between the two mechanisms. Example graphs are given in Figs. 2(a)–(c).

For the cross-section integrated over $\boldsymbol{q}_1$ and $\boldsymbol{q}_2$, the only leading-power contributions comes from SPS. Suppressed by Λ^2/Q^2 are two types of graphs in addition to DPS:

- Graphs with a collinear twist-two distribution (i.e., a parton density) for one proton and a collinear twist-four distribution for the other one, as in Fig. 2(d). We refer to this as the twist-four mechanism in the following.
- Graphs with a collinear twist-three distribution for each proton, as in Fig. 2(e). This will be referred to as the twist-three mechanism.

As was already noted in Ref. 4 (see also Chapter 3), the integration over $\boldsymbol{q}_1$ and $\boldsymbol{q}_2$ forces all hard interactions to occur at the same transverse position in the SPS/DPS interference, which thus becomes a special case of the twist-three mechanism. By contrast, in TMD factorization the graphs in Fig. 2(d) and 2(e) are suppressed by Λ^2/Q^2 compared with the SPS/DPS interference in Fig. 2(c).

In an unpolarized proton, the number of possible collinear twist-three distributions is severely restricted by helicity conservation, and only distributions with a quark and an antiquark of opposite helicity are allowed.[5] Such distributions do not have any cross talk with gluon distributions. One can hence expect them to lack the small x enhancement of quark or gluon DPDs, so that there is some justification for neglecting them (in the same spirit as neglecting the parton-type interference distributions mentioned in Section 2). Notice that in TMD factorization, the twist-three distributions occurring in the SPS/DPS interference are not subject to restrictions from parton helicity conservation: since all three parton fields are at different transverse positions, orbital angular momentum can compensate a mismatch of parton helicities in this case.

A special class of graphs for the twist-four mechanism, shown in Fig. 3, has been associated with "rescattering" in Ref. 6 (see also Chapter 10). Each propagator marked by a bar in the figure has a denominator of the form $ax - b + i\epsilon$, where x is a loop variable and a, b are fixed by external kinematics. Keeping the pole part of each propagator and neglecting the principal value part of the integration puts the two lines on shell, and the process looks like one $2 \to 2$ scattering followed by a second one. The calculation of this two-pole part in terms of two unpolarized $2 \to 2$ partonic cross-sections is indeed correct if in the twist-four distribution the quantum numbers are coupled such that partons with equal momentum fractions are unpolarized and form a color singlet. However, it is not obvious that the pole parts of the loop integrations should dominate over the principal value contributions in general kinematics. This may happen for jets with very large rapidities.[7] We also emphasize

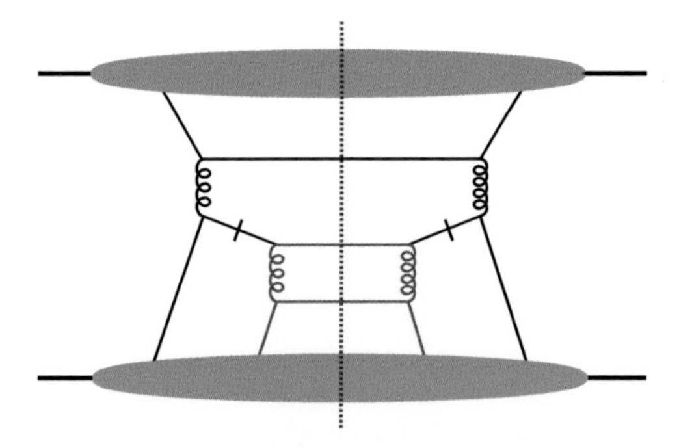

Fig. 3 A graph for three-jet production by the twist-four mechanism that has been associated with "rescattering" in kinematics where the lines marked by a bar are on shell. All three partons in the final state are understood to have large transverse momenta.

that the partons marked by bars do *not* physically propagate over distances much larger than $1/Q$. Technically speaking, their propagator poles can be avoided by a complex contour deformation in the loop integrals, and physically speaking one finds that the "rescattering" of Fig. 3 does not correspond to a classically allowed scattering process.[3] It is therefore inappropriate to associate final- or initial-state parton showers to these partons.

4. Short-Distance Splitting and Double Counting

At small inter-parton distances, the dominant contribution to a DPD comes from perturbative splitting of one parton into two, as depicted in Fig. 4(a). Let us for now concentrate on collinear DPDs. At leading order in α_s, the contribution of the $1 \to 2$ splitting mechanism is easily computed and reads

$$ {}^{R}F_{a_1 a_2}(x_1, x_2, \boldsymbol{y})\big|_{\text{spl,pt}} = \frac{1}{y^2} \frac{\alpha_s}{2\pi^2} \, {}^{R}P_{a_0 \to a_1 a_2}\left(\frac{x_1}{x_1 + x_2}\right) \frac{f_{a_0}(x_1 + x_2)}{x_1 + x_2}, \tag{9} $$

where f_{a_0} is an unpolarized PDF and $P_{a_0 \to a_1 a_2}$ a splitting function. The $1/y^2$ behavior can be deduced already by dimensional counting. Note that this mechanism gives strong color and spin correlations: chirality conservation for massless quarks

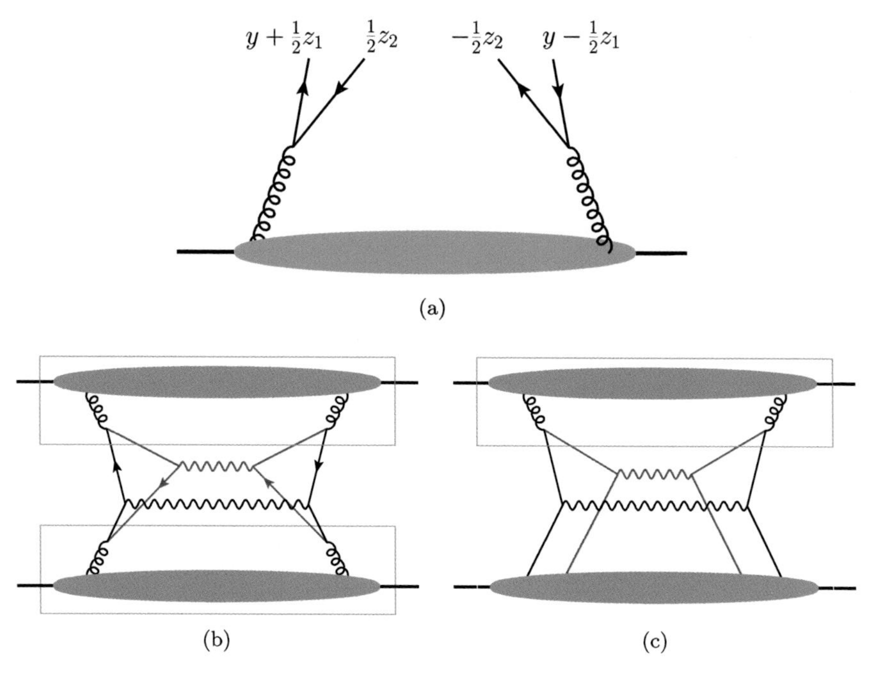

Fig. 4 Splitting provides a short-distance contribution to DPDs (a) and gives rise to 1v1 (b) and 2v1 (c) graphs for the DPS cross-section. The boxes represent double parton distributions; partons emerging from them have virtualities much smaller than Q.

results for instance in complete anti-alignment of the quark and antiquark helicities in $g \to q\bar{q}$.

DPDs also contain an "intrinsic" short-distance part, where the two partons may be thought of as part of the "intrinsic" wave function of the proton. This part is related to a twist-four distribution and only diverges logarithmically at small y. Finally, one may have a $1 \to 2$ splitting only in the amplitude or only in its conjugate: this contribution involves a collinear twist-three distribution and lacks small x enhancement, as discussed in Section 3.

Inserting the short-distance form (9) into the cross-section formula (1), we see that without the function Φ the integral over $\boldsymbol{y}$ would be power divergent. This power divergence is associated with so-called 1v1 (1 versus 1 parton) diagrams, in which there are $1 \to 2$ splittings in both protons as shown in Fig. 4(b). Note that this is the same graph as in Fig. 2(b), which represents a loop correction in the leading-power SPS mechanism. The difference is that in Fig. 4(b) the quark virtualities are understood to be much smaller than Q, whereas in Fig. 2(b) they are of order Q. The small y divergence in the DPS formula without Φ is not present in reality: it arises from using DPS approximations in the small y region where they are not valid. It should hence be removed and replaced with the appropriate SPS expression, in a manner that avoids double counting between SPS and DPS. The analogous double counting problem for multi-jet production has already been noticed some time ago.[8]

A short-distance divergence in the $\boldsymbol{y}$ integral also appears for so-called 2v1 (2 versus 1 parton) diagrams as in Fig. 4(c), where a $1 \to 2$ splitting takes place in only one proton. This divergence is only logarithmic, and it corresponds to the overlap of the DPS contribution with the twist-four mechanism shown in Fig. 2(d). The importance of the 2v1 mechanism has been emphasized in Refs. 9–13.

In the following we describe a solution to these problems that was elaborated in Ref. 5. Different approaches have been presented in Refs. 9–12,14, the one of Refs. 9 and 10 being reviewed in Chapter 5. A detailed comparison between them is given in Ref. 5. The formalism described here resolves the double counting problem, retains the concept of double parton distributions that have a field theoretic definition, and permits the study of higher order contributions in a practical way. The other approaches just mentioned do not possess all these features.

The first step is to insert the function $\Phi(y\nu)$ into the cross-section formula (1). (We insert the square of Φ for consistency with the TMD case.) This function regulates the divergences just discussed by removing the region $y \ll 1/\nu$ from what we *define* to be DPS. It must satisfy $\Phi(u) \to 0$ for $u \to 0$ and $\Phi(u) \to 1$ for $u \gg 1$. Suitable forms are $\Phi(u) = 1 - \exp(-u^2/4)$, or a hard cutoff $\Phi(u) = \Theta(u - b_0)$ with $b_0 = 2e^{-\gamma_E}$ chosen to simplify analytic expressions.

To avoid double counting between DPS and SPS, and between DPS and the twist-four mechanism, we introduce subtraction terms in the overall cross-section:

$$\sigma_{\text{tot}} = \sigma_{\text{DPS}} - \sigma_{\text{1v1,pt}} + \sigma_{\text{SPS}} - \sigma_{\text{2v1,pt}} + \sigma_{\text{tw4}}. \tag{10}$$

The subtraction terms depend on ν in such a way that the dependence on this unphysical parameter cancels on the right-hand side (to the order of perturbative accuracy of the calculation). Note that σ_{SPS} and σ_{tw4} do not depend on ν. In particular, σ_{SPS} is simply calculated in the usual way with no modifications. The 1v1 subtraction $\sigma_{\text{1v1,pt}}$ is constructed in a simple way by replacing the DPDs in the cross-section formula (1) by the perturbative splitting approximation (9) or its equivalent at higher orders in α_s. Similarly, $\sigma_{\text{2v1,pt}}$ is obtained by replacing one of the two DPDs by its splitting approximation and the other one by its intrinsic short-distance part. There are some subtleties in choosing adequate scales μ in these distributions, especially if the scales Q_1, Q_2, Q_h of the two DPS subprocesses and of the SPS subprocess are very different, but we will not dwell on this here.

An appropriate choice for the scale ν is the minimum of Q_1 and Q_2. With this choice, σ_{DPS} does contain short-distance contributions for which the DPS approximations are not valid, but these contributions are removed by the subtraction terms in the overall cross-section. This is quite similar to choosing factorization scales $\mu \sim Q$ in collinear PDFs and DPDs: the parton distributions then contain virtualities up to the hard scale Q, but double counting is avoided by subtractions in the hard-scattering cross-sections.

Let us demonstrate how the prescription works. At small $y \sim 1/Q$, one has $\sigma_{\text{DPS}} \approx \sigma_{\text{1v1,pt}} + \sigma_{\text{2v1,pt}}$ by construction (the product of the intrinsic parts of each DPD gives a power suppressed contribution at small y, so that the absence of a subtraction term $\sigma_{\text{2v2,pt}}$ is no problem). One thus has $\sigma_{\text{tot}} \approx \sigma_{\text{SPS}} + \sigma_{\text{tw4}}$, as is appropriate for the short-distance region. The dependence on the unphysical cutoff scale ν cancels between DPS and the subtraction terms. At $y \gg 1/Q$, the dominant contribution to σ_{SPS} comes from 1v1 type loops in the region where the DPS approximations are valid, such that $\sigma_{\text{SPS}} \approx \sigma_{\text{1v1,pt}}$. Similarly, we have $\sigma_{\text{tw4}} \approx \sigma_{\text{2v1,pt}}$. As a result we obtain $\sigma_{\text{tot}} \approx \sigma_{\text{DPS}}$, as appropriate. The construction just explained is a special case of the general subtraction formalism discussed in Chapter 10 of Ref. 1.

For the scale choice $\nu \sim \min(Q_1, Q_2)$, one can show that the combination $\sigma_{\text{tw4}} - \sigma_{\text{2v1,pt}}$ in Eq. (10) is subleading compared to σ_{DPD} by a logarithm $\log(Q/\Lambda)$, where Λ is an infrared scale. This combination can hence be dropped at leading logarithmic order, which is of great practical benefit since the computation of the twist-four contribution is technically quite involved. For the same scale choice, one finds that σ_{DPD} includes the appropriate resummation of large DGLAP logarithms in the 2v1 graphs.

In order to estimate the theoretical uncertainty from missing higher order terms in this framework, one can vary the parameters μ_1, μ_2 and ν, similar to how one varies only μ in the single scattering case. Note that the variation in ν of the DPS term alone provides an order-of-magnitude estimate of SPS graphs containing a double box as in Fig. 2(b), since it involves the same PDFs, overall coupling constants and kinematic region (small y, corresponding to large transverse momenta and virtualities of internal lines). An alternative estimate is provided by the double counting subtraction term $\sigma_{\text{1v1,pt}}$. Therefore, a small ν variation of σ_{DPS} compared to its central value indicates that $\sigma_{\text{1v1,pt}}$ and the corresponding loop contribution to σ_{SPS} are negligible compared to σ_{DPS}. Several scenarios where the ν variation is reduced in this way were found in Ref. 5, for instance when the parton pairs in the relevant DPDs cannot be produced in a single leading-order splitting (e.g., $u\bar{d}$), or when low x values are probed in the DPDs. In such cases, one may justifiably neglect the appropriate perturbative order of σ_{SPS}, together with the 1v1 subtraction term. Such processes and kinematic regions are the most promising ones to make useful calculations and measurements of DPS, especially because there are only few cases for which SPS is computed at the order containing the double box (essentially only double electroweak gauge boson production).

Now let us turn to the TMD case, where the pattern of overlaps and divergences is somewhat different. In particular, the ultraviolet divergences in σ_{DPS} associated with 1v1 graphs become logarithmic rather than a power. This is related to the fact that DPS and SPS have the same power behavior in the small q_T region. The inter-parton distance $\boldsymbol{y}_+ = \boldsymbol{y} + \frac{1}{2}(\boldsymbol{z}_1 - \boldsymbol{z}_2)$ in the amplitude and its counterpart $\boldsymbol{y}_- = \boldsymbol{y} - \frac{1}{2}(\boldsymbol{z}_1 - \boldsymbol{z}_2)$ in the complex conjugate amplitude (see Fig. 4(a)) are independent variables now. When one of these distances is small, the TMDs are dominated by perturbative $1 \to 2$ splitting for the corresponding parton pair. The divergent behavior of σ_{DPS} in the region where both y_+ and y_- go to zero corresponds to the overlap of DPS with the SPS double box graph in Fig. 2(b). The region where only y_- goes to zero corresponds to an overlap with the SPS/DPS interference graph in Fig. 2(c).

It is clear then that in this case the DPS term must be regulated when either y_+ or y_- go to zero, as is done in Eq. (2). The SPS/DPS interference terms must also be regulated for small y_+ or y_-, since they overlap with SPS again. Subtraction terms must be included as appropriate to remove the double counting, as elaborated in Ref. 5.

5. Collinear DPDs: Evolution

The twist-two operators in the definition of DPDs contain ultraviolet divergences that require renormalization. This leads to the familiar DGLAP evolution equations of ordinary PDFs, and to corresponding equations for collinear DPDs. Taking different scales μ_1, μ_2 for the partons with momentum fractions x_1 and x_2, we have a homogeneous evolution equation

$$\frac{\partial}{\partial \log \mu_1^2} {}^{R}F_{a_1 a_2}(x_1, x_2, \boldsymbol{y}; \mu_1, \mu_2, \zeta) = \sum_{b_1} \int_{x_1}^{1-x_2} \frac{dx_1'}{x_1'} \, {}^{R}P_{a_1 b_1}\left(\frac{x_1}{x_1'}; \mu_1, \frac{x_1 \zeta}{x_2}\right) \times {}^{R}F_{b_1 a_2}(x_1', x_2, \boldsymbol{y}; \mu_1, \mu_2, \zeta) \tag{11}$$

in μ_1 and its analogue for μ_2. The two parton pairs with momentum fractions x_1 or x_2 evolve separately. Note that in the color singlet sector, both ${}^1F$ and ${}^1P$ are ζ independent, and ${}^1P$ is the same DGLAP evolution kernel as for ordinary PDFs.

The interplay of DGLAP evolution with the splitting mechanism described in Section 4 has important consequences when one or both momentum fractions x_1, x_2 are small.[5] It can change the $1/y^2$ dependence of the splitting contribution (9) into a much flatter y dependence, which increases the contribution of the region $y \gg 1/\nu$ in the 1v1 and 2v1 cross-sections. The size of the effect depends on kinematics and on the parton types involved.

The Fourier integral that converts $F(x_i, \boldsymbol{y})$ into a momentum space DPD $F(x_i, \boldsymbol{\Delta})$ has a logarithmic divergence at small y from the splitting contribution, which requires additional ultraviolet renormalization.[3] In the following, we concentrate on color singlet distributions ($R = 1$) and on equal scales $\mu_1 = \mu_2$. One way to define $F(x_i, \boldsymbol{\Delta})$ is to perform the Fourier transform in $D = 4 - 2\epsilon$ dimensions and use ordinary $\overline{\text{MS}}$ renormalization for the splitting divergence. The resulting evolution equation has an additional inhomogeneous term, which at LO in α_s reads

$$\frac{\partial}{\partial \log \mu^2}\, {}^1F_{a_1 a_2}(x_1, x_2, \boldsymbol{\Delta}; \mu) = \{\text{homogeneous terms}\} + \frac{\alpha_s(\mu)}{2\pi}\, {}^1P_{a_0 \to a_1 a_2} \times \left(\frac{x_1}{x_1 + x_2} \right) \frac{f_{a_0}(x_1 + x_2; \mu)}{x_1 + x_2}, \tag{12}$$

with the same kernel ${}^1P_{a_0 \to a_1 a_2}$ as in Eq. (9). The homogeneous terms have the same form as in the evolution of $F(x_i, \boldsymbol{y})$, with $\boldsymbol{y}$ replaced by $\boldsymbol{\Delta}$. This inhomogeneous evolution has been discussed extensively in the literature.[15–19]

Whilst the scheme presented here requires position space DPDs $F(x_i, \boldsymbol{y})$ for computing cross-sections, the momentum space DPDs have a property that makes their study worthwhile. At $\boldsymbol{\Delta} = \boldsymbol{0}$, unpolarized momentum space DPDs satisfy sum rules[18] for the momentum and the flavor quantum numbers of one of the two partons:

$$\begin{aligned} \sum_{a_2 = q, \bar{q}, g} \int_0^{1-x_1} dx_2\, x_2\, {}^1F_{a_1 a_2}(x_1, x_2, \boldsymbol{0}) &= (1 - x_1)\, f_{a_1}(x_1), \\ \int_0^{1-x_1} dx_2\, \left[{}^1F_{a_1 q}(x_1, x_2, \boldsymbol{0}) - {}^1F_{a_1 \bar{q}}(x_1, x_2, \boldsymbol{0}) \right] &= N_{a_1 q}\, f_{a_1}(x_1), \end{aligned} \tag{13}$$

where $N_{a_1 q}$ is a combinatorial factor. The validity of these sum rules for $\overline{\text{MS}}$ renormalized distributions can be shown to all orders in perturbation theory.[20] A relation between DPDs in momentum and position space can be established by defining distributions

$${}^1F_{\Phi}(x_1, x_2, \boldsymbol{\Delta}; \mu, \nu) = \int d^2\boldsymbol{y}\, e^{i\boldsymbol{\Delta}\boldsymbol{y}}\, \Phi(y\nu)\, {}^1F(x_1, x_2, \boldsymbol{y}; \mu), \tag{14}$$

where the logarithmic singularity at small y is removed by the same regulator function Φ used in the cross-section. These DPDs and the $\overline{\text{MS}}$ renormalized ones differ only by the treatment of the ultraviolet region, so that their difference can be computed in perturbation theory.[5] To order α_s, one finds that ${}^{1}F(x_1, x_2, \boldsymbol{\Delta}; \mu) - {}^{1}F_{\Phi}(x_1, x_2, \boldsymbol{\Delta}; \mu, \mu)$ at $\boldsymbol{\Delta} = \mathbf{0}$ is a calculable function of $x_1/(x_1 + x_2)$ times the inhomogeneous term in Eq. (12).

We note that the factorization formula (1) can be rewritten in terms of $\int d^2\boldsymbol{\Delta}\, {}^{R}F_{\Phi}(\bar{x}_1, \bar{x}_2, -\boldsymbol{\Delta})\, {}^{R}F_{\Phi}(x_1, x_2, \boldsymbol{\Delta})$. This has been used to show that, at leading logarithmic accuracy, the collinear 2v2 and 2v1 cross-sections given in Refs. 9–13 are consistent with the formalism presented here.[5]

6. Soft Gluons and Sudakov Logarithms

The proof of the DPS factorization formulas (1) and (2) proceeds in close analogy to the case of SPS. Here we only sketch the steps that lead to the construction of DPDs and their evolution equations in rapidity, referring to Refs. 21 and 22 for details. One starts by showing that graphs contributing to the cross-section at leading power in Λ/Q factorize into hard, collinear and soft subgraphs, as depicted for the double Drell–Yan process in Fig. 5(a). In the hard-scattering subgraphs H_1 and H_2 all internal lines are far off shell, the subgraphs A and B involve only momenta collinear to one of the incoming protons, and the subgraph S describes the exchange of soft gluons between the right-moving partons in A and the left-moving ones in B.

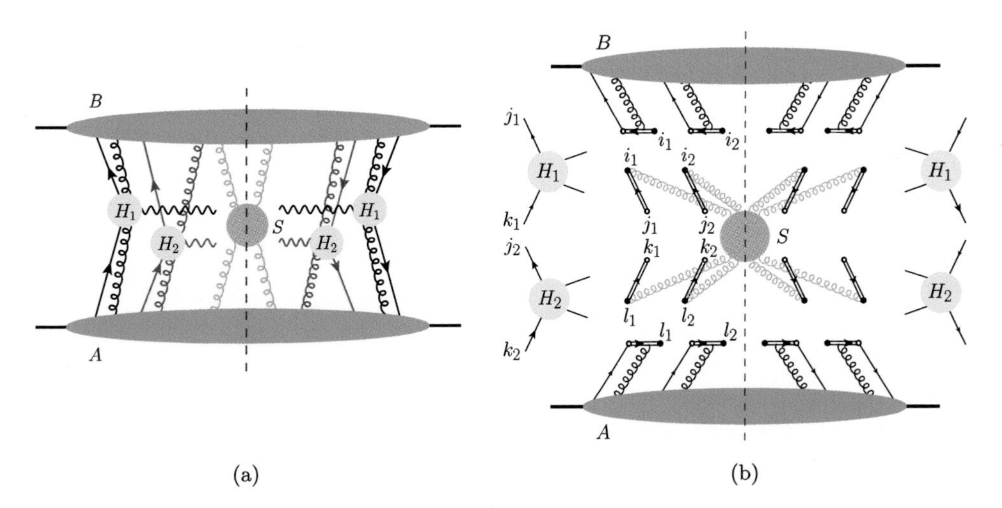

Fig. 5 (a) Factorized graph for double Drell–Yan production. (b) Graph for double dijet production after the Grammer–Yennie approximation and Ward identities have been applied. $i_1, i_2, \ldots, l_2$ are color indices.

To obtain a useful factorization formula, soft gluons must be decoupled from A and B. To achieve this, one performs a Grammer–Yennie approximation. For one gluon with momentum ℓ flowing from S into A, this reads

$$S_\mu(\ell)\, A^\mu(\ell) \approx S^-(\ell)\, \frac{v_R^+}{\ell^- v_R^+ + i\epsilon}\, \ell^- A^+(\tilde{\ell}) \approx S_\mu(\ell)\, \frac{v_R^\mu}{\ell v_R + i\epsilon}\, \tilde{\ell}_\nu A^\nu(\tilde{\ell}), \tag{15}$$

where $v_R = (v_R^+, v_R^-, \mathbf{0})$ specifies a direction with large positive rapidity $Y_R = \frac{1}{2}\log|v_R^+/v_R^-|$. In the first step, we have retained only the largest component of A^μ and replaced $\ell = (\ell^+, \ell^-, \boldsymbol{\ell})$ with $\tilde{\ell} = (0, \ell^-, \boldsymbol{\ell})$ in that factor, which involves plus momenta much larger than ℓ^+. The second step is more delicate and will be discussed in Section 7. One can now apply a Ward identity to $\tilde{\ell}_\nu A^\nu(\tilde{\ell})$, which removes the gluon attachment from A. The soft gluon emerging from S then couples to a Wilson line along the direction v_R. Naively, one would take v_R lightlike, i.e., set $v_R^- = 0$, but this would lead to so-called rapidity divergences of loop integrals inside S. Instead of a finite rapidity Y_R, one may use other methods to regulate these divergences, two of which have been applied to DPS in Refs. 23 and 24.

In full analogy, one can replace soft gluons coupling to B by gluons coupling to a Wilson line along a direction v_L with large negative rapidity. The soft factor S can now be written as the vacuum expectation value of Wilson line operators and is thus decoupled from A and B.

A similar argument is used for longitudinally polarized collinear gluons exchanged between A and H_1 or H_2, removing them from the hard scatters and coupling them to Wilson lines along v_L. These Wilson lines appear in the proton matrix element $\langle p|\mathcal{O}_{a_1}\mathcal{O}_{a_2}|p\rangle$ associated with the factor A. The same steps are followed for gluons exchanged between B and H_1 or H_2, resulting in Wilson lines along v_R in the matrix element associated with B.

The factors in the cross-section formula are tied together by color indices in a way that is shown in Fig. 5(b) for double dijet production. We first discuss the production of color singlet particles in H_1 and H_2. In this case, the soft factor is contracted with $\delta^{j_1 k_1}\delta^{j_2 k_2}$ and analogous factors for the index pairs on the right-hand side of the final-state cut.

It is useful to project the collinear and soft factors on color representations R, as was explained for DPDs in Section 2. The collinear factors then become vectors ${}^{R}\!A$, ${}^{R}\!B$ in the space of color representations, and the soft factor for color singlet production becomes a matrix ${}^{RR'}\!S$. In the cross-section we have the combination $\sum_{RR'} {}^{R}\!B\; {}^{RR'}\!S\; {}^{R'}\!A$.

The soft matrix in the cross-section depends on the rapidity difference $Y = Y_R - Y_L$ of the Wilson lines along v_R and v_L. This dependence is given by a Collins–Soper equation

$$\frac{\partial}{\partial Y}\, {}^{RR''}\!S(Y) = \sum_{R'} {}^{RR'}\!\widehat{K}\; {}^{R'R''}\!S(Y) \tag{16}$$

with a rapidity independent matrix $\widehat{K}$. Introducing an intermediate rapidity Y_C between Y_R and Y_L, one can write $S(Y)$ as the product of two matrices, one depending on $Y_R - Y_C$ and the other on $Y_C - Y_L$. Combining one of them with A and the other with B, one obtains DPDs $F_A(Y_C)$ and $F_B(Y_C)$. In the DPDs, one can take the limits $Y_L \to -\infty$ and $Y_R \to \infty$ without encountering rapidity divergences. The proton matrix element in Eq. (4) then contains lightlike Wilson lines. The final cross-section formula involves the sum $\sum_R {}^R F_B \, {}^R F_A$ with color-dependent DPDs, as anticipated in Section 2. Note that A, B and S are non-perturbative quantities since they involve low virtualities. Eliminating S by defining distributions F_A and F_B thus represents a significant simplification.

The construction sketched so far is common to TMD and collinear factorization. However, ultraviolet renormalization works differently in the two cases, which we now discuss in turn.

TMD factorization. It is useful to express the rapidity dependence of the DPDs in terms of boost invariant quantities, $\zeta = 2x_1 x_2 (p^+)^2 \, e^{-2Y_C}$ for F_A and an analogue $\bar{\zeta}$ for F_B. We concentrate on F_A from now on and omit the subscript A. Restoring the dependence on all other variables, we have a Collins–Soper equation

$$\frac{\partial}{\partial \log \zeta} \, {}^R F(x_i, \boldsymbol{z}_i, \boldsymbol{y}; \mu_i, \zeta) = \frac{1}{2} \sum_{R'} {}^{RR'} K(\boldsymbol{z}_i, \boldsymbol{y}; \mu_i) \, {}^{R'} F(x_i, \boldsymbol{z}_i, \boldsymbol{y}; \mu_i, \zeta) \tag{17}$$

with a matrix kernel K related to $\widehat{K}$ in Eq. (16). Its μ_1 dependence is given by

$$\frac{\partial}{\partial \log \mu_1} \, {}^{RR'} K(\boldsymbol{z}_i, \boldsymbol{y}; \mu_i) = -\delta_{RR'} \, \gamma_K(\mu_1), \tag{18}$$

whilst for the DPD we have

$$\frac{\partial}{\partial \log \mu_1} \, {}^R F(x_i, \boldsymbol{z}_i, \boldsymbol{y}; \mu_i, \zeta) = \gamma_F(\mu_1, x_1 \zeta / x_2) \, {}^R F(x_i, \boldsymbol{z}_i, \boldsymbol{y}; \mu_i, \zeta) \tag{19}$$

with

$$\gamma_F(\mu, \zeta) = \gamma_F(\mu, \mu^2) - \frac{1}{2} \gamma_K(\mu) \, \log \frac{\zeta}{\mu^2}. \tag{20}$$

Here $\gamma_K(\mu)$ and $\gamma_F(\mu, \mu^2)$ depend on μ via $\alpha_s(\mu)$. Analogues of Eqs. (18) and (19) hold for the μ_2 dependence. Note that the kernel K and the anomalous dimensions γ_K and γ_F differ for quarks and for gluons, but are independent of parton polarization and quark flavor.

The system of evolution equations can be solved analytically (provided the matrix K can be diagonalized analytically). The solution exponentiates Sudakov double logarithms, controlled by γ_K, and single logarithms going with γ_F and K. Except for the scaling of ζ by x_1/x_2 or x_2/x_1, the double logarithms have the same form as for single TMDs. When F_A is multiplied with F_B in the cross-section, logarithms of ζ and $\bar{\zeta}$ turn into logarithms of the invariant masses Q_1 and Q_2 in the two hard-scattering subprocesses.

Collinear factorization. In collinear factorization, the soft factor simplifies considerably. Using color algebra, one can show that the general soft factor shown in Fig. 5(b) is the same as the one for color singlet production, which is contracted with $\delta^{j_1 k_1} \delta^{j_2 k_2}$ etc. Moreover one finds that the soft matrix ${}^{RR'}S$ is diagonal. The color matrix algebra in the construction of DPDs thus becomes trivial. The product of factors in the cross-section can then be written as

$$\sum_R {}^R H_1 \, {}^R H_2 \, {}^R F_B \, {}^R F_A . \tag{21}$$

If a color singlet state is produced, then all color projections ${}^R H$ are equal, otherwise they differ. Multiplying ${}^R H$ with a flux factor, one obtains the subprocess cross-sections ${}^R\hat{\sigma}$ in the collinear factorization formula (1). Collinear DPDs depend on ζ as

$$\frac{\partial}{\partial \log \zeta} \, {}^R F(x_i, \boldsymbol{y}; \mu_i, \zeta) = \frac{1}{2} \, {}^R J(\boldsymbol{y}; \mu_i) \, {}^R F(x_i, \boldsymbol{y}; \mu_i, \zeta) \tag{22}$$

with

$$\frac{\partial}{\partial \log \mu_1} \, {}^R J(\boldsymbol{y}; \mu_i) = - \, {}^R \gamma_J(\mu_1) \tag{23}$$

and an analogous equation for the μ_2 dependence. The DGLAP kernels in the evolution equation (11) depend on ζ via

$${}^R P_{ab}(x; \mu, \zeta) = {}^R P_{ab}(x; \mu, \mu^2) - \frac{1}{4} \delta_{ab} \delta(1-x) \, {}^R \gamma_J(\mu) \, \log \frac{\zeta}{\mu^2} . \tag{24}$$

The ζ dependence of the DPDs can be given in analytical form. It contains exponentiated double logarithms controlled by γ_J and single logarithms going with the kernel J. In the color singlet sector, the soft factor is trivial, ${}^{11}S = 1$, and correspondingly one has ${}^1 J = {}^1 \gamma_J = 0$. In physical terms, the effects of soft gluon exchange cancel in this case. As a result, color non-singlet DPDs are suppressed by Sudakov logarithms, whereas color singlet DPDs are not.[23,25]

7. Glauber Gluons and Factorization

A crucial step in showing that soft gluon exchange between the subgraphs A and B in Fig. 5(a) can be subsumed into the vacuum expectation value of Wilson lines is to establish the absence of contributions from the so-called Glauber region. In the following, we restrict ourselves to color singlet production, since it is the only context in which this issue has been studied. The arguments apply both to collinear and TMD factorization.

For gluons leaving the soft subgraph S in Fig. 5(a), there are in fact two distinct momentum regions that contribute to the cross-section at leading power. The first one can be called the "central soft" region, where all components of the momentum ℓ have comparable size, $|\ell^+| \sim |\ell^-| \sim |\boldsymbol{\ell}|$. In this region, the second step of the Grammer–Yennie approximation (15) is valid: we have $\ell^- A^+ \approx \tilde{\ell}_\nu A^\nu$ because $\boldsymbol{\ell A}$ is power suppressed compared to $\ell^- A^+$. The second one is the Glauber region, which is characterized by $|\ell^+\ell^-| \ll \boldsymbol{\ell}^2$. Gluons in this region mediate small-angle scattering of a right-moving parton on a left-moving one. In the Glauber region, we can have $|\boldsymbol{\ell A}| \sim |\ell^- A^+|$, so that the Grammer–Yennie approximation fails. This presents a serious obstacle to factorization.

Of course, the soft momentum ℓ in a graph is not held fixed but integrated over. For many types of soft gluon attachment, the integration over ℓ^+ or ℓ^- (or both) can be deformed away from the real axis into the complex plane in such a way that one has $|\ell^+\ell^-| \sim \boldsymbol{\ell}^2$ on the deformed integration contour and thus avoids the Glauber region. This is only possible when the poles in ℓ^+ or ℓ^- of the propagators depending on ℓ do not obstruct the deformation. In such cases, the contribution from ℓ in the Glauber region can be validly subsumed into the contribution from a collinear or a central soft region, where Grammer–Yennie approximations can be applied to achieve factorization.

Examples of soft attachments for which we may deform the momentum out of the Glauber region are given by the gluons with momenta ℓ_1 and ℓ_4 in Fig. 6. In fact, for all of the "novel" types of soft attachment that appear only when we consider DPS rather than SPS, such a deformation is possible. Note that the sign of $i\epsilon$ in the denominators of Eq. (15) is chosen precisely such that it does not obstruct these contour deformations. As a consequence, the Wilson lines in the construction of soft factors and DPDs are past pointing (in light-cone coordinates), as in Eq. (6).

The one type of soft attachment for which the propagator poles obstruct a deformation out of the Glauber region is exemplified by the gluons with momenta ℓ_2 and ℓ_3 in Fig. 6. This is an attachment between a right-moving and a left-moving spectator parton after the two hard scatters (where "after" refers to the topology of graphs and not to the time coordinate in some reference frame). Of course, such exchanges occur already in SPS. As shown for instance in Ref. 26, the contribution from the Glauber region cancels to leading power when one sums over all final state cuts of a given graph. This requires that the cross-section is differential

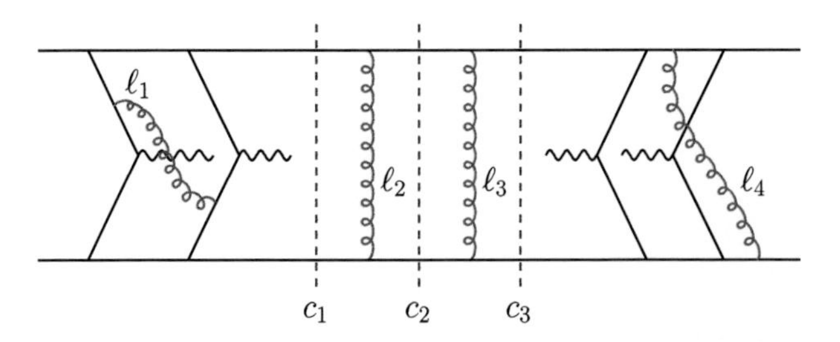

Fig. 6 Graph for double Drell–Yan production with several soft gluons exchanged between left and right moving fast partons. The three possible final state cuts of the graph are denoted by c_1, c_2 and c_3.

only in properties of the hard scattering products, but fully inclusive over the remaining particles. The same argument applies to DPS. The principle ensuring this cancellation is unitarity: spectator scattering does affect details of the final state, but its net effect is zero if the observable is not sensitive to the spectator momenta. The DPS cross-sections (1) and (2) satisfy this requirement. On the other hand, one can show that Glauber gluon exchange does break factorization for observables depending on the momenta of the spectator partons (or better, of the "beam jets" into which these partons hadronize).

The argument just sketched works for simple graphs, essentially at the level of single soft gluon exchange. To demonstrate Glauber cancellation at all orders, a more powerful technique is needed, based on the light-front ordered version of QCD perturbation theory (LCPT). This argument was given for DPS in Ref. 21, generalizing the treatment of the SPS case in Ref. 1. In the LCPT picture, one sees again that from the point of view of the Glauber gluons, single and double hard scattering look rather similar, and that the troublesome "final state" poles obstructing the deformation out of the Glauber region cancel after the sum over of final state cuts. Again, a unitarity argument is used to achieve this cancellation.

8. Perturbative Transverse Momenta

TMD factorization is applicable when the typical size q_T of the transverse momenta $\boldsymbol{q}_1, \boldsymbol{q}_2$ is much smaller than the hard scale Q. This includes the multi-scale regime $\Lambda \ll q_T \ll Q$. In this situation, the small parameter in the general power counting of Section 3 becomes q_T/Q rather than Λ/Q.

The Fourier exponent $e^{-i(\boldsymbol{q}_1\boldsymbol{z}_1+\boldsymbol{q}_2\boldsymbol{z}_2)}$ in the TMD cross-section (2) limits the distances z_1 and z_2 to typical size $1/q_T$. Double TMDs can then be computed in terms of perturbative subprocesses at scale q_T and of collinear matrix elements expressing the physics at scale Λ, which significantly increases the predictive power of theory. The distance y is not restricted in this way, and there are in fact two different regimes for DPS.

If $y \sim 1/\Lambda$ is of hadronic size, then the mechanism generating perturbative transverse momenta is the emission of partons, described by ladder graphs as in Fig. 7(a). For the DPDs we then have

$$\begin{aligned} {}^{R}F_{a_1a_2}(x_i, \boldsymbol{z}_i, \boldsymbol{y}; \mu_i, \zeta) = \sum_{b_1b_2} & {}^{R}C_{a_1b_1}(x_1', \boldsymbol{z}_1; \mu_1, x_1\zeta/x_2) \\ & \underset{x_1}{\otimes} {}^{R}C_{a_2b_2}(x_2', \boldsymbol{z}_2; \mu_2, x_2\zeta/x_1) \underset{x_2}{\otimes} {}^{R}F_{b_1b_2}(x_i', \boldsymbol{y}; \mu_i, \zeta) \end{aligned} \tag{25}$$

with convolution products $\otimes$ as in Eq. (11). This is the same mechanism as in SPS (a prominent example is Drell–Yan production with $\Lambda \ll q_T \ll Q$), and the short-distance coefficients ${}^{1}C$ are the same as the ones for single TMDs. The relation (25)

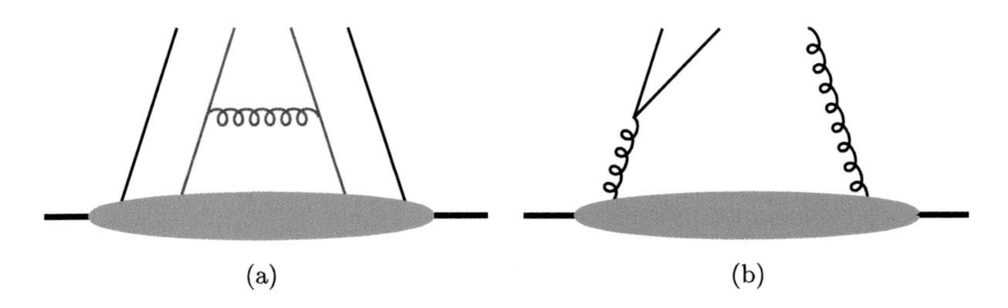

Fig. 7 (a) Ladder graph for a DPD. (b) Splitting contribution for a twist-three TMD.

can be understood in terms of an operator product expansion, with the operators in Eq. (5) being expanded for small z_1, z_2 while keeping y large.

The second regime for DPS is when y is of the same size as $z_i \sim 1/q_T$. The operator product expansion for the double TMD then involves three types of terms, already presented in Section 4. The four partons at small relative distances can originate from a collinear PDF via splitting, as in Fig. 4(a), from a collinear twist-four distribution without any splitting, or from a collinear twist-three distribution with parton splitting only in the amplitude or its conjugate. As already noted, the latter case requires chiral-odd distributions and lacks small x enhancement.

In the DPS cross-section, we have 1v1, 2v1 and 2v2 contributions from the four combinations of splitting and intrinsic contributions to the two DPDs. It is important to note that these have different power behavior in q_T, namely

$$\frac{Q^4\, d\sigma}{d^2\boldsymbol{q}_1\, d^2\boldsymbol{q}_2} \underset{yq_T\sim 1}{\sim} \begin{cases} \alpha_s^2/q_T^2 & \text{for 1v1,} \\ \alpha_s\Lambda^2/q_T^4 & \text{for 2v1,} \\ \Lambda^4/q_T^6 & \text{for 2v2,} \end{cases} \qquad \frac{Q^4\, d\sigma}{d^2\boldsymbol{q}_1\, d^2\boldsymbol{q}_2} \underset{y\Lambda\sim 1}{\sim} \Lambda^2/q_T^4, \tag{26}$$

where we have also specified the behavior of the contribution from $y \sim 1/\Lambda$. Although this contribution, as well as the 2v1 part at $y \sim 1/q_T$ are suppressed by Λ^2/q_T^2 compared with 1v1, it makes sense to keep them since they have a stronger small x enhancement and involve fewer powers of α_s. Explicit expressions for the different terms, including the Sudakov factors resulting from ζ evolution, are given in Ref. 22. We note that Eq. (26) holds if $|\boldsymbol{q}_1 + \boldsymbol{q}_2| \sim |\boldsymbol{q}_1| \sim |\boldsymbol{q}_2|$ are all of order q_T. Other regimes have been discussed in Refs. 3, 9 and 10.

To obtain the physical cross-section, one must combine DPS with SPS and the SPS/DPS interference, as discussed in Section 4. The TMDs in these contributions can be expressed in terms of collinear matrix elements as well. For SPS, they are just the ordinary PDFs. For the interference term, one has contributions with collinear twist-three distributions (lacking small x enhancement) and contributions with a PDF and a short-distance splitting only on one side of the final state cut, as shown in Fig. 7(b). Overall, one thus finds that — if collinear twist-three distributions are neglected — the only parton distributions needed for TMD factorization in the regime $\Lambda \ll q_T \ll Q$ are collinear DPDs and ordinary PDFs.

9. Status of Factorization

Significant progress has been made towards establishing factorization formulas for DPS processes at the same level of rigor as for SPS. In fact, many of the results we have sketched can even be extended to the case of three or more hard scatterings in a rather straightforward manner. However, a description of the color structure becomes rather cumbersome in this case, as does the discussion of perturbative splitting and double counting with other mechanisms. To conclude this overview, we list what in our opinion are major remaining open issues in DPS factorization.

No all-order proof is available for the nonabelian Ward identities required for decoupling soft gluons from the collinear factors (see Section 6). Examples at lowest order have been given in Ref. 3. It may be possible to adapt the proof of the corresponding Ward identities in single Drell–Yan production[27] to DPS, but this has not been worked out.

A crucial ingredient for constructing DPDs is the evolution equation (16) of the soft matrix S, for which no general proof has been given yet. For small distances y and z_i, one can calculate S in perturbation theory and easily finds that Eq. (16) is valid at one loop.[3] Its validity at two loops is corroborated by the calculation in Ref. 24 (which uses a different regulator for rapidity divergences). An all-order proof has recently been put forward in Ref. 28, but it is currently not clear whether it applies to the rapidity regulator employed in the present work. We also note that the construction sketched below Eq. (16) requires S to be positive semidefinite. There is no general proof for this, but it can be motivated by perturbative arguments.[22]

A technical problem in the construction of soft factors are gluons that couple only to Wilson lines along one direction. Such so-called Wilson line self-interactions are divergent for Wilson lines of infinite length. It is easy to see that they cancel in the cross-section by construction, but one must also show that they cancel in the individual parton distributions in the factorization formula. Some discussion for SPS is given in Sec. 13.7 of Ref. 1, but it would be desirable to have a more explicit solution to this problem, before applying it to the case of DPS.

Finally, the cancellation of Glauber gluon exchange has only been shown for DPS processes producing colorless particles.[21] An extension of this argument to the production of colored particles, relevant, e.g., for jet production, has not even been worked out for SPS, as far as we know. Such an extension may be possible for collinear factorization, whereas for TMD factorization there are strong arguments that this cannot even be done for SPS.[2]

Many of the subtleties in DPS factorization, such as the presence of parton correlations and the perturbative splitting mechanism, are by now quite well understood on the theory side. Their phenomenological importance, however, remains to be quantified for many interesting cases. This opens a wide field of studies for the future.

Acknowledgments

J. G. acknowledges financial support from the European Community under the FP7 Ideas program QWORK (contract 320389). The figures in this contribution were produced with JaxoDraw.[29]

Chapter 3

Inclusive and Exclusive Cross-Sections, Sum Rules

D. Treleani and G. Calucci*

Dipartimento di Fisica dell'Università di Trieste and INFN, Sezione di Trieste, Strada Costiera 11, Miramare-Grignano, I-34151 Trieste, Italy

In a simplified model of multiple parton interactions the inclusive cross-sections, of processes with large momentum transfer exchange, acquire the statistical meaning of factorial moments of the distribution in multiplicity of interactions, while more exclusive cross-sections, which can provide complementary information on the interaction dynamics, become experimentally viable. Inclusive and exclusive cross-sections are linked by sum rules, which can be tested experimentally.

1. Introduction

The relevance of multiple parton interactions (MPIs) in high energy hadronic collisions was apparent since the very beginning of the description of large p_t processes by pQCD.[30–33] To recall some of the main ideas in the matter, we will review a simplified model of MPIs, where a cutoff in the transverse momentum exchange is introduced, to separate the hard from the soft component of the interaction, and one exploits the very different properties of hard and soft interactions in the transverse coordinates space. Being localized in a smaller region, the hard component of the interaction can in fact be disconnected, in such a way that each disconnected component may be treated as an independent hard sub-interaction, which constitutes the characteristic feature of the MPIs. In the actual model, each different hard sub-interaction is treated perturbatively, without however distinguishing between quarks and gluons and between different states of spin and charge. In addition only MPIs, where each disconnected hard sub-interaction is initiated by two partons, are taken into account.

*Now retired.

The picture of MPIs obtained in the model satisfies various non-trivial properties, which will be discussed in detail.

By reviewing the kinematics of the process, in the next section we will outline the geometrical features of the double parton interaction (DPI) inclusive cross-section, we will introduce the "effective cross-section" and we will show that interference terms between DPIs and single parton interactions (SPIs) are suppressed. In the following section, we will generalize the cross-section to the case of an increasingly large number of MPIs and we will write the expression of the total contribution of hard interactions to the inelastic cross-section in the simplest case, where all multi-parton correlations can be neglected. The case where two-body correlations play an important role will be discussed by means of a suitable functional formalism in the successive section. The possibility of introducing explicit expressions, not only for the inclusive but also for a definite set of exclusive cross-sections, will be outlined in the last section, where we will discuss also the sum rules, linking inclusive and exclusive cross-sections.

2. Double Scattering

The DPI inclusive cross-section is obtained by the unitarity diagram in Fig. 1, in the limit of large c.m. energy and large momentum transfer exchange in the two partonic interactions, which generate the final states with overall momenta P and Q,[30] while the remnants of the hadron carry momenta $\underline{A}$ and $\underline{B}$. The soft blobs ϕ_A and ϕ_B represent the hadron bound state and the virtual lines attached to ϕ are

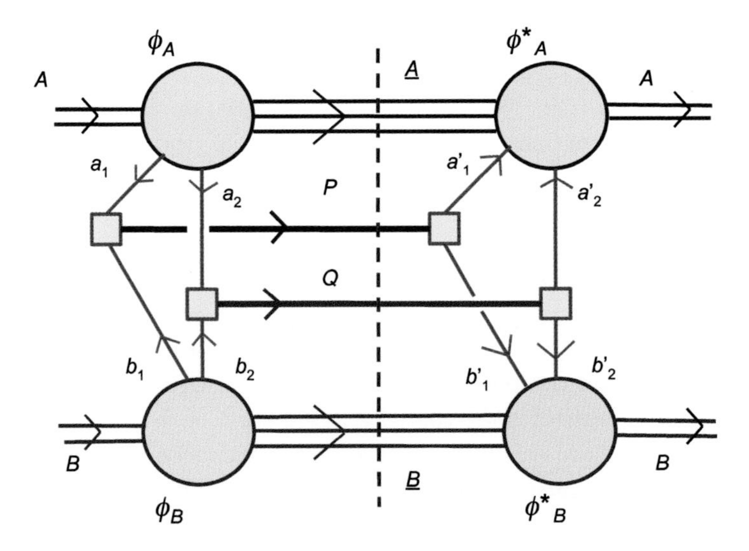

Fig. 1 Unitarity diagram for the double parton scattering cross-section. P and Q represent the total four-momenta of the partonic states produced in the two interactions, e.g., of two pairs of large p_t jets.

characterized by transverse momenta and off-shell scales of the order of the hadron mass. In the hadron–hadron c.m. frame, the corresponding light cone + components are of $\mathcal{O}\sqrt{s}$ in the upper part of the diagram and the light cone − components are of $\mathcal{O}\sqrt{s}$ in the lower part of the diagram, while the initial transverse components do not grow with the c.m energy $\sqrt{s}$. Because of momentum conservation, the overall transverse momenta of P and Q are of order of the transverse momenta of initial state partons, while the transverse momenta of the final state partons generated by the two hard interactions are large.

The diagram is characterized by five independent loops. As loop variables one may choose

$$P, \quad Q, \quad \delta = (a_1 - a_2)/2, \quad \delta' = (a_1' - a_2')/2, \quad \underline{A}. \tag{1}$$

Energy–momentum conservation in the interaction vertices give

$$P = a_1 + b_1 = a_1' + b_1', \quad Q = a_2 + b_2 = a_2' + b_2', \tag{2}$$

and one has

$$\begin{aligned} a_1 &= (A - \underline{A})/2 + \delta, & b_1 &= P - (A - \underline{A})/2 - \delta, \\ a_2 &= (A - \underline{A})/2 - \delta, & b_2 &= Q - (A - \underline{A})/2 + \delta, \\ a_1' &= (A - \underline{A})/2 + \delta', & b_1 &= P - (A - \underline{A})/2 - \delta', \\ a_2' &= (A - \underline{A})/2 - \delta', & b_2 &= Q - (A - \underline{A})/2 + \delta'. \end{aligned} \tag{3}$$

One may introduce the light cone components

$$(a_1)_- = (a_1^2 + a_{1,t}^2)/(a_1)_+, \quad (b_1)_+ = (b_1^2 + b_{1,t}^2)/(b_1)_- \tag{4}$$

and, keeping into account that the soft blobs allow virtualities and transverse momenta only of the order of the hadronic scale R (in coordinates space), one has

$$a_1^2 \simeq a_{1,t}^2 \simeq b_1^2 \simeq b_{1,t}^2 = \mathcal{O}(1/R^2), \tag{5}$$

so, in conclusion, we get

$$\begin{aligned} (a_1)_+ &= \mathcal{O}(\sqrt{s}), & (a_1)_- &= \mathcal{O}(1/R^2\sqrt{s}), \\ (b_1)_+ &= \mathcal{O}(1/R^2\sqrt{s}), & (b_1)_- &= \mathcal{O}(\sqrt{s}). \end{aligned} \tag{6}$$

A similar argument holds for the primed variables and for the variables with index 2. One obtains

$$\begin{aligned} P_+ &\simeq (a_1)_+ \pm \mathcal{O}(1/R^2\sqrt{s}) \simeq (a_1')_+ \pm \mathcal{O}(1/R^2\sqrt{s}), \\ P_- &\simeq (b_1)_- \pm \mathcal{O}(1/R^2\sqrt{s}) \simeq (b_1')_- \pm \mathcal{O}(1/R^2\sqrt{s}), \\ Q_+ &\simeq (a_2)_+ \pm \mathcal{O}(1/R^2\sqrt{s}) \simeq (a_2')_+ \pm \mathcal{O}(1/R^2\sqrt{s}), \\ Q_- &\simeq (b_2)_- \pm \mathcal{O}(1/R^2\sqrt{s}) \simeq (b_2')_- \pm \mathcal{O}(1/R^2\sqrt{s}), \end{aligned} \tag{7}$$

and, as shown in Fig. 2, one has

$$k = (a_1 + a_2)/2 = (a_1' + a_2')/2, \quad \delta = (a_1 - a_2)/2, \quad \delta' = (a_1' - a_2')/2$$

which imply

$$\begin{aligned} \delta_- &\approx \delta'_- = \mathcal{O}(1/R^2\sqrt{s}), \\ \delta_+ &\approx \delta'_+ = \frac{1}{2}(P - Q)_+ \pm \mathcal{O}(1/R^2\sqrt{s}). \end{aligned} \tag{8}$$

The loop integrations over $\delta_\pm$, $\delta'_\pm$ are therefore restricted to a range of $\mathcal{O}(1/R^2\sqrt{s})$. The integrations on δ_- hence involve only the upper vertex ϕ_A and the propagators of the lines with momenta a_1 and a_2, whose "minus" components are also of $\mathcal{O}(1/R^2\sqrt{s})$. One needs in fact to keep into account only of the kinematical variables which grow as $\sqrt{s}$ in the hard interaction vertices; while the lower vertex ϕ_B and the propagators of the lines with momenta b_1 and b_2, whose "minus" components are of $\mathcal{O}(\sqrt{s})$, are practically constant for variations of δ_- of $\mathcal{O}(1/R^2\sqrt{s})$. Conversely for the integration on δ_+, while similar arguments hold for the integrations on δ'_-, δ'_+.

The dependence on the initial state is thus through quantities like

$$\psi(a_{1t}, a_{2t}, a_{1+}, a_{2+}, \underline{A}_-) \equiv \int \frac{\phi_A(a_1, a_2, \underline{A})}{a_1^2 a_2^2} \frac{d\delta_-}{2\pi} \tag{9}$$

while the values of a_{1+}, a_{2+}, $\underline{A}_-$ are determined by the final state observables P_+, Q_+, $\underline{A}^2$.

All different initial state quantities ψ are linked through the integrations on the transverse momenta a_{1t}, a_{2t}, etc. To deal with the transverse momentum integrations, one may introduce the two-dimensional Fourier transform

$$\begin{aligned} &\tilde{\psi}(s_1, s_2,\ a_{1+}, a_{2+}, \underline{A}_-) \\ &\quad = \int d^2a_{1t}d^2a_{2t} \frac{e^{i(\mathbf{s_1}\cdot\mathbf{a_{1t}}+\mathbf{s_2}\cdot\mathbf{a_{2t}})}}{(2\pi)^2} \psi(a_{1t}, a_{2t}, a_{1+}, a_{2+}, \underline{A}_-). \end{aligned} \tag{10}$$

Keeping into account the conservation constraints in the vertices, the flow of transverse momenta may be expressed as shown in Fig. 2. The discontinuity of the diagram is thus given by a Fourier integral over all independent transverse momenta. The argument of the Fourier exponential coming from ψ and ψ^* is

$$\begin{aligned} &(\mathbf{k}+\delta)\cdot\mathbf{s}_1 + (\mathbf{k}-\delta)\cdot\mathbf{s}_2 + (-\mathbf{k}-\delta+\mathbf{P})\cdot\mathbf{s}_1' + (-\mathbf{k}+\delta+\mathbf{Q})\cdot\mathbf{s}_2' \\ &\quad - (\mathbf{k}+\delta')\cdot\mathbf{s}_1'' - (\mathbf{k}-\delta')\cdot\mathbf{s}_2'' - (-\mathbf{k}-\delta'+\mathbf{P})\cdot\mathbf{s}_1''' - (-\mathbf{k}+\delta'+\mathbf{Q})\cdot\mathbf{s}_2''' \\ &= (\mathbf{k}+\delta)\cdot(\mathbf{S}+\mathbf{s}/2) + (\mathbf{k}-\delta)\cdot(\mathbf{S}-\mathbf{s}/2) + (-\mathbf{k}-\delta+\mathbf{P})\cdot(\mathbf{S}'+\mathbf{s}'/2) \\ &\quad + (-\mathbf{k}+\delta+\mathbf{Q})\cdot(\mathbf{S}'-\mathbf{s}'/2) - (\mathbf{k}+\delta')\cdot(\mathbf{S}''+\mathbf{s}''/2) - (\mathbf{k}-\delta')\cdot(\mathbf{S}''-\mathbf{s}''/2) \\ &\quad - (-\mathbf{k}-\delta'+\mathbf{P})\cdot(\mathbf{S}'''+\mathbf{s}'''/2) - (-\mathbf{k}+\delta'+\mathbf{Q})\cdot(\mathbf{S}'''-\mathbf{s}'''/2), \end{aligned} \tag{11}$$

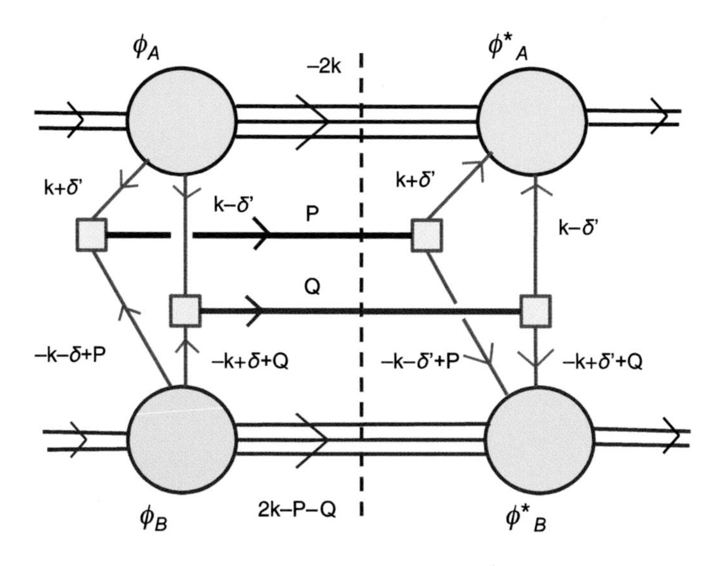

Fig. 2 Transverse momenta flow in the double parton scattering diagram.

where $\mathbf{s}_1, \mathbf{s}_2, \mathbf{s}'_1, \mathbf{s}'_2$ etc. are the transverse coordinates conjugate to the transverse momenta in Fig. 2 and $\mathbf{S}, \mathbf{s}, \mathbf{S}', \mathbf{s}'$ etc. are the centres of mass and the relative transverse coordinates of $(\mathbf{s}_1, \mathbf{s}_2)$, $(\mathbf{s}'_1, \mathbf{s}'_2)$ etc.

The integrations on transverse momenta give:

$$\begin{aligned}
&d^2\delta \rightarrow \mathbf{s} = \mathbf{s}', \\
&d^2\delta' \rightarrow \mathbf{s}'' = \mathbf{s}''', \\
&d^2\mathbf{P} \rightarrow \mathbf{S}' + \mathbf{s}'/2 - \mathbf{S}''' - \mathbf{s}'''/2 = 0, \\
&d^2\mathbf{Q} \rightarrow \mathbf{S}' - \mathbf{s}'/2 - \mathbf{S}''' + \mathbf{s}'''/2 = 0, \\
&d^2\mathbf{k} \rightarrow \mathbf{S} = \mathbf{S}'',
\end{aligned} \tag{12}$$

which imply

$$\mathbf{s} = \mathbf{s}' = \mathbf{s}'' = \mathbf{s}''' \quad \text{and} \quad \mathbf{S} = \mathbf{S}'',\ \mathbf{S}' = \mathbf{S}'''. \tag{13}$$

The cross-section may hence be expressed as a function of $|\tilde{\psi}|^2$. More precisely one may define the two-body parton distribution

$$\Gamma(x_1, x_2; \mathbf{s}) \equiv \int |\tilde{\psi}(\mathbf{s}_1, \mathbf{s}_2, a_{1+}, a_{2+}, \underline{A}_-)|^2 d^2\mathbf{S} dA_-, \tag{14}$$

where $x_i = (a_i)_+/A_+$, and the cross-section is proportional to the convolution

$$\int \Gamma_A(x_1, x_2; \mathbf{s}) \Gamma_B(x'_1, x'_2; \mathbf{s}) d^2\mathbf{s}, \tag{15}$$

which shows that the cross-section is given by the incoherent superposition of two different hard collisions, where the two interacting pairs are localized at the same relative transverse distance $\mathbf{s}$. The hard part of the interaction is therefore disconnected in two different parts, which are physically separated in the transverse coordinates space by the distance $\mathbf{s}$, which is of the order of the hadron size, namely much larger as compared with the transverse size of the two regions where the hard interactions take place.

One should point out that this conclusion relies on the assumption that $\Gamma_A(x_1, x_2; \mathbf{s})$ is not singular for $\mathbf{s} \to 0$ or, in other words, on the assumption that $\Gamma_A(x_1, x_2; \mathbf{s})$ is characterized by a non-perturbative scale of the order of the hadron mass at small $\mathbf{s}$. On the other hand, a singular behavior of Γ at small $\mathbf{s}$ would necessarily induce a large relative momentum between the final states with overall momenta P and Q, while, in the case of DPIs, the configurations one is looking for, between the states with momenta P and Q, are those characterized by a relative momentum of the order of the hadron mass.

The final expression of the DPI cross-section σ_D is as follows[30]:

$$\frac{d\sigma_D|_{ij}}{\prod_{k=1,2} dx_k dx'_k} = \frac{1}{1+\delta_{ij}} \int_{p_t^c} \Gamma_A(x_1, x_2; \mathbf{s}) \hat{\sigma}_i(x_1 x'_1) \hat{\sigma}_j(x_2 x'_2) \Gamma_B(x'_1, x'_2; \mathbf{s}) d^2\mathbf{s}$$
$$\equiv \frac{1}{1+\delta_{ij}} \frac{\sigma_i(x_1 x'_1)\sigma_j(x_2 x'_2)}{\sigma_{\text{eff}}}, \tag{16}$$

where the two partonic interactions are labeled with the indices i, j and $1+\delta_{ij}$ is the symmetry factor for the case of two identical interactions. The transverse momenta of the produced partons are integrated with the cutoff p_t^c, $\hat{\sigma}_{i,j}$ are the partonic cross-sections and $\sigma_{i,j}$ the hadron–hadron inclusive cross-sections. The last line here above defines the "effective cross-section" σ_{eff}, which summarizes in a single quantity all unknowns in the process.

3. Interference Term

A interference diagram between an SPI and a DPI is shown in Fig. 3, with the corresponding flow of momenta. P and Q represent the total four-momenta of the partonic states produced in the two interactions, e.g., of two pairs of large p_t jets.

It may be useful to look at the picture of the interference process in transverse space. When following the same line of reasoning of the previous section, the amplitude is expressed by the Fourier transform of the following combination, of functions of transverse coordinates of the initial state interacting partons:

$$\int \tilde{\varphi}_A(\mathbf{s}) \tilde{\psi}^*_A(\mathbf{s}_1, \mathbf{s}_2) \tilde{\varphi}_B(\mathbf{s}') \tilde{\psi}^*_B(\mathbf{s}'_1, \mathbf{s}'_2) d^2\mathbf{s}\, d^2\mathbf{s}_1 d^2\mathbf{s}_2 d^2\mathbf{s}' d^2\mathbf{s}'_1 d^2\mathbf{s}'_2, \tag{17}$$

where $\tilde{\psi}_{A,B}$ have been defined in the previous section, $\tilde{\varphi}_{A,B}$ are the analogous quantities describing the soft components on the left side of the cut in Fig. 3 and the dependence on the longitudinal variables is implicit.

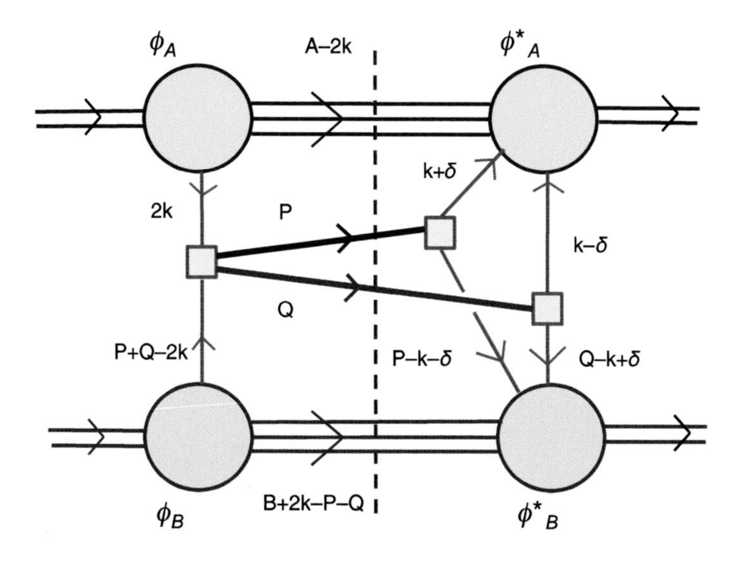

Fig. 3 Flow of momenta in a interference diagram.

The argument of the Fourier exponential is

$$2\mathbf{k}\cdot\mathbf{s}+(\mathbf{P}+\mathbf{Q}-2\mathbf{k})\cdot\mathbf{s}'-(\mathbf{k}+\delta)\cdot\mathbf{s}_1-(\mathbf{k}-\delta)\cdot\mathbf{s}_2-(\mathbf{P}-\mathbf{k}-\delta)\cdot\mathbf{s}_1'$$
$$-(\mathbf{Q}-\mathbf{k}+\delta)\cdot\mathbf{s}_2', \tag{18}$$

where $\mathbf{s}$, $\mathbf{s}_1, \mathbf{s}_2, \mathbf{s}', \mathbf{s}_1', \mathbf{s}_2'$ are the transverse coordinates conjugate to the transverse momenta shown in Fig. 3.

The integrations on transverse momenta give

$$\begin{aligned} d^2\mathbf{P} &\to \mathbf{s}'-\mathbf{s}_1'=0,\\ d^2\mathbf{Q} &\to \mathbf{s}'-\mathbf{s}_2'=0,\\ d^2\delta &\to \mathbf{s}_1-\mathbf{s}_2-\mathbf{s}_1'+\mathbf{s}_2'=0,\\ d^2\mathbf{k} &\to 2\mathbf{s}-2\mathbf{s}'-\mathbf{s}_1-\mathbf{s}_2+\mathbf{s}_1'+\mathbf{s}_2'=0, \end{aligned} \tag{19}$$

which imply

$$\mathbf{s}'=\mathbf{s}_1'=\mathbf{s}_2' \quad \text{and} \quad \mathbf{s}=\mathbf{s}_1=\mathbf{s}_2$$

Expression (18) thus reduces to

$$\int \tilde{\varphi}_A(\mathbf{s})\tilde{\psi}_A^*(\mathbf{s},\mathbf{s})\tilde{\varphi}_B(\mathbf{s}')\tilde{\psi}_B^*(\mathbf{s}',\mathbf{s}')d^2\mathbf{s}\,d^2\mathbf{s}'$$
$$=\left[\int \tilde{\varphi}_A(\mathbf{s})\tilde{\psi}_A^*(\mathbf{s},\mathbf{s})d^2\mathbf{s}\right]\cdot\left[\int \tilde{\varphi}_B(\mathbf{s}')\tilde{\psi}_B^*(\mathbf{s}',\mathbf{s}')d^2\mathbf{s}'\right],$$

which shows that the two interactions are localized in the same region in transverse space.

The overlap of the two amplitudes in Fig. 3 restricts considerably the kinematical configurations available to the final state. In fact, the amplitude on the left-hand side of the cut in Fig. 3 depends, in particular, on $\mathbf{P} - \mathbf{Q}$. On the other hand, the difference between the two partons' transverse momenta entering the soft blob ϕ_B^*, in the amplitude on the right-hand side of the cut in Fig. 3, is $(\mathbf{P} - \mathbf{k} - \delta) - (\mathbf{Q} - \mathbf{k} + \delta) = \mathbf{P} - \mathbf{Q} - 2\delta$ and the difference between the two partons' transverse momenta entering the soft blob ϕ_A^* is $(\mathbf{k} + \delta) - (\mathbf{k} - \delta) = 2\delta$. So, when δ is small, the relative momentum of the two lines entering the soft blow ϕ_B^* is about $\mathbf{P} - \mathbf{Q}$ and when $2\delta \approx \mathbf{P} - \mathbf{Q}$, the relative momentum of the two lines entering the soft blow ϕ_A^* is about $\mathbf{P} - \mathbf{Q}$: As one may notice, by looking at Fig. 3, the soft blob in the amplitude on the left-hand side of the cut limits $\mathbf{P} + \mathbf{Q}$ to values of the order of $1/R$, with R the hadron radius, without putting a similar bound on the difference $\mathbf{P} - \mathbf{Q}$, which is only governed by the hard process. As in the case of the diagonal contribution to the cross-section in Fig. 2, the difference $\mathbf{P} - \mathbf{Q}$ is, on the contrary, limited to values of the order of $1/R$ by the amplitude on the right-hand side of the cut. The final state configurations generated by the two amplitudes in Fig. 3 are thus very different and, in a kinematical regime where the single and double parton scattering integrated cross-sections are comparable, the effect is a sizable reduction of the contribution of the interference term to the cross-section.[4]

One should point out that the same phase space restriction holds for the DPI diagonal term discussed in the previous section (see Chapter 2 and Ref. 3). The reduction of the final state phase-space available to DPIs, as compared to SPIs, originates from the non-perturbative scale, of order $\mathcal{O}(R)$, characterizing the process. By dimensional reasons (cf. Eq. 16) DPIs are thus suppressed at large p_t^c and, conversely, DPIs grow faster, as compared to SPIs, when p_t^c decreases. The c.m. energy and the cut p_t^c determine the lower limit of the initial parton's fractional momenta in the process and thus the initial parton flux. At large c.m. energies and fixed p_t^c, the parton flux of DPIs at low x grows about as the square of the parton flux of SPIs, which may compensate the phase space restriction, in such a way that the integrated rate of DPIs can also exceed the integrated rate of SPIs. The flux factor of the interference term grows less rapidly and, in comparison with the diagonal terms, its contribution to the cross-section is thus suppressed.

4. The Simplest Poissonian Model

A simple generalization of the expression of the DPI cross-section, where all correlations are neglected, leads to a Poissonian distribution of MPIs, with average number depending on the value of the impact parameter,[34] which is a basic feature of most Monte Carlo simulation codes of high energy hadronic interactions.[35–37]

One may start by introducing in the SPI cross-section the three-dimensional parton density $D(x, b)$, namely the average number of partons with a given momentum fraction x and with transverse coordinate b (the dependence on flavour and on the resolution of the process is understood) and making the simplifying assumption $D(x,b) = G(x)f(b)$, where $G(x)$ is the usual parton distribution function and $f(b)$ is normalized to one. The SPI inclusive cross-section is thus given by

$$\sigma_S = \int_{p_t^c} G_A(x)\hat{\sigma}(x,x')G_B(x')dxdx' \tag{20}$$

$$= \int_{p_t^c} G_A(x)f_A(b)\hat{\sigma}(x,x')G_B(x')f_B(b-\beta)d^2bd^2\beta dxdx', \tag{21}$$

where integrations are done with the lower limit in the exchanged transverse momenta p_t^c.

The expression allows a simple geometrical interpretation. Given the large momentum exchange, the partonic interaction is localized in the overlap volume of the two hadrons and one may identify with b and with $b-\beta$ the transverse coordinates of the two colliding partons, β being the impact parameter of the hadronic collision. In the case of a DPI, neglecting all correlations in the multi-parton distributions, one has

$$\Gamma(x_1,x_2;r) = G(x_1)G(x_2)F(r) \equiv G(x_1)G(x_2)\int f(b)f(b-r)d^2b \tag{22}$$

and, if the two interactions are identical, the DPI cross-section σ_D in Eq. (16) is thus expressed by

$$\begin{aligned}\sigma_D &= \frac{1}{2!}\int_{p_t^c} G_A(x_1)G_A(x_2)f_A(b)f_A(b-r)\hat{\sigma}(x_1,x_1')d^2bdx_1dx_1' \\ &\quad\times \hat{\sigma}(x_2,x_2')G_B(x_1')G_B(x_2')f_B(b')f_B(b'-r)d^2b'dx_2dx_2'd^2r \\ &= \frac{1}{2!}\int_{p_t^c} G_A(x_1)f_A(b_1)\hat{\sigma}(x_1,x_1')G_B(x_1')f_B(b_1-\beta)d^2b_1dx_1dx_1' \\ &\quad\times G_A(x_2)f_A(b_2)\hat{\sigma}(x_2,x_2')G_B(x_2')f_B(b_2-\beta)d^2b_2dx_2dx_2'd^2\beta \\ &= \int \frac{1}{2!}\left(\int_{p_t^c} G_A(x)f_A(b)\hat{\sigma}(x,x')G_B(x')f_B(b-\beta)d^2bdxdx'\right)^2 d^2\beta,\end{aligned} \tag{23}$$

where the following change of variables has been made: $b_1 = b$, $b_1 - \beta = b'$, $b_2 = b - r$ and $b_2 - \beta = b' - r$.

The expression may be readily generalized to the case of the inclusive cross-section of N identical partonic interactions σ_N:

$$\sigma_N = \int \frac{1}{N!}\left(\int_{p_t^c} G_A(x)f_A(b)\hat{\sigma}(x,x')G_B(x')f_B(b-\beta)d^2bdxdx'\right)^N d^2\beta.$$

The cross-sections σ_N may violate unitarity when p_t^c is small. One may however notice that the integrand here above is dimensionless and that it may thus be understood as the probability to have N-partonic collisions in a inelastic event. The unitarity problem is thus solved by normalizing the integrand. The probability $P_N(\beta)$, of having N partonic interactions in a hadronic collisions with impact parameter β, is thus given by

$$\begin{aligned}P_N(\beta) &\equiv \frac{\left(\sigma_S F(\beta)\right)^N}{N!} e^{-\sigma_S F(\beta)}, \quad \text{where} \\ \sigma_S F(\beta) &\equiv \int_{p_t^c} G_A(x)\hat{\sigma}(x,x')G_B(x')dxdx' \int f_A(b)f_B(b-\beta)d^2b.\end{aligned} \tag{24}$$

By summing all probabilities one obtains in this way the hard cross-section σ_{hard}, namely the contribution to the inelastic cross-section due to all events with *at least* one partonic interaction with momentum transfer exchange larger than the cutoff p_t^c:

$$\begin{aligned}\sigma_{\text{hard}} &= \sum_{N=1}^{\infty} \int P_N(\beta)d^2\beta = \sum_{N=1}^{\infty} \int d^2\beta \frac{\left(\sigma_S F(\beta)\right)^N}{N!} e^{-\sigma_S F(\beta)} \\ &= \int d^2\beta [1 - e^{-\sigma_S F(\beta)}].\end{aligned} \tag{25}$$

For large β the overlap function $F(\beta)$ goes to zero. When σ_S is large, σ_{hard} thus gives a measure of the size of the overlap region.

The cross-section σ_{hard} depends on the choice of the cutoff p_t^c. By considering the case of two different cutoffs in transverse momenta, p_1 and p_2, with $p_1 < p_2$ and writing $\sigma_S F(\beta) = T(\beta)$, one identifies in T a softer component T_S, when $p_1 < p_t < p_2$, and a harder component T_H, when $p_2 < p_t$, in such a way that $T = T_H + T_S$. One may thus look for the MPI contribution to the total inelastic cross-section due to the harder component T_H. By making in Eq. (25) the replacement

$$T^N \Rightarrow \sum_{K=1}^{N} \binom{N}{K} T_H^K \times T_S^{N-K} = (T_H + T_S)^N - T_S^N$$

and, summing over N, without considering the color degrees of freedom and even without distinguishing between quarks and gluons, one obtains in this way $\sigma_{\text{hard}}(p_2)$, namely the contribution to the inelastic cross-section due to all events with at least one partonic interaction with momentum transfer exchange larger than p_2. Actually:

$$\begin{aligned}\sigma_{\text{hard}}(p_2) &= \int d^2\beta \left[\sum_{N=1}^{\infty} \frac{(T_H + T_S)^N}{N!} e^{-(T_H+T_S)} - \sum_{N=1}^{\infty} \frac{T_S^N}{N!} e^{-(T_H+T_S)}\right] \\ &= \int d^2\beta [e^{(T_H+T_S)} \times e^{-(T_H+T_S)} - e^{T_S} \times e^{-(T_H+T_S)}] \\ &= \int d^2\beta [1 - e^{-T_H(\beta)}] = \int d^2\beta \sum_{N=1}^{\infty} \frac{T_H(\beta)^N}{N!} e^{-T_H(\beta)}.\end{aligned} \tag{26}$$

As defined, σ_{hard} thus satisfies a natural consistency requirement: it is insensitive to the softer cutoff p_1 and, when replacing p_t^c with p_2, the expression in Eq. (25) is the same as the expression in Eq. (26).

While the hard component of the inelastic cross-section is given by a Poissonian distribution of MPIs, with the average number depending on the impact parameter of the hadronic collision, the inclusive cross-section σ_N acquires a well-defined meaning in terms of factorial moment of the distribution in multiplicity of partonic collisions. By working out the average number of collisions one in fact obtains:

$$\langle N\rangle\sigma_{\text{hard}} = \int d^2\beta \sum_{N=1}^{\infty} \frac{N\left[\sigma_S F(\beta)\right]^N}{N!} e^{-\sigma_S F(\beta)} = \int d^2\beta \sigma_S F(\beta) = \sigma_S,$$

which is precisely the expression of the single scattering inclusive cross-section. More in general, one may write:

$$\begin{aligned}\frac{\langle N(N-1)\ldots(N-K+1)\rangle}{K!}\sigma_{\text{hard}} &= \int d^2\beta \sum_{N=K}^{\infty} \frac{N(N-1)\ldots(N-K+1)}{K!} P_N(\beta) \\ &= \int d^2\beta \frac{1}{K!}\left[\sigma_S F(\beta)\right]^K = \sigma_K.\end{aligned} \tag{27}$$

Unitarity corrections thus cancel out in all inclusive cross-sections σ_K, which is an explicit proof of the validity of the cancellation of Abramovsky, Gribov and Kancheli (AGK)[38] in the actual uncorrelated model of MPIs.

It's worth pointing out a few features:

- One may define two different sets of MPI cross-sections: the inclusive cross-sections, given by the factorial moments of the distribution in multiplicity of partonic collisions, and the exclusive cross-sections, namely the different contributions to σ_{hard} in Eq. (25). Both sets of cross-sections are expressed fully explicitly in terms of the same quantity, the average number of collisions at a fixed impact parameter, $\sigma_S F(\beta)$.
- All inclusive cross-sections are divergent for $p_t^c \to 0$. The relation with the multiplicity of collisions shows that the divergence is due to the factorial moments. In other words the cause of the divergence is the number of collisions, which become very large at low p_t.
- While all inclusive cross-sections become increasingly large at low p_t, because of the increasingly large number of partonic interactions, all exclusive cross-section, where the number of hard interactions is kept fixed, become, on the contrary, smaller and smaller at low p_t.

5. A Functional Approach

A rather general approach to MPIs is by a functional formalism.[39]

One may introduce the exclusive n-parton distributions $W_n(u_1 \dots u_n)$, namely the probabilities to have the hadron in a configuration with n partons with coordinates u_i, which represent the variables (b_i, x_i), being b the transverse partonic coordinate and x the corresponding fractional momentum. The scale for the distributions is given by the cutoff p_t^c, that defines the separation between soft and hard collisions, and the distributions are symmetric in the variables u_i. The generating functional is defined by

$$\mathscr{Z}[J] = \sum_n \frac{1}{n!} \int J(u_1) \dots J(u_n) W_n(u_1 \dots u_n) du_1 \dots du_n. \tag{28}$$

The conservation of the probability implies the normalization condition $\mathscr{Z}[1] = 1$. The probabilities of the various configurations, namely the exclusive distributions, are the coefficients of the expansion of $\mathscr{Z}[J]$ for $J = 0$. The coefficients of the expansion of $\mathscr{Z}[J]$ for $J = 1$ give the many-body densities, i.e., the inclusive distributions:

$$D_1(u) = \left. \frac{\delta \mathscr{Z}}{\delta J(u)} \right|_{J=1}, \quad D_2(u_1, u_2) = \left. \frac{\delta^2 \mathscr{Z}}{\delta J(u_1) \delta J(u_2)} \right|_{J=1} \quad \dots . \tag{29}$$

Correlations, which describe how much the distribution deviates from a Poissonian, are obtained by the expansion of the logarithm of the generating functional, $\mathscr{F}[J] \equiv \ln \mathscr{Z}[J]$, for $J = 1$:

$$\begin{aligned} \mathscr{F}[J] = & \int D_1(u)[J(u) - 1] du \\ & + \sum_{n=2}^{\infty} \frac{1}{n!} \int C_n(u_1 \dots u_n)[J(u_1) - 1] \dots [J(u_n) - 1] du_1 \dots du_n. \end{aligned}$$

Obviously one has $\mathscr{F}[1] = 0$ and, in the Poissonian case, $C_n \equiv 0, n \geq 2$.

A general expressions of the semi-hard cross-section, which takes into account of all possible MPIs, is

$$\begin{aligned} \sigma_{\text{hard}} = & \int d\beta \int \sum_n \frac{1}{n!} \frac{\delta}{\delta J(u_1)} \cdots \frac{\delta}{\delta J(u_n)} \mathscr{Z}_A[J] \\ & \times \sum_m \frac{1}{m!} \frac{\delta}{\delta J'(u'_1 - \beta)} \cdots \frac{\delta}{\delta J'(u'_m - \beta)} \mathscr{Z}_B[J'] \\ & \times \left. \left\{ 1 - \prod_{i=1}^{n} \prod_{j=1}^{m} [1 - \hat{\mathrm{P}}_{i,j}(u, u')] \right\} \prod du du' \right|_{J=J'=0}, \end{aligned} \tag{30}$$

where β is the impact parameter between the two interacting hadrons A and B and $\hat{\mathrm{P}}_{i,j}$ is the probability for the parton i (of hadron A) to have a hard interaction with the parton j (of hadron B).

Analogously to the case of nucleus–nucleus interactions,[40] the cross-section is obtained in this way by summing all contributions due to all different hadronic configurations (the sums over n and m) and, for each pair of values n and m, one has a contribution to $\sigma_{\rm hard}$ when at least one hard interaction takes place. The interaction probability is here fully determined by the two-body interaction probabilities $\hat{\rm P}_{i,j}$, in such a way that the term in curly brackets in Eq. (30) represents the probability to have at least one interaction. In the cross-section both disconnected interactions with $n = m$ and connected interactions, also with $n \neq m$, are included. One may focus on multiple disconnected interactions, each initiated by two partons. Only the terms with $n = m$ have thus to be taken into account. One may write the term in curly brackets as

$$S \equiv 1 - \exp \sum_{ij} \ln(1 - \hat{\rm P}_{ij}) = 1 - \exp\left[-\sum_{ij}\left(\hat{\rm P}_{ij} + \frac{1}{2}\hat{\rm P}_{ij}\hat{\rm P}_{ij} + \cdots\right)\right], \tag{31}$$

where all repeated indices have to be removed. Only of the first term of the expansion of the logarithm thus contributes and one has to make the following substitution:

$$S \Rightarrow 1 - \exp\left[-\sum_{ij}\hat{\rm P}_{ij}\right] \Rightarrow \sum_{ij}\hat{\rm P}_{ij} - \frac{1}{2}\sum_{ij}\sum_{k\neq i, l\neq j}\hat{\rm P}_{ij}\hat{\rm P}_{kl}\ldots. \tag{32}$$

The resulting cross-section is expressed in a rather compact way:

$$\begin{aligned}\sigma_{\rm hard}(\beta) &= \exp(\partial_J)\cdot\exp(\partial_{J'})[1 - \exp(-\partial_J\cdot\hat{\rm P}\cdot\partial_{J'})]\mathscr{Z}_A[J]\mathscr{Z}_B[J']|_{J=J'=0}\\ &= [1 - \exp(-\partial_J\cdot\hat{\rm P}\cdot\partial_{J'})]\mathscr{Z}_A[J]\mathscr{Z}_B[J']|_{J=J'=1},\end{aligned} \tag{33}$$

where all convolutions are understood. In the simplest non-trivial case all correlations C_n with $n > 2$ can be neglected and the cross-section simplifies to:

$$\begin{aligned}\sigma_{\rm hard}(\beta) = &\left[1 - \exp\left\{-\int du du' \partial_J \hat{\rm P}(u,u')\partial_{J'}\right\}\right]\\ &\cdot \exp\left\{\int D_A(u)J(u)du + \frac{1}{2}\int C_A(u,v)J(u)J(v)dudv\right\}\\ &\cdot \exp\left\{\int D_B(u)J'(u)du + \frac{1}{2}\int C_B(u,v)J'(u)J'(v)dudv\right\}\Bigg|_{J=J'=0},\end{aligned} \tag{34}$$

which can be worked out explicitly. One obtains[39]

$$\sigma_{\rm hard}(\beta) = 1 - \exp\left[-\sum_n \frac{a_n}{2} - \sum_n \frac{b_n}{2n}\right], \tag{35}$$

where

$$a_n = (-1)^{n+1} \int D_A(u_1)\hat{\mathrm{P}}(u_1, u_1')C_B(u_1', u_2')\hat{\mathrm{P}}(u_2', u_2)C_A(u_2, u_3)$$
$$\cdots \hat{\mathrm{P}}(u_n, u_n')D_B(u_n') \prod_{i=1}^{n} du_i du_i' + A \leftrightarrow B,$$
$$b_n = (-1)^{n+1} \int C_A(u, u_1)\hat{\mathrm{P}}(u_1, u_1')C_B(u_1', u_2')$$
$$\cdots \hat{\mathrm{P}}(u_n, u_n')C_B(u_n', u')\hat{\mathrm{P}}(u', u)dudu' \prod_{i=1}^{n} du_i du_i'. \tag{36}$$

Remarkably, *also in the general case, where all correlations C_n are taken into account*, by working out the factorial moments of the distribution in the number of collisions, one obtains the same result of the simplest Poissonian model.

From Eq. (33) one may in fact express the hard cross-section as a sum of MPIs:

$$\begin{aligned}\sigma_{\mathrm{hard}}(\beta) &= [1 - \exp(-\partial_J \cdot \hat{\mathrm{P}} \cdot \partial_{J'})]\mathscr{L}_A[J]\mathscr{L}_B[J']|_{J=J'=1} \\ &= \sum_{N=1}^{\infty} \frac{(\partial_J \cdot \hat{\mathrm{P}} \cdot \partial_{J'})^N}{N!} \mathrm{e}^{-\partial_J \cdot \hat{\mathrm{P}} \cdot \partial_{J'}} \mathscr{L}_A[J]\mathscr{L}_B[J']|_{J=J'=1},\end{aligned} \tag{37}$$

and the average number of collisions is

$$\begin{aligned}\langle N \rangle \sigma_{\mathrm{hard}}(\beta) &= \sum_{N=1}^{\infty} \frac{N(\partial_J \cdot \hat{\mathrm{P}} \cdot \partial_{J'})^N}{N!} \mathrm{e}^{-\partial_J \cdot \hat{\mathrm{P}} \cdot \partial_{J'}} \mathscr{L}_A[J]\mathscr{L}_B[J']|_{J=J'=1} \\ &= \partial_{J_1} \cdot \hat{\mathrm{P}} \cdot \partial_{J_1'} \sum_{N=0}^{\infty} \frac{(\partial_J \cdot \hat{\mathrm{P}} \cdot \partial_{J'})^N}{N!} \mathrm{e}^{-\partial_J \cdot \hat{\mathrm{P}} \cdot \partial_{J'}} \mathscr{L}_A[J]\mathscr{L}_B[J']|_{J=J'=1} \\ &= (\partial_{J_1} \cdot \hat{\mathrm{P}} \cdot \partial_{J_1'})\mathscr{L}_A[J]\mathscr{L}_B[J']|_{J=J'=1} \\ &= \int D_A(x_1; b_1)\hat{\sigma}(x_1 x_1')D_B(x_1'; b_1 - \beta)dx_1 dx_1' d^2 b_1 \\ &\equiv \sigma_S(\beta),\end{aligned} \tag{38}$$

where $\hat{\sigma}(x_1 x_1')$ is the parton–parton cross-section, integrated with $p_t > p_t^c$. Given the localization of the interactions in transverse space, the parton–parton interaction probability has in fact been represented as a δ-function of the transverse coordinates: $\hat{\mathrm{P}}(u, u') = \hat{\sigma}(x, x')\delta(\mathbf{b} - \mathbf{b}')$. Analogously

$$\begin{aligned}\frac{\langle N(N-1)\rangle}{2!}\sigma_{\mathrm{hard}}(\beta) &= \frac{1}{2!}\int D_A(x_1 x_2; b_1 b_2)\hat{\sigma}(x_1 x_1')\hat{\sigma}(x_2 x_2') \\ &\quad \times D_B(x_1' x_2'; b_1 - \beta, b_2 - \beta)dx_1 dx_1' d^2 b_1 dx_2 dx_2' d^2 b_2, \\ &\equiv \sigma_D(\beta)\end{aligned} \tag{39}$$

and, in general,

$$
\begin{aligned}
&\frac{\langle N(N-1)\dots(N-K+1)\rangle}{K!}\sigma_{\text{hard}}(\beta) \\
&= \frac{1}{K!}\int D_A(x_1\dots x_K;b_1\dots b_K)\hat{\sigma}(x_1x_1')\dots\hat{\sigma}(x_Kx_K') \\
&\quad\times D_B(x_1'\dots x_K';b_1-\beta\dots b_K-\beta)dx_1dx_1'd^2b_1\dots dx_Kdx_K'd^2b_K \\
&\equiv \sigma_K(\beta),
\end{aligned}
\tag{40}
$$

which shows that *when considering the case, where all connected interactions are neglected and where each partonic interaction is initiated by two partons*, for any choice of multiparton distributions, the inclusive cross-sections are given by the factorial moments of the distribution in the number of partonic collisions.

The validity of the cancellation of AGK[38] is thus proved explicitly in the actual model of MPIs on rather general grounds.

6. Exclusive Cross-Sections, Sum Rules

With the simplifying assumptions here above, in *pp* collisions the inclusive cross-sections are thus given by the factorial moments of the distribution in the number of MPIs. The most basic information on the distribution in the number of collisions, the average number, corresponds to the single scattering inclusive cross-section of the pQCD parton model and, analogously, the K-parton scattering inclusive cross-section σ_K corresponds to the Kth factorial moment of the distribution in the number of collisions.

As already pointed out, a way alternative to the set of factorial moments, to provide the whole information of the distribution, is represented by the set of the different terms of the probability distribution of multiple collisions. In addition to the set of the inclusive cross-sections σ_K, one may thus consider the set of the exclusive cross-sections $\tilde{\sigma}_N$, which correspond to the different terms of the probability distribution and which represent the cross-sections where one selects the events where *only* a given number N of collisions are present.[41,42] The following relations thus hold:

$$
\sigma_{\text{hard}} \equiv \sum_{N=1}^{\infty} \tilde{\sigma}_N, \quad \sigma_K \equiv \sum_{N=K}^{\infty} \binom{N}{K} \tilde{\sigma}_N, \tag{41}
$$

which may be also understood as a set of sum rules connecting the inclusive and the exclusive cross-sections.

While the non-perturbative input to the inclusive cross-section σ_K is given by the K-parton distributions of the hadron structure, as implicit in Eqs. (25) and (37), the non-perturbative input to the exclusive cross-sections is given by an infinite set of multi-parton distributions. On the other hand, the request of being in a perturbative regime limits the number of partonic collisions and the sum rules in Eq. (41) can be saturated by a few terms, in such a way that the exclusive cross-sections can be expressed by finite combinations of inclusive cross-sections.

A particular case, already considered in the previous section, where all correlations C_n with $n > 2$ are negligible, can be worked out fully explicitly. The exponential in Eq. (35) represents the probability of no interaction at a given impact parameter β and all exclusive cross-sections can be obtained from the argument of the exponential.

Consider the interaction probability

$$1 - \prod_{i,j=1}^{N} (1 - \hat{\mathrm{P}}_{ij}), \tag{42}$$

where N is the maximal number of possible partonic interactions between two given initial partonic configurations and each index assumes a given value only once, in such a way that possible connected interactions are not included. The probability of having only a single interaction is

$$\left(-\frac{\partial}{\partial g}\right) \prod_{i,j=1}^{N} (1 - g\hat{\mathrm{P}}_{ij}) \Bigg|_{g=1} = \sum_{kl} \hat{\mathrm{P}}_{kl} \prod_{ij \neq kl}^{N} (1 - g\hat{\mathrm{P}}_{ij}) \Bigg|_{g=1}, \tag{43}$$

the probabilities of a double and of a triple interaction are

$$\begin{aligned} \frac{1}{2!}\left(-\frac{\partial}{\partial g}\right)^2 \prod_{i,j=1}^{N} (1 - g\hat{\mathrm{P}}_{ij}) \Bigg|_{g=1} &= \frac{1}{2!} \sum_{kl} \sum_{rs} \hat{\mathrm{P}}_{kl}\hat{\mathrm{P}}_{rs} \prod_{ij \neq (kl,rs)}^{N} (1 - g\hat{\mathrm{P}}_{ij}) \Bigg|_{g=1}, \\ \frac{1}{3!}\left(-\frac{\partial}{\partial g}\right)^3 \prod_{i,j=1}^{N} (1 - g\hat{\mathrm{P}}_{ij}) \Bigg|_{g=1} &= \frac{1}{3!} \sum_{kl} \sum_{rs} \sum_{tu} \hat{\mathrm{P}}_{kl}\hat{\mathrm{P}}_{rs}\hat{\mathrm{P}}_{tu} \prod_{ij \neq (kl,rs,tu)}^{N} (1 - g\hat{\mathrm{P}}_{ij}) \Bigg|_{g=1}, \end{aligned} \tag{44}$$

while the corresponding expressions for the exclusive cross-sections are

$$\begin{aligned} \left(-\frac{\partial}{\partial g}\right) e^{-X(g)} \Bigg|_{g=1} &= X'(g)\, e^{-X(g)}|_{g=1}, \\ \frac{1}{2!}\left(-\frac{\partial}{\partial g}\right)^2 e^{-X(g)} \Bigg|_{g=1} &= \frac{1}{2!}\{[X'(g)]^2 - X''(g)\}\, e^{-X(g)}|_{g=1}, \\ \frac{1}{3!}\left(-\frac{\partial}{\partial g}\right)^3 e^{-X(g)} \Bigg|_{g=1} &= \frac{1}{3!}\{X'''(g) + [X'(g)]^3 - 3X'(g)X''(g)\}\, e^{-X(g)}|_{g=1}, \end{aligned} \tag{45}$$

where $X = \frac{1}{2}(\sum a_n + \sum b_n/n)$ depends on g through a_n and b_n (cf. Eq. (36)) by the replacement $\hat{\mathrm{P}} \to g\hat{\mathrm{P}}$. If however one is interested in expression where some kinematical variables are free, i.e., not integrated, the substitution to be performed is: $\hat{\mathrm{P}}(u, u') \to g(u, u')\hat{\mathrm{P}}(u, u')$ and the subsequent derivatives become functional derivatives $\delta/\delta g(u,u')$; we keep symbols like X', X'' also for the functional derivatives.

It is convenient to expand X and its derivatives in the number of elementary collisions

$$X = X_1 + X_2 + X_3 + \cdots, \text{where:}$$

$$X_1 = \int D_A(u)\hat{\mathrm{P}}(u,u')D_B(u')dudu',$$

$$X_2 = -\frac{1}{2}\left[\left[\int D_A(u_1)\hat{\mathrm{P}}(u_1,u_1')C_B(u_1',u_2')\hat{\mathrm{P}}(u_2',u_2)D_A(u_2)\prod_{i=1}^{2} du_i du_i' + A \leftrightarrow B\right]\right.$$

$$-\frac{1}{2}\int C_A(u_1,u_2)\hat{\mathrm{P}}(u_1,u_1')C_B(u_1',u_2')\hat{\mathrm{P}}(u_2',u_2)\prod_{i=1}^{2} du_i du_i', \qquad (46)$$

$$X_3 = \int D_A(u_1)\hat{\mathrm{P}}(u_1,u_1')C_B(u_1',u_2')\hat{\mathrm{P}}(u_2',u_2)C_A(u_2,u_3)\hat{\mathrm{P}}(u_3,u_3')$$

$$\times D_B(u_3')\prod_{i=1}^{3} du_i du_i'.$$

It should be pointed out that, when expanding at a fixed order in powers of $\hat{\mathrm{P}}$, one could easily introduce the triple correlations T in the multi-parton distributions by including in X_3 terms like $\int D_A(u_1)\hat{\mathrm{P}}(u_1,u_1')D_A(u_2)\hat{\mathrm{P}}(u_2,u_2')$ $D_A(u_3)\hat{\mathrm{P}}(u_3,u_3')T_B(u_1',u_2',u_3')\prod du_i du_i'$ and similar combinations involving the double correlation C.[41]

The derivatives at $g = 1$ can be worked out.[41] By expanding Eq. (45) and its derivatives in the number of elementary collisions and disregarding all terms beyond the third order in $\hat{P}$, one obtains the following expressions:

$$\begin{aligned} \tilde{\sigma}_1' &= (X_1' + X_2' + X_3')(1 - X_1 - X_2 + X_1 \cdot X_1/2), \\ 2 \times \tilde{\sigma}_2'' &= (X_1' \cdot X_1' + 2X_1' \cdot X_2' - X_2'' - X_3'')(1 - X_1), \\ 3 \times \tilde{\sigma}_3''' &= \frac{1}{2}(X_3''' + X_1' \cdot X_1' \cdot X_1' - 3X_1' \cdot X_2''), \end{aligned} \qquad (47)$$

where $\tilde{\sigma}_1'$ etc. are the exclusive cross-sections, differential in the different partonic variables.

The integrated exclusive cross-sections are

$$\begin{aligned} \tilde{\sigma}_1 &= X_1 - X_1^2 - X_1X_2 + X_1^3/2 + 2X_2 - 2X_2X_1 + 3X_3, \\ 2 \times \tilde{\sigma}_2 &= X_1^2 + 4X_1X_2 - 2X_2 - 6X_3 - X_1^3 + 2X_1X_2, \\ 3 \times \tilde{\sigma}_3 &= 3X_3 + (X_1)^3/2 - 3X_1X_2. \end{aligned} \qquad (48)$$

A representation of the inclusive cross-section σ_S', at all orders in the number of elementary collisions $\hat{\sigma}$, and of the exclusive cross-sections $\tilde{\sigma}_1'$, at order $\hat{\sigma}^2$, is shown in Fig. 4. The yellow circles represent the inclusive one-body parton distribution, the

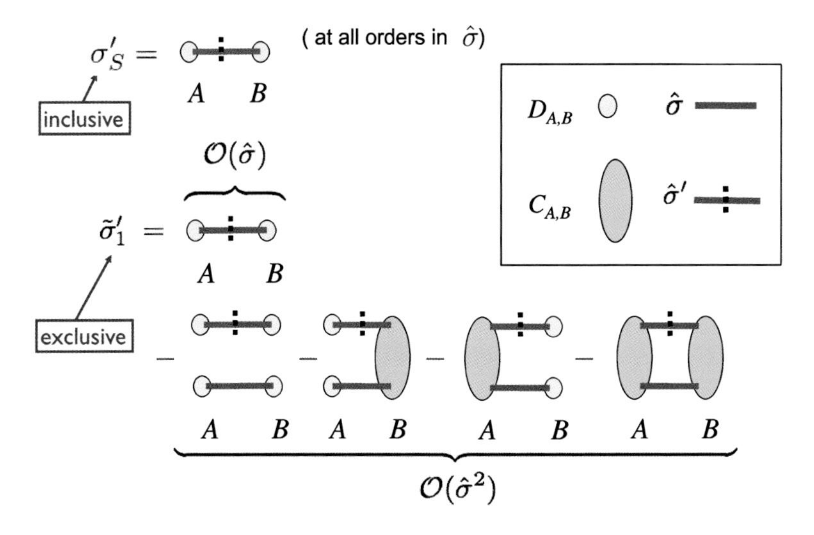

Fig. 4 Symbolic representation of the inclusive and of the exclusive cross-sections σ'_S and $\tilde{\sigma}'_1$ at order $\hat{\sigma}^2$ (see the main text).

green oval the two-body parton correlation, the red line the elementary interaction, differential when the red line is crossed by black points.

The sum rules of Eq. (41) are satisfied as follows:

$$\begin{aligned}
\tilde{\sigma}_1 + 2\times\tilde{\sigma}_2 + 3\times\tilde{\sigma}_3 &= X_1 - X_1^2 + 2X_2 + X_1^2 - 2X_2 - X_1X_2 + X_1^3/2 - 2X_2X_1 \\
&\quad + 3X_3 + 4X_1X_2 - 6X_3 - X_1^3 + 2X_1X_2 + 3X_3 + (X_1)^3/2 \\
&\quad - 3X_1X_2 = X_1 \equiv \sigma_S, \qquad (49) \\
2\times\tilde{\sigma}_2 + 6\times\tilde{\sigma}_3 &= X_1^2 - 2X_2 \\
&\quad + 4X_1X_2 - 6X_3 - X_1^3 + 2X_1X_2 + 6X_3 + (X_1)^3 - 6X_1X_2 \\
&= X_1^2 - 2X_2 \equiv 2\times\sigma_D \\
6\times\tilde{\sigma}_3 &= 6X_3 + (X_1)^3 - 6X_1X_2 \equiv 3!\times\sigma_T,
\end{aligned}$$

where σ_S, σ_D and σ_T are respectively the single, double and triple parton scattering inclusive cross-sections. Explicitly

$$\sigma_S = X_1 = \int D_A\hat{\mathrm{P}}D_B,$$

$$\sigma_D = \frac{1}{2}[X_1^2 - 2X_2] = \frac{1}{2}\left[\int D_A\hat{\mathrm{P}}D_B \cdot D_A\hat{\mathrm{P}}D_B + \int D_A\hat{\mathrm{P}}C_B\hat{\mathrm{P}}D_A\right.$$

$$+\int D_B\hat{\text{P}}C_A\hat{\text{P}}D_B+\int C_A\hat{\text{P}}C_B\hat{\text{P}}\Big]$$
$$=\frac{1}{2}\int [D_AD_A+C_A]\hat{\text{P}}\hat{\text{P}}[D_BD_B+C_B], \tag{50}$$

where $[DD+C]\equiv D_2$ is the two-body parton distribution, as defined in Eq. (29) and the arguments of the two $\hat{\text{P}}$s here above are, of course, different. An analogous expression may be written for σ_T.

The relations (49) may be inverted

$$\begin{aligned}\tilde{\sigma}_1&=\sigma_S-2\sigma_D+3\sigma_T,\\ \tilde{\sigma}_2&=\sigma_D-3\sigma_T,\\ \tilde{\sigma}_3&=\sigma_T,\end{aligned} \tag{51}$$

which allow expressing the scale parameters characterizing the double and triple parton collisions in terms of the single scattering inclusive cross-section σ_S and of the exclusive cross-sections $\tilde{\sigma}_1$ and $\tilde{\sigma}_2$:

$$\begin{aligned}\sigma_D&=\sigma_S-\tilde{\sigma}_1-\tilde{\sigma}_2=\frac{1}{2}\frac{\sigma_S^2}{\sigma_{\text{eff}}},\\ \sigma_T&=\frac{1}{3}(\sigma_S-\tilde{\sigma}_1-2\tilde{\sigma}_2)=\frac{1}{6}\sigma_S^3\frac{1}{\tau\sigma_{\text{eff}}^2},\end{aligned} \tag{52}$$

where the scale factor of the triple parton scattering cross-section has been characterized by the dimensionless parameter τ (see Chapter 9 and Ref. 43).

7. Concluding Remarks

The present model of MPIs is characterized by different non-trivial features and its extension to a more general case would represent an important step towards a comprehensive understanding of MPI dynamics. In the model one can in fact prove that:

- The cancellation of AGK[38] holds for each MPI inclusive cross-section, both in the uncorrelated case and when correlations in the many-parton distributions are taken into account, Eqs. (38)–(40). In both cases the MPI inclusive cross-sections are thus given by the factorial moments of the distribution in multiplicity of the partonic interactions.
- A consequence is that each MPI inclusive cross-section can be safely evaluated at a given order in the number of partonic collisions, since unitarity corrections are not going to spoil the calculation.
- In addition to the inclusive cross-sections one may introduce the exclusive cross-sections, corresponding to the case where only a given number of hard interactions take place. Inclusive and exclusive cross-sections are linked by sum rules, Eq. (41),

which allow to evaluate also the exclusive cross-sections, at a given order in the number of partonic collisions, e.g., Eq. (51).

- Inclusive and exclusive cross-sections are measured independently and their comparison can thus provide an additional handle for the determination of the non-perturbative parameters, which characterize the MPIs, e.g., Eq. (52), and the related unknown non-perturbative properties of the hadron structure.

We like to end showing by an example what we may learn on correlations from high-energy proton–deuteron collisions.

As already noted, a simple and efficient tool to study many-parton dynamics is given by the effective cross-section, Eq. (16), $\sigma_{\rm eff} = {\sigma_S}^2/(2 \cdot \sigma_D)$ and its generalizations. Although $\sigma_{\rm eff}$ is related to the size of the hadron, there are possible concurrent effects, which can be exemplified by two extreme situations:

(1) Configurations with high multiplicity are frequent so many double (and multiple) collisions are produced, e.g., the parton number follows a negative-binomial distribution instead of a Poissonian distribution, then $\sigma_{\rm eff}$ becomes small.
(2) Partons are strictly correlated so that if a pair collides, another pair also collides. Again $\sigma_{\rm eff}$ becomes small.

Observing multiple collisions both on free nucleons and on nucleons bound in light nuclei helps in separating the different form of correlations: Consider a double hard scattering in proton–deuteron collision.[44] There are events where the projectile interacts twice with one nucleon of the target, the other bound nucleon being a spectator, so nothing new may be learned, and there are events where the projectile interacts with both bound nucleons. In the latter case, together with the spatial correlations of the partons, the deuteron wave function plays a relevant role and, since the deuteron wave function is well known, a new window may be opened on the actual content of parton pairs in the nucleon.

Chapter 4

Parton Correlations in Double Parton Scattering

T. Kasemets* and S. Scopetta†

*Nikhef and Department of Physics and Astronomy, VU University Amsterdam, De Boelelaan 1081, NL-1081 HV Amsterdam, The Netherlands

†Department of Physics and Geology, University of Perugia and INFN, Sezione di Perugia, Via A. Pascoli, I-06123, Perugia, Italy

Double parton scattering events are directly sensitive to the correlations between two partons inside a proton and can answer fundamental questions on the connections between the proton constituents. In this chapter, the different types of possible correlations, our present knowledge of them, and the processes where they are likely to be important, are introduced and explained. The increasing integrated luminosity at the LHC and the refinements of the theory of double parton scattering, lead to interesting prospects for measuring, or severely constraining, two-parton correlations in the near future.

1. Introduction

The study of double parton scattering (DPS) events can open up a window to see, for the first time, how the constituents of the proton are connected to each other. The correlations between the properties of two partons in one proton can be directly probed, measuring how two partons inside the proton affect one another. So far, only indirect tests of these correlations have been possible, studying for example, by means of electromagnetic interactions, how the collective behavior of the constituents sums up to give the proton spin.

This allows us to answer questions such as: How does the probability to find one parton in a certain spin state affect the probability to find the second parton in the same spin state? In this chapter we will look at two quarks or gluons inside the proton, explain the different ways they can be connected to one another, describe the state-of-the-art of the field as well as the perspectives for future studies of these correlations. This will be possible in processes where DPS forms a major contribution, such as same-sign double-W production.

Assuming factorization (see Chapter 2), the DPS cross-section for the production of final states A and B takes the form (see Chapters 2 and 3 and Ref. 30)

$$d\sigma_{\mathrm{DPS}}^{AB} = \frac{m}{2} \sum_{abcd,R} \int d^2\boldsymbol{y}\; {}^{R}F_{ac}(x_1, x_2, \boldsymbol{y})\, {}^{R}F_{bd}(x_3, x_4, \boldsymbol{y})\; d^{R}\hat{\sigma}_{ab}^{A}\, d^{R}\hat{\sigma}_{cd}^{B}\,, \tag{1}$$

where $m = 1$ if $A = B$, $m = 2$ otherwise, R denotes the different possible color representations and a, b, c, d label simultaneously the species (parton-type and flavor) and polarization of the partons contributing to the production of the final states. In Eq. (1), $d\hat{\sigma}$ represents the differential partonic cross-section (for example, differential in the rapidities of the produced particles). The functions F are the double parton distributions (dPDFs), encoding the probability to find the two interacting partons, with longitudinal fractional momenta x_1, x_2 at a relative transverse distance $\boldsymbol{y}$ inside the proton. They depend additionally on factorization scales $\mu_{A(B)}$, and for $R \neq 1$, on a rapidity scale (see Chapter 2). If extracted from data, as noticed a long time ago,[45] dPDFs would offer for the first time the opportunity to investigate two-parton correlations. This would be a two-body property, carrying information which is different and complementary to that encoded in one-body distributions, such as generalized parton distributions (GPDs).[46] This is illustrated in Fig. 1.

For cross-sections differential also in the net transverse momenta of each of the two hard interactions, the dPDFs are replaced in the factorization theorem by the double transverse momentum-dependent parton distributions (dTMDs). These distributions depend on two additional transverse vectors and allow for a number of further correlations, for example between the spin and transverse momenta of the partons. They are interesting also from a more theoretical point of view, with the rich color structure in combination with the non-trivial dependence on the soft gluon exchanges.[22] In the region where the two net transverse momenta are small, DPS and single parton scattering (SPS) both contribute to the cross-section at the same power, which makes it promising for DPS extractions. However, for simplicity we will focus on the dPDFs during the rest of this chapter.

In the following, we will have a closer look at what is currently known about the different correlations, describe the effects expected in cross-sections and the prospects for their measurement. The chapter is structured as follows: In the next section, we will look at the correlations between the kinematic variables x_i and $\boldsymbol{y}$ of the dPDFs. In Section 3, we will focus instead on the correlations between color, spin, flavor and fermion number of the two partons. In Section 4 we will summarize and give an outlook to what we consider are the most promising future directions.

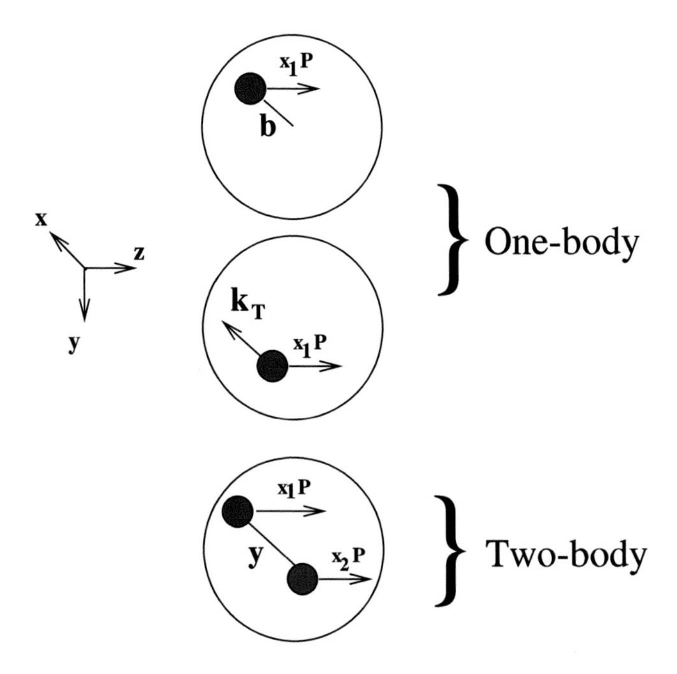

Fig. 1 From top to bottom: a pictorial representation of an impact parameter-dependent parton distribution, i.e., the Fourier transform of a GPD when the momentum transfer is purely transverse; a transverse momentum-dependent parton distribution (TMD); a dPDF, which, at variance with the two previous cases, is a two-body distribution.

2. Kinematic Correlations

As stressed in the introduction, the two-body information encoded in dPDFs is different and complementary to that described by one-body parton distributions. Nevertheless, a connection between dPDFs, presently largely unknown, and one-body quantities can be obtained by making a number of assumptions on the dPDFs. First, all color representations different from the color singlet are neglected (i.e., only $R = 1$ is considered), together with all possible correlations between spins, flavors and fermion-numbers. Thereafter, correlations between x_1 and x_2 are neglected. The dPDFs then take the form

$$F_{jk}(x_1, x_2, \mathbf{y}) = \int d^2\mathbf{b} F_j(x_1, \mathbf{b} + \mathbf{y}) F_k(x_2, \mathbf{b}), \tag{2}$$

where $F_i(x, \mathbf{b})$ is a parton distribution dependent on the impact parameter $\mathbf{b}$, the transverse distance of the parton from the transverse centre of mass of the hadron.[46] This function is the Fourier transform of a GPD in a process where the momentum transfer is transverse. Neglecting moreover correlations between x_1, x_2

and $\mathbf{b}$, one can write

$$F_i(x, \mathbf{b}) = f_i(x)G(\mathbf{b}), \tag{3}$$

where $f_i(x)$ is a parton distribution function (PDF) and the transverse profile $G(\mathbf{b})$ has been assumed to be equal for all parton species. One should notice that Eq. (3) has been found to fail in all model calculations of GPDs (see, e.g., Refs. 47 and 48), as well as in the first analyses of data from deeply virtual Compton scattering.[49] The assumptions described above are often used to infer properties of dPDFs from those of single particle distributions. The relations Eqs. (2) and (3) have been introduced and critically discussed in a mean field approach in Refs. 9 and 82.

Since dPDFs are largely unknown, and only sum rules relating them to PDFs are available,[10,18,20,50,51] model calculations can be very useful and have been performed. Models are usually developed at low energy, but are able to reproduce some relevant features of nucleon parton structure. Since in models the number of degrees of freedom is fixed, they can be predictive in particular in the valence region, at x larger than, say, 0.1. In such model calculations, the factorized structures in Eqs. (2) and (3) do not arise. Relevant correlations between x_1 and x_2, violating Eq. (2), and between x_1, x_2 and $\mathbf{b}$, violating Eq. (3), have been found in the valence region in a variety of approaches. This result was obtained, for example, in a modified version of the simplest bag model,[52] in constituent quark models[53,54] in a valon model[55] and in dressed quark models.[56]

In particular, in Ref. 54 a light-front (LF) Poincaré covariant approach, reproducing the essential sum rules of dPDFs without ad hoc assumptions and containing natural two-parton correlations, has been described. An example of the information that model calculations can provide is shown in Fig. 2, where the effect of the breaking of the factorization between longitudinal and transverse variables is emphasized.

It is crucial to explore if the breaking of the properties, Eqs. (2) and (3), found in the valence region, survive at LHC kinematics, dominated by low-x and high energy scales. As a matter of fact, model estimates are valid in general at a low scale Q_0, the so-called hadronic scale. The results of the calculations should therefore be evolved using perturbative QCD (pQCD) in order to compare them with data taken at a momentum scale $Q > Q_0$, according to a well-established procedure, proposed already in Refs. 57 and 58. The evolution of dPDFs has been studied for a long time. The first studies were performed in the late 1970s/early 1980s[15,16] with much theoretical progress being made in recent years — for a detailed discussion on this topic we invite the reader to look at Chapter 2. One should notice that, even if a factorized structure of the dPDF were valid at a given scale, the different evolution properties of dPDFs and PDFs would break it at a different scale, generating perturbative correlations. These correlations have been discussed in a largely model-independent way in Ref. 59, incorporating the homogeneous evolution equations. The evolution tends to pull the average transverse separation in quark and gluon distributions towards a common value,

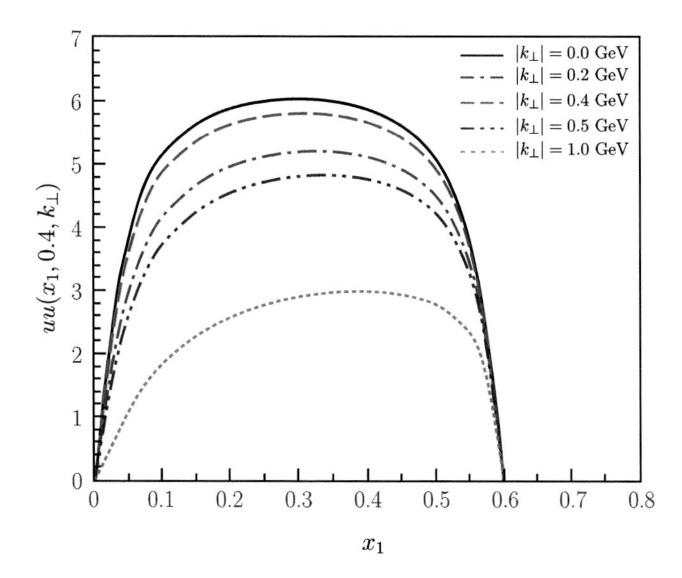

Fig. 2 The distribution $uu(x_1, x_2, \mathbf{k}_\perp)$, Fourier transform of the dPDF $F_{uu}(x_1, x_2, \mathbf{b})$, for the proton, for $x_2 = 0.4$, according to the LF model calculation of Ref. 54, at the low momentum scale of the model. If it were possible to factorize the dependence on the longitudinal momenta x_1, x_2 and that on the transverse variable $\mathbf{k}_\perp$, the distributions would have the same symmetric shape for the different values of $\mathbf{k}_\perp$.

but this is a relatively slow process and differences can remain up to high scales. Similarly, correlations between the momentum fractions and the transverse separation present at a low scale can remain in large-scale processes, as described here below.

The interplay of perturbative and non-perturbative correlations between different kind of partons has been described also using homogeneous QCD evolution applied to the results of the correlated LF model.[60] It was found that their effect tends to be washed out at low-x for the valence, flavor non-singlet distributions, while they can affect singlet distributions in a sizable way. This different behavior can be understood in terms of a delicate interference of non-perturbative correlations, generated by the dynamics of the model, and perturbative ones, generated by the model-independent evolution procedure.

Concerning the correlation between the $\mathbf{y}$ and x_1, x_2 dependences in dPDFs, some qualitative understanding can be inferred from studies of hard exclusive processes, involving $f_i(x, \mathbf{b})$ of a single parton inside the proton. In particular, measurements of $\gamma p \to J/\psi p$ at HERA[61,62] indicate a logarithmic dependence $\langle \mathbf{b}^2 \rangle = \text{const} + 4\alpha' \log(1/x)$ with $\alpha' \approx 0.15\ \text{GeV}^{-2} = (0.08\,\text{fm})^2$ for gluons with $x \simeq 10^{-3}$. Studies of nucleon form factors[63] and calculations of Mellin moments $\int dx\, x^n f_i(x, \mathbf{b})$ with $n = 0, 1, 2$ in lattice QCD[64] indicate that for x above 0.1 the decrease of $\langle \mathbf{b}^2 \rangle$ with x is even stronger. Although this is one-body information, one could wonder whether the correlations between the $\boldsymbol{y}$ dependence and x_1, x_2 in double parton distributions could follow the behavior of the one-body quantity, with the $\mathbf{b}$ distribution becoming more narrow with increasing x. If this is the case, important consequences could be expected for multi-parton interactions.[65] The production of hard final states requires relatively large momentum fractions of the partons entering the corresponding hard interaction. This would favor small values of $\mathbf{b}$, which is the transverse distance of the parton from the transverse centre of the proton. The collision would therefore be rather central and thus the transverse interaction area for the colliding protons would be rather large, a fact which in turn favors additional interactions.

Such correlations may have a sizable impact, e.g., on the underlying event activity in Z production, as shown in a study with Pythia 8.[66]

3. Quantum-Number Correlations

Two partons inside a single proton can have their quantum numbers correlated. Perhaps the most straightforward example comes from the valence sector of the proton. If we, for one interaction, extract one valence up quark from the proton, it is natural to expect that the chance to find another valence up quark in the proton is reduced. It seems reasonable to expect such effects to be sizable at relatively large momentum fractions and to reduce as the density of partons increases towards small momentum fractions. This phenomenon naturally fits into the dPDFs, F_{ab}, of two partons a and b inside a proton.

We will focus here on another type of correlation and interference which occurs at the quantum level, and for which we reserve the label *quantum-number correlations*. This includes correlations and interferences in color, spin, flavor and fermion number.[23,33,67] Understanding how this occurs in double parton scattering, but not in single parton scattering, is not complicated. From a diagram such as the one in Fig. 3, we can see that two quarks *leave* the right-moving proton (represented by the lower green ellipse) on the left side of the final-state cut and two quarks *return* to the proton on the right side of the cut. The quantum numbers of the two quarks in the amplitude have to sum up to the quantum numbers in the conjugate amplitude, which still leaves room for the two quarks in the amplitude to individually have different quantum numbers from their partners in the conjugate amplitude.

In particular, this allows for quantum number interferences, which is another way of viewing the correlations. If we take color as example (even though, as we will see, it might not have the largest impact), and couple each parton in the amplitude with its partner in the conjugate amplitude (i.e., parton with the same longitudinal momentum fraction x_i) we have two possible combinations: $3 \otimes \bar{3} = 1 \oplus 8$. Repeating this with the other pair we obtain

$$(3 \otimes \bar{3}) \otimes (3 \otimes \bar{3}) = (1 \otimes 1) \oplus (1 \otimes 8) \oplus (8 \otimes 1) \oplus (8 \otimes 8) = 1 \oplus 1 \oplus \ldots, \quad (4)$$

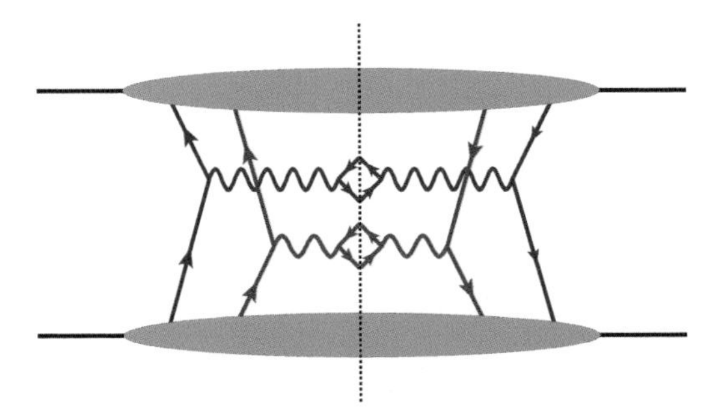

Fig. 3 Double vector boson production. In contrast to single parton scattering, only the *sum* of the quantum numbers of the partons leaving the protons on the left and returning on the right-hand side of the final-state cut have to match.

where the "..." refer to combinations that do not produce a total color singlet. The requirement that the sum of the quantum numbers on the left and right side of the final-state cut have to be equal amounts to the requirement that when coupling all four partons, we need to obtain a color singlet. We therefore see that for the quark case we can obtain the singlet in two ways: either by coupling two individual color singlet pairs or by coupling two color octet pairs. This results in two independent double quark distributions for the two color states in Eq. (1), labeled as ${}^{R}F$ with $R = 1, 8$. In the cross-section, both distributions contribute and color-singlet production is proportional to ${}^{1}F{}^{1}F + {}^{8}F{}^{8}F$ (with the normalization of the distributions as in Ref. 3). The color-octet term has hard interactions with color interferences between the amplitude and conjugate, i.e., it is a genuine quantum effect which can never appear in a single hard scattering. Under the assumption of zero correlations between the two hard interactions, no such interference could take place and the octet distributions would vanish.

Similar to the color, also the spin of the two partons can be correlated and give rise to a large number of different polarized dPDFs. There can be interferences in flavor, for example between up and down quarks in double-W boson production. This type of interference is illustrated by the diagrams in Fig. 4. Furthermore, there can be interference in fermion number between quarks, antiquarks and gluons, with examples being given in Fig. 5. It is interesting to note that spin correlations leading to distributions of transversely polarized quarks and linearly polarized gluons have a rather unique signature. They induce a dependence on the azimuthal angle (for example between the Z-boson decay planes) and lead to azimuthal spin asymmetries in unpolarized proton scattering.[68] It is important to realize that experimental extractions of DPS signals are based on Monte Carlo generators which assume a flat azimuthal distributions, which might no longer be true in the presence of correlations.

The result of all the correlations is a flora of independent double parton distributions, of which we have little knowledge and no experimental extractions.

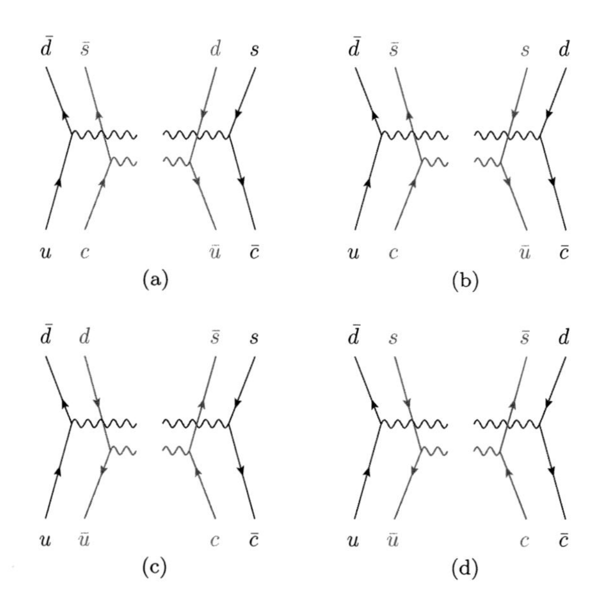

Fig. 4 Flavor interference in double W production. Two possible processes are shown for W^+W^+ production in (a, b), and for W^+W^- production in (c, d). Figure from Ref. 68. q and $\bar{q}$ labels partons corresponding to a quark field or a conjugate quark field in the relevant dPDF. Graphs (b) and (d) have flavor interference only for the proton at the bottom, while graphs (a) and (c) come with flavor interference distributions for both protons.

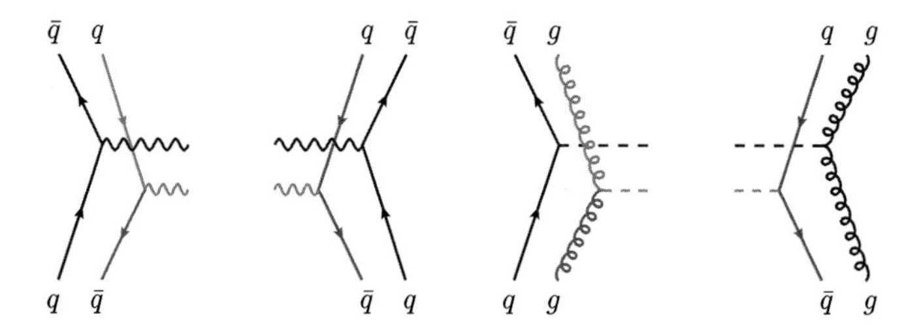

Fig. 5 Fermion number interference examples for double Drell–Yan (left) and double Higgs production (right). q and $\bar{q}$ labels partons corresponding to a quark field or a conjugate quark field in the relevant dPDF and g labels a gluon field.

One might question what predictive power we have, and can hope to obtain. The answer to this question leads us into a discussion of what we know about the different correlation effects, when they are likely to play an important role and when we believe they can be safely neglected. The information available to this end comes from two main categories of studies. The first studies the distributions in different types of hadron models, or derives theoretical bounds, and attempts to quantify the size of the correlations. The second examines how the perturbatively calculable evolution of these distributions influences their shapes and sizes.

3.1. *Models and bounds*

There are a couple of different hadron model calculations which consider different quantum-number correlations. Focus has been on the polarization of the partons (apart from the kinematic correlations already discussed) and large correlations have been observed. For quark and antiquark distributions, large spin correlations were found in the MIT bag model[52] and in light-front constituent quark models.[54] The domain of validity of these models is principally the region of large momentum fractions, and thus they serve best as initial conditions to the double DGLAP evolution equations. This was done in Ref. 54 with the observation that the spin correlations are sizable even after evolution. Within a dressed-quark model of the mixed quark–gluon distributions, the spin correlations were observed to be large for certain polarization types, such as two longitudinally polarized partons and the combination of a transversely polarized quark and an unpolarized gluon.[56] In addition to model calculations, theoretical upper bounds on the correlations, including spin, flavor, fermion number and color, have been derived from the probability interpretation (or positivity) of dPDFs.[69,70]

3.2. *Evolution*

The dPDFs evolve according to a double ladder version of the DGLAP evolution equations, i.e., a double DGLAP evolution (see Chapter 2). Cross talk between the ladders is suppressed by the large distance $\boldsymbol{y}$ separating the two partons, which is typically of the size of the proton. The evolution starts at a scale of the order of $1/|\boldsymbol{y}|$ and evolves up to the scale of the respective hard interaction.[5] This evolution generically leads to a reduction of the correlations between the two partons and decreases the importance of the two interference/correlation dPDFs. However, the rate at which this occurs varies significantly for the different types of correlations and the momentum fractions of the partons.

If we allow for a slight oversimplification, the current state of knowledge can be summarized in a short paragraph: The color correlations are Sudakov suppressed and expected to be small in large-scale processes (see Chapter 2 and Refs. 23 and 25). This can be understood from the fact that those correlations require color information to travel over the large distance $\boldsymbol{y}$ inside the proton. Therefore, for processes above $Q_i^2 \sim 100\ \mathrm{GeV}^2$ they are expected to play a minor role.

Gluon polarizations at low momentum fractions (where DPS is most relevant) are also quite rapidly suppressed through the evolution. This suppression can be understood from the gluon splitting kernels: The unpolarized gluon splitting kernel at small x goes as $1/x$, leading to the large increase of the gluon density for small momentum fractions (as is well known from single parton distributions). The polarized splitting kernels on the other hand go as x^0 for longitudinal polarization and x for linearly polarized gluons. The quark polarizations on the other hand can remain sizable up to high scales.[59] Figure 6 shows two examples of the

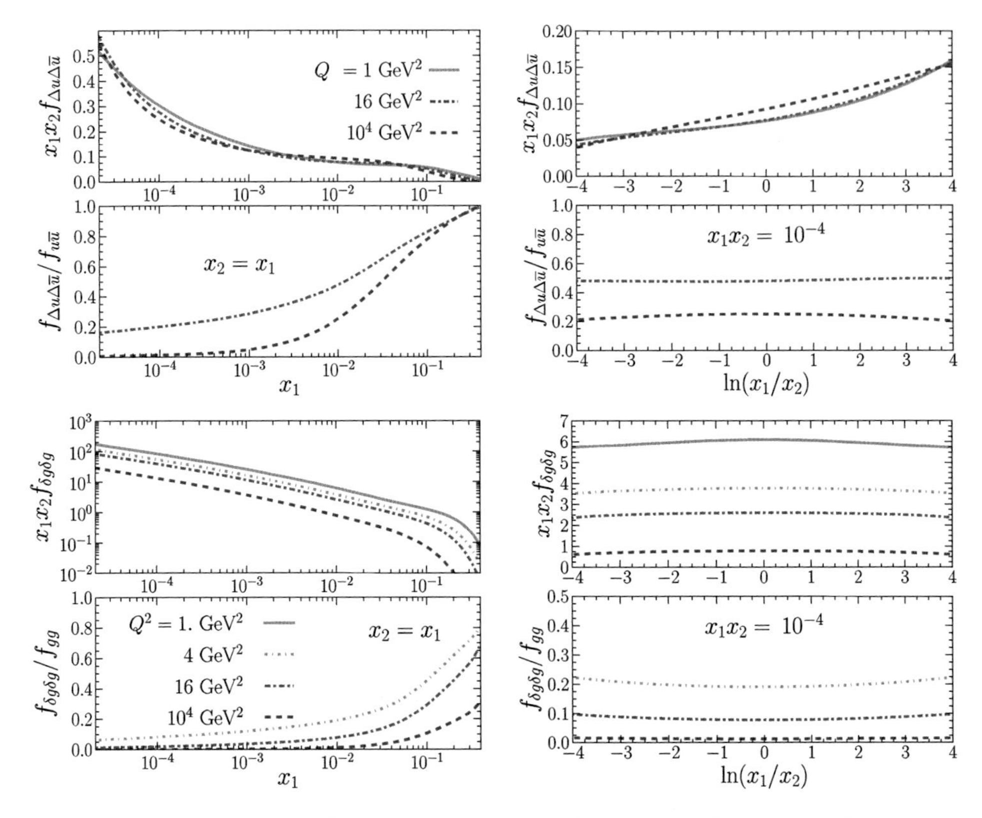

Fig. 6 Evolution of longitudinally polarized up-quarks (top) and linearly polarized gluons (bottom), either as a function of $x_1 = x_2$ (left) or as a function of $\log x_1/x_2$ (right). At the initial scale of 1 GeV, the polarization is maximized (equal to the unpolarized distribution). Lower panels show ratio of polarized over unpolarized distributions. Figure from Ref. 59.

suppression for the most suppressed gluon polarization and the least suppressed quark polarization, starting with maximal polarization (i.e., polarized equal to unpolarized) at the input scale of 1 GeV. Fermion-number interference is expected to be small at large scales, since the interference distributions do not mix with the gluon distributions (which drives the evolution at small to moderate x) under leading order evolution and always involve color interference. Flavor interference on the other hand is still relatively unexplored, but also does not mix with the gluon distributions.

4. Prospects

We have seen that two-parton correlations are very interesting properties of the non-perturbative proton structure, and they can be relevant in specific DPS channels. So far, it has been challenging to observe them at the LHC and extract dPDFs from data. While waiting for precise data expected from LHC at high luminosity in the near future, one could look for signatures of the presence of correlations in an extracted quantity, the so-called effective cross-section, σ_{eff}. Let us introduce now this quantity. Since dPDFs are largely unknown, it has been useful to describe DPS cross-sections independently of dPDFs, through the approximation

$$d\sigma_{\text{DPS}}^{AB} \simeq \frac{m}{2} d\sigma_{\text{SPS}}^{A} \frac{d\sigma_{\text{SPS}}^{B}}{\sigma_{\text{eff}}}, \tag{5}$$

where $d\sigma_{\text{SPS}}^{A(B)}$ is the SPS cross-section with final state $A(B)$:

$$d\sigma_{\text{SPS}}^{A(B)} = \sum_{i,k} f_i(x_1) f_k(x_3)\, d\hat{\sigma}_{ik}^{A(B)}(x_1, x_3)\,. \tag{6}$$

The physical meaning of Eq. (5) is that, once the process A has occurred with cross-section $d\sigma_{\text{SPS}}^{A}$, the ratio $d\sigma_{\text{SPS}}^{B}/\sigma_{\text{eff}}$ represents the probability of process B to occur. So far, a constant value of σ_{eff} has been assumed in the experimental analyses performed. In this way, different collaborations have extracted values of σ_{eff}, analyzing events with different final states and with different centre-of-mass energies of the hadronic collisions. The results have large error bars and their central values vary in the range 2–20 mb (see, for example, Figs. 8 and 9 in Ref. 71). However, these numbers are to be taken with caution as the different extractions rely on different assumptions, for example, with regards to the SPS cross-sections. It is interesting to realize that the approximations leading to Eq. (5), with a constant σ_{eff}, from Eq. (1), are the same leading the dPDF to its full factorized form. As a matter of fact, by inserting Eqs. (2) and (3) into Eq. (1), one obtains σ_{eff} from Eq. (5) and (6) as follows:

$$\sigma_{\text{eff}}^{-1} = \int d^2\mathbf{y}\, [T(\mathbf{y})]^2, \tag{7}$$

with the quantity

$$T(\mathbf{y}) = \int d^2\mathbf{b}\, G(\mathbf{b}+\mathbf{y})\, G(\mathbf{b}), \tag{8}$$

controlling the double parton interaction rate. The fact that σ_{eff} does not show any dependence on parton fractional momenta, hard scales or parton species, is clearly a consequence of the assumptions in Eqs. (2) and (3). If those assumptions were relaxed σ_{eff} would explicitly depend on scales and flavors, and on all momentum fractions, and would be a complicated average (with x_i dependent weights) of all the correlations described by the double parton distributions. One could therefore analyze data looking for such a dependence. Besides, model calculations show that correlations in momentum fractions cannot be treated separately from those involving also $\mathbf{y}$: the way in which the dPDF differs from the product of single parton densities changes with $\mathbf{y}$.[54,72] Using model calculations without the assumptions leading to Eqs. (2) and (3), σ_{eff} was found to depend non-trivially on longitudinal momenta. In particular, this was obtained in the LF constituent quark model,[73] as well as in a holographic approach.[74] Very recently, the LF model calculation of dPDFs has been used to evaluate the cross-section for same-sign W boson pair production, a promising channel to look for signatures of double parton interactions at the LHC. In this way, the average value of the DPS cross-section was found to be in line with previous estimates which make use of a constant $\sigma_{\rm eff}$ as an external parameter, not necessary in this approach. The novel obtained dependence on longitudinal momenta addresses the possibility to observe two-parton correlations, in this channel, in the next LHC runs.[75] An example of these results is shown in Fig. 7.

Since in the DPS cross-section the dependence upon $\mathbf{y}$ is integrated over, a direct test of the breaking of Eq. (3) in DPS, addressing correlations between $\mathbf{y}$ and x_1, x_2 in dPDFs, appears difficult at the moment. An indirect test of these correlations is expected from future measurements at Jefferson Lab, COMPASS and at a possible future electron–ion collider,[77] where at least a detailed picture of the one-body distribution $F_a(x, \mathbf{b})$, should be at hand.

As for the correlations between quantum numbers described in the previous section, their impact on cross-sections has been studied only in a limited number

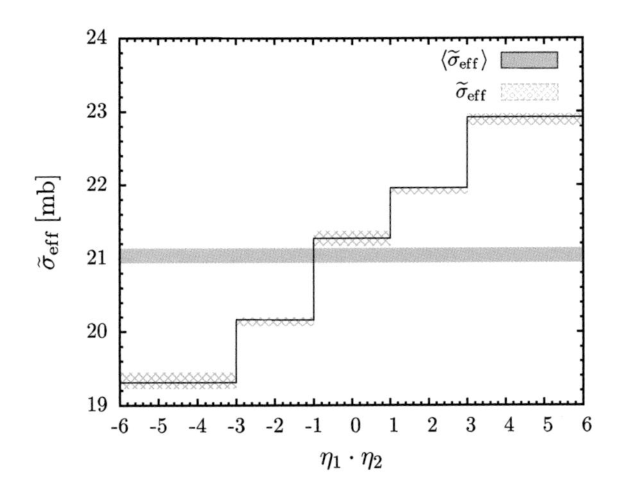

Fig. 7 The quantity $\tilde{\sigma}_{\rm eff} = \frac{m}{2} d\sigma^A_{\rm SPS} d\sigma^B_{\rm SPS}/d\sigma^{AB}_{\rm DPS}$, for the production of two W bosons with the same sign, in the kinematics of the CMS measurements of Ref. 76. SPS and DPS cross-sections are calculated using PDFs and dPDFs obtained in the LF model.[73] No factorized structure has been assumed for dPDFs. In this way, a dependence on the longitudinal variable $\eta_1 \cdot \eta_2$ is clearly predicted and could be tested in future analyses. $\eta_{1,2}$ are the pseudorapidities of the detected muons in the final state, naturally related to the longitudinal parton momenta. Figure from Ref. 75, where further details can be found. Figure reprinted with permission. Copyright 2017 by the American Physical Society.

of cases. For the production of two D^0 mesons, as measured by LHCb,[78] the low masses of the final states allows for a large impact on the size of the cross-section from longitudinally polarized gluons, reaching a contribution of up to 50% of the unpolarized.[79] This is an example of the importance of further exploratory studies of DPS to find channels and phase space regions in which two-parton correlations are more pronounced and easily measured. In this sense, input is expected also from proton–nucleus scattering, where the DPS contribution is known to be enhanced.[80]

There are several elements working together to provide a promising near future for DPS in general, and measurement of correlations in particular:

(1) The continuous refinements of the DPS theory, including for example a scheme to combine, without double-counting, the SPS and DPS cross-sections described in Chapter 2.
(2) The increasing integrated luminosity collected by the experiments at the LHC.
(3) The improved precision to which the SPS cross-sections are known.

Combined, this provides good reasons to further develop the theory for DPS, motivation for phenomenological studies of the effects correlations have on actual observables, and good prospects for interesting experimental results to confront the theory with in the upcoming years.

Double TMDs enter cross-sections when the transverse momenta of for example two vector bosons are measured and small. In this region, there is no factorization theorem without considering both single and double parton scattering. The formalism to treat this region in both single and double parton scattering,[22] will allow for interesting prospects to investigate the correlations in DPS, including those between the transverse momenta of the two partons. The experimental searches have now measured same-sign double-W production,[76] often put forward as the cleanest signal for DPS. Interesting results are expected also in channels where the separation between single and double parton scattering is less straightforward. An increased precision on both DPS and SPS sides will lead to a situation where the double parton distributions are the main unknown. Using differential calculations and resummation at high logarithmic accuracy, for example in double boson production, the combination of DPS and SPS will be important and comparisons to data will enable extractions of dPDFs and interparton correlations, or experimentally constrain them.

In summary, the increased luminosity will allow for more differential measurements. Moving towards a theory that allow for more complete phenomenological explorations, simultaneously treating both SPS and DPS, provides the basis for our belief that inter-parton correlations might soon be an experimentally established fact, or a heavily constrained hypothesis.

Acknowledgments

Many discussions with F.A. Ceccopieri, L. Fanò, J. Gaunt, M. Rinaldi, M. Traini, D. Treleani, V. Vento are gratefully acknowledged. We thank S. Cotogno and T. van Daal for useful comments on the manuscript. TK acknowledges support from the Alexander von Humboldt Foundation and the European Community under the "Ideas" program QWORK (contract 320389).

Chapter 5

Multiparton pp and pA Collisions: From Geometry to Parton–Parton Correlations

Boris Blok* and Mark Strikman†

**Department of Physics, Technion — Israel Institute of Technology, Haifa, Israel*

†*Physics Department, Pennsylvania State University, University Park, USA*

We derive expressions for the cross-section of the multiparton interactions (MPI) based on the analysis of the relevant Feynman diagrams. We express the MPI cross-sections through the double (triple,...) generalized parton distributions (GPDs). In the mean field approximation for the double GPDs, the answer for the double parton cross-section is expressed through the integral over two gluon–nucleon form factor which was measured in the exclusive DIS vector meson production. We explain under what conditions the derived expressions for MPI correspond to an intuitive picture of hard interactions in the impact parameter representation. The mean field approximation in which correlations of the partons are neglected fails to explain the double parton interaction data. Perturbative-QCD-induced correlations enhance the cross-section by a factor of 1.5–2 for large $p_\perp$ and $0.001 < x < 0.1$ explaining the current data. We argue that in the small x kinematics ($10^{-4} \leq x \leq 10^{-3}$) where effects of perturbative correlations diminish, a diffractive nonperturbative mechanism kicks in and generates positive correlations comparable in magnitude with perturbative ones. We explain how our technique can be used for calculations of MPI in the proton–nucleus scattering. The interplay of hard interactions and underlying event is discussed, as well as different geometric pictures for each of the MPI mechanisms — pQCD, non-perturbative correlations and mean field. Predictions for value of $\sigma_{\rm eff}$ for various processes and a wide range of kinematics are given. We show that together different MPI mechanisms give good description of experimental data, both at Tevatron, and LHC, including the central kinematics studied by ATLAS and CMS detectors, and forward (heavy flavors) kinematics studied by LHCb.

1. Introduction

It is widely realized now that hard *multiple parton interactions* (MPIs) occur with a probability of the order one in typical inelastic LHC proton–proton pp collisions. Indeed the ratio of the integral of the inclusive jet cross-section with transverse momenta $p_\perp \geq$ few GeV and $\sigma_{\rm in}(NN)$ gives the average multiplicity of hard collisions (dijet production) larger than one, see, e.g., Refs. 35 and 81. Hence MPIs play an important role in the description of inelastic pp collisions. MPIs were first introduced in the eighties[30,32] and in the last decade became a subject of a number of theoretical studies, see, e.g., Refs. 3,9,10,13,14,18,21,23,80 and 82–96 and references therein.

Also, in the past several years, a number of Double Parton Scattering (DPS) measurements in different channels were carried out,[78,97–106] while many Monte Carlo (MC) event generators now incorporate MPIs.[36,36,66,107–115]

The DPS cross-section for (say) the production of two pairs of jets is traditionally parametrized as

$$\frac{d\sigma(4\to 4)}{d\Omega_1 d\Omega_2} = \frac{1}{\sigma_{\rm eff}}\frac{d\sigma(2\to 2)}{d\Omega_1}\frac{d\sigma(2\to 2)}{d\Omega_2}, \qquad (1)$$

where Ω_i is the phase volume for production of a pair of jets where $\sigma_{\rm eff}$ is a priori function of x_i, p_{t_i}, flavour and spin. Here p_t is a hard transverse scale of the jets. Initially it was conjectured[30] that parameter $\sigma_{\rm eff}$ is related to the total inelastic cross-section of the hadron–hadron interactions.

Later on within the framework of the geometric picture implemented in the Monte Carlo models $\sigma_{\rm eff}$ was written as a convolution of the four single parton impact parameter distributions, $f(\rho_i)$ assuming that these distributions do not depend on x and on flavor, cf. Fig. 1.

$$\frac{1}{\sigma_{\rm eff}} = \int d^2\rho_i d^2 b f(\rho_1)f(\rho_2)f(\rho_3)f(\rho_4)\delta(\rho_1-\rho_3-\mathbf{b})\delta(\rho_2-\rho_4-\mathbf{b}). \qquad (2)$$

One can see from Eq. (2) that the factor $\sigma_{\rm eff}$ characterizes the transverse area occupied by the partons participating in two hard collisions. It also includes the effect of possible longitudinal correlations between the partons. (In our calculations and some of Pythia latest versions, the distributions are always x-dependent, see Appendix).

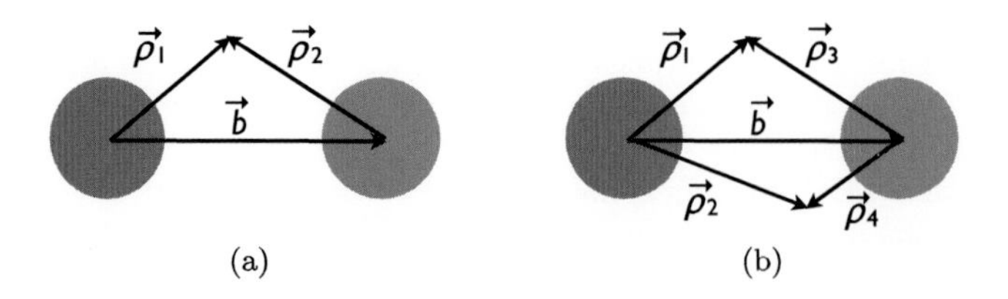

Fig. 1 Geometry of one and two hard collisions in impact parameter picture.

Parameters of the transverse distribution were chosen to reproduce the MPI data obtained at the Tevatron which reported $\sigma_{\text{eff}} \approx 15$ mb.[97–99]

Further study used the QCD factorization theorem for the exclusive vector meson production to extract one particle Generalized Parton Distributions (GPD) $f(\rho, x|Q^2)$ from the photo/electro production data. Under assumption that partons in colliding nucleons are not correlated, a much larger $\sigma_{\text{eff}} \geq 30$ mb was found.[116] This strongly suggested that significant parton–parton correlations are present in nucleons.

In this chapter, we will summarize our studies of the mechanisms which generate perturbative and non-perturbative correlations between the partons and allow to explain many features of the data. In particular we explain the geometry of MPI and show that the MPI cross-section receives contributions from the mean field contribution, perturbative QCD (pQCD) and non-perturbative mechanisms, connected with non-factorizable initial conditions. Each of these three mechanisms corresponds to its different range of impact parameters (with the mean field one being most central). Together they lead to a good agreement of experimental MPI cross-sections.

The paper is organized as follows.

In Section 2 we present the geometrical picture of MPI and explain that hard collisions on average involve much smaller impact parameters than minimum bias inelastic collisions.

In Section 3 we review the parton level calculation of the DPS using Feynman diagram analysis which allows to express the DPS cross-section through the convolution of two double generalized parton distributions (GPDs). (This convolution breaks down beyond the parton model, see below.) The double GPDs in the mean field approximation are expressed through a product of single GPDs which are extracted from the studies of the exclusive vector meson production.

In Section 4 we analyze contribution to the DPS of the correlation mechanism induced by the pQCD evolution. General expressions are derived both for the cross-section differential in jet imbalances δ_{ij} and the cross-section integrated over δ_{ij}.

The numerical results for the contribution of pQCD correlation mechanism are presented in Section 5. We find that the perturbative mechanism may enhance the DPS rates at large $p_\perp$ (large virtualities) and $x \sim 10^{-2}$–10^{-3} by a factor 1.5–2 allowing to explain the observed rates for a number of DPS processes.

In Section 6 we argue that a new soft mechanism of the parton–parton correlations becomes important for $x \leq 10^{-3}$ which is due to presence of multi-Pomeron exchanges. We explain that this mechanism is relevant for the understanding of the rate of minijet production and the production of two D-mesons in the forward kinematics studied by LHCb.[78,104–106]

In Section 7 we apply our technique to calculate the rate of MPI in proton–nucleus collisions taking into account pQCD corrections to the parton model approximation[80] and finding that pQCD corrections further increase the rate of MPI in pp and pA scattering.[93]

In Section 8 we consider several consequences of the different impact parameter localization of the minimum bias and hard collisions. In particular we explain that the b-space unitarity leads to the requirement that jet production cross-section should be suppressed is compared to the pQCD result even at large impact parameters.

Our conclusions are presented in Section 9.

2. Transverse Picture of Multiparton Interactions

2.1. *Impact parameter distribution in hard collisions*

A natural framework for visualization of the MPI is the impact parameter representation of the collision. Indeed, in the high energy limit angular momentum conservation implies that the impact parameter b becomes a good quantum number. Also the hard collisions are localized in the transverse plane at the relative distances $\sim 1/Q$ where Q is transverse momentum transfer (a hard scale of the problem). Combined, they lead to an intuitive picture of the MPI.

To describe the transverse geometry of the pp collisions with production of a dijet, it is convenient to consider probability to find a parton with given x and transverse distance ρ from the nucleon transverse centre of mass, $f_i(x_i, \rho_i)$. This quantity allows a formal operator definition, and it is referred to as the diagonal GPD. It is related to non-diagonal GPDs which enter in the description of the exclusive meson production (see the appendix for discussion of the information on ρ dependence of GPDs which is available from the studies of the exclusive vector meson production in the DIS).

The inclusive cross-section in the Leading Twist (LT) pQCD regime does not depend on the transverse structure of the colliding hadrons — the cross-section is expressed through the convolution of parton densities. Indeed, we can write

$$\begin{aligned}\sigma_h &\propto \int d^2b d^2\rho_1 d^2\rho_2 \delta(\rho_1 + b - \rho_2) f_1(x_1, \rho_1) f_2(x_2, \rho_2)\sigma_{2\to 2} \\ &= \int d^2\rho_1 d^2\rho_2 f_1(x_1, \rho_1) f_2(x_2, \rho_2)\sigma_{2\to 2} \\ &= D_1(x_1, Q_1^2) D_2(x_2, Q_2^2)\sigma_{2\to 2}. \end{aligned} \quad (3)$$

Here at the last step we used the relation between diagonal GPD and PDF $D_j(x, Q^2)$: $\int d^2\rho f_j(x, \rho, Q^2) = D_j(x, Q^2)$.

At the same time, as soon as one wants to describe the structure of the final state in production of say dijets, it is important to know whether a hard process occurs at different average impact parameters than in the minimum bias interactions. It turns out that at the LHC energies a dijet trigger selects, in average, a factor of two smaller impact parameters than in the minimum bias events. For example this influences the Underlying Event (UE) structure of the collision. This implies that the multijet activity, energy flow should be much stronger in these events than in the minimum bias events. Obviously, the magnitude of the enhancement does depend on the transverse distribution of partons and on the correlation between the partons in the transverse plane. The information on the transverse distribution of partons has been extracted from experimental measurements, and is summarized in the appendix.

In the case of collisions with N hard subprocesses, the interaction picture corresponds to a pairwise localization of N partons of each of the nucleons at short distances (Fig. 1(b)), leading to the cross-section of collision of hadrons a and b proportional to (if correlations between partons are neglected)

$$\sigma_h^{(N)} \propto \int d^2b \prod_{i=1}^{i=N} d\rho_i d\rho_i' \delta(\rho_i + b - \rho_i') f_a(\rho_i, Q_i) f_b(\rho_i', Q_i). \tag{4}$$

The geometric pairwise overlap with N partons of hadrons a and b nearby pairwise provides a geometric factor L^{2-2N} in the cross-section for N hard collisions, where L is the linear scale proportional to the transverse linear scale of the colliding hadrons, i.e., L^2 is the transverse area occupied by partons, σ_{eff}. Equation (4) includes correlations between partons both on the hadronic distance scale and local correlations due to the QCD evolution. In the case of perturbative correlations when two partons of one of the colliding nucleons are close together the overlap factor is enhanced as compared to the uncorrelated case, see discussion in Section 3.

Using the information on the transverse spatial distribution of partons in the nucleon, one can obtain the distribution over impact parameters in pp collisions with hard parton–parton processes.[116] It is given by the overlap of two parton wave functions as depicted in Fig. 1.

The probability distribution of pp impact parameters in events with a given hard process, $P_2(x_1, x_2, b|Q^2)$, is given by the ratio of the cross-section at given b and the cross-section integrated over b. As a result

$$\begin{aligned} P_2(x_1, x_2, b|Q^2) \equiv \int d^2\rho_1 \int d^2\rho_2 \, \delta^{(2)}(\boldsymbol{b} - \boldsymbol{\rho}_1 + \boldsymbol{\rho}_2) \\ \times F_{2g}(x_1, \rho_1|Q^2) \, F_{2g}(x_2, \rho_2|Q^2), \end{aligned} \tag{5}$$

which obviously satisfies the normalisation condition

$$\int d^2b \, P_2(x_1, x_2, b|Q^2) = 1. \tag{6}$$

Here the two gluon–nucleon form F_{2g} is defined as

$$F_{2g}(x_1, \rho_1|Q^2) = f(x_1, \rho_1|Q^2)/D(x, Q^2). \tag{7}$$

(For detailed discussion of parametrisation and properties of F_{2g}, see the appendix.) This distribution represents an essential tool for phenomenological studies of the underlying event in pp collisions,[116,117] see discussion in Section 7.

For the dipole and exponential parameterisation of F_{2g} (see Eq. (A.4) in the appendix), Eq. (5) leads to (for $x \equiv x_1 = x_2$)

$$P_2(x, b|Q^2) = \begin{cases} (4\pi B_g)^{-1} \exp[-b^2/(4B_g)], \\ [m_g^2/(12\pi)]\,(m_g b/2)^3\, K_3(m_g b), \end{cases} \tag{8}$$

where the parameters B_g and m_g are taken at the appropriate values of x and Q^2. Since B_g increases with a decrease of x, the distribution in b depends on the x values of the colliding partons and their virtualities, however this effect is pretty small for production of jets at central rapidities, see, e.g., Figs. 4 and 5 in Ref. 117.

Comment: A word of caution is necessary here. The transverse distance b for dijet events is defined as the distance between the transverse centres of mass of two nucleons. It may not coincide with b defined for soft interactions where soft partons play an important role. For example, if we consider dijet production due to the interaction of two partons with $x \sim 1$, $\rho_1, \rho_2 \sim 0$ since the transverse centre of mass coincides with transverse position of the leading parton in the $x \to 1$ limit. As a result, b for the hard collision will be close to zero. On the other hand the rest of the partons may interact in this case at very different transverse coordinates. As a result, such configurations may contribute to the inelastic pp cross-section at much larger b for the soft interactions. However for the parton collisions at $x_1, x_2 \ll 1$ the recoil effects are small and so two values of b should be close.

2.2. *Impact parameter distribution in minimum bias collisions*

The derived distribution $P_2(x, b|Q^2)$ should be compared to the distribution of the minimum bias inelastic collisions which can be expressed in terms of the profile function of the pp elastic amplitude $\Gamma(s, b)$ ($\Gamma(s, b) = 1$ if the interaction is completely absorptive at given b)

$$P_{\text{in}}(s, b) = [1 - |1 - \Gamma(s, b)|^2]/\sigma_{\text{in}}(s), \tag{9}$$

where $\int d^2b\, P_{\text{in}}(s, b) = 1$.

Our numerical studies indicate that the impact parameter distributions with the jet trigger (Eq. (8)) are much more narrow than that in minimum bias inelastic events at the same energy (Eq. (9)) — see Fig. 2, and that b-distribution for events with a dijet trigger is a very weak function of the p_T of the jets or their rapidities, for example, for the case of the pp collisions at $\sqrt{s} = 13$GeV the median value of b, $b_{\text{median}} \approx 1.2$ fm and $b_{\text{median}} \approx 0.65$ fm for minimum bias and dijet trigger events respectively.[117]

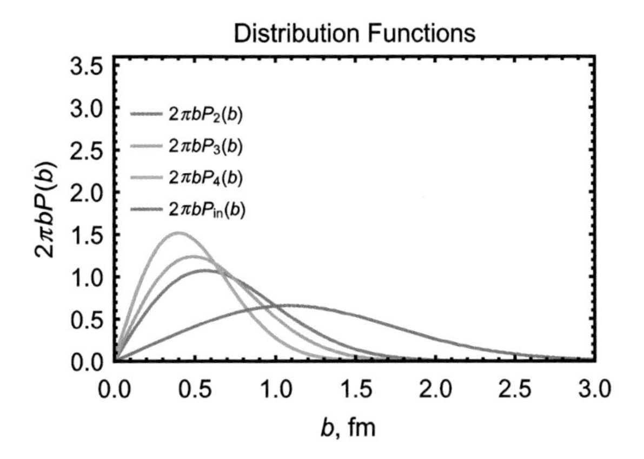

Fig. 2 Normalized probabilities of minimum bias, inclusive two, and four parton collisions and collision involving three partons as a function of the impact parameter.

Note here that in many experimental analyses the minimum bias cross-section is defined as the inelastic non-diffractive cross-section. Since inelastic diffraction is a peripheral process in *pp* scattering, $\sigma_{\text{min.bias}}$ defined this way corresponds to somewhat smaller b than the ones given by Eq. (9).

For $N \geq 2$ dijet processes,

$$b_{\text{median}}(N) \approx \frac{1}{\sqrt{N}} b_{\text{median}}(N = 1). \tag{10}$$

Hence inclusive $N \geq 2$ processes are dominated by collisions at very small impact parameters where gluon fields of two nucleons strongly overlap: $b_{\text{median}} < 2r_g^{(N)}(x)$ (here $r_g^{(N)}(x) \geq 0.4$ fm is the transverse radius of the gluon distribution in nucleons), cf. Fig. 2.

Since the large impact parameters give the dominant contribution to σ_{in}, our analysis indicates that there are two pretty distinctive classes of *pp* collisions — large b collisions which are predominantly soft and central collisions with strongly enhanced rate of hard collisions. We refer to this pattern as the two transverse scale picture of *pp* collisions at collider energies.[116]

3. GPD and Mean Field Approach to MPI

Description of the MPI is a multi-scale problem. This is not only because the separate parton–parton interactions may differ in hardness. More importantly, *each* single hard interaction possesses *two* very different hardness scales. The distinctive feature of the DPS is that it produces two pairs of nearly back-to-back jets, so that in the collision of partons 1 and 3 the first (larger) scale is given by the invariant mass of the jet pair, $Q^2 = 4J_{1\perp}^2 \simeq 4J_{3\perp}^2$, while the second scale is the magnitude of the *total transverse momentum* of the pair: $\delta^2 = \delta_{13}^2$. It is important to stress that in the MPI physics there is *no factorization* in the usual sense of the word. The cross-sections do not factorize into the product of the hard parton interaction cross-sections and the multi-parton distributions depending on momentum fractions x_i and the hard scale(s).[a] A general approach to double (multi) hard interactions has been developed in Ref. 82. It turned out that the *transverse momentum* of the parton in the w.f. and that of its counterpart in the conjugated w.f. are indeed necessarily different, with their difference $\mathbf{\Delta}$ being conjugate to the relative transverse distance between the two partons in the hadron. This has led to the introduction of the new object — the *generalized double parton distribution*, ${}_2$GPD, which depends on a new momentum parameter $\mathbf{\Delta}$[9,82] (see also Chapters 2 and 3).

3.1. *Generalized two-parton distribution*

3.1.1. *${}_2GPD$ and their connection to the wave functions*

In Refs. 9 and 82 we have shown that the QFT description of the double hard parton collisions calls for introduction of ${}_2$GPD. Defined in the momentum space, it characterizes two-parton correlations inside hadron:[82] $D_h(x_1, x_2, Q_1^2, Q_2^2; \mathbf{\Delta})$. Here the index h refers to the hadron, x_1 and x_2 are the light-cone fractions of the parton momenta, and Q_1^2, Q_2^2 the corresponding hard scales. As has been mention above, the two-dimensional vector $\mathbf{\Delta}$ is the Fourier conjugate to the relative distance between the partons 1 and 2 in the impact parameter plane. The distribution obviously depends on the parton species; we suppress the corresponding indices for brevity.

The ${}_2$GPD are expressed through multiparton light cone wave functions as

$$
\begin{aligned}
D(x_1, x_2, p_1^2, p_2^2, \vec{\Delta}) &= \sum_{n=3}^{\infty} \int \frac{d^2k_1}{(2\pi)^2} \frac{d^2k_2}{(2\pi)^2} \theta(p_1^2 - k_1^2)\theta(p_2^2 - k_2^2) \\
&\quad \times \int \prod_{i\neq 1,2} \frac{d^2k_i}{(2\pi)^2} \int_0^1 \prod_{i\neq 1,2} dx_i\,(2\pi)^3 \delta\left(\sum_{i=1}^{i=n} x_i - 1\right) \delta\left(\sum_{i=1}^{i=n} \mathbf{k}_i\right) \\
&\quad \times \psi_n(x_1, \mathbf{k}_{1t}, x_2, \mathbf{k}_{2t}, ., \mathbf{k}_{it}, x_i..)\psi_n^+ \\
&\quad \times (x_1, \vec{k_{1t}} + \vec{\Delta}, x_2, \vec{k_{2t}} - \vec{\Delta}, x_3, \mathbf{k}_{3t}, \ldots).
\end{aligned}
\tag{11}
$$

Note that this distribution is diagonal in the space of all partons except the two partons involved in the collision. Here ψ is the parton wave function normalized to one in the usual way. An appropriate summation over color and Lorentz indices is implied.

[a] One may try to restore factorization by adding $1 \otimes 1$ contribution, although this contribution has different singularity structure and is not double collinearly enhanced. For progress in this direction see Chapter 2. In this review we treat $1 \otimes 1$ contribution as part of the loop correction to conventional multijet creation processes $1 + 1 \to \cdots$.

The double hard interaction cross-section (and, in particular, that of production of two dijets) can be expressed through the convolution of ${}_2$GPDs.

The *effective interaction area* $\sigma_{\rm eff}$ defined in Eq. (1) is given by the convolution of the ${}_2$GPDs of incident hadrons over the transverse momentum parameter $\mathbf{\Delta}$ normalized by the product of single parton inclusive pdfs[b]

$$\frac{1}{\sigma_{\rm eff}} \equiv \frac{\int \frac{d^2\mathbf{\Delta}}{(2\pi)^2} D_{h_1}(x_1, x_2, Q_1^2, Q_2^2; \mathbf{\Delta}) D_{h_2}(x_3, x_4, Q_1^2, Q_2^2; -\mathbf{\Delta})}{D_{h_1}(x_1, Q_1^2) D_{h_1}(x_2, Q_2^2) D_{h_2}(x_3, Q_1^2) D_{h_2}(x_4, Q_2^2)}. \tag{12}$$

Equation (12) (and similar expressions for any number of MPI) can be rewritten in the transverse coordinate representation and corresponds to the transverse geometry depicted in Fig. 1 with $\mathbf{\Delta}$ Fourier conjugated to the difference of transverse coordinates of partons: $\rho_1 - \rho_3$.

${}_2$GPDs enter also the expressions for the differential distributions in the jet transverse momentum imbalances δ_{ik} (the integral over δ_{ik} is the "total" DPS cross-section — Eq. (12)). In the inclusive case, the hardness parameters of the ${}_2$GPDs are given by the jet transverse momenta Q_i^2, while for the differential distributions, by the jet imbalances δ_{ik}^2. The corresponding formulas derived in the leading collinear approximation of pQCD can be found in Ref. 9. It is worth emphasizing here that the DPS cross-section *does not factorize* into the product of the hard parton interaction cross-sections and the two two-parton distributions depending on momentum fractions x_i and the hard scales, Q_1^2, Q_2^2.

Note that one can introduce in the same way the N-particle GPD, ${}_N GPD$, which can be probed (say) by the production of N pairs of jets.[82] In this case, the first N arguments k_i are shifted by $\overrightarrow{\Delta_i}$ subject to the constraint $\sum_i \overrightarrow{\Delta_i} = 0$. So the cross-section is proportional to

$$\sigma_{2N} \propto \int \prod_{i=1}^{i=N} \frac{d\overrightarrow{\Delta}_i}{(2\pi)^2} D_a(x_1, \ldots, x_N, \overrightarrow{\Delta}_1, \ldots, \overrightarrow{\Delta}_N)$$
$$\times D_b(x'_1, \ldots, x'_N, \overrightarrow{\Delta}_1, \ldots, \overrightarrow{\Delta}_N) \delta\left(\sum_{i=1}^{i=N} \overrightarrow{\Delta}_i\right). \tag{13}$$

The N-parton GPDs are expressed through multiparton wave functions analogously to Eq. (11).

The above approach allows to take into account consistently the perturbative mechanism of two-parton correlation when the two partons emerge from *perturbative splitting* of one parton taken from the hadron wave function since one needs to separate these correlations from the $2 \to 4$ mechanism of the jet production.

In perturbative scenario the production of the parton pairs is concentrated at much smaller transverse distances between partons. As a result, the corresponding

[b]We do not consider here spin correlation effects, and in kinematics under investigation. Also, color interference effects are negligible in the kinematics considered in this review.[13,92]

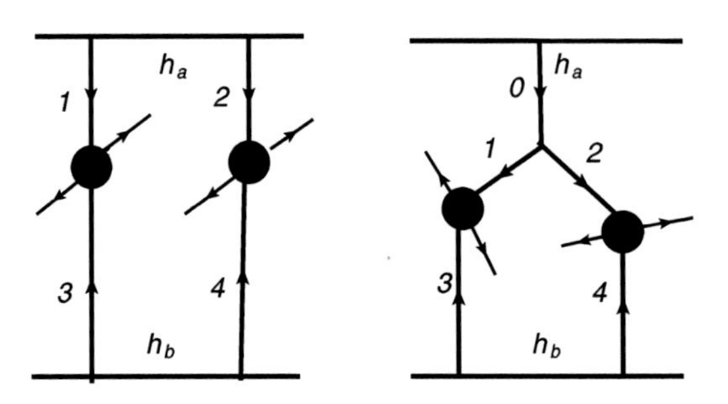

Fig. 3 Sketch of the two considered DPS mechanisms: $2 \otimes 2$ (left) and $1 \otimes 2$ (right) mechanism.

contribution to ${}_2$GPD turns out to be practically independent of Δ^2 in a broad range of Δ — up to the hard scale(s) characterizing the hard process under consideration (Δ^2 only affects the lower limit of the transverse momentum integrals in the parton cascades, resulting in a mild logarithmic dependence). The weak dependence on Δ results in a distribution over impact parameters for DPS which is intermediate between the mean field contribution and dijet b-distributions, cf. Fig. 2. Given essentially different dependence on Δ, one has to treat the two contributions separately by casting the ${}_2$GPD as a sum of two terms depicted in Fig. 3:

$$D_h(x_1, x_2, Q_1^2, Q_2^2; \boldsymbol{\Delta}) = {}_{[2]}D_h(x_1, x_2, Q_1^2, Q_2^2; \boldsymbol{\Delta}) + {}_{[1]}D_h(x_1, x_2, Q_1^2, Q_2^2; \boldsymbol{\Delta}). \tag{14}$$

Here the subscript ${}_{[2]}$ in ${}_{[2]}D$ marks the mechanism in which two partons originate from the hadron (Fig. 3(left)) while subscript ${}_{[1]}$ in the term ${}_{[1]}D$ denotes mechanism in which one parton perturbatively splitting into two (Fig. 3(right)).

Let us stress that it follows from the above formulas that in the impact parameter space these GPDs have a probabilistic interpretation. In particular, they are positively definite in the impact parameter space, see the discussion in Ref. 10 (we consider only the diagonal case when initial and final wave functions coincide).

3.2. *Modeling ${}_{[2]}D$: The mean field approach*

To proceed with quantitative estimates, one needs a model for the non-perturbative two-parton distributions in a proton. A priori, we know next to nothing about them. The first natural step to take is an *approximation of independent partons*/the mean field approximation (as already discussed in Section 2). It allows one to relate ${}_2$GPD with known objects, namely[82]

$${}_{[2]}D(x_1, x_2, Q_1^2, Q_2^2; \Delta) \simeq f(x_1, Q_1^2; \Delta^2) f(x_2, Q_2^2; \Delta^2). \tag{15}$$

Here f is the non-forward parton correlator (known as generalized parton distribution, GPD) that determines, e.g., hard vector meson production at HERA and enters in our case in the diagonal kinematics in x ($x_1 = x_1'$).

Modeling ${}_2$GPD using Eq. (15) has its limitations. First of all, it does not respect the obvious restriction $D(x_1 + x_2 > 1) = 0$. So, x_i have to be taken not too large (say, $x_i \ll 0.5$). Actually, the neglect of correlations is likely to be a good approximation only at much smaller $x \leq 0.1$. In any case, currently one can extract GPDs only from the theoretical analysis of the hard exclusive amplitude like $\gamma_L^* + N \to VM + N$ for $x < 0.05$.

There is an additional caveat — in the vector meson production two gluons in t-channel carry different light cone fractions while in the case of the scattering amplitude x's in the $|in\rangle$ and $\langle out|$ state are equal. Also, in the vector meson production modulus squared of the amplitude enters while in the MPI case we deal with the imaginary part of the zero angle amplitude. As a result a simple connection between the gluon GPD and the observed cross-section exists only if the virtualities are large enough and x is small enough, $x \leq 0.1$.

On the other hand, x_i should not be *too small* to stay away from the region of the Regge–Gribov phenomena where there are serious reasons for parton correlations to be present at the non-perturbative level (see the discussion in Ref. 85 and in Section 6).

Thus, we expect that x-range where the mean field NP model Eq. (15) is applicable for ${}_2$GPD is $10^{-1} \geq x_i \geq 10^{-3}$.

The GPDs can be parameterized as

$$f(x_1, Q_1^2; \Delta^2) \simeq D(x_1, Q_1^2) \times F_{2g}(x_1, \Delta^2, Q^2), \tag{16}$$

with D being the usual single parton distribution functions and F being the so-called two-gluon form factor of the hadron. The latter is a non-perturbative object; it falls fast with the "momentum transfer" Δ^2. In our following numerical studies we will use the model of the two gluon form factor extracted from the data on exclusive J/ψ photoproduction. This analysis is summarized in the appendix.

Using parameterization of Eq. (8), one finds [82,116]

$$\frac{1}{\sigma_{\text{eff}}} = \int \frac{d^2\Delta}{(2\pi)^2} F_{2g}^4(x, \Delta) = (8\pi B_g(x)))^{-1} \approx 32 \text{ mb}, \tag{17}$$

for $x \sim 0.01$ for the exponential parameterization fit and practically the same number, $\frac{m_g^2}{28\pi}$, for the dipole fit with B_g related to m_g^2 according to Eq. (A.5). (Here and below we do not write explicit dependence of σ_{eff} on x.) Numerically Eq. (17) leads to approximately a factor of two smaller production cross-section than the one observed at the Tevatron at $x \geq 0.01$ for four jet and 3 jet +photon DPS. Since the two gluon form factor decreases faster with t with decrease of x, the mean field model leads to increase of σ_{eff} with energy for the central rapidities and fixed p_t. Note that the two-exponential parameterization of transverse parton density used in a number of versions of Pythia which described experimental values of σ_{eff} strongly contradicts the data on the J/ψ photoproduction, see, e.g., Fig. 3 in Ref. 118.

Using Eq. (13) and exponential parameterization of GPD one can also find the effective cross-section for n hard collisions in mean field approach:

$$\frac{1}{\sigma_{\rm eff}^{(n)}} = \frac{1}{(2\pi)^{N-1}} \prod_{i=1}^{i=N} \frac{1}{B_i + B_i'} \frac{1}{\sum_{i=1}^{i=N} 1/(B_i + B_i')}. \tag{18}$$

Here $B_i \equiv B(x_i), B_i' \equiv B(x_i')$ for N dijet process with x_i are Bjorken fractions for hadron a, and x_i' are Bjorken fractions for colliding hadron b. For $N = 2$ we get the familiar result:[90]

$$\frac{1}{\sigma_{\rm eff}} = \frac{1}{2\pi} \frac{1}{B_1 + B_1' + B_2 + B_2'}. \tag{19}$$

The particular case of this formula for $N = 3$ was recently considered in Ref. 43.

4. pQCD Correlations

4.1. $1 \otimes 2$ *DPS process*

Actually, the non-perturbative (NP) and perturbative (PT) contributions *do not* enter the physical DPS cross-section in the arithmetic sum Eq. (14), driving one even farther from the familiar factorization picture based on universal (process independent) parton distributions. As explained Ref. 9, a double hard interaction of two pairs of partons that *both* originate from PT splitting of a single parton from each of the colliding hadrons, does not produce back-to-back dijets in a double collinear approximation (their contribution has only single collinear enhancement, although numerically, being of leading twist, can be numerically important, cf. Ref. 5). In fact, such an eventuality corresponds to a one-loop correction to the usual $2 \to 4$ jet production process and should not be looked upon as a multi-parton interaction. The term ${}_{[1]}D_{h_1} \times {}_{[1]}D_{h_2}$ has to be excluded from the product $D_{h_1} \times D_{h_2}$, the conclusion we share with Gaunt and Stirling.[84]

So, we are left with two sources of genuine two-parton interactions: four-parton collisions described by the product of (PT-evolved) ${}_2$GPDs of NP origin ($2 \otimes 2$),

$${}_{[2]}D_{h_1}(x_1, x_2, Q_1^2, Q_2^2; \mathbf{\Delta})\, {}_{[2]}D_{h_2}(x_3, x_4, Q_1^2, Q_2^2; -\mathbf{\Delta}), \tag{20}$$

and three-parton collisions due to an interplay between the NP two-parton correlation in one hadron and the two partons emerging from a PT parton splitting in another hadron ($1 \otimes 2$), described by the combination

$$\begin{aligned} &{}_{[2]}D_{h_1}(x_1, x_2, Q_1^2, Q_2^2; \mathbf{\Delta})\, {}_{[1]}D_{h_2}(x_3, x_4, Q_1^2, Q_2^2; -\mathbf{\Delta}) \\ &\quad + {}_{[1]}D_{h_1}(x_1, x_2, Q_1^2, Q_2^2; \mathbf{\Delta})\, {}_{[2]}D_{h_2}(x_3, x_4, Q_1^2, Q_2^2; -\mathbf{\Delta}). \end{aligned} \tag{21}$$

Given that ${}_{[2]}D$ falls fast at large Δ, a mild logarithmic Δ-dependence of ${}_{[1]}D$ can be neglected in the product in Eq. 21.

4.2. *Composition of the* $1 \otimes 2$ *DPS cross-section*

In order to derive the DPS cross-section, one has to start with examination of the double differential transverse momentum distribution and then integrate it over jet imbalances δ_{ik}. Why is this step necessary? The parton distribution $D(x, Q^2)$ — the core object of the QCD-modified parton model — arises upon logarithmic integration over the transverse momentum up to the hard scale, $k_\perp^2 < Q^2$. Analogously, the ${}_2$GPD $D(x_1, x_2, Q_1^2, Q_2^2; \mathbf{\Delta})$ embeds *independent integrations* over parton transverse momenta $k_{1\perp}^2$, $k_{2\perp}^2$ up to Q_1^2 and Q_2^2, respectively. However, the $1 \otimes 2$ DPS cross-section contains a specific contribution ("short split", see below) in which the transverse momenta of the partons 1 and 2 are strongly correlated (nearly opposite). This pattern does not fit into the structure of the pQCD evolution equation for ${}_2$GPD where $k_{1\perp}$ and $k_{2\perp}$ change independently. Given this subtlety, a legitimate question arises whether the expression for the integrated $1 \otimes 2$ cross-section (Eq. (21)) based on the notion of the two-parton distribution ${}_{[1]}D$ takes the short split into account. The expression for the differential distribution over jet imbalances was derived in Ref. 9 in the leading collinear approximation of pQCD. It resembles the "DDT formula" for the Drell–Yan spectrum[119] and contains two derivatives of the product of ${}_2$GPDs that depend on the corresponding δ_{ik} as hardness scales, and the proper Sudakov form factors depending on the ratio of Q_i^2 and δ_{ik}^2.

In particular, in the region of *strongly ordered* imbalances,

$$\frac{\pi^2 d\sigma^{\mathrm{DPS}}}{d^2\delta_{13}\, d^2\delta_{24}} \propto \frac{\alpha_s^2}{\delta_{13}^2\, \delta_{24}^2}; \qquad \delta_{13}^2 \gg \delta_{24}^2,\ \delta_{13}^2 \ll \delta_{24}^2, \tag{22}$$

the differential $1 \otimes 2$ cross-section reads

$$\begin{aligned} \frac{\pi^2 d\sigma_{1\otimes 2}}{d^2\delta_{13}\, d^2\delta_{24}} = \frac{d\sigma_{\mathrm{part}}}{d\hat{t}_1\, d\hat{t}_2} \frac{d}{d\delta_{13}^2} \frac{d}{d\delta_{24}^2} \Bigg\{ \int \frac{d^2\mathbf{\Delta}}{(2\pi)^2} \\ \times {}_{[1]}D_{h_1}(x_1, x_2, \delta_{13}^2, \delta_{24}^2; \mathbf{\Delta})\ {}_{[2]}D_{h_2}(x_3, x_4, \delta_{13}^2, \delta_{24}^2; -\mathbf{\Delta}) \\ \times S_1\left(Q_1^2, \delta_{13}^2\right) S_3\left(Q_1^2, \delta_{13}^2\right) \cdot S_2\left(Q_2^2, \delta_{24}^2\right) S_4\left(Q_2^2, \delta_{24}^2\right) \Bigg\} \\ + \{h_1 \leftrightarrow h_2\}. \end{aligned} \tag{23}$$

The differential distribution for the $2 \otimes 2$ DPS mechanism has a similar structure, see Eq. (25) of Ref. 9.

In addition to Eqs. (22) and (23) there is another source of double collinear enhancement in the differential $1 \otimes 2$ cross-section. It is due to the kinematical region where the two imbalances nearly compensate each other,

$$\delta'^2 = (\delta_{13} + \delta_{24})^2 \ll \delta^2 = \delta_{13}^2 \simeq \delta_{24}^2, \tag{24}$$

and the dominant integration region is complementary to that of Eq. (22):

$$\frac{\pi^2 d\sigma_{\rm short}^{\rm DPS}}{d^2\delta_{13}\, d^2\delta_{24}} \propto \frac{\alpha_{\rm s}^2}{\delta'^2\, \delta^2}; \qquad \delta'^2 \ll \delta^2. \tag{25}$$

This enhancement characterizes the set of $1 \otimes 2$ graphs in which accompanying radiation has transverse momenta not exceeding δ'.

In this situation, the parton that compensates the overall imbalance, $\mathbf{k}_\perp = -\delta'$ is radiated off the incoming, quasi-real, parton legs. At the same time, the virtual partons after the core splitting "0"$\to$ "1"+"2" enter their respective hard collisions without radiating any offsprings on the way.

The $1 \to 2$ splitting occurs close to the hard vertices, therefore the name "short split" (aka "endpoint contribution"[9]).

A complete expression for the differential distribution in the jet imbalances due to a short split was derived in the leading collinear approximation (Eq. (27) of Ref. 9):

$$\begin{aligned}
\frac{\pi^2\, d\sigma_{\rm short}^{\rm DPS}}{d^2\delta_{13}\, d^2\delta_{24}} &= \frac{d\sigma_{\rm part}}{d\hat{t}_1\, d\hat{t}_2} \cdot \frac{\alpha_{\rm s}(\delta^2)}{2\pi\, \delta^2} \sum_c P_c^{(1,2)}\!\left(\frac{x_1}{x_1+x_2}\right) \\
&\quad \times S_1(Q_1^2, \delta^2)\, S_2(Q_2^2, \delta^2) \\
&\quad \times \frac{d}{d\delta'^2} \left\{ S_c(\delta^2, \delta'^2) \frac{D_{h_1}^c(x_1+x_2, \delta'^2)}{x_1+x_2} S_3(Q_1^2, \delta'^2) S_4(Q_2^2, \delta'^2) \right. \\
&\quad \left. \times \int \frac{d^2\boldsymbol{\Delta}}{(2\pi)^2}\, {}_{[2]}D_{h_2}(x_3, x_4, \delta'^2, \delta'^2; -\boldsymbol{\Delta}) \right\} + \{h_1 \leftrightarrow h_2\}.
\end{aligned} \tag{26}$$

The short split becomes less important when the scales of the two hard collisions are different. Indeed, the logarithmic integration over δ^2 is kinematically restricted from above, $\delta^2 < \delta^2_{\rm max} \simeq \min\{Q_1^2, Q_2^2\}$. As a result, in the kinematics where transverse momenta of jets in one pair are much larger than in the second pair, e.g., $Q_1^2 \gg Q_2^2$, the contribution of the short split is suppressed as

$$\sigma_{\rm short}^{(3\to 4)} / \sigma^{(3\to 4)} \propto S_1(q_1^2, q_2^2)\, S_3(q_1^2, q_2^2) \ll 1 \quad (Q_1^2 \gg Q_2^2).$$

Here S_1 and S_3 are the double logarithmic Sudakov form factors of the partons "1" and "3" that enter the hard interaction with the larger hardness scales. The short split induces a strong correlation between jet imbalances which is worth trying to look for experimentally.

The relative weight of the short split depends on the process under consideration. For most DPS processes in the kinematical region we have studied, it typically provides 10–15% of the pQCD correlation contribution. However, it becomes more important when the nature of the process favors parton splitting. In particular, this is the case for the double Drell–Yan pair production where the short split contribution reaches 30–35%. On the contrary, the short split turns out to be practically negligible for the same-sign double W-meson production.[10]

Thus, for the integrated DPS cross-section we obtain two contributions to the effective interaction area:

$$\frac{\prod_{i=1}^{4} D(x_i)}{\sigma_4} = \int \frac{d^2\mathbf{\Delta}}{(2\pi)^2}\ {}_{[2]}D_{h_1}(x_1, x_2, Q_1^2, Q_2^2; \mathbf{\Delta})\ {}_{[2]}D_{h_2}(x_3, x_4, Q_1^2, Q_2^2; -\mathbf{\Delta}),$$

$$\begin{aligned}\frac{\prod_{i=1}^{4} D(x_i)}{\sigma_3} &= \int \frac{d^2\mathbf{\Delta}}{(2\pi)^2}[{}_{[2]}D_{h_1}(x_1, x_2, Q_1^2, Q_2^2; \mathbf{\Delta})\,{}_{[1]}D_{h_2}(x_3, x_4, Q_1^2, Q_2^2) \\ &\quad + {}_{[1]}D_{h_1}(x_1, x_2, Q_1^2, Q_2^2)\,{}_{[2]}D_{h_2}(x_3, x_4, Q_1^2, Q_2^2; \mathbf{\Delta})],\end{aligned} \tag{27}$$

and we refer the reader to Ref. 9 for explicit expression for ${}_1D$ as an integral in terms of DGLAP kernels and PDFs.

It is worth noting here that our analysis demonstrates that a compact and intuitively clear expression containing the product of the ${}_2$GPDs${}_{[2]}$D and ${}_{[1]}$D in Eq. (27) is valid only for the *integrated* $1 \otimes 2$ cross-section. While the expression for differential distributions (26) contains three terms corresponding to $2 \otimes 2$ mechanism and long and short splits in $1 \otimes 2$ mechanism, the expression for full cross-section (27) is entirely expressed through GPD and does not contain short split contribution explicitly. This phenomenon was discussed in detail in Ref. 95. The basic idea is that the differential distribution for the long split in Eq. (26) contains the singular term $\sim \delta(\delta_{13}^2 - \delta_{24}^2)$. This can be seen by explicit double differentiation in Eq. (27). This term clearly must not be taken into account while calculating the differential distributions, since it corresponds to the short split kinematics of Eq. (25). That is, in this kinematic region the short split is working and the last term in Eq. (26) correctly describes the differential distribution. The fake singular term has to be subtracted. However while this fake singular term is concentrated in an area zero region of the phase space, its integral over disbalances gives rise to a total cross-section which remarkably coincides with the short split contribution, i.e., Eq. (27) does not include short splits explicitly, but correctly sums all contributions to full cross-section, non-trivially avoiding double-counting.

Finally let us stress that we consider all hard scales, including jet disbalances to be sufficiently hard ($\gg \Lambda_{QCD}$), so we can calculate all differential distributions consistently within the DGLAP formalism (see Ref. 119). The case when some of the scales can become soft was considered using transverse momentum distributions in Refs. 3 and 22.

4.3. *Modeling $_1D$ terms*

Turning to the $1 \otimes 2$ term, we neglect a mild logarithmic Δ-dependence of $_{[1]}D$ in (27) and use the model of section 3B for $_{[2]}D$ to obtain

$$\sigma_3{}^{-1} \simeq \frac{7}{3} \cdot \left[\frac{_{[1]}D(x_1,x_2)}{D(x_1)D(x_2)} + \frac{_{[1]}D(x_3,x_4)}{D(x_3)D(x_4)}\right] \times \sigma_4{}^{-1}, \tag{28}$$

where we used dipole parameterization for the two gluon form factor:

$$\int \frac{d^2\boldsymbol{\Delta}}{(2\pi)^2} F_{2g}^2(\Delta^2) = \frac{m_g^2}{12\pi}.$$

Very similar results are obtained for the exponential parameterization. We will parameterize the result in terms of the ratio

$$R \equiv \frac{\sigma_{1\otimes 2}}{\sigma_{2\otimes 2}} = \frac{\sigma_4}{\sigma_3}. \tag{29}$$

For the effective interaction area,

$$\sigma_{\rm eff}^{-1} = \sigma_4^{-1} + \sigma_3^{-1}, \tag{30}$$

we have

$$\sigma_{\rm eff} = \frac{\sigma_{\rm eff}^{\rm mean\ field}}{1+R}, \tag{31}$$

where $\sigma_{\rm eff}^{\rm mean\ field}$ is the mean field value of $\sigma_{\rm eff}$, obtained using either dipole or exponential fit for F_{2g}. The difference between the values for $\sigma_{\rm eff}$ obtained using these two fits is within the current experimental errors of the J/ψ data. In the numerical simulations for DPS below we use the dipole fit, that works slightly better for values of Bjorken x corresponding to hard DPS, while for the underlying event (UE) we used the exponential fit, that works slightly better for small x relevant for the UE. The difference however is of the order of several percent and can be neglected.

Within the framework of the NP two-parton $_2$GPD model, Eq. (15), there is only one free parameter Q_0^2. The DPS theory developed in this section can be applied to various processes and holds in a range of energies and different kinematical regions. Therefore, having fixed the Q_0^2 value, say, from the Tevatron data, one can consider all other applications (in particular, to the LHC processes) as parameter-free theoretical predictions.

4.4. *Analytical estimate of pQCD correlations*

The PT parton correlations cannot be neglected. Indeed, let us choose a scale Q_0 that separates NP and PT physics to be sufficiently low, so that parton cascades due to the evolution between Q_0 and Q_i^2 are well developed. To get a feeling of the relative importance of the PT correlation, as well as to understand its dependence on x and the ratio of scales, Q^2 vs. Q_0^2, the following lowest order PT estimate can be used.

Imagine that at the scale Q_0 the nucleon consisted of n_q quarks and n_g gluons ("valence partons") with relatively large longitudinal momenta, so that triggered partons with $x_1, x_2 \ll 1$ resulted necessarily from PT evolution. In the first logarithmic order, $\alpha_s \log(Q^2/Q_0^2) \equiv \xi$, the inclusive spectrum can be represented as

$$D \propto (n_q C_F + n_g N_c)\xi,$$

where we suppressed x-dependence as irrelevant. If both gluons originate from the same "valence" parton, then

$$_{[1]}D \propto \frac{1}{2} N_c \xi \cdot D + (n_q C_F^2 + n_g N_c^2)\xi^2, \tag{32}$$

while independent sources give:

$$\begin{aligned} _{[2]}D &\propto \left(n_q(n_q-1)C_F^2 + 2n_q n_g C_F N_c + n_g(n_g-1)N_c^2\right)\xi^2 \\ &= D^2 - \left(n_q C_F^2 + n_g N_c^2\right)\xi^2. \end{aligned} \tag{33}$$

Hence

$$\frac{_{[2]}D(x_1, x_2)}{D(x_1)D(x_2)} - 1 \simeq \frac{N_c}{2(n_q C_F + n_g N_c)}. \tag{34}$$

The correlation is driven by the gluon cascade — the first term in Eq. (32) — and is not small (being of the order of unity). It gets diluted when the number of independent "valence sources" at the scale Q_0^2 increases. This happens, obviously, when x_i are taken smaller. On the other hand, for large $x_i \sim 0.1$ and increasing, the effective number of more energetic partons in the nucleon is about two and decreasing, so that the relative importance of the $1 \otimes 2$ processes grows.

We conclude that the relative size of PT correlations is of the order one, provided $\xi = \mathcal{O}(1)$.

5. Numerical Results for DPS

5.1. *Calculation framework*

We consider in this chapter hard DPS with $p_t > 10$–15 GeV, where the $1 \otimes 2$ mechanism gives the dominant correction to the mean field results. (The pattern for small p_t relevant for UE is discussed in the next section.) In numerical calculations we used the GRV92 parameterization of gluon and quark parton distributions in the proton.[120] We have checked that using more advanced GRV98 and CTEQ6L parameterizations does not change the numerical results. The explicit GRV92 parameterization is speed efficient and allows one to start the PT evolution at rather small virtuality scales. The combination $(Q_0^2 + \Delta^2)$ was used as the lower cutoff for the logarithmic transverse momentum integrals in the parton evolution, which induced a mild (logarithmic) Δ-dependence on top of the relevant power of the two-gluon form factor $F_{2g}(\Delta^2)$.

To quantify the role of the $1 \otimes 2$ DPS subprocesses, we calculated the ratio R defined in Eq. (29) in the kinematical region $10^{-3} \leq x_i \leq 10^{-1}$ for Tevatron ($\sqrt{s} = 1.8$–$1.96\,\text{TeV}$) and LHC energies ($\sqrt{s} = 7\,\text{TeV}$). We chose to consider three types of ensembles of colliding partons:

(1) $u(\bar{u})$ quark and three gluons which is relevant for "photon plus 3 jets" CDF and D0 experiments,
(2) four gluons (two pairs of hadron jets),
(3) $u\bar{d}$ plus two gluons, illustrating W^+jj production.
(4) $u\bar{d}$ plus $d\bar{u}$, corresponding to the W^+W^- channel.

5.2. *Perturbative* $1 \otimes 2$ *correlation at the Tevatron*

5.2.1. *CDF experiment*

In Fig. 4 we show the profile of the $1 \otimes 2$ to $2 \otimes 2$ ratio R for the $\gamma +$3jets process in the kinematical domain of the CDF experiment.[97] The calculation was performed for the dominant "Compton scattering" channel of the photon production: $g(x_2) + u(\bar{u})(x_4) \to \gamma + u(\bar{u})$. The longitudinal momentum fractions of two gluons producing second pair of jets are x_1 and x_3. The typical transverse momenta were taken to be $p_{\perp 1,3} \simeq 5$ GeV for the jet pair, and $p_{\perp 2,4} \simeq 20$ GeV for the photon–jet system. In Fig. 4, R is displayed as a function of rapidities of the photon–jet, $\eta_2 = \frac{1}{2}\ln(x_2/x_4)$, and the 2-jet system, $\eta_1 = \frac{1}{2}\ln(x_1/x_3)$.

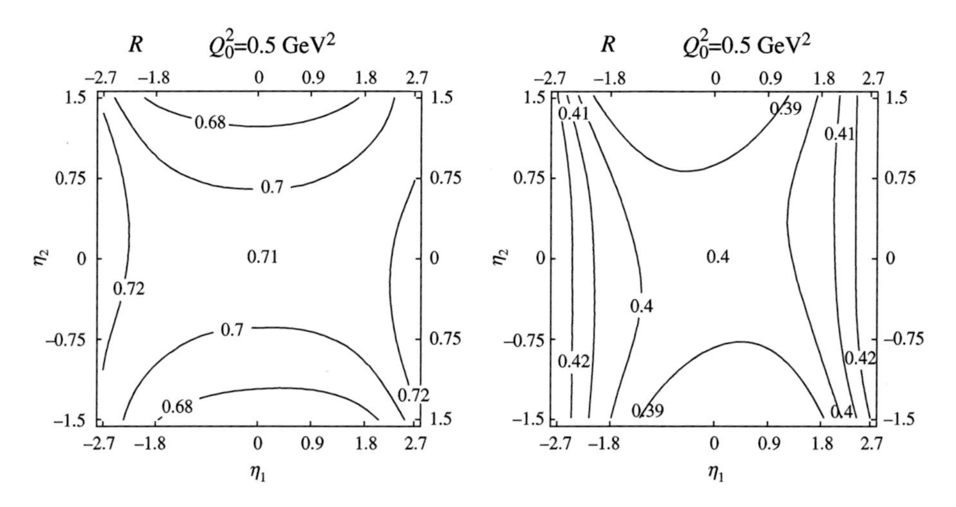

Fig. 4 The $1 \otimes 2/2 \otimes 2$ ratio, Eq. (29), in the CDF kinematics for the process $p\bar{p} \to \gamma + 3\,\text{jets} + X$.

We observe that the enhancement factor lies in the ballpark of $1+R \sim 1.5-1.8$. Processed through Eq. (31), it translates into $\sigma_{\rm eff} \simeq 18-21$ mb. This expectation has to be compared with the CDF finding $\sigma_{\rm eff} = 14.5 \pm 1.7\, ^{+1.7}_{-2.3}$ mb. A recent reanalysis of the CDF data points at an even smaller value: $\sigma_{\rm eff} = 12.0 \pm 1.4\, ^{+1.3}_{-1.5}$ mb.[112] Both these values are significantly smaller than our estimate and the result of D0 experiment discussed in the next subsection.

From Fig. 4 one also observes that the R factor and hence $\sigma_{\rm eff}$ exhibits a very mild x-dependence at these fixed hardness values.

5.2.2. *D0 experiment*

The ratio R is practically constant in the kinematical domain of the D0 experiment which studied photon+3 jets production[98,99] and is very similar to that of the CDF experiment shown above in Fig. 4. So, for the D0 kinematics we instead display in Fig. 5 the enhancement factor $1+R$ as a function of $p_\perp$ of the secondary jet pair for photon transverse momenta 10, 20, 30, 50, 70, and 90 GeV (from bottom to top). The corresponding prediction for $\sigma_{\rm eff}$ is shown in Fig. 6 in comparison with the D0 findings. Both the absolute value and the hint of decrease of $\sigma_{\rm eff}$ with increase of $p_\perp$ look satisfactory.

5.3. *LHC energies*

In Fig. 7 we show the $1 \otimes 2$ to $2 \otimes 2$ ratio for production of two pairs of back-to-back jets with transverse momenta 50 GeV produced in collision of gluons at the LHC energy of $\sqrt{s} = 7$ TeV (the pattern for higher LHC energies is very similar).

Dependence on the hardness parameters of the DPS process of double gluon–gluon collisions is illustrated in Fig. 8. For the sake of illustration, we have chosen

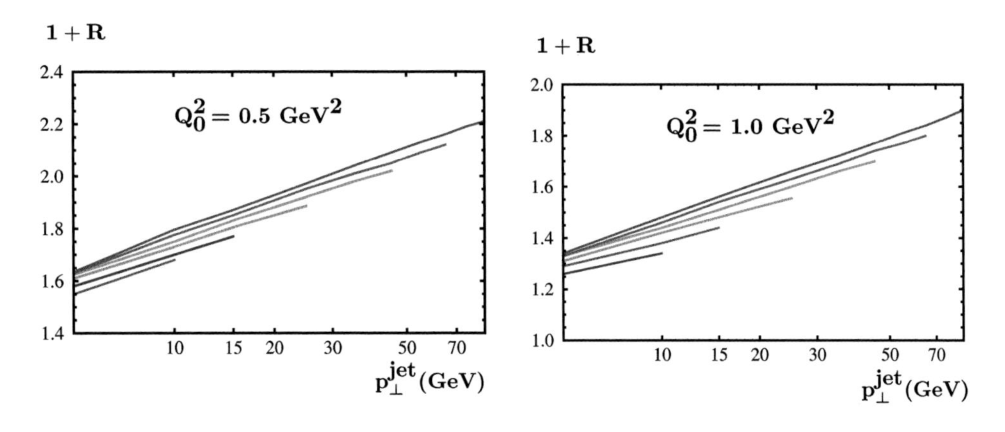

Fig. 5 Left: Central rapidity photon+3 jets production in $u(\bar{u})$ — gluon collisions in the D0 kinematics for $Q_0^2 = 0.5$ GeV^2. Right: for $Q_0^2 = 1$ GeV^2.

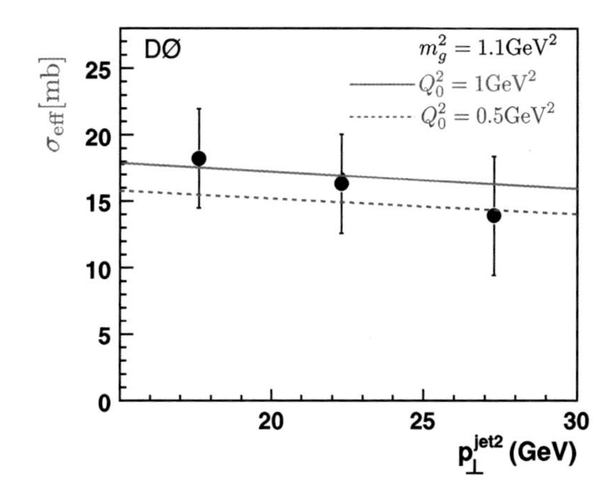

Fig. 6 σ_{eff} as a function of the hardness of the second jet in the kinematics of the D0 experiment[98,99] for $p_{\perp\gamma} = 70$ GeV.

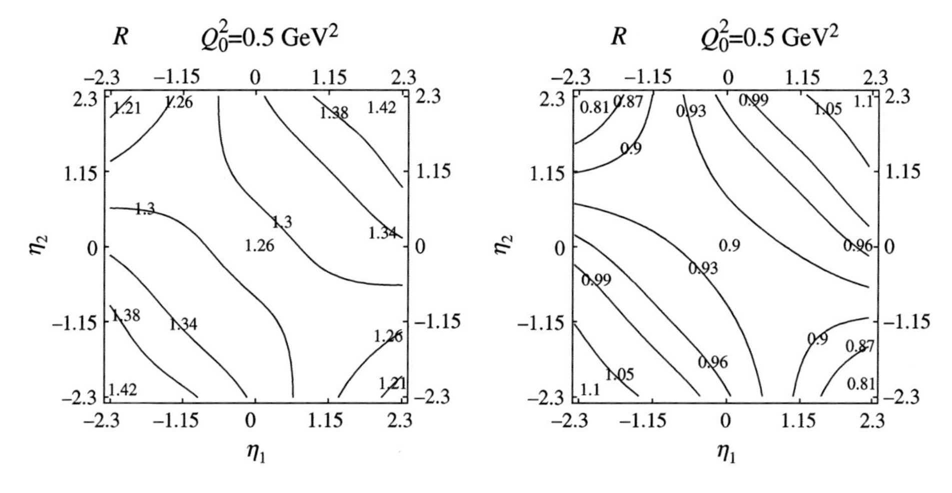

Fig. 7 Rapidity dependence of the R factor for two pairs of $p_\perp = 50$ GeV jets produced in gluon–gluon collisions.

the value of the $p_\perp$ cutoff parameter, varied $Q_0^2 = 0.5, 1, 2\,\text{GeV}^2$, and calculated the σ_{eff} as a function of transverse momenta of the second dijet.[90]

For considered $\sqrt{s}, p_\perp$ range, R increases by about 15–25% with increase of the hardness of one of the jet pairs. This corresponds to approximately 10% drop of σ_{eff}.

Finally, in Fig. 9 we show the rapidity profile of the R ratio for the process of production of the vector boson, $u\bar{d} \to W^+$, accompanied by an additional pair of (nearly back-to-back) jets with transverse momenta $p_\perp = 30$ GeV produced in a gluon–gluon collision.

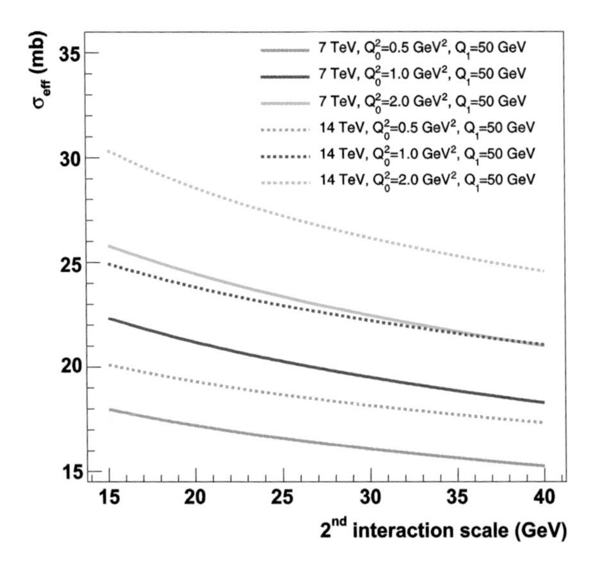

Fig. 8 σ_{eff} for two dijets in DPS at the LHC.

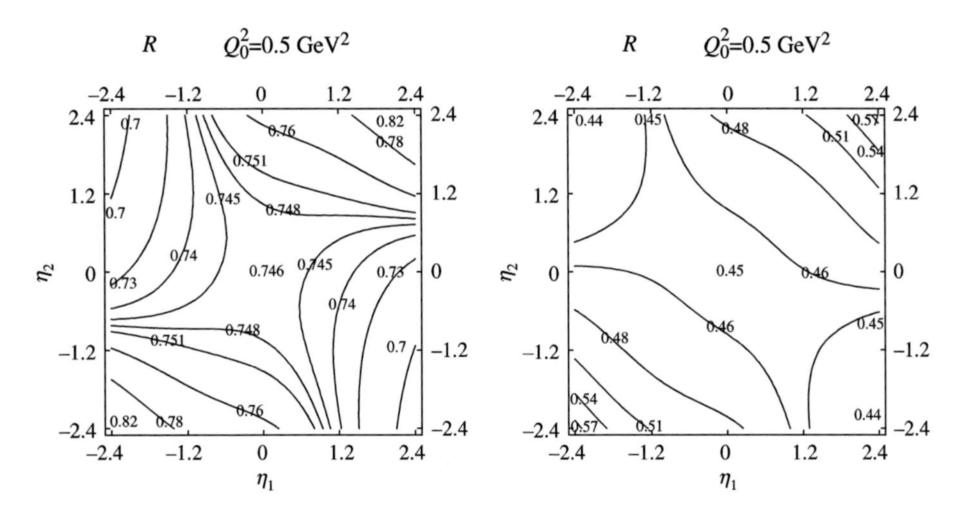

Fig. 9 Ratio R for production of W plus a pair of $p_\perp \simeq 30$ GeV gluon jets.

It is worth noticing that the effect of perturbatively induced parton–parton correlations is maximal for equal rapidities of the W and the jet pair, and slowly decreases with increase of difference of rapidities. This feature is more pronounced when the cutoff parameter Q_0^2 is taken larger. In this case the PT correlation becomes smaller and, at the same time, exhibits a stronger rapidity dependence.

The recent ATLAS study[102] reported for this process the value $\sigma_{\text{eff}} = 15 \pm 3^{+5}_{-3}$ mb which is consistent with the expected enhancement due to

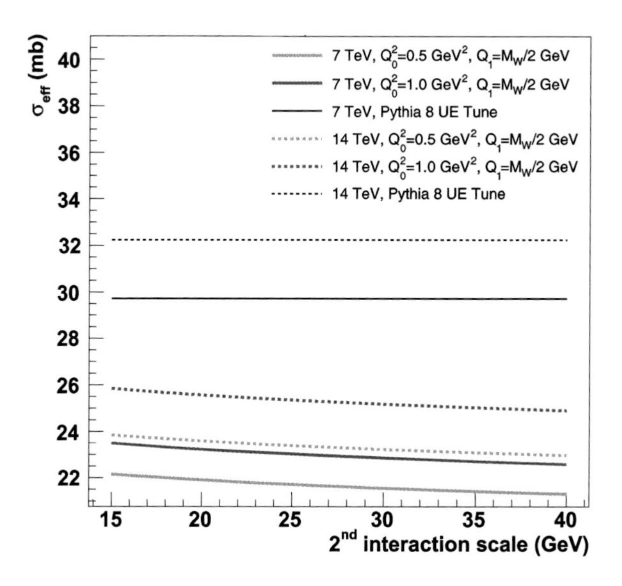

Fig. 10 $\sigma_{\rm eff}$ for Wjj processes as function of a transverse scale of a dijet.

contribution of the 12 DPS channel, see Eq. (29). The characteristic feature of our approach is that $\sigma_{\rm eff}$ depends both on the longitudinal fractions and transverse scale. For example, consider Wjj processes: Figure 10 presents the dependence of $\sigma_{\rm eff}$ on the transverse momenta of jets of the second pair, $p_\perp \equiv p_{\perp 2}$.

We observe that within the considered kinematic range, R increases by about 15–25% with increase of the hardness of one of the jet pairs. This corresponds to approximately 10% decrease of $\sigma_{\rm eff}$. It is worth noticing that the effect of perturbatively induced parton–parton correlations is maximal for equal rapidities of the W and the jet pair, and diminishes with increase of the rapidity interval between W and $2j$. This feature is more pronounced when the cutoff parameter Q_0^2 is taken larger, so that the pQCD correlation becomes smaller and, at the same time, exhibits a stronger rapidity dependence.

Theoretical derivation of the effective interaction area $\sigma_{\rm eff}$ ("effective cross section") in Refs. 9, 10, 82 and 85 relied on certain assumptions and approximations. Our approach to perturbative QCD effects in DPS developed in Ref. 9 was essentially probabilistic. In particular, we did not discuss the issue of possible interference between $1 \otimes 2$ and $2 \otimes 2$ two-parton amplitudes. One can argue that such eventuality should be strongly suppressed. Indeed, spatial properties of accompanying radiation produced by so different configurations make them unlikely to interfere, since in the $2 \otimes 2$ mechanism a typical transverse distance between two partons from the hadron w.f. is of order of the hadron size, while in the $1 \otimes 2$ case it is much smaller and is determined by a hard scale. Moreover, we disregarded potential contributions from non-diagonal interference diagrams that are due to crosstalk between partons in the amplitude and the conjugated amplitude. Such contributions appear to be negligible in the kinematic region under consideration.[13]

Finally, our prediction for the DPS cross-sections was based on a model assumption of the absence of NP two-parton correlations in the proton. This assumption is the simplest guess. One routinely makes it due to the lack of any firsthand information about such correlations. In Ref. 85 we have pointed out a source of genuine non-perturbative two-parton correlations that should come into play for very small x values, $x \ll 10^{-3}$, and estimated its magnitude via inelastic diffraction in the framework of the Regge–Gribov picture of high energy hadron interactions. The theory of small x NP correlations and their role are discussed in the next chapter.

In order to be able to reliably extract the DPS physics, one has to learn how to theoretically predict contribution of two parton collision with production of two hard systems (four jets in particular). This is the dominant channel, and it is only in the back-to-back kinematics that the $2\otimes 2$ and $1\otimes 2$ DPS processes become competitive with it. Among first subleading pQCD corrections to the $1\otimes 1$ amplitude, there is a loop graph that looks like a two-by-two parton collision. However this resemblance is deceptive. Unlike the $2\otimes 2$ and $1\otimes 2$ contributions, this specific correction does not depend on the spatial distribution of partons in the proton (information encoded in σ_{eff}), it is not power enhanced in the region of small transverse momenta of hard systems, and therefore does not belong to the DPS mechanism.[9,82,84] Treating the amplitude corresponding to splitting of two incoming partons at the one-loop level, corresponds to the two-loop accuracy for the cross-section. Until this accuracy is achieved, the values of σ_{eff} extracted by experiments should be considered as tentative.

Our first conclusion is that in the kinematical region explored by the Tevatron and the LHC experiments, the x-dependence of σ_{eff} turns out to be rather mild. This by no means implies, however, that σ_{eff} can be looked upon as any sort of a universal number. On the contrary, we see that the presence of the perturbative correlation due to the $1\otimes 2$ DPS mechanism results in the dependence of σ_{eff} not only on the parton momentum fractions x_i and on the hardness parameters, but also on the type of the DPS process.

For example, in the case of golden DPS channel of production of two same sign W bosons,[18] the discussed mechanism leads to expectation of significantly larger σ_{eff} than for, say, W plus two jets process. Indeed, the comparison of the values of R for central production of two gluon jet pairs, Wjj and W^+W^+ (with jet transverse momenta $p_\perp \simeq M_W/2$), gives ($\sqrt{s} = 7\,\text{TeV}$, $\eta_1 = \eta_2 = 0$)

$$\begin{aligned} R(jj+jj) &= 1.18\ (0.81), \\ R(W+jj) &= 0.75\ (0.45), \\ R(W^+W^+) &= 0.49\ (0.26), \end{aligned} \tag{35}$$

for $Q_0^2 = 0.5$ (1.0) GeV^2. As a result of the different magnitude of the perturbative correlation contribution for different processes, the effective interaction areas $\sigma_{\rm eff}$ comes out to be significantly different for the three processes:

$$\begin{aligned} jj + jj &: \sigma_{\rm eff} = 14.5\text{–}20\,\text{mb}, \\ W + jj &: \sigma_{\rm eff} = 20\text{–}23.5\,\text{mb}, \\ W^+W^+ &: \sigma_{\rm eff} = 21.5\text{–}25.4\,\text{mb}. \end{aligned} \tag{36}$$

In all cases the effective cross-section is smaller for lower Q_0^2 due to more developed perturbative parton cascades.

In contrast to the W^+W^+ channel, the *double Drell–Yan process* favors the $1\otimes2$ mechanism, $g \to u\bar{u}$. As a result, the effective interaction area in this case turns out to be significantly smaller. For example, for the central production of two Z bosons at $\sqrt{s} = 7$TeV we find

$$R(ZZ) = 1.03\ (0.73), \text{ corresponding to } \quad \sigma_{\rm eff}(ZZ) = 15.9\text{–}18.5\,\text{mb}. \tag{37}$$

The results for $\sigma_{\rm eff}$ for higher LHC energies are quite close (within the accuracy of measurements), cf. Figs. 8 and 10, and have similar pattern.

We mentioned above that an important feature of the $1 \otimes 2$ mechanism is its dependence on the hardness of the process. With increase of Q_i^2, the $1 \otimes 2$ to $2 \otimes 2$ ratio R is predicted to increase rather rapidly, resulting in smaller values of $\sigma_{\rm eff}$. At the same time, with decrease of the $p_\perp$ of the jets this contribution decreases. We have seen above, that such a trend is consistent with the D0 data for $x \geq 10^{-2}$, Fig. 6.

6. Non-factorized Contribution to ${}_2D$ at the Initial Q_0 Scale

6.1. *Basic ideas*

There is an additional contribution to the DPS at small x which is related to the soft dynamics. It was first discussed in Ref. 85, and in a more detail in Ref. 92. It was demonstrated in Ref. 92 that soft dynamics leads to positive correlations between partons at small x which have to be included in the calculation of the DPS cross-section. These soft correlations can be calculated using the connection between correlation effects in MPI and inelastic diffraction. The emerging non-factorized contribution to ${}_2$GPD is calculated at the initial scale Q_0^2 that separates soft and hard physics and which we consider as the starting scale for the DGLAP evolution. One expects that for this scale the single parton distributions at small x are given by the soft Pomeron and soft Reggeon exchange.[121]

The diagrams of Figs. 11 and 12 lead to a simple expression for the non-factorizable/correlated contribution (see Ref. 92 for details). For the correlated

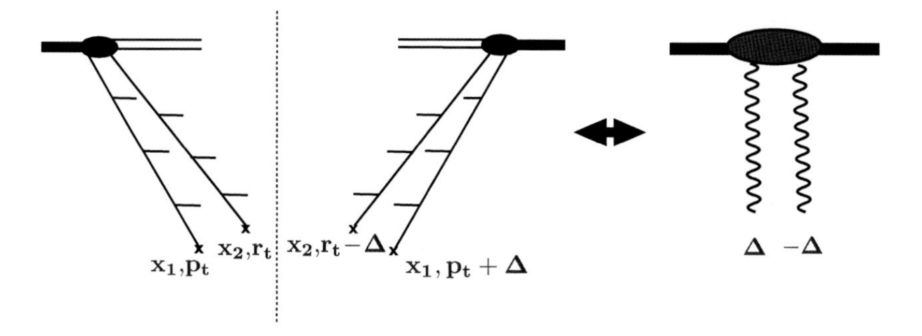

Fig. 11 $_2$GPD as a two Pomeron exchange.

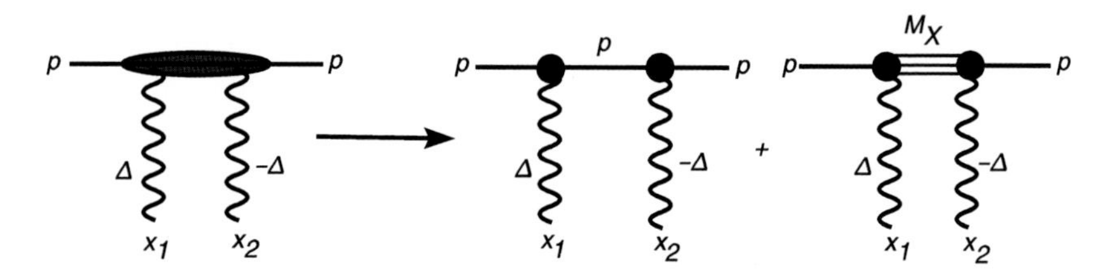

Fig. 12 $2I\!P$ contribution to $_2D$ and corresponding Reggeon diagrams.

contribution, we have

$$
\begin{aligned}
{}_2D(x_1, x_2, Q_0^2)_{nf} &= c_{3I\!P} \int_{x_m/a}^{1} \frac{dx}{x^2} D(x_1/x, Q_0^2) D(x_2/x, Q_0^2) \left(\frac{1}{x}\right)^{\alpha_{I\!P}} \\
&\quad + c_{I\!PI\!PI\!R} \int_{x_m/a}^{1} \frac{dx}{x^2} D(x_1/x, Q_0^2) D(x_2/x, Q_0^2) \left(\frac{1}{x}\right)^{\alpha_{\mathrm{R}}}.
\end{aligned} \tag{38}
$$

Here $x_m = \max(x_1, x_2)$. We also introduced an additional factor of $a = 0.1$ in the limit of integration over x (or, equivalently, the limit of integration over diffraction masses M^2) to take into account that the Pomeron exchanges occupy at least two units in rapidity, i.e., $M^2 < 0.1 \cdot \min(s_1, s_2)$ $(s_{1,2} = m_0^2/x_{1,2})$, or $x > \max(x_1, x_2)/0.1$, where $m_0^2 = m_N^2 = 1$ GeV^2 is the low limit of integration over diffraction masses. Here $c_{3I\!P}$ and $c_{I\!PI\!PI\!R}$ are normalized three Pomeron and Pomeron–Pomeron–Reggeon vertices. We determine $c_{3I\!P}$ and $c_{I\!PI\!PI\!R}$ from the HERA data[122] for the ratio of inelastic and elastic diffraction at $t = 0$: $\omega \equiv \frac{\frac{d\sigma_{\mathrm{in.\ dif.}}}{dt}}{\frac{d\sigma_{\mathrm{el}}}{dt}}|_{t=0} = 0.25 \pm 0.05$, and from analysis of diffraction for large x carried in Ref. 123. which shows that $c_{I\!PI\!PI\!R} \sim 1.5 c_{3I\!P}$. We are considering here relatively low energies (relative large x) and a rather modest energy interval. Hence we neglect energy dependence of $c_{3I\!P}$. Numerically, we obtain $c_{3I\!P} = 0.075 \pm 0.015, c_{I\!PI\!PI\!R} \sim 0.11 \pm 0.03$ for $Q_0^2 = 0.5$ GeV^2 and $c_{3I\!P} = 0.08 \pm 0.015$ and $c_{I\!PI\!PI\!R} = 0.12 \pm 0.03$ for $Q_0^2 = 1.$ GeV^2, using the Pomeron intercept values given below. Note that the intercept of the Pomeron that splits into 2 (region between two blobs in Fig. 11) is always 1.1 for $t = 0$, i.e., this Pomeron is by definition soft, and the intercept of the Reggeon is 0.5.

For the parton density in the ladder we use:[92] $xD(x, Q_0^2) = \frac{1-x}{x^{\lambda(Q_0^2)}}$, where the small x intercept of the parton density λ is taken from the GRV parameterization[124] for the nucleon gluon pdf at Q_0^2 at small x. Numerically $\lambda(0.5\text{GeV}^2) \sim 0.27$, $\lambda(1.0\text{GeV}^2) \sim 0.31$. A more elaborate treatment of the parton structure of the soft pomeron is possible including quark degrees of freedom, cf. Ref. 125.

Consider now the $t = -\Delta^2$ dependence of the above expressions. The t-dependence of the factorized contribution to ${}_2D_f$ is given by

$$F(t) = F_{2g}(x_1, t) \cdot F_{2g}(x_2, t) = \exp((B_{\text{el}}(x_1) + B_{\text{el}}(x_2))t/2), \tag{39}$$

where F_{2g} is the two gluon–nucleon form factor. The t-dependence of the non-factorized term (38) is given by the t-dependence of the inelastic diffraction: $\exp(B_{\text{in}}t)$. Using the exponential parameterization $\exp(B_{\text{in}}t)$ for the t-dependence of the square of the *inelastic vertex* $pM_X I\!P$, the experimentally measured ratio of the slopes $B_{\text{in}}/B_{\text{el}} \simeq 0.28$[126] translates into the absolute value $B_{\text{in}} = 1.4-1.7\,\text{GeV}^2$.

The evolution of the initial conditions (38) is given by

$$\begin{aligned} {}_2D(x_1, x_2, Q_1^2, Q_2^2)_{\text{nf}} &= \int_{x_1}^{1} \frac{dz_1}{z_1} \int_{x_2}^{1} \frac{dz_2}{z_2} G(x_1/z_1, Q_1^2, Q_0^2) G(x_2/z_2, Q_2^2, Q_0^2) \\ &\quad \times {}_2D(z_1, z_2, Q_0^2)_{\text{nf}}, \end{aligned} \tag{40}$$

where $G(x_1/z_1, Q_1^2, Q_0^2)$ is the conventional DGLAP gluon–gluon kernel[119] which describes evolution from Q_0^2 to Q_1^2. In our calculation we neglect initial sea quark densities in the Pomeron at scale Q_0^2 (obviously the Pomeron does not receive contributions from the valence quarks). Let us define the ratio of evolved non-factorized (correlated) and factorized (un-correlated) terms at $t = -\Delta^2 = 0$ to be K:

$$K(x_1, x_2, Q_1^2, Q_2^2) \equiv \frac{{}_2D(x_1, x_2, Q_1^2, Q_2^2, Q_0^2)_{\text{nf}}}{D(x_1, Q_1^2) D(x_2, Q_2^2)}. \tag{41}$$

We refer the reader to Ref. 96 for the numerical calculation of this ratio.

6.2. σ_{eff} in the central kinematics

The enhancement coefficient is now given by the

$$R = R_{\text{pQCD}} + R_{\text{soft}}, \tag{42}$$

where R_{pQCD} corresponds to the contribution of $1 \otimes 2$ pQCD mechanism (Fig. 3 (right)) calculated in Ref. 10 (see Section 4), while the expression for R_{soft} is given by

$$R_{\text{soft}} = \frac{4K}{1 + B_{\text{in}}/B_{\text{el}}} + \frac{K^2 B_{\text{el}}}{B_{\text{in}}} + K R_{\text{pQCD}} B_{\text{el}}/B_{\text{in}}, \tag{43}$$

where we calculate all factors for $x_1 = x_2 = x_3 = x_4 = \sqrt{4Q^2/s}$, with s being invariant energy of the collision. We present our numerical results in Figs. 13 and 14:

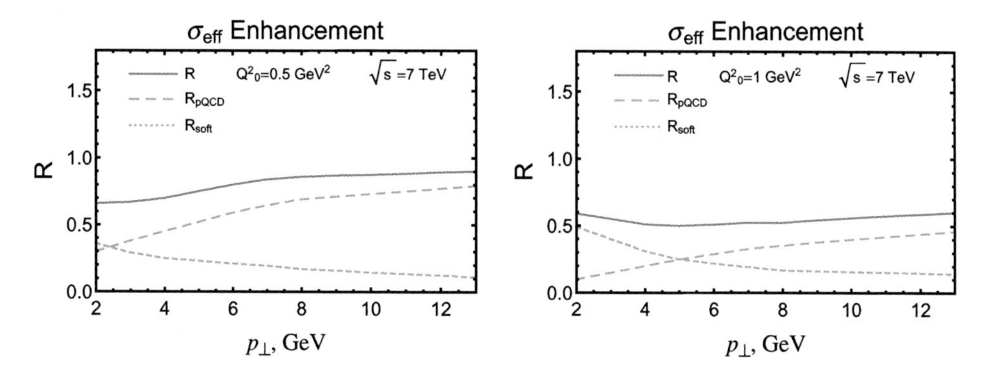

Fig. 13 σ_{eff} as a function of the transverse scale $p_\perp$ for $Q_0^2 = 0.5$ (left), and $Q_0^2 = 1$ GeV^2 (right) in the central kinematics. We present the mean field, the mean field plus $1 \otimes 2$ mechanism and total σ_{eff} for $\sqrt{s} = 13$ TeV.

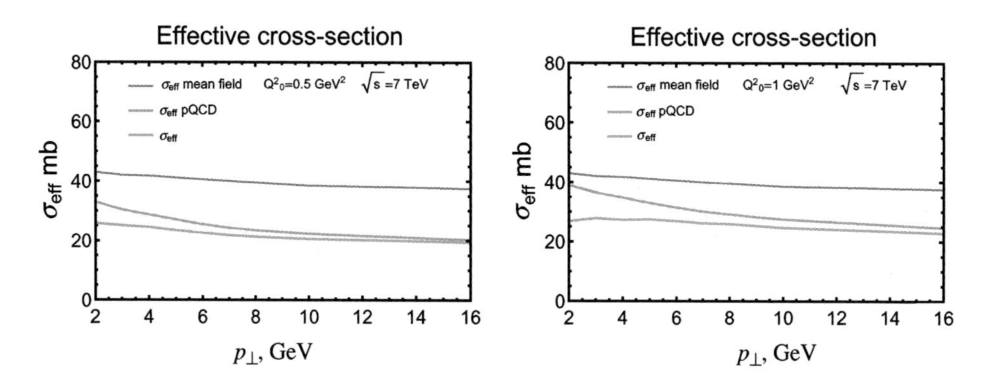

Fig. 14 R for different Q_0^2 and $\sqrt{s} = 13$ TeV.

In addition, in order to illustrate the pattern of the σ_{eff} behavior in both UE and DPS, in Fig. 13 we give the example of the $p_\perp$ dependence of σ_{eff} for the transverse momenta region 2–50 GeV for $Q_0^2 = 0.5$ GeV^2 (for $Q_0^2 = 1$ GeV^2 the behavior is very similar).

We also studied the energy dependence of σ_{eff} for fixed transverse momenta $p_\perp$ on s in the UE kinematic region in the energy region from Tevatron to LHC. We find that σ_{eff} slowly increases with s and practically flattens out at the top of the LHC energies, see Fig. 15 (right).

In order to understand the evolution of σ_{eff} at higher incident energies for given transverse scale we would need the information on the x dependence of the two-gluon form factor for small $x \leq 10^{-4}$ and of the inelastic diffraction which are likely to come from the current analyses of the J/ψ diffractive production in the ultraperipheral collisions at the LHC.

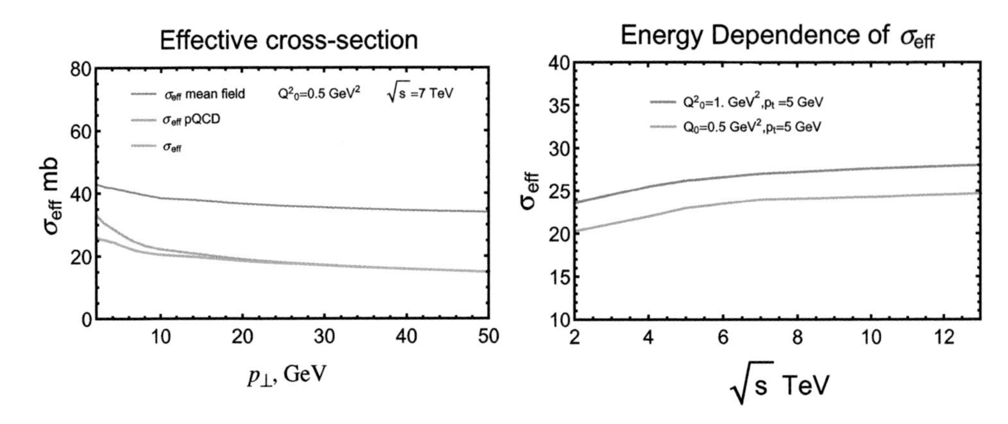

Fig. 15 Left: σ_{eff} for the entire transverse momenta region ($Q_0^2 = 0.5$ GeV^2). Right: The characteristic energy dependence of σ_{eff} on c.m.s. energy $\sqrt{s}$.

Our current estimates of non-factorizable contribution should be considered as semiquantitative due to the large uncertainties in diffraction parameters as well as the use of the "effective" values for the reggeon/pomeron parameters which include screening corrections (effects of cuts) very roughly. Nevertheless, our results indicate a number of basic features of soft nonperturbative parton–parton correlations which are relevant for the central LHC dynamics.

(i) For large transverse momenta, relevant for hard DPS scattering, soft effects are small and essentially negligible, contributing only 5% to the enhancement coefficient R if we start evolution at the scale $Q_0^2 = 0.5$ GeV^2, and 10–15% for the starting scale of 1 GeV^2, for $p_\perp \sim$ 15–20 GeV. Thus they do not influence detailed hard DPS studies described in the previous sections. Our results also indicate that the characteristic transverse momentum p_{t0}, for which soft correlations constitute given fixed fraction of the enhancement factor R rapidly increase with s.

(ii) The soft non-factorizable contributions may contribute significantly in the underlying event dynamics, especially at the scales $p_\perp =$ 2–4 GeV where they are responsible for about 50% of the difference between mean field result and full prediction for σ_{eff} for $Q_0^2 = 0.5$ GeV^2 case. If we would start evolution at $Q_0^2 = 1$ GeV^2, soft effects would dominate up to scale $p_\perp \sim 4$ GeV. In the UE the account of the soft contribution leads to stabilization of the results for σ_{eff}, and to its slower decrease with increase of $p_\perp$ than in the approximation in which only perturbative correlations, i.e., the $1 \otimes 2$ mechanism is included. These values for σ_{eff} for UE, especially for scales 2–4 GeV are very close to the ones used by Pythia.

We see that the new framework gives a reasonable description of the data over the full transverse momenta range, with weaker dependence of the quality of the fit on the starting point of the evolution Q_0 than in Ref. 90.

(iii) The evolution of $\sigma_{\rm eff}$ with transverse scale is stabilized for the UE regime, as shown in Fig. 13 leading to an almost plateau like behavior of $\sigma_{\rm eff}$ with a slight decrease with increase of the transverse scale.

(iv) The inclusion of the soft correlations stabilizes the incident energy dependence of $\sigma_{\rm eff}$. It changes only slightly between 3.5 TeV and 6.5 TeV proton collision energies for the same transverse scale for small $p_\perp$. In other words, the increase of the soft correlations compensates the decrease of the relative pQCD contribution with an increase of the collision energy due to decrease of effective x_i.

We refer the reader to the original papers[95,96] for more details.

One of the processes which is sensitive to the non-factorizable contribution is production of double open charm in the forward kinematics which was recently studied in the LHCb experiment. We find that the mean field approximation for the double parton GPD, which neglects parton–parton correlations, underestimates the observed rate by a factor of two. The enhancement due to the perturbative QCD correlation $1 \otimes 2$ mechanism which explains the rate of double parton interactions at the central rapidities is found to explain 60–80% of the discrepancy. We find[95] that non-factorized contributions to the initial conditions for the DGLAP collinear evolution of the ${}_2$GPD discussed above play an important role in this kinematics. Combined, the two correlation mechanisms provide a good description of the rate of double charm production reported by the LHCb[78] with the result weakly sensitive to the starting point of the QCD evolution. At the same time we cannot reproduce small values of $\sigma_{\rm eff}$ for the double J/ψ channel reported by LHCb which may indicate a more complicated pQCD mechanism of charmonium production.

In addition, let us note that it was suggested in Ref. 79 that strong gluon spin correlations may be present in the nucleon wave function. Such correlation may lead to an increase the double charm production cross-sections by up to 60%.

7. MPI in Proton–Nucleus Scattering

Above we considered the case of *pp* scattering. The same formalism is applicable for collision of any two hadrons and in particular for the proton–nucleus scattering. Cross-sections of the MPI processes for the *pA* case were first calculated in the parton model approximation in Ref. 80. It was demonstrated that the double, triple, and other MPI are strongly enhanced in the proton collisions with heavy nuclei due to a possibility of hard collisions occurring simultaneously on 2 $(3,\ldots)$ nucleons at the same impact parameter. It was also emphasized that a comparison of the MPI in pp and *pA* scattering would allow to study longitudinal correlations of partons in nucleons.[80] Further, it was demonstrated in Ref. 93 that corrections to the parton model expression in the mean field approximation are much smaller than in the

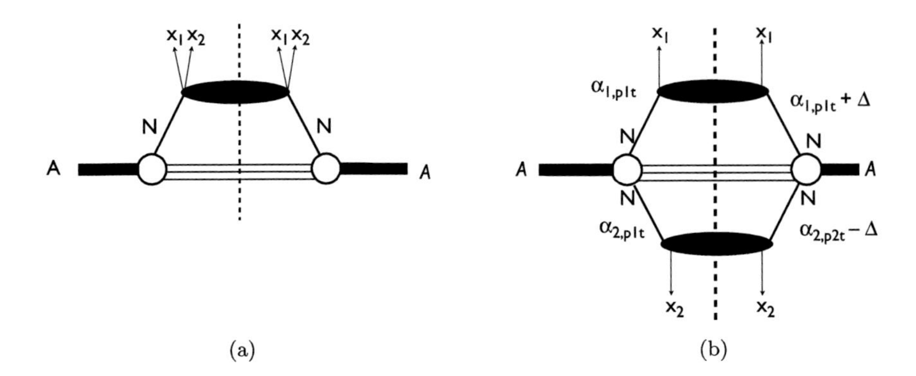

Fig. 16 Impulse approximation and two nucleon contributions to DPS in pA scattering.

pp case. The color interference effects in pA scattering were discussed in Ref. 127. Analysis of Ref. 92 indicates that these effects are very small.

The technique we described in Section 2 allows to perform the calculation in a very compact form.[93] We will focus on the case of DPS. In this case we have two contributions — the impulse approximation, corresponding to two partons of the nucleus involved in the collision belonging to the same nucleon (Fig. 16(a)), σ_1:

$$\sigma_1 = A\sigma_{NN}, \tag{44}$$

and two different nucleons, σ_2 (Fig. 16(b)). Since the b-dependence of the nuclear density is much more gradual than that for the nucleon σ_2 is not sensitive to the transverse distance between the partons of the nucleon and hence proportional to the double parton distribution (i.e., ${}_2$GPD at $\mathbf{\Delta} = 0$). The probability to pick up two partons at similar impact parameters in the heavy nucleus is $\propto A^{4/3}$, leading to $\sigma_2/\sigma_1 \propto A^{1/3}$.

For simplicity we restrict the discussion to the case of $x_i^{(A)} \geq 0.01$ where interference effects corresponding to $x_1^{(A)}(x_2^{(A)})$ belonging to one nucleon in the $|in\rangle$ state and to another nucleon in the $\langle out|$ state may be neglected. For analysis of the effect of the leading twist nuclear shadowing see Ref. 92.

Similar to the *pp* case, the expression for σ_2 is proportion to the integral over Δ of the product of the ${}_2$GPDs of proton and nucleus. The ${}_2$GPD form factor of the nucleus is

$${}_2\mathrm{GPD}_A(\Delta) \propto F_A(\Delta, -\Delta) \cdot F_{2g}^2(x, \Delta). \tag{45}$$

Here $F_A(\Delta, -\Delta)$ is the two-body nuclear form factor. In the mean field approximation for the nucleus wave function:

$$F_A(\Delta, -\Delta) = A^2 F_A^2(\Delta) \approx A^2 \exp(-R_A^2 \Delta^2/3), \tag{46}$$

where $F_A(\Delta)$ is the nucleus single-body form factor normalized to one at $\Delta = 0$. Since the Δ^2 dependence of $F_A^2(\Delta)$ is much stronger than that of $F_{2g}^4(x, \Delta)$ the later can be neglected in the integral

$$\int d^2\Delta_2 \mathrm{GPD}(\Delta, N) \cdot_2 \mathrm{GPD}(\Delta, A) = \int d^2\Delta A^2 F_A^2(\Delta). \tag{47}$$

It is convenient at this point to switch to the impact parameter representation using $\int d^2\Delta A^2 F_A^2(\Delta) = \int d^2 b T_A^2(b)$, where $T_A(b)$ is the nuclear thickness function normalized to A: $\int d^2 b T_A(b) = A$ leading to

$$\frac{\sigma_4(x_1', x_2', x_1, x_2)}{d\hat{t}_1 d\hat{t}_2} = \sigma_4(NN) \cdot \underbrace{\int T(b) d^2 b}_{A} + \frac{f_p(x_1', x_2')}{f_p(x_1') f_p(x_2')} \frac{d\sigma_2(x_1', x_1)}{d\hat{t}_1} \frac{d\sigma_2(x_2', x_2)}{d\hat{t}_2} \underbrace{\int T^2(b) d^2 b}_{\propto A^{4/3}}. \tag{48}$$

In the mean field approximation for NN scattering we use in this chapter, the $1\otimes 2$ contribution is strongly suppressed in pA scattering: $R_{pA}/R_{NN} \approx 1/5$.[93]

We find for the ratio of double to single scattering terms:

$$\sigma_2/\sigma_1 = 1.1 \left(\frac{\sigma_{\mathrm{eff}}}{15\mathrm{mb}}\right) \cdot \left(\frac{A}{40}\right)^{.39} (1 + R_{NN}/5), \tag{49}$$

for $A \geq 40$. Hence for the typical hard kinematics: $\sigma_{\mathrm{eff}} \sim$ 15–20 mb, $R_{NN} = 0.8$ we find the enhancement of DPS as compared to the impulse approximation result:

$$1 + \sigma_2/\sigma_1 \approx 3.5\text{–}4.2. \tag{50}$$

for the lead nucleus. Note that Eq. (48) is valid for fixed b as well. Hence we expect that for collisions with heavy nuclei at small b the ratio σ_2/σ_1 is enhanced by an additional factor of ~ 1.5. Thus one can look for the effect of MPI in pA scattering by comparing central and peripheral pA collisions.

Much larger enhancement is expected for higher order MPI.[80] However only chance to observe such rare events would be for rather small $p_\perp$ and small x_A where leading twist shadowing effects would significantly reduce elementary cross-sections.

8. Soft–Hard Interplay in *pp* Collisions at the LHC

8.1. *Underlying event and transverse geometry*

The *pp* LHC data already provide important tests of the transverse geometry of *pp* collisions described in Sec. 2.

Let us first consider production of a hadron (minijet) with momentum $p_\perp$. The observable of interest here is the transverse multiplicity, defined as the multiplicity of particles with transverse momenta in a certain angular region perpendicular to the transverse momentum of the trigger particle or jet (the standard choice is the interval $60° < |\Delta\phi| < 120°$ relative to the jet axis; see Ref. 128

and Chapter 11 for an illustration and discussion of the experimental definitions). In the central collisions one expects a much larger transverse multiplicity due to the presence of multiple hard and soft interactions. At the same time the enhancement should be a weak function of $p_\perp$ in the region where main contribution is given by the hard mechanism.[116,117] The predicted increase and eventual flattening of the transverse multiplicity agrees well with the pattern observed in the existing data. At $\sqrt{s} = 0.9\,\text{TeV}$ the transition occurs approximately at $p_{T,\text{crit}} \approx 4\,\text{GeV}$, at $\sqrt{s} = 1.8\,\text{TeV}$ at $p_{T,\text{crit}} \approx 5\,\text{GeV}$, and at $p_{T,\text{crit}} = 6$–$8\,\text{GeV}$ for $7\,\text{TeV}$.[129,130] Note also that $p_{T,\text{crit}}$ is smaller for the single hadron trigger than for a jet trigger since the leading hadron carries a fraction $\sim$0.6–0.7 of the jet momentum, see comparison the CMS jet data and ALICE single hadron data in Fig. 3 of Ref. 131.

One possible interpretation is that the minimum $p_\perp$ at which particle production due to hard collisions starts to dominate significantly increases with the collision energy. Another is that for the small $p_\perp$ one selects events with fewer MPI collisions due to the cutoff on minimum $p_\perp$ which becomes stronger with increase of the incident energy. Both these effects are likely to be related to the onset of the high gluon density regime in the central *pp* interactions since an increase of incident energy leads to partons in the central *pp* collisions propagating through stronger and stronger gluon fields.

Many further tests of the discussed picture were suggested in Ref. 117. They include (i) checking that the transverse multiplicity does not depend on rapidities of the jets, (ii) studying the multiplicity at $y < 0$ for events with jets at $y_1 \sim y_2 \sim 2$. This would allow one to check that the transverse multiplicity is universal and that multiplicity in the away and the towards regions is similar to the transverse multiplicity for $y \leq 0$. (iii) Studying whether transverse multiplicity is the same for quark and gluon induced jets. Since the gluon radiation for production of $W^\pm, Z$ is smaller than for the gluon dijets, a subtraction of the radiation effect mentioned below is very important for such comparisons.

Note that the contribution of the jet fragmentation to the transverse cone as defined in the experimental analyses is small but not negligible especially at smaller energies ($\sqrt{s} = 0.9$ TeV). It would be desirable to use a more narrow transverse cone, or subtract the contribution of the jets fragmentation. Indeed, the color flow contribution[132] leads to a small residual increase of the transverse multiplicity with $p_\perp$. However the jet fragmentation effect depends on p_T rather than on $\sqrt{s}$. Hence it does not contribute to the growth of the transverse multiplicity, which is a factor of ~ 2 between $\sqrt{s} = 0.9$ TeV and $\sqrt{s} = 7.0$ TeV. In fact, a subtraction of the jet fragmentation contribution would somewhat increase the rate of the increase of the transverse multiplicity in the discussed energy interval. This allows one to obtain a value for the lower limit of the rate of the increase of the multiplicity in the central ($\langle b \rangle \sim 0.6$ fm) *pp* collisions of $s^{0.17}$. It is a bit steeper than the s dependence of multiplicity in the central heavy-ion collisions.

8.2. *Correlation of soft and hard multiplicities*

It was demonstrated recently[131] that the rates of different hard processes observed in jet production by CMS and in J/ψ, D-meson production by ALICE normalized to the average hard process multiplicity, R universally depend on the underlying event charged-particle multiplicity normalized to the average charged-particle multiplicity at least until it becomes four times higher than average. Note here that the recoil jet multiplicity has to be subtracted from the underlying multiplicity.

It is worth emphasizing here, that similarity between R in the CMS and ALICE measurements is highly non-trivial as the rapidity intervals used for determination of N_{ch} differ by a factor of ~ 3.

The ratio of the inclusive rate of hard signals at fixed b to the average one in bulk events is given as follows[133]:

$$R(b) = P_2(b)\sigma_{\mathrm{in}}. \tag{51}$$

The median of the distribution over N_{ch} should roughly correspond to the median of the distribution over impact parameters. For the studied inelastic sample $\sigma_{\mathrm{in}} \approx$ 55 mb. Using parameterization of $P_2(b)$ given by Eq. (8) we find $R_{\mathrm{median}} \approx 2$ which agrees well with the data.

The relation (51) breaks down when the multiplicity becomes very large starts to select events with $b \sim 0$ and $R(0) \approx 4$ is reached. For larger R, b stays close to zero, while the high multiplicity trigger starts to select events with number of hard collisions larger than average one for $b \sim 0$.

In this limit large fraction of the total multiplicity originates from gluon emission in processes associated with minijet production. So one can expect that in this limit $N_{\mathrm{ch}}/\langle N_{\mathrm{ch}}\rangle$ is proportional to the number of the hard collisions, N, and leading to the linear dependence between N and $N_{\mathrm{ch}}/\langle N_{\mathrm{ch}}\rangle$. This expectation is consistent with the data.

An interesting question is whether high multiplicity events originate from tail of distribution over number of hard collisions at $b \sim 0$ or from some correlated configurations. In Ref. 133 it was suggested that for the highest observed multiplicities (which occur with probability $\sim 10^{-4}$–10^{-5}) fluctuations of the gluon density are important.

8.3. *Unitarity and consistency in multiple hard collisions*

One of the important observations of the MC models is that to reproduce the data one needs to suppress production of minijets. As discussed in Chapter 10, Pythia[113] introduces the energy-dependent suppression factor

$$R(p_\perp) = p_T^4/(p_\perp^2 + p_0^2(s))^2, \tag{52}$$

with $p_0(\sqrt{s} = 7\ \mathrm{TeV}) \approx 3\ \mathrm{GeV}/c$, corresponding to $R(p_T = 4\ \mathrm{GeV}/c) = 0.4$. In Herwig[36] a cutoff of similar magnitude is introduced of the form $\theta(p_\perp - p_0'(s))$.

A complimentary way to see that a mechanism of the suppression has to exist follows from the analysis of the restrictions related to the value of the total inelastic cross-section at a fixed impact parameter.[134,135] (Note here that the large inclusive cross-section of production of minijets which exceeds the total inelastic cross-section does not violate the S-channel unitarity since it effectively measures multiplicity of minijet production — see Chapter 3.)

It is possible to rewrite the cross-section of the production of minijets as a series of positive terms $\sigma_i = \int d^2b\tilde{P}_i(b)$, where $\tilde{P}_i(b)$ is the probability that in the collision at fixed b exactly i minijet pairs are produced. The total probability of inelastic interaction at given b is expressed through the elastic scattering amplitude (Eq. (9)).

Unitarity in the b space leads to the condition

$$P_{\text{hard}}(b) = \sum_{i=1}^{\infty} \tilde{P}_i(b) \leq P_{\text{in}}(s, b). \tag{53}$$

As we discussed in Section 2 distribution over b for generic inelastic collisions is much broader than for hard binary collisions and that of binary collisions is much broader than of MPI events (Fig. 2). As a result in an MPI model without non-perturbative correlations one finds that for $b \geq 1.2$ fm $\tilde{P}_1(b) \gg \tilde{P}_2(b)$. Consequently, for such b inclusive cross-section for production of minijets and $P_{\text{hard}}(b)$ practically coincide. As a result for such b the analysis does not depend on the details of modeling. Numerical studies[134,135] indicate that to satisfy inequality Eq. (53) one needs to suppress production of minijets in the momentum range similar to that introduced in the Pythia[113] and Herwig[36] models. In fact one may need an even stronger cutoff than the one found in Refs. 134 and 135. Indeed in Eq. (53) we did not take into account that at the LHC inelastic diffraction contributes a significant fraction, $\sim$15–20%, of σ_{in} and it is predominantly due to events with no minijet production (remember that even in DIS where absorptive effects are small diffraction constitutes a small fraction of the small x cross-section ($\lesssim$20%)). Since for small b interaction is essentially black and hence diffraction is impossible, the main contribution of diffraction to $P_{\text{in}}(b)$ should concentrate at $b \geq 1.2$ fm, leading to a need for even stronger cutoff.

A dynamical mechanism for a strong cutoff for the interaction at large impact parameters is not clear yet. Indeed, typical x_1, x_2 for hard collisions are 10^{-2}–10^{-3} for which pQCD work well at HERA. Also, colliding partons of the nucleon "1" ("2") propagate typically though much smaller gluon densities of the nucleon "2" ("1") so if the suppression is due to the high gluon density effects one would expect a very strong dependence of the cutoff on impact parameter. Alternatively, one would have to introduce very strong correlations for partons at the nucleon periphery.

Understanding the origin of this phenomenon is a challenge for future studies.

9. Conclusion

We developed the momentum space technique for describing MPI based on introduction of double (triple, ...) parton GPDs. It allows one to effectively introduce both the mean field approximation which is constrained by the data on single parton GPDs and develop the framework for including perturbative and non-perturbative correlations between the partons. We find that perturbative correlations enhance the high $p_{\perp}$ DPS rates bringing theoretical predictions into a fair agreement with most of the experiments. In the underlying event kinematics, an additional NP mechanism becomes significant whose strength was estimated based on information about double Pomeron exchange. The NP mechanism largely compensates the increase of σ_{eff} expected in the mean field approximation due to the increase of the gluon distribution radius with decrease of x.

Taken together these three mechanisms provide a good description of experimental data on MPI in the entire kinematical domain, including forward heavy flavor production observed in LHCb (see Chapter 8).

The dijet production and even more so MPI occur at smaller impact parameters than the soft interactions giving leading to explanation of some of the regularities of UE and to a conclusion that a strong suppression of minijet production even in peripheral collisions is necessary for explaining *pp* data.

Further studies are necessary in order to go beyond the leading log approximation as well as to understand dynamical mechanism of the suppression of the minijet production. It would be desirable to find a way to distinguish the scenario presented here with a low Q^2 scale starting point of the pQCD evolution and weak NP correlations at $x > 10^{-3}$ and a scenario where the NP correlations are present at $x \sim 10^{-2}$ while pQCD evolution starts at significantly higher scale.

Next, further work may be needed to describe recent experimental data on J/Ψ production in central kinematics.[136–139]

Here we focused on the $x_i < 0.1$ domain. The large x region is certainly of much interest for understanding the nucleon structure. For example, strong quark–antiquark correlations may arise[140] from dynamical chiral symmetry breaking. Also, one expects significant correlations between valence quarks. They could be studied in the forward DPS for example in the production of two forward pions.[141]

Acknowledgements

We thank our coauthors M. Azarkin, Yu. Dokshitzer, L. Frankfurt, P. Gunnellini, T. Rogers, A. Stasto, D. Treleani, C. Weiss, and U. Wiedemann for numerous discussions and insights. We thank V. Belyaev for discussions of the LHCb charm data. We thank J. Gaunt, F. Hautmann, T. Kasemets, and S. Scopetta for reading the manuscript and valuable comments. M.S.'s research was supported by the US

Department of Energy Office of Science, Office of Nuclear Physics under Award No. DE-FG02-93ER40771.

Appendix

The QCD factorization theorem for exclusive vector meson (VM) production[142] states that in the leading twist approximation the differential cross-section of the process $\gamma_L^* + p \to VM + p$ is given by the convolution of the hard block, meson wave function and generalized gluon parton distribution, $g(x_1, x_2, t|Q^2)$, where x_1, x_2 are the longitudinal momentum fractions of the emitted and absorbed gluon (we discuss here only the case of small x which is of relevance for the LHC kinematics). Of a particular interest is the generalized parton distribution (GPD) in the "diagonal" case, $g(x, t|Q^2)$, where $x_1 = x_2$ and denoted by x, and the momentum transfer to the nucleon is in the transverse direction, with $t = -\Delta_\perp^2$ (we follow the notation of Refs. 116 and 117). This function reduces to the usual gluon density in the nucleon in the limit of zero momentum transfer, $g(x, t = 0|Q^2) = g(x|Q^2)$. Its two-dimensional Fourier transform

$$g(x, \rho|Q^2) \equiv \int \frac{d^2\Delta_\perp}{(2\pi)^2}\, e^{i(\boldsymbol{\Delta}_\perp \boldsymbol{\rho})}\, g(x, t = -\boldsymbol{\Delta}_\perp^2|Q^2) \tag{A.1}$$

describes the one-body density of gluons with given x in transverse space, with $\rho \equiv |\boldsymbol{\rho}|$ measuring the distance from the transverse centre-of-momentum of the nucleon, and is normalized such that $\int d^2\rho\, g(x, \rho|Q^2) = g(x|Q^2)$. It is convenient to separate the information on the total density of gluons from their spatial distribution and parameterize the GPD in the form

$$g(x, t|Q^2) = g(x|Q^2)\, F_{2g}(x, t|Q^2), \tag{A.2}$$

where the latter function satisfies $F_{2g}(x, t = 0|Q^2) = 1$ and is known as the two-gluon form factor of the nucleon. Its Fourier transform describes the normalized spatial distribution of gluons with given x,

$$F_{2g}(x, \rho|Q^2) \equiv \int \frac{d^2\Delta_\perp}{(2\pi)^2}\, e^{i(\boldsymbol{\Delta}_\perp \boldsymbol{\rho})}\, F_{2g}(x, t = -\boldsymbol{\Delta}_\perp^2|Q^2), \tag{A.3}$$

with $\int d^2\rho\, F_{2g}(x, \rho|Q^2) = 1$ for any x.

The QCD factorization theorem predicts that the t-dependence of the VM production should be a universal function of t for fixed x (up to small DGLAP evolution effects). Indeed the t-slope of the J/ψ production is practically Q^2 independent, while the t-slope of the production light vector mesons approaches that of J/ψ for large Q^2. The t-dependence of the measured differential cross-sections of exclusive processes at $|t| < 1\,\text{GeV}^2$ is commonly described either by an exponential, or by a dipole form inspired by analogy with the nucleon elastic form factors. Correspondingly, we consider here two parameterizations of the two-gluon form factor:

$$F_{2g}(x,t|Q^2) = \begin{cases} \exp(B_g t/2), \\ (1 - t/m_g^2)^{-2}, \end{cases} \tag{A.4}$$

where the parameters B_g and m_g are functions of x and Q^2. The two parametrisations give very similar results if the functions are matched at $|t| = 0.5\,\text{GeV}^2$, where they are best constrained by present data (see Fig. 3 of Ref. 143); this corresponds to[117]

$$B_g = 3.24/m_g^2. \tag{A.5}$$

The analysis of the HERA exclusive data leads to

$$B_g(x) = B_{g0} + 2\alpha_g' \ln(x_0/x), \tag{A.6}$$

where $x_0 = 0.0012, B_{g0} = 4.1 \, (^{+0.3}_{-0.5}) \, \text{GeV}^{-2}, \alpha_g' = 0.140 \, (^{+0.08}_{-0.08}) \, \text{GeV}^{-2}$ for $Q_0^2 \sim 3$ GeV^2. For fixed x, $B(x,Q^2)$ slowly decreases with increase of Q^2 due to the DGLAP evolution.[116] The uncertainties in parentheses represent a rough estimate based on the range of values spanned by the H1 and ZEUS fits, with statistical and systematic uncertainties added linearly. This estimate does not include possible contributions to α_g' due to the contribution of the large size configurations in the vector mesons and changes in the evolution equation at $-t$ comparable to the intrinsic scale. Correcting for these effects may lead to a reduction of α_g' and hence to a slower increase of the area occupied by gluons with decrease of x.

Chapter 6

Phenomenology of Final States with Jets

Paolo Gunnellini

Deutsches Elektronen-Synchroton DESY,
Notkestraße 85, 22607 Hamburg, Germany

Final states with jets are discussed from a phenomenological point of view, with focus on analysis selection and sensitive observables. Measurements on proton–proton collision published by the different experiments and results relevant for the estimation of double parton scattering contributions are discussed.

1. Double Parton Scattering with Jets in the Final States

Abundant channels which can be used for double parton scattering (DPS) detection at hadronic colliders are the ones with at least one pair of jets in the final state. This is due to the fact that the dijet production cross-section at low and medium transverse momentum (p_{T}) is much higher than the cross-section of any other process in proton–proton interactions. In fact, for any given hard interaction in hadronic collisions, there is a high probability that a jet pair is produced by a secondary hard interaction. The primary hard scattering may create high-scale objects, such as a light-, a b-, or a top-quark pair, a W or a Z boson, a photon plus a jet or even a Higgs boson. Most of these final states with the aforementioned high-scale objects accompanied by a dijet pair have been widely investigated by the experimental collaborations at various collision energies.

One of the simplest and most intuitive scenario is the one involving four jets. The four jets might be produced by two different processes: a single parton scattering (SPS), represented in Fig. 1 with a pictorial view and with the specific Feynman diagrams, and a double parton scattering (DPS), shown in the same way in Fig. 2. In SPS events, the four jets are produced by a single chain, with all jets coming from a single hard scattering and associated hard radiation; in DPS, instead, the two pairs of jets in the final state are produced in two separate chains. These two

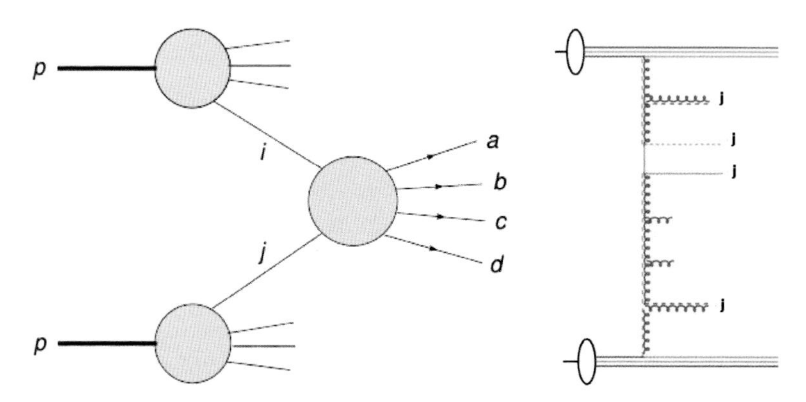

Fig. 1 A SPS event in a pp collision, with four jets in the final state. In the left plot, a sketch of the active partons (i, j) producing the four objects, (a, b, c, d), in a single chain, is represented, while in the right plot, the Feynman diagram of the process is shown.

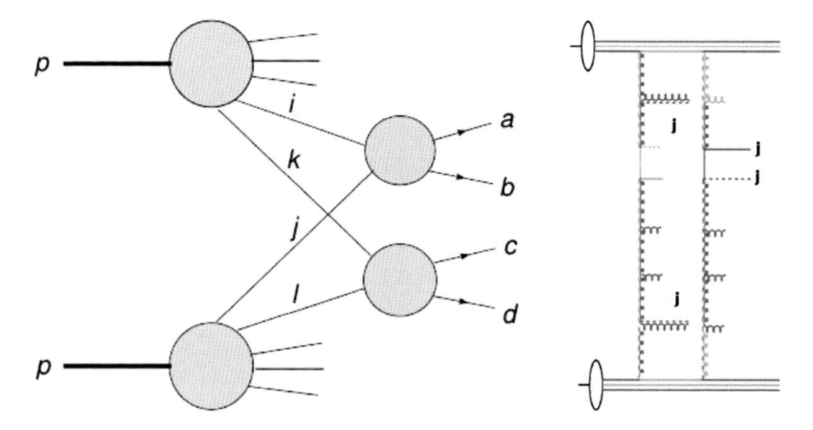

Fig. 2 A DPS event in a pp collision, with four jets in the final state. In the left plot, a sketch of the active partons (i, j) and (k, l) producing the four objects, (a, b) and (c, d), in two different chains, is represented, while in the right plot, the Feynman diagram of the process is shown.

production mechanisms translate into different configurations and topologies, with a higher correlation expected among the jets created via SPS.

Thanks to their high cross-section, final states with jets were historically the first ones to be considered for DPS contributions. Evidence for DPS contributions in final states with jets has been observed in various channels and at different energies. Final states involving four jets were already measured in the 1980–1990s in proton–antiproton ($p\bar{p}$) collisions by the AFS collaboration[144] and by the UA2 collaboration[145] at the CERN $p\bar{p}$ collider, as well as in $\gamma + 3$ jets measured by the DØ[146] and CDF[97] collaborations at the Tevatron collider. More recently, the ATLAS[71] and CMS[100,147] collaborations have contributed with various four-jet scenarios, and the DØ collaboration with the $\gamma + \text{b/c}$ jet+2 jet final state.[148] A more extensive review of the different measurements is provided in Section 4.

Conclusions from this collection of measurements were that a contribution from DPS is crucial to describe the measured distributions. This contribution is generally quantified by quoting a value for σ_{eff}, which might be thought of as related to the transverse size of the proton or to the overlapping area between the colliding protons. Under the assumption of complete factorization of the two hard scatterings, the cross-section of two simultaneous processes, A and B, occurring within the same hadronic collision is given by the formula,

$$\sigma_{A,B}^{\text{DPS}} = \frac{1}{1+\delta_{AB}} \frac{\sigma_A \sigma_B}{\sigma_{\text{eff}}}, \tag{1}$$

where σ_A and σ_B are the single inclusive partonic cross-sections for processes A and B and δ_{AB} is the Kronecker delta included in order to have the denominator equal to 2 if A and B are the same processes and equal to 1 if A and B refer to different ones. All σ_{eff} measurements published so far, are based on the assumption of Eq. (1). Even if this approximation is rather simplicistic, an extraction of σ_{eff} with this assumption for different physics processes at different energies is very meaningful: if values of σ_{eff} turn out to be different for the various channels, it would be a clear indication that this simplified model needs to be improved, e.g., by introducing correlations between the partons.

An experimental analysis aiming for a DPS-signal extraction generally starts from sensitivity studies[149] of final-state observables which might help to separate the SPS background from the DPS contributions. In the following section, observables which are useful to discriminate SPS and DPS processes are presented.

2. DPS-Sensitive Observables

A separation between the two contributions on an event-by-event basis is not feasible, since it is not possible to identify whether a single event was produced by a single or a double chain process. What is indeed possible, is an estimation of the contributions of the two production channels by looking at observables sensitive to DPS, which investigate correlations between the final-state objects: distributions measured in data should be a combination of different fractions of SPS and DPS contributions. The general analysis strategy is based on the identification of two physical objects, one expected to originate from the primary hard scattering and the other from the secondary hard interaction. In a final state containing four jets, the most natural choice for the two objects are two jet pairs. In a $\gamma + 3$ jet scenario, a dijet pair and a γ + 1-jet system are generally considered. In several phenomenological studies, it has been shown that many observables might be sensitive to DPS signals and they have been used in experimental measurements. In early measurements of four-jet final states,[144] the minimum of the transverse momentum imbalance was first considered:

$$I = \frac{1}{2}(\mathbf{p}_{T,1}^2 + \mathbf{p}_{T,2}^2), \tag{2}$$

where $\mathbf{p}_{T,1}$ and $\mathbf{p}_{T,2}$ are the vectorial transverse momenta of the two dijet systems, which in each event minimize I. A similar variable can also be defined, obtained by first boosting the four-jet system to rest in the transverse plane and then calculating

$$I_b = \mathbf{p}_T^2, \tag{3}$$

where $\mathbf{p}_T$ is the minimum total transverse momentum of each dijet system after the boost. This variable was referred to as "boosted imbalance" in Ref. 145. The boosted-imbalance spectrum has the advantage of being insensitive to the transverse momentum of the four-jet system, possibly produced by additional partonic hard emissions. For both I and I_B, DPS events are expected to contribute mostly at low values, while SPS processes have a broader distribution. Another DPS-sensitive observable, with very similar features as I, is

$$2\mathrm{S}^2 = \min\left(\frac{|\mathbf{p}_{T,i} + \mathbf{p}_{T,j}|^2}{|\mathbf{p}_{T,i}| + |\mathbf{p}_{T,j}|} + \frac{|\mathbf{p}_{T,k} + \mathbf{p}_{T,l}|^2}{|\mathbf{p}_{T,k}| + |\mathbf{p}_{T,l}|}\right), \tag{4}$$

where the different $\mathbf{p}_T$ refers to the vectorial transverse momenta of the jets and the minimization is performed over the three possible permutations of the jet pairs. As I and I_b, S should be close to zero for DPS events. The S variable, since it is a normalized p_{T} balance, is less affected by the uncertainties related to jet energy calibration, which cancel out in the ratio. A normalized p_{T} balance can be studied also between only two of the selected jets and can be defined as

$$\Delta_{ij}^{p_{\mathrm{T}}} = \frac{|\mathbf{p}_{T,i} + \mathbf{p}_{T,j}|}{|\mathbf{p}_{T,i}| + |\mathbf{p}_{T,j}|}. \tag{5}$$

DPS-sensitive observables might also be the ones investigating the position of jet pairs in the transverse and longitudinal plane. One can measure the difference in azimuthal angle, $\Delta\phi$, or in rapidity, Δy, between two of the four selected jets. These variables were used, for instance, by the ATLAS four-jet measurement[71] (see Section 4.3). Jets from DPS are expected to be in a back-to-back configuration in the transverse plane, with $\Delta\phi$ very close to π and with a Δy distribution more flat with respect to SPS processes.

Other DPS-sensitive measurements also considered a global variable, labeled as ΔS, relative to the configuration of the selected physics objects in the transverse plane. The definition of ΔS needs the identification of two systems, e.g., two dijet pairs in four-jet events. By doing that, one can measure the difference in azimuthal angle between the planes of the two systems, through the following expression:

$$\Delta S = \arccos\frac{\mathbf{p}_T(\mathrm{jet}_1, \mathrm{jet}_2) \cdot \mathbf{p}_T(\mathrm{jet}_3, \mathrm{jet}_4)}{|\mathbf{p}_T(\mathrm{jet}_1, \mathrm{jet}_2)| \cdot |\mathbf{p}_T(\mathrm{jet}_3, \mathrm{jet}_4)|}, \tag{6}$$

where $\mathbf{p}_T(\text{jet}_1, \text{jet}_2)$ and $\mathbf{p}_T(\text{jet}_3, \text{jet}_4)$ are the vectorial transverse momenta of the two dijet systems. The ΔS variable was used in measurements of both four-jet and $\gamma + 3$ jet[a] final states performed by CDF, DØ and CMS. A variable carrying the same information as ΔS but with a different definition was used by the ATLAS four-jet measurement:[71]

$$\Delta\Phi = |\phi_{ij} - \phi_{kl}|, \tag{7}$$

where ϕ_{ij} is the difference in azimuthal angle between two of the selected jets. For both ΔS and ΔΦ, DPS events are expected to have a flat contribution on these observables, while the region at $\sim\pi$ is mostly filled by SPS processes.

Other novel variables which might be used for investigating DPS contributions but were not yet used in experimental measurements for DPS extraction are listed in the following:

- Δx: difference in longitudinal momentum fraction between two interacting partons in the collision.[b]
- Total sum of the energy (or transverse energy) deposited in the calorimeters.
- Total sum of transverse momenta of the selected physics objects.

Phenomenological studies have shown that observables, such as ΔS or ΔΦ, which include information from the whole final state, are more sensitive to contributions from DPS than the ones which focus on the kinematic of only one object pair, such as $\Delta_{ij}^{p_T}$. Additionally, using more than one variable in the DPS extraction, as done in the ATLAS and CMS jet measurements, is also able to increase the analysis discrimination power. A larger sensitivity might also be reached by performing double differential studies, for instance by investigating the listed variables in bins of particle multiplicity.

An important aspect for reaching a large DPS sensitivity is the set of selection cuts applied to the considered final state. Theoretical studies have shown that a requirement of large rapidity separation, i.e., $|y| > 6.0$, between jets in the final state might be able to increase the sensitivity to DPS contributions.[151] However, experimental measurements with such rapidity separation are hard to perform since current instruments detect jets only up to $|\eta| \sim 4.5$. For larger $|\eta|$ values, if calorimetric coverage is available, jet reconstruction becomes very challenging because of the large particle flux produced in this phase space.

[a]In $\gamma + 3$ jet final states, one of the jets in Eq. (6) is replaced by the photon.

[b]Note that a measurement using Δx introduces a bias if one wants to investigate the x dependence of the σ_{eff} value.[150]

3. Analysis Strategy

To measure observables sensitive to the DPS contribution allows the possibility to extract a value of $\sigma_{\rm eff}$, which quantifies the amount of DPS for the considered final states. All $\sigma_{\rm eff}$ measurements performed so far are based on the approximation implemented in Eq. (1), of complete factorization and independence of the two hard scatterings occurring within the same collision.

The most common extraction methods are the so-called *template method* and the *inclusive-fit method*. In the following, the two methods are discussed.

In the approach of the template method, templates for background and signal are built and their relative fractions are determined through a fitting procedure: the results correspond to the combination of background and signal which best describe the DPS-sensitive observables. The background template includes events coming from SPS events, while in the signal one, DPS processes are contained. With the fraction of DPS signal contained in the data, $\sigma_{\rm eff}$ can be obtained with a simple formula which also accounts for selection efficiencies and cross-section values of the selected processes.[102,103] For instance, for a four-jet final state, with two dijet pairs defined in two phase spaces A and B, Eq. (1) becomes

$$\sigma_{4j}^{\rm DPS} = \frac{1}{1+\delta_{AB}} \frac{\sigma_{2j}^{A}\sigma_{2j}^{B}}{\sigma_{\rm eff}}, \tag{8}$$

where σ_{2j}^{A} and σ_{2j}^{B} are the cross-sections for dijet events in the considered phase spaces. The approach followed by the ATLAS measurement is based on the expression for the DPS cross-section,

$$\sigma_{4j}^{\rm DPS} = f_{\rm DPS} \cdot \sigma_{4j}, \tag{9}$$

where σ_{4j} is the inclusive cross-section for four-jet events, including both DPS and SPS contributions, and $f_{\rm DPS}$ represents the fraction of DPS events in the considered four-jet final states. The expression for $\sigma_{\rm eff}$, thus, might be translated to:

$$\sigma_{\rm eff} = \frac{1}{1+\delta_{AB}} \frac{1}{f_{\rm DPS}} \frac{\sigma_{2j}^{A}\sigma_{2j}^{B}}{\sigma_{4j}}. \tag{10}$$

In case of a partial overlap between the A and B phase spaces, the symmetry factor needs to be adjusted to take that into account. The size of the overlap is proportional to the ratio between σ_{2j}^{B} and σ_{2j}^{A} (assuming B is included within A),

$$\frac{1}{1+\delta_{AB}} \to 1 - \frac{\gamma}{2}, \tag{11}$$

where

$$\gamma = \frac{\sigma_{2j}^{B}}{\sigma_{2j}^{A}} \tag{12}$$

The logic behind the adjustment is as follows: assuming that the phase space B is completely included in phase space A, the expression of the DPS cross-section, defined in Eq. (1), can be rewritten as

$$\sigma_{\text{eff}} \cdot \sigma_{\text{DPS}} = \sigma_{2j}^{A-B} \cdot \sigma_{2j}^{B} + \frac{1}{2}(\sigma_{2j}^{B})^2. \tag{13}$$

The first term on the right-hand side of Eq. (13) relates two mutually exclusive cross-sections and therefore the symmetry factor in this terms is set to one. The second term contains two equivalent cross-sections, thus its symmetry factor is set to half. The cross-sections in Eq. (13) may be expressed in terms of σ_{2j}^{A} and the fractional overlap γ,

$$\sigma_{2j}^{B} = \gamma\sigma_{2j}^{A}, \tag{14}$$

$$\sigma_{2j}^{A-B} = (1-\gamma)\sigma_{2j}^{A}. \tag{15}$$

If one introduces these relations in Eq. (13), one gets

$$\sigma_{\text{eff}} \cdot \sigma_{\text{DPS}} = \left(1 - \frac{\gamma}{2}\right) \sigma_{2j}^{A} \cdot \sigma_{2j}^{B}. \tag{16}$$

To extract σ_{eff}, it is thus necessary to measure three different cross-sections, σ_{2j}^{A}, σ_{2j}^{B} and σ_{4j}, and estimate f_{DPS}. The cross-sections are generally measured by defining different samples from data, while f_{DPS} is the output of the template method. Even if using the template method and within the same theoretical assumptions, the approach followed by the measurements performed at the DØ and CDF collaborations was slightly different. It relied on the estimation of the rates of "pure" DPS events and of events with the same signature as the DPS event but with two separate collisions, occurring within the same bunch crossing.

The other method which can be used for a determination of a σ_{eff} value is the so-called *inclusive-fit* method, which was described and developed in Ref. 152. This method relies on inclusive tunes of predictions simulated with MC generators. It similarly uses measured differential cross-sections, as a function of DPS-sensitive observables, but it does not try to separate background and signal contributions: it rather fits the variables inclusively. The free parameters of the fit are the parameters which govern the amount of DPS within the considered MC event generator. However, the estimation of σ_{eff} is based on the assumptions of the multiparton interaction model implemented in the MC event generator used for the fit. The output of the inclusive method is the set of parameters which best describes the fitted data. This corresponds to a certain value of σ_{eff} for that considered MPI model.

4. Summary of Available Measurements

In this section, a review of the published measurement with jets in the final states is presented.

4.1. *Early measurements in proton–antiproton collisions*

The first DPS-sensitive measurement of a four-jet final state is dated back to 1986, performed by the AFS collaboration. Proton–antiproton collision data were collected at a centre-of-mass energy $\sqrt{s} = 63$ GeV corresponding to a total integrated luminosity (L) of 10 pb^{-1}. Events were selected by applying a lower cutoff of 31 GeV on the total transverse energy detected in the calorimeters. Out of the detected energy deposits, four clusters are identified as jets through a simple jet-finding algorithm in η–ϕ plane and required to have $p_{\text{T}} > 4$ GeV. The imbalance I in transverse momentum, defined in Eq. (2), and its corresponding "boosted" version I_b, defined in Eq. (3), were measured for extracting the signal from DPS and resulted in a value of $\sigma_{\text{eff}} = 5$ mb. No experimental uncertainty was provided for that.

The UA2 collaboration also performed a measurement of a four-jet final state with $\text{p}\bar{\text{p}}$ collision data corresponding to an integrated luminosity of 7.6 pb^{-1}, collected in 1988 and 1989 at the CERN $\text{p}\bar{\text{p}}$ collider running at $\sqrt{s} = 630$ GeV. The four jets were identified through their energy deposition in the calorimeters. The event selection required exactly four jets with a transverse energy above 15 GeV in the pseudorapidity range $|\eta| < 2$. In order to select events well contained in the calorimetry system, an upper threshold of 20 GeV on the missing transverse energy was applied. Events were collected through single and dijet triggers and after all analysis requirements, around 10,000 events were selected. To study the amount of DPS signal contributing to the considered final state, the 2S^2 variable, defined in Eq. (4) was measured. A good description of the data could be obtained with predictions from SPS diagrams only, and a lower limit of $\sigma_{\text{eff}} > 8.3$ mb was obtained at 95% confidence level.

4.2. *Measurements with jets at the Tevatron*

The CDF and DØ experiments published various measurements for different final states in $\text{p}\bar{\text{p}}$ collisions:

- four-jet at $\sqrt{s} = 1.8$ TeV ($L = 325$ nb^{-1}) by CDF,[153]
- $\gamma + 3$ jet at $\sqrt{s} = 1.8$ TeV ($L = 16.0$ pb^{-1}) by CDF,[97]
- $\gamma + 3$ jet at $\sqrt{s} = 1.96$ TeV ($L = 8.7$ fb^{-1}) by DØ.[146]

For the σ_{eff} determination, all analyses investigated the ΔS observable, as defined in Eq. (6). In four-jet final states, the jets, clustered through the anti-k_T algorithm with a cone size $R = 0.7$, were ordered in p_{T} and then associated in pairs according to that. In $\gamma + 3$ jet final states, the first system was formed by the photon and the highest-p_{T} jet and the second one by the two remaining jets. Jets were selected with a nominal threshold for the jet p_{T} equal to 25 GeV in $|\eta| < 3.5$ for the four-jet measurement and equal to 6 GeV in $|\eta| < 3.5$ for the γ+3 one. In the latter final state, the γ was selected with $p_{\text{T}} > 16$ GeV in $|\eta| < 0.9$. Figure 3 shows

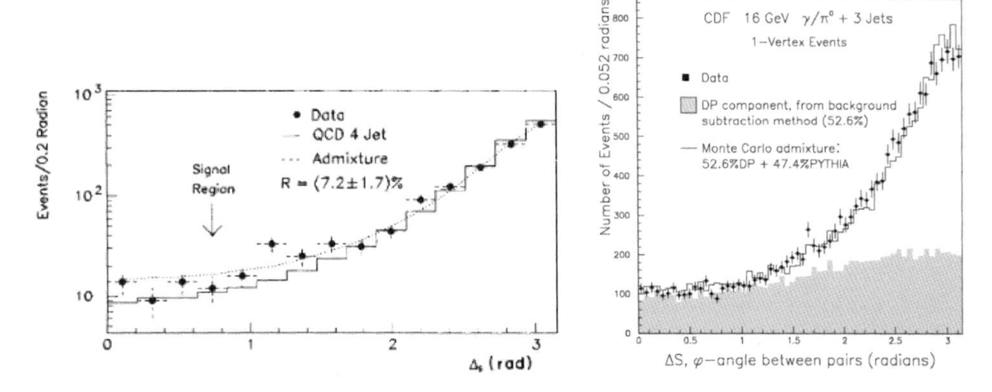

Fig. 3 Normalized distributions of the ΔS observable measured by the CDF Collaboration in a four-jet[153] (left) and a $\gamma+3$ jet[97] (right) final state. Figures reprinted with permission. Left: Copyright 1993 by the American Physical Society. Right: Copyright 1997 by the American Physical Society.

the ΔS observable in the four-jet and in the $\gamma + 3$-jet final states measured by the CDF experiment at $\sqrt{s} = 1.8$ TeV. The DPS fractions extracted through the template fit are also shown in the figures, together with the curves obtained with the combination of SPS and DPS templates. Fractions of 7.2% and 57.6% were obtained for, respectively, the four-jet and the $\gamma + 3$-jet final states, which correspond to $\sigma_{\text{eff}}(4\text{j}) = 12.1^{+10.7}_{-5.4}$ mb and $\sigma_{\text{eff}}(\gamma + 3\text{ jets}) = 14.5 \pm 1.7$ (stat.) $^{+1.7}_{-2.3}$ (syst.).

In order to differentiate the light and heavy-flavour (HF) sectors and investigate a possible flavour dependence, the DØ Collaboration performed the measurement of two different scenarios with a γ and three jets in the final state: an inclusive one, without any requirement on the flavour of the selected jets, and an HF one, with the requirement of at least one of three jets to originate from a charm or a bottom quark. The photon was selected with $p_{\text{T}} > 26$ GeV in $|\eta| < 1.0$ or $1.5 < |\eta| < 2.5$, while the jets were required to have $p_{\text{T}} > 15$ GeV in $|\eta| < 2.5$. For the second and the third jet, an additional upper threshold was set to 35 GeV, in order to increase the DPS sensitivity of the selection. In Fig. 4, the normalized cross-sections as a function of ΔS are shown for the two scenarios. A fraction of DPS events equal to 21% and 17% is obtained in the, respectively, $\gamma + 3$-jet and the corresponding heavy-flavour final state. Combinations of SPS and DPS templates, scaled with the corresponding fitted fractions, describe well the measured data points. The measured DPS fractions translate to values of σ_{eff} of 12.7 ± 0.2 (stat.) ± 1.3 (syst.) and 14.6 ± 0.6 (stat.) ± 3.2 (syst.), respectively.

Additionally, in order to investigate the dependence of the DPS contributions as a function of the scale of the secondary hard process, an analysis investigating a similar $\gamma + 3$-jet final state was performed by the DØ Collaboration by applying different jet p_{T} thresholds, in a range between 15 and 25 GeV.[146] No evidence

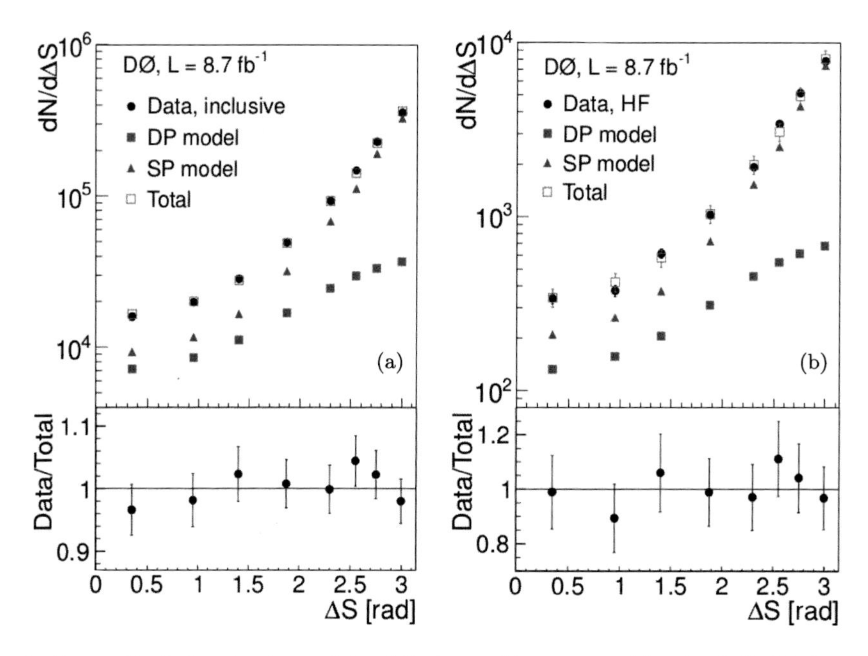

Fig. 4 Normalized distributions of the ΔS observable measured by the DØ Collaboration in an inclusive $\gamma + 3$ jet[148] (left) and heavy-flavour $\gamma + 3$ jet[148] (right) final state. Points corresponding to SP and DP templates are also represented, as well as the fitted combination of the two. Figure reprinted with permission. Copyright 2014 by the American Physical Society.

for different $\sigma_{\rm eff}$ values was observed, due to the large systematical uncertainties affecting the measurements.

4.3. *Measurements with jets in ATLAS*

For a four-jet measurement, the ATLAS collaboration used a dataset corresponding to 37 pb^{-1} at $\sqrt{s} = 7$ TeV. For the determination of $\sigma_{\rm eff}$, three various samples of events were used, two dijet samples and one four-jet sample. Jets were reconstructed through the anti-k_T clustering algorithm with a cone size $R = 0.6$ and were required to have $p_{\rm T} > 20$ GeV in the pseudorapidity range $|\eta| < 4.4$. The dijet samples had at least two jets in the final state, while the four-jet one has at least four jets. In one of the dijet samples and in the four-jet sample, the highest-$p_{\rm T}$ (leading) jet was required to have a transverse momentum greater than 42.5 GeV to comply with the requirements of the available jet triggers. Only the four jets with the highest transverse momentum were considered in the selection but events with additional jets in the final state were not rejected. This corresponds to an inclusive four-jet selection. For the quantitative measurement of the DPS signal, $\Delta^{p_{\rm T}}$, $\Delta\phi$, $\Delta\Phi$, and Δy between pairs of jets were considered. Templates representing the SPS background and the DPS signal were defined. In particular, based on Monte Carlo

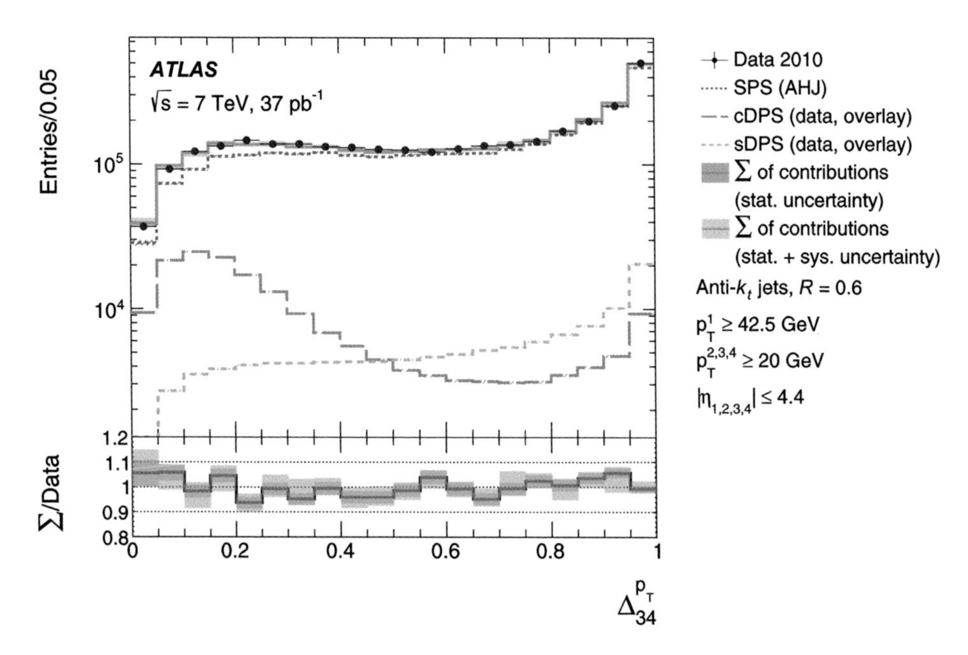

Fig. 5 Normalized distribution of Δ^{p_T} measured in a four-jet final state measured by ATLAS[71] Collaborations. The data points are compared to the sum of the background and signal templates, scaling by the respective fractions obtained through the template method.

studies, double parton processes contribute in two ways to the considered four-jet final state. In one contribution, referred to as complete-DPS (cDPS), the secondary scatter produces two of the four leading jets in the event. In the second contribution, referred to as semi-DPS (sDPS), three of the selected four jets are produced in the hardest scatter, and the other jet comes from a secondary scatter. The SPS sample was extracted from a multijet sample of MC simulated events, while the sDPS and cDPS samples were constructed by overlaying events from data. The aforementioned DPS-sensitive observables obtained for the different samples were given as input in the training of an artificial neural network, which was then applied to the measured quantities. The choice of using a neural network with variables relative to all possible jet combinations is motivated by the fact that, in such a way, no association of the selected jets in pairs is needed. Figure 5 shows the normalized p_T balance between the two lowest-p_T jets. The sum of the contributions from the three templates, scaled by the respective fractions, reproduces the measured data points well.

From a template fit to the output of the neural network, the fraction of DPS events was estimated to be $f_{\rm DPS} = 0.084^{+0.009}_{-0.012}$ (stat.)$^{+0.054}_{-0.036}$ (syst.), which translates to a value of $\sigma_{\rm eff} = 14.9^{+1.2}_{-1.0}$ (stat.)$^{+5.1}_{-3.8}$ (syst.). Systematic effects dominate the total experimental uncertainty on the $\sigma_{\rm eff}$ value, in particular the effect on the jet p_T reconstruction.

4.4. *Measurements with jets in CMS*

Two different measurements with four jets selected in the final state were published by CMS at $\sqrt{s} = 7$ TeV:

- an exclusive final state with two "hard" jets and two "soft" jets, referred to as "4j"[100] in the following;
- an inclusive final state with two jets originated from b-quarks and two other jets (without any requirement on the flavour of the originating quark), referred to as "2b2j"[147] in the following.

Both measurements used a dataset delivered during low-luminosity runs collected in 2010 characterized by a small number of overlapping interactions (pile-up). In the 4j scenario, exactly four jets clustered with the anti-k_T algorithm with cone size R = 0.5 were selected: two of them, the "hard" jets, were required to have $p_{\rm T} > 50$ GeV and the other two ("soft") had $p_{\rm T} > 20$ GeV. All four were selected in $|\eta| < 4.7$. If additional jets with $p_{\rm T} > 20$ GeV were measured, the event was rejected. A symmetric $p_{\rm T}$ threshold of 20 GeV was instead applied to the jets in the 2b2j scenario. Two of the jets, labeled as "b-jets", were required to contain a b-quark within the jet cone and were selected in $|\eta| < 2.4$, where information of the tracker could be used for tagging the b-jets. The other two jets were selected in $|\eta| < 4.7$. In this case, presence of additional jets within the selected phase space was not rejected and the leading b-jets and the leading other jets were considered. Three DPS-sensitive observables are measured:

(1) ΔS (see Eq. (6)), which measures the azimuthal angle of the two selected dijet planes ("hard" and "soft" jet pairs for the 4j analysis, and "b-jet" and "other-jet" pairs for the 2b2j one);
(2) $\Delta^{p_{\rm T}}$ (see Eq. (5)), which measures the normalized $p_{\rm T}$ balance between the "soft" jets and between the "other" jets in, respectively, the 4j and 2b2j scenario;
(3) $\Delta\phi$, measuring the difference in azimuthal angle between the same pairs as for $\Delta^{p_{\rm T}}$ for the two analyses;

Figure 6 shows the ΔS variable measured by CMS in the 4j and 2b2j scenarios. The data were compared to various predictions from Monte Carlo event generators which simulate the hard process at a leading or next-to-leading order in the strong coupling within the matrix element (ME) calculation and include contributions of the parton shower (PS). The outcome of both measurements was that predictions are able to reproduce the DPS-sensitive observables only if they include contributions from multiparton interactions. If multiparton interactions are switched off in the simulation, the regions of the phase space where a DPS signal is expected are significantly underestimated.

The inclusive-fit method was applied to the two CMS four-jet measurements as displayed in Fig. 7, which shows a visual example of the method tested on the

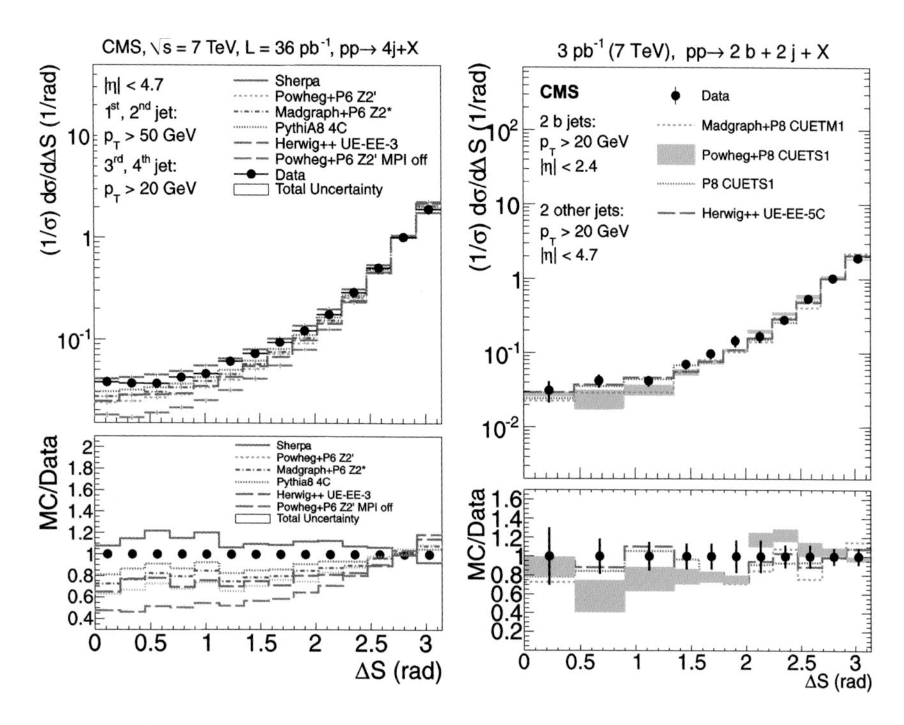

Fig. 6 Normalized distributions of the ΔS observable measured in the 4j (left) and in the 2b2j (right) final states, measured by the CMS[100,147] collaboration at $\sqrt{s} = 7$ TeV. The data points are compared to various predictions from Monte Carlo event generators with and without the simulation of the MPI.

ΔS observable in the 4j final state.[100] For this study, predictions from the Pythia 8 MC event generator were considered. The blue band represents the envelopes of the predictions obtained by the variations of the MPI parameters. The red curve is the result of the fit to the data.

The following results for $\sigma_{\rm eff}$ were obtained for the 4j and the 2b2j final states:

$$\sigma_{\rm eff}(4j) = 19.0^{+4.6}_{-3.0} \text{ mb}, \tag{17}$$

$$\sigma_{\rm eff}(2b2j) = 23.2^{+3.3}_{-2.5} \text{ mb}. \tag{18}$$

The uncertainties on the $\sigma_{\rm eff}$ value are mainly due to the experimental inaccuracy of the jet transverse momenta.

It was observed that the set of MPI parameters obtained from the fits to DPS-sensitive observables was not able to describe underlying-event data.[101,154] This shows the difficulties of the current models to describe both the soft and the hard spectrum of the MPI contributions with the same set of parameters. Siodmok *et al.*[110] released a set of parameters for the Herwig++ event generator, which is able to correctly predict underlying-event data while implementing a value of $\sigma_{\rm eff} \sim 15$ mb, which is closer to the experimental measurements. However, it

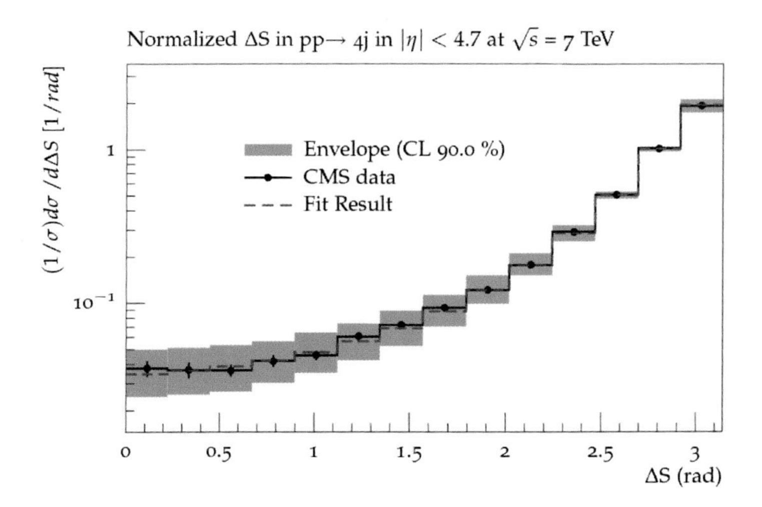

Fig. 7 Envelopes of the predictions simulated with the Pythia 8 Monte Carlo event generator obtained with different parameters on the correlation observable ΔS in the four-jet scenario measured by the CMS collaboration, together with the measured data.[100]

was shown[149] that the differential DPS-sensitive distributions are not optimally reproduced with this set of parameters. Phenomenological studies,[90,91] including a dependence of σ_{eff} on the scale of the hard interactions and on the longitudinal momentum fraction of the interacting partons, were able to fix this incompatibility, outside the simple scheme used in Eq. (1).

In order to investigate the same channel widely investigated by the CDF and DØ experiments, the CMS collaboration also performed the measurement of the $\gamma + 3$ jet final state,[155] by measuring similar observables as done in the 4j and 2b2j analyses. Two different p_{T} thresholds were set for the event selection: one of 75 GeV for a photon and the leading jet, and one of 20 GeV for the two additional jets. All physics objects were selected in the pseudorapidity region $|\eta| < 2.5$. However, the selected phase space was not able to show a large sensitivity to DPS contributions and no σ_{eff} measurement was performed.

5. Discussion About σ_{eff} Determination

So far, in most of the experimental measurements aiming for extracting a DPS contribution, the template method has been adopted. The differences among the various measurements have been mostly the definition of background and signal templates and the fit procedure. For instance, the DØ, CDF, and ATLAS measurements mainly rely on data-driven signal definitions;[97,146,148] DPS-like events are selected from two independent collisions recorded in data, carefully corrected for possible contamination from pile-up. The physics objects of the final state are then selected pair-wise in the two different events, mimicking, thus, the occurrence of a DPS in the same collision.

The background template, instead, relies on events generated on a MC basis, by switching off the simulation of MPI. The CMS and ATLAS collaborations refined this method by comparing results obtained from additional template definitions. For instance, for the signal definition, DPS events are also evaluated from MC simulations, e.g., Pythia 8,[113] which give the possibility of generating two hard scatterings within the same collision. The SPS background, instead, is built based on simulated events, by generating a higher-order ME (like, for instance, by including also 2→3 or 2→4 diagrams) and setting an upper scale for the MPI, not simply switching them off: this is a better solution, since a scenario without MPI is, indeed, too unrealistic and, especially in final states with jets, the contribution of MPI is relevant for the determination of the jet transverse momenta. In fact, the MPI not only might produce "genuine" jets from additional hard interactions, but also add an offset amount of energy to the existing jets. The upper scale chosen for the MPI depends on the p_{T} threshold of the physics objects of the final state, which are expected to come from the second hard scattering: the Pythia 8, Herwig 6 + Jimmy[36,156] and Sherpa[157] event generators have been used for the background definition, within the CMS and ATLAS measurements. The mentioned generators allow the possibility of excluding part of the MPI spectrum from the simulated events, while, for instance, Pythia 6,[158] which was used by CMS for the estimation of the model dependence, does not contain this option.

From these templates, the DPS fraction was estimated in the measurements performed at Tevatron,[97,148,153,159] by fitting only one sensitive observable. The four-jet measurement performed by ATLAS[71] used the output of an artificial neural network, based on many DPS-sensitive observables.

From the experience gained with the available measurements, a suggestion for general criteria for the definition of background and signal templates can be summarized as follows:

(1) use more than one MC generator in order to properly estimate the systematic uncertainty;
(2) make sure that signal and background cover the full phase space: this translates into the fact that the background template is not a sample with MPI switched off, but rather a sample with a second interaction below a given scale;
(3) select an inclusive scenario, rather than an exclusive one, namely set a physics selection which allows any number of additional interactions, and not only two;
(4) investigate the dependence of the background template on the generator used for its definition: it might happen that the inclusion of a higher number of partons in the ME fills a similar region of the phase space as the DPS signal; if this is the case, a choice of an ME which guarantees stability of the obtained results has to be taken.

Note that different assumptions used for background and signal definitions might bring to different results for $\sigma_{\rm eff}$.[c]

The inclusive-fit method, by fitting the variables inclusively, removes possible ambiguities related to the choice of the templates, which might affect the $\sigma_{\rm eff}$ extraction. Note that the procedure can be tested for any model implemented in an MC generator which has tunable parameters. A simple estimation of the model dependence of the measurement is thus feasible. An important point of the method is also that predictions obtained with the new parameter settings can be tested on other measured DPS-sensitive observables, in order to check if the agreement with the data globally improves after the fit, and on other measurements sensitive to MPI occurring at lower scales, in order to test the predictivity power of the extracted set of parameters. The inclusive-fit method has been applied to the four-jet[100] and two b- and two-jet[147] final states, measured by CMS and can be further tested against any DPS-sensitive measurement.

6. Summary and Outlook

Multijet scenarios sensitive to DPS contributions have been measured at various collision energies, investigating several observables and scales. So far, no significant evidence of any energy dependence could be observed for the value of $\sigma_{\rm eff}$, due to the large experimental uncertainty affecting the measurements. In the future, specific efforts will be necessary to decrease the jet energy scale and the pile-up effects at low transverse momenta which drive the systematic uncertainties in jet measurements. At the moment, within the available models, it appears very difficult to describe both measurements sensitive to the underlying event and soft multiparton interactions and measurements sensitive to DPS.[152] This tension in the description of hard- and soft-scale processes within the same framework may indicate some deficiency of the whole model, which could be improved, e.g., by introducing correlations among the initial partons inside the protons.

Even though measurements with jets are not the most sensitive ones to DPS signals among the possible ones at hadron colliders, due to the overwhelming background of SPS processes, they are produced with very high statistical accuracy.

[c]A striking example of the issues appearing for a $\sigma_{\rm eff}$ value extracted with the template method is the series of paper published about the CDF measurement of the $\gamma + 3$ jet final state. After the measurement of $\sigma_{\rm eff}$ published by CDF and pointing to a value of 14.5 mb,[97] a phenomenological study performed by Treleani[160] was carried out in order to "adapt"the value of $\sigma_{\rm eff}$ to an inclusive scenario, namely by taking into account also the possibility of having more than two hard scatterings within the same collision; this study resulted into a $\sigma_{\rm eff}$ value of $\sim$10.3 mb. After that, a further adjustment was published by Siodmok *et al.*, by correcting for the assumption of not knowing experimentally how many hard scatterings occur in the considered collisions.[112] Thus, the $\sigma_{\rm eff}$ value went up to 12.0 mb. Further details can be found in the related papers.

Thanks to that, they offer a unique opportunity of performing multidifferential measurements, which allow a systematic investigation of the dependence of σ_{eff} on the scales of the two hard scatterings, the longitudinal momentum fractions of the interacting partons and the flavour of the outgoing partons. Measuring DPS-sensitive observables for jet final states which apply different phase space requirements might shed light on these important points.

Chapter 7

Phenomenology of Final States with Massive Vector Bosons

Orel Gueta

Deutsches Elektronen-Synchrotron (DESY), Platanenallee 6, 15738 Zeuthen, Germany

Massive vector bosons (VB) are considered ideal for the study of double parton scattering (DPS). Various methodologies have been proposed throughout the years to determine the DPS contribution to final states involving VB. These have led to recent LHC measurements of DPS. The corresponding phenomenological studies, analysis procedures and experimental results are described in detail. Potential future studies are suggested, emphasizing the inherent advantages of VB final states in DPS measurements.

1. Introduction

The prospect of studying the production of massive vector bosons (VB) in double parton scattering (DPS) was envisioned long ago. The process of two simultaneous Drell–Yan (DY) annihilations was identified as a valuable test of the constituent picture well before the first experimental observation of DPS.[161,162] It was estimated that the DPS contribution to the production of two high-mass virtual photons might be competitive with the contribution from a single interaction, allowing to study the structure of the hadron without the need to reconstruct jets in the final state. The contribution from double DY was even used in an attempt to explain the excess in the observed rate of four-muon events in πp and pp interactions.[163]

Soon after the experimental observations of the intermediate W and Z vector bosons,[164–167] various studies were proposed to learn about their decay modes, production mechanisms and couplings. In particular, it was suggested to study the triple gauge coupling in hadron–hadron collisions where a pair of gauge bosons is produced[168] (see Fig. 1(a)). In this context, the DPS mechanism producing a pair of gauge bosons was considered irreducible background, the contribution of which was

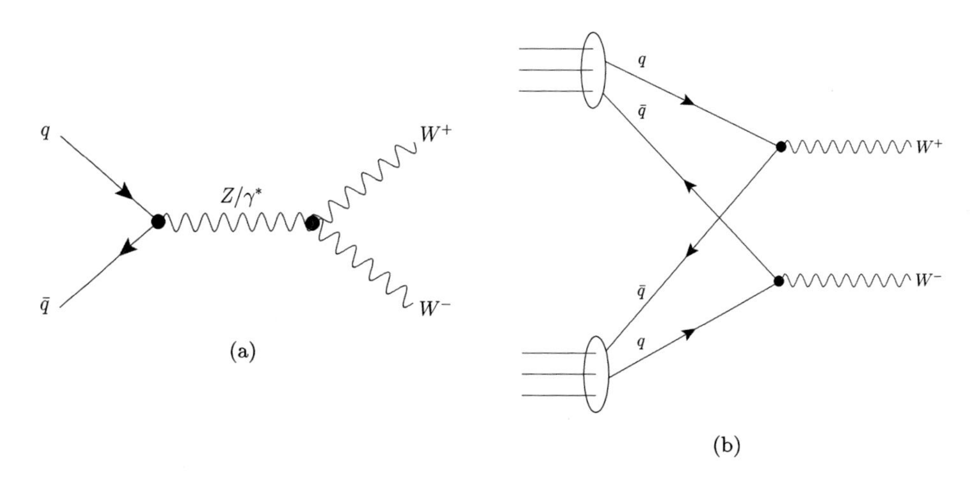

Fig. 1 Examples of leading-order Feynman diagrams for W^+W^- production in (a) an SPS via the triple gauge coupling and (b) DPS.

estimated to be negligible at Sp$\bar{p}$S energies. An example of leading-order Feynman diagrams for W^+W^- production in DPS is shown in Fig. 1(b).

The first evidence of DPS in the four-jet final state, observed by the AFS experiment,[162] led to a number of phenomenological studies of DPS involving massive vector bosons. Some studies focused on the benefits of studying DPS as a way to gain insight into the structure of the hadron, while others treated the DPS contribution as background to other processes. The common thread is the need to identify DPS and estimate its contribution as precisely as possible. A summary of some of the studies performed so far is given in the next section.

2. Phenomenological Studies

2.1. *DPS with VB final states at the Tevatron*

With the start of the Tevatron, the DPS contribution to the $W+2$j final state was discussed in the context of background to rare physical processes. Several interesting processes share the same experimental signature, such as

- $p\bar{p} \to H \to W^*(\to \ell\nu) + W(\to q\bar{q})$;
- $p\bar{p} \to t\bar{t} \to b + \bar{b} + W(\to \ell\nu) + W(\to q\bar{q})$;
- $p\bar{p} \to W(\to \ell\nu) + W(\to q\bar{q})$.

At the time, these processes had not yet been observed. To improve the chances of their observation, a method to suppress the DPS background was proposed. In a DPS production of $W+2$j, the W boson is the result of a DY annihilation of a $q\bar{q}$ pair. It is therefore expected to be produced with low transverse momentum, ${p_T}^W \approx 0$. On the other hand, a W produced in the processes listed above typically has ${p_T}^W \approx m_W$. Requiring a minimum ${p_T}^W$ threshold on the leptonically decaying

W was then suggested as a way to reduce the DPS background to negligible values.[169]

The DPS contribution to the $V+2\text{j}$ final state was also discussed as background to QCD studies performed with $V + n\text{j}$ events at the Tevatron. Its contribution was computed to be about 10% of the higher order QCD processes for typical kinematic requirements at the Tevatron.[170] As more advanced studies of QCD were conducted, in particular α_s measurements, it became clear that more precise estimates of the DPS contribution would be beneficial. The following ratio was suggested as a possible observable with which to estimate the DPS contribution to $V+n\text{j}$ events:[170]

$$R_{3/2} = \frac{\sigma_{V+3\text{j}}}{\sigma_{V+2\text{j}}}. \tag{1}$$

The parameter $\sigma_{V+n\text{j}}$ is the inclusive cross-section for the production of a W or Z boson in association with n jets. As seen in Fig. 2 for $Z+n\text{j}$ production, a significant enhancement in $R_{3/2}$ is expected due to DPS in the range $5 < p_{\text{T}}{}^{\text{jet}} < 15$ GeV. The study demonstrated that low-p_{T} QCD measurements are particularly affected by the contribution of multi-parton interactions. This is considered one of the difficulties of performing BFKL[171,172] studies in hadronic collisions.

The interest in DPS as a way to investigate the spatial distribution of partons within the proton, as well as background to other processes, prompted a number

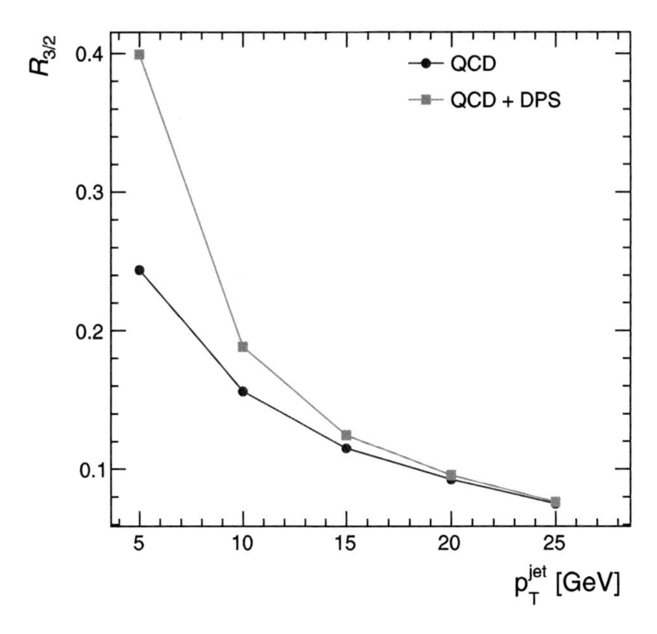

Fig. 2 The ratio $R_{3/2}$ in $Z + n\text{j}$ production, defined in Eq. (1), as a function of the jet minimum transverse momentum, $p_{\text{T}}{}^{\text{jet}}$. Black circles represent the ratio for the QCD cross-sections and red squares stand for their sum with the DPS contribution. Plot adapted from Fig. 9 in Ref. 170. Figure reprinted with permission. Copyright 1998 by the American Physical Society.

of DPS measurements at the Tevatron. CDF performed the first in the four-jet final state,[153] the same final state used in the previous measurements by the AFS and UA2 collaborations.[145,162] Soon after the measurement was published, it was noted that while final states consisting only of jets offer large cross-sections, they suffer from large experimental uncertainties and severe backgrounds.[173] The study in Ref. 173 explored the prospect of DPS measurements in the $\gamma + 3\text{j}$, $2\gamma + 2\text{j}$ and $2\ell + 2\text{j}$ final states, where 2ℓ represents the leptonic decay products of a VB. It predicted that the fraction of DPS (f_{DPS}) in $\gamma + 3\text{j}$ would be a few percent higher than in the four-jet final state. In the $2\gamma + 2\text{j}$ and $2\ell + 2\text{j}$ final states, f_{DPS} was predicted to be 3–7 times larger than in the four-jet final state, respectively. The relative cross-sections of the three processes were estimated to be

- $\sigma(\gamma + 3\text{j})/\sigma(4\text{j}) \sim 10^{-3}$,
- $\sigma(2\gamma + 2\text{j})/\sigma(4\text{j}) \sim 10^{-6}$,
- $\sigma(2\ell + 2\text{j})/\sigma(4\text{j}) \sim 10^{-6}$.

Despite the small relative cross-sections of the latter two processes, it was suggested in Ref. 173 that the large increase in f_{DPS} makes a σ_{eff} measurement possible. DØ collaboration recently (2016) published a measurement in the $2\gamma + 2\text{j}$ final state,[174] which suggests that a measurement in the $2\ell + 2\text{j}$ final state is also feasible, especially considering that f_{DPS} is expected to be higher and the separation between DPS and single parton scattering (SPS) is easier because of the superior energy and angular resolution of lepton reconstruction. However, despite the phenomenological studies suggesting to study DPS with VB in the final state at the Tevatron, no such measurements have so far been published.

2.2. *DPS with VB final states at the LHC*

The prospect of the LHC sparked renewed interest in DPS. The high energy available at the LHC and the high density of partons accessible with low fractions of the proton momenta enhance the probability of DPS producing high-p_{T} and massive particles. This led to a growing number of publications discussing the DPS contribution to hadronic collisions at the LHC.

2.2.1. *DPS involving VB as background at the LHC*

Early on it was noted that fulfilling the discovery potential of the LHC necessitates an accurate estimation of the DPS background to various physics channels.[175] Prior to the discovery of the Higgs boson, one of the promising channels for discovery was $p+p \to WH + X$ with $W \to \ell\nu$ and $H \to b\bar{b}$. The DPS production of a W boson in association with two b jets was considered as possible background to this channel.[176] However, a later study demonstrated that the DPS contribution to the $b\bar{b}$ invariant mass spectrum is below 120 GeV,[177] effectively proving that DPS does not pose a hindrance for Higgs searches and measurements in the WH channel.

Other than a discovery machine, the LHC is a top-quark factory. The most precise measurements of the characteristics of the top quark to date are expected at the LHC. With this in mind, the background to the $t\bar{t}$ final state from DPS production of a VB in association with jets was estimated and found to be small.[178]

The σ_{eff} values used in the above estimations were taken from measurements in the four-jet and $\gamma + 3\text{j}$ final states since at the time those were the only ones available. In addition, the large uncertainties of $\sim$20% on the measured σ_{eff} values resulted in sizeable uncertainties in the estimated DPS contributions. To test the assumption that σ_{eff} is process independent and obtain more accurate determinations of σ_{eff}, a number of studies suggested to measure DPS in a variety of final states.

2.2.2. *Proposed studies of DPS with VB at the LHC*

A more precise estimation of the effect of DPS can be obtained by determining σ_{eff} in accurately measured final states. At the LHC, it is experimentally easier to collect data and perform measurements in leptonic final states, such as W boson production. Vector boson production is therefore a reasonable candidate for DPS measurements, taking into account the following: the longitudinal momentum fraction in proton–proton interactions is approximated by $x \sim 2p_{\text{T}}/\sqrt{s}$. Processes producing particles with p_{T} of a few GeV in the central region of the detector probe x values of the order 10^{-3}. The high density of partons in the proton at these x values leads to an increase in the probability for multiple partonic interactions in the same collision. In the case of VB production, the x values probed are higher, of the order 10^{-2}, and the parton density is smaller. To study DPS in this case, with a relatively small amount of data, the cross-section of the second process has to be large.

This last point led to a number of suggestions for final states to perform measurements in. Seeing that the largest cross-section at the LHC is for jet production, the $W+n\text{j}$ and $Z+n\text{j}$ with $n = 2, 3, 4$ final states were considered.[178,179] In particular, following the study of DPS as background to WH, the $W + b\bar{b}$ final state was suggested for the determination of σ_{eff}.[177]

Taking advantage of the topologies of the events was shown to help in separating between DPS and SPS. For example, requiring a large difference in azimuthal angle between the pair of jets in $W+2\text{j}$ events was shown to enhance the DPS contribution for both light and b jets. In the case of $W + 4\text{j}$, it was found that the DPS fraction can be enhanced by requiring at least two jets with $|\Delta\eta| > 3.8$.[178] The same conclusions were reached in $Z + n\text{j}$ events, but there the DPS cross-section is about an order of magnitude smaller.[179] Nonetheless, the simpler Z reconstruction through two charged leptons makes this final state attractive for DPS studies. A detailed description of a number of DPS-sensitive observables is given in Section 2.2.3.

Same-sign W pair production is considered the ideal channel for the determination of σ_{eff}. It was suggested well before the start of the LHC.[175] However, the event rate is too low to perform the measurement within the first few years of the LHC. Nevertheless, a number of studies were performed, pointing out the potential benefit from exploring correlations between the interactions in this final state.[180]

2.2.3. *DPS-sensitive observables*

Over the years, many different observables were proposed for their potential to differentiate between DPS and SPS in VB final states. The underlying assumption in all of them is that the two interactions taking place in a DPS are largely uncorrelated.

Applying this to the case of DPS $W+2\text{j}$ production, the kinematics of the W and dijet systems are considered decorrelated. Therefore, the momenta of the two jets must compensate each other in the transverse plane, orienting them back-to-back in azimuthal angle. This renders the azimuthal separation between the jets,

$$\Delta\phi(\text{jet1},\text{jet2}) = |\phi(\text{jet}_1) - \phi(\text{jet}_2)|, \tag{2}$$

a potentially useful observable. The variable $\Delta\phi(\text{jet1},\text{jet2})$ is expected to peak around π in DPS events. The transverse momentum of the W in the SPS production channel must balance that of the jets. In a DPS, low values of ${p_{\text{T}}}^W$ are more likely. Based on this effect, ${p_{\text{T}}}^W$ was suggested to be used to suppress the DPS contribution to W + jets events.[169] Combining these two principles of low ${p_{\text{T}}}^W$ values and the expected orientation of the jets, the vectorial sum of the transverse momenta of the jets may be used,

$$\Delta p_{\text{T}}(\text{jet1},\text{jet2}) = |\mathbf{p}_{\text{T}}(\text{jet}_1) + \mathbf{p}_{\text{T}}(\text{jet}_2)|. \tag{3}$$

Low values of $\Delta p_{\text{T}}(\text{jet1},\text{jet2})$ are expected when ${p_{\text{T}}}^W$ is small and the jets are back-to-back.[102] The following observable incorporates the decay products of the W as well:

$$S_{p_{\text{T}}}^{W+2\text{j}} = \frac{1}{\sqrt{2}}\sqrt{\left(\frac{|\mathbf{p}_{\text{T}}(\text{jet}_1) + \mathbf{p}_{\text{T}}(\text{jet}_2)|}{|\mathbf{p}_{\text{T}}(\text{jet}_1)| + |\mathbf{p}_{\text{T}}(\text{jet}_2)|}\right)^2 + \left(\frac{|\mathbf{p}_{\text{T}}(\ell) + \mathbf{p}_{\text{T}}(E_{\text{T}}^{\text{miss}})|}{|\mathbf{p}_{\text{T}}(\ell)| + |\mathbf{p}_{\text{T}}(E_{\text{T}}^{\text{miss}})|}\right)^2}, \tag{4}$$

where $E_{\text{T}}^{\text{miss}}$ is the missing transverse energy in the event, representing the transverse momentum of the neutrino. The $S_{p_{\text{T}}}^{W+2\text{j}}$ variable is inspired by the observable constructed in the CDF DPS analysis in the four-jet final state.[153] It was shown to have sensitivity to DPS in the $W + b\bar{b}$ final state.[177]

The observables above, given for $W + 2\text{j}$, are relevant also for the $Z + 2\text{j}$ final state. As the Z boson decays to two charged leptons, observables incorporating the VB kinematics like $S_{p_{\text{T}}}^{Z+2\text{j}}$ provide even better sensitivity to DPS.

Studies have shown that combining two observables, such as $\Delta\phi(\text{jet1},\text{jet2})$ and $S_{p_{\text{T}}}^{W+2\text{j}}$, offers superior sensitivity to DPS.[103,177] Taking this further, a multi-variate analysis in the form of an artificial neural network was employed in a study of DPS production of $V + 2\mathrm{j}$ as background to WH.[181] Among other observables, including those mentioned above, the following variables were considered:

$$\Delta\eta(\mathrm{jet1},\,\mathrm{jet2}) = |\eta(\mathrm{jet}_1) - \eta(\mathrm{jet}_2)|, \tag{5}$$

$$\Delta\eta(V,\mathrm{jet12}) = |\eta(V) - (\eta(\mathrm{jet}_1) + \eta(\mathrm{jet}_2))|, \tag{6}$$

$$\Delta\Phi = \Delta\phi(\mathbf{p}_\mathrm{T}(V),\quad \mathbf{p}_\mathrm{T}(\mathrm{jet}_1,\mathrm{jet}_2)). \tag{7}$$

All three were shown to be sensitive to the presence of DPS. The $\Delta\eta(\mathrm{jet1},\,\mathrm{jet2})$ and $\Delta\eta(V,\mathrm{jet12})$ distributions are wider in $V + 2\mathrm{j}$ DPS events than in WH events. The azimuthal angle between the VB and the dijet system, $\Delta\Phi$, is expected to be uniform in DPS, while peaking around π in WH events. The $\Delta\Phi$ observable, albeit with a different definition, was used in the CMS DPS measurement in the $W + 2\mathrm{j}$ final state (see Section 3.6).

3. $W + 2\mathrm{j}$ Measurements

Following the surge of interest in MPI at the LHC, both ATLAS and CMS considered DPS measurements as key parts of their scientific program. Considering the various phenomenological studies performed prior to the start of the LHC indicating the benefits of determining σ_eff in processes involving vector bosons; and the large cross-section for producing a W boson with or without jets; the $W + 2\mathrm{j}$ final state was a natural candidate for the first DPS measurement.

Figure 3(a) shows examples of Feynman diagrams of the two interactions taking place in the production of a W boson with two jets in a DPS. In one interaction, a W boson is produced in the annihilation of a $q\bar{q}$ pair and decays leptonically. This provides a clean experimental tag for the event. In the second interaction, a dijet is produced via gluon–gluon scattering. The $W + 2\mathrm{j}$ final state can also be produced in an SPS, as shown in Fig. 3(b), constituting an irreducible background to DPS.

3.1. *Analysis strategy*

Both ATLAS and CMS determined σ_eff using the phenomenological formula,

$$\sigma_\mathrm{eff} = \frac{\hat{\sigma}_{W+0\mathrm{j}} \cdot \hat{\sigma}_{2\mathrm{j}}}{\hat{\sigma}^\mathrm{DPS}_{W+2\mathrm{j}}} = \frac{1}{f_\mathrm{DPS}} \cdot \frac{\hat{\sigma}_{W+0\mathrm{j}} \cdot \hat{\sigma}_{2\mathrm{j}}}{\hat{\sigma}^\mathrm{total}_{W+2\mathrm{j}}}. \tag{8}$$

The $\hat{\sigma}_{W+0\mathrm{j}}$ and $\hat{\sigma}_{W+2\mathrm{j}}$ are the particle-level cross-sections for the production of a W boson with zero or two jets respectively. The superscripts denote either the inclusive or the DPS production mechanism of the $W + 2\mathrm{j}$ final state. The $\hat{\sigma}_{2\mathrm{j}}$ is the particle-level dijet cross-section and f_DPS is the DPS fraction,

$$f_\mathrm{DPS} = \frac{N^\mathrm{DPS}_{W+2\mathrm{j}}}{N^\mathrm{total}_{W+2\mathrm{j}}}, \tag{9}$$

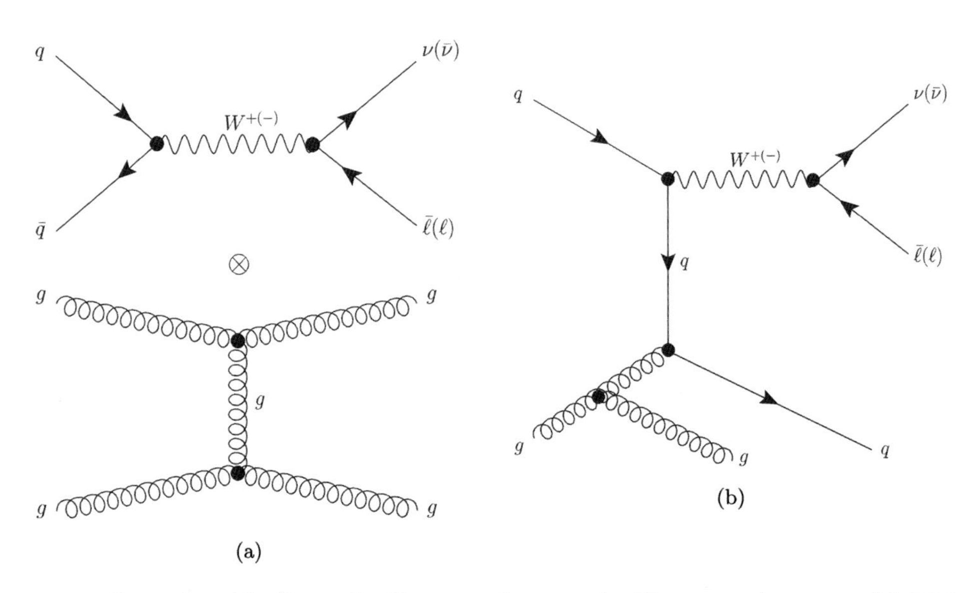

Fig. 3 Examples of leading-order Feynman diagrams for $W + 2\text{j}$ production in (a) DPS and (b) SPS.

where $N_{W+2\text{j}}$ is the number of inclusive or DPS $W + 2\text{j}$ events, as denoted by the superscript. The main challenge in DPS measurements in this (and other) final state is the extraction of f_{DPS}.

It is impossible to extract DPS candidate events on an event-by-event basis. As a result, both ATLAS and CMS adopted the template fit approach to estimate the DPS contribution. In this approach, DPS-sensitive observables are constructed and fitted to a combination of SPS and DPS templates,

$$\mathscr{D} = (1 - f_{\text{DPS}})\mathscr{H}_{\text{SPS}} + f_{\text{DPS}}\mathscr{H}_{\text{DPS}}, \tag{10}$$

with f_{DPS} as the free parameter in the fit. The distributions of the observable in the data, the SPS and the DPS templates are denoted by $\mathscr{D}$, $\mathscr{H}_{\text{SPS}}$ and $\mathscr{H}_{\text{DPS}}$ respectively. Details on the SPS and DPS templates are given in Section 3.4.

The second observable necessary to determine σ_{eff} in Eq. (8) is the ratio of cross-sections,

$$R = \frac{\hat{\sigma}_{W+0\text{j}}}{\hat{\sigma}_{W+2\text{j}}}. \tag{11}$$

Both experiments took advantage of the fact that the cross-sections in the ratio were measured using the same sample. This simplified the measurement of R through partial cancellations of reconstruction corrections and uncertainties.

The last observable necessary for the determination of σ_{eff} is the $\hat{\sigma}_{2\text{j}}$ cross-section. The kinematic requirements for the selection of the dijet sample play a crucial role in the measurement. As the dijet cross-section rapidly drops with p_{T}, reducing the p_{T} threshold for jets significantly enhances the DPS contribution.

The first data taking period of the LHC presented an opportunity to perform the measurement under optimal conditions. The number of multiple proton–proton interactions per bunch crossing (pile-up) increased gradually during the first year of operation, averaging to approximately 0.4. This allowed to use a relatively low jet-p_{T} threshold of 20 GeV in the selection of the dijet sample and reduced the effect of pile-up on the measurement.

3.1.1. *Exclusive final state*

Various phenomenological studies[42,112,160] suggested that assuming factorization in the form of Eq. (8), leads to a σ_{eff} which is independent of the processes, phase-space and parton distribution functions only if the cross-sections are inclusive. That is, the measurements of $\hat{\sigma}_{W+2\mathrm{j}}$ and $\hat{\sigma}_{2\mathrm{j}}$ should be performed while including two or more jets in the final state. However, both experiments estimated the corrections to σ_{eff} from using an exclusive selection of one W boson and exactly two jets to be of the order of a few percent. The use of an exclusive selection proved to be beneficial in identifying DPS. At leading order, requiring exactly two jets and assuming independent interactions, the jets are in the back-to-back topology and uncorrelated with the W boson. The DPS-sensitive observables were constructed based on this expected topology.

3.2. *Data samples*

3.2.1. *ATLAS selection*

ATLAS used for the measurement the LHC data set collected in 2010 of pp collisions at $\sqrt{s} = 7$ TeV, corresponding to approximately 36 pb^{-1} of integrated luminosity. Electron and muon triggers were used to collect candidate events of W bosons decaying leptonically, $W \to e\nu$ and $W \to \mu\nu$. Electrons and muons were required to have ${p_{\mathrm{T}}}^{\ell} > 20$ GeV, $|\eta^{e}| < 2.47$, $|\eta^{\mu}| < 2.4$ and fulfill standard quality and isolation requirements.[102,182–184] To select events with a W boson, in addition to requiring exactly one lepton (e or μ) in the event, events were required to have $E_{\mathrm{T}}^{\mathrm{miss}} > 25$ GeV and $m_{\mathrm{T}} > 40$ GeV. The transverse mass, m_{T}, is defined as

$$m_{\mathrm{T}} = \sqrt{2{p_{\mathrm{T}}}^{\ell} E_{\mathrm{T}}^{\mathrm{miss}}(1 - \cos\Delta\phi_{\ell,E_{\mathrm{T}}^{\mathrm{miss}}})}, \tag{12}$$

where $\Delta\phi_{\ell,E_{\mathrm{T}}^{\mathrm{miss}}}$ is the angle between $\mathbf{p}_{\mathrm{T}}^{\ell}$ and $\mathbf{E}_{\mathrm{T}}^{\mathrm{miss}}$.

Jets were identified using the anti-k_t algorithm,[185] with a radius parameter $R = 0.4$, and required to have $p_{\mathrm{T}} > 20$ GeV and $|\eta| < 2.8$. ATLAS used the rapidity of the jets rather than the pseudorapidity, but for consistency throughout this chapter, η is used. Jets were rejected if they overlapped with selected leptons within a cone of radius 0.5. Tracking information was used where available to reject additional jets from pile-up interactions. Using these requirements, ATLAS selected three samples,

- a $W + 0j$ sample, consisting of events with a reconstructed W boson and no jets with $p_T > 20$ GeV;
- a $W + 2j$ sample, consisting of events with a reconstructed W boson and exactly two jets with $p_T > 20$ GeV;
- a dijet sample, consisting of events with exactly two jets with $p_T > 20$ GeV.

3.2.2. *CMS selection*

To select $W + 2j$ events, CMS used the larger data set collected in 2011, corresponding to approximately 5 fb^{-1} of pp collisions at $\sqrt{s} = 7$ TeV. Muon triggers were used to collect candidate events of W bosons decaying via the muonic channel. Despite using only the $W \to \mu\nu$ channel, the statistical uncertainties in the CMS measurement were significantly smaller than in the ATLAS measurement thanks to the larger data set used.

A sample of dijet events was also used in the CMS measurement. Similarly to ATLAS, CMS took advantage of the low pile-up conditions of the 2010 data set and used it to select the dijet sample.

The selection requirements imposed by CMS were similar to those used in the ATLAS measurement. A summary of the requirements enforced in the two measurements is presented in Table 1.

3.3. *Simulated samples*

Studying the kinematic properties of simulated $W + 2j$ events, both experiments found that Monte Carlo generators which generate only $2 \to 1$ and $2 \to 2$ processes do not reproduce the kinematic distributions of the data. Such generators produce most of the additional jets during parton showering, leading to a softer jet p_T spectrum than that measured in data. This difference was seen mainly in the DPS-sensitive region. Therefore, it was concluded that the generators used to produce

Table 1. Summary of the main selection requirements in the ATLAS and CMS DPS measurements in the $W + 2j$ final state.

Requirement	ATLAS	CMS
W decay channel	$W \to e\nu$, $W \to \mu\nu$	$W \to \mu\nu$
Lepton kinematics	${p_T}^{\ell} > 20$ GeV, $\lvert\eta^e\rvert < 2.47$, $\lvert\eta^\mu\rvert < 2.4$	${p_T}^{\mu} > 35$ GeV, $\lvert\eta^\mu\rvert < 2.1$
W reconstruction	$E_T^{miss} > 25$ GeV, $m_T > 40$ GeV	$E_T^{miss} > 30$ GeV, $m_T > 50$ GeV
Jet anti-k_t radius parameter	$R = 0.4$	$R = 0.5$
Jet kinematics	$p_T > 20$ GeV, $\lvert\eta\rvert < 2.8$	$p_T > 20$ GeV, $\lvert\eta\rvert < 2.0$

events for the SPS sample must include a proper implementation of additional hard radiation. Otherwise, the effect of missing hard radiation might be interpreted as a DPS contribution. Subsequently, to produce W boson signal events, the experiments used matrix-element generators interfaced to multi-purpose generators for hadronization and parton showering.

ATLAS produced $W+n$j events using the Alpgen[186] matrix-element generator, interfaced to Herwig[156] for parton showering and hadronization and Jimmy[36] for the addition of MPI. This combination of generators is referred to as A+H+J in the following. The CTEQ6L1[187] parton distribution functions were used with the AUET2 tune.[188]

CMS produced two samples of $W + n$j events with the MadGraph 5[189,190] matrix-element generator using the CTEQ6L[187] PDF set. One of the samples was interfaced to Pythia 6[158] for parton showering and hadronization using the Z2 tune.[191] The other was showered and hadronized with Pythia 8[192] using the 4C tune.[114] The sample showered with Pythia 6 was used for detector-level studies and to obtain distributions of the data corrected for detector effects. The MadGraph 5 + Pythia 8 sample was used to construct the DPS and SPS templates.

Further checks and estimations of systematic uncertainties were performed by both experiments using additional samples of W events produced with other Monte Carlo event generators. Potential sources of physics background to the W signal were modeled by a combination of simulated samples. The background contribution from multi-jet production was estimated from data. The dominant source of background in the ATLAS measurement was found to be from multi-jet production, while in the CMS measurement it was determined to come from top-quark production and Drell–Yan processes. A likely reason for the difference between the experiments is the use of the $W \to e\nu$ decay channel in the ATLAS measurement. ATLAS estimated the multi-jet background contribution to be more than double in the electron channel than in the muon channel. In both measurements, the contribution from background sources was subtracted from the DPS-sensitive distributions in the data.

3.4. *Templates*

The fit to extract f_{DPS} is performed using templates representing the SPS and DPS contribution. ATLAS and CMS extracted the SPS template from the simulated samples of $W+n$j events since it is impossible to directly generate SPS events. Monte Carlo generators incorporate MPI in their description of hard hadronic interactions. Some of the outgoing partons from these additional interactions can materialize as high-p_{T} jets, while the majority add extra soft hadronic activity to the original hard scattering in the event. The SPS sample cannot be obtained by simply switching off MPI as it would lead to an incomplete description of the hadronic interaction. Instead, only events with MPI resulting in high-p_{T} jets which satisfy the DPS signal definition of the measurement should be excluded from the SPS template.

To extract SPS events, ATLAS introduced a threshold on the p_{T} of outgoing partons, $p_{\mathrm{T}}^{\mathrm{max}}$, and rejected events containing two or more partons from MPI with $p_{\mathrm{T}} > p_{\mathrm{T}}^{\mathrm{max}}$. CMS found the effect of $p_{\mathrm{T}}^{\mathrm{max}}$ on their measurement to be negligible and instead selected SPS events by rejecting all events having MPI partons with $|\eta| < 2.0$. More details on the $p_{\mathrm{T}}^{\mathrm{max}}$ studies are given in Section 3.7.

The experiments took different approaches for the DPS template. ATLAS used DPS-sensitive observables constructed solely from the kinematics of the dijet system (see Section 3.5). To reduce the dependence on the modeling of dijet production, the dijet sample selected from data was used as the DPS template. CMS produced the DPS sample by randomly mixing dijet events simulated using Pythia 8 with $W + 0\mathrm{j}$ events selected from the MadGraph 5 + Pythia 8 $W + n\mathrm{j}$ sample. Both approaches embody the underlying assumption of the measurement that the two interactions in a DPS are uncorrelated.

3.5. *DPS-sensitive observables — ATLAS*

ATLAS studied various observables for their sensitivity to distinguish DPS production of $W + 2\mathrm{j}$ from the SPS production mechanism. The observables suggested in previous phenomenological studies were considered (see Section 2.2.3), among them those which rely on the kinematics of the W boson, reconstructed using $E_{\mathrm{T}}^{\mathrm{miss}}$. The reconstruction of $E_{\mathrm{T}}^{\mathrm{miss}}$ suffered from experimental inaccuracies at the time, leading ATLAS to decide not to use observables which incorporate the W kinematics. Therefore, observables based solely on the kinematics of the jets were considered. The p_{T} distribution of the individual jets and the azimuthal angle between them were studied. The former was ruled out because it offered limited differentiating power due to its dependency on the jet energy calibration. The latter was found to be sensitive to various systematic effects, in particular pile up and the underlying event. Two observables quantifying the balance in p_{T} of the two jets were used instead,

$$\Delta_{\mathrm{jets}} = |\mathbf{p}_{\mathrm{T}}(\mathrm{jet}_1) + \mathbf{p}_{\mathrm{T}}(\mathrm{jet}_2)| \tag{13}$$

and

$$\Delta_{\mathrm{jets}}^{n} = \frac{|\mathbf{p}_{\mathrm{T}}(\mathrm{jet}_1) + \mathbf{p}_{\mathrm{T}}(\mathrm{jet}_2)|}{|\mathbf{p}_{\mathrm{T}}(\mathrm{jet}_1)| + |\mathbf{p}_{\mathrm{T}}(\mathrm{jet}_2)|}. \tag{14}$$

The distributions of $\Delta_{\mathrm{jets}}^{n}$ and Δ_{jets} in the inclusive $W + 2\mathrm{j}$ A+H+J sample, and the SPS sample extracted from it, are shown in Fig. 4. The effect of DPS is seen in the distributions of both observables, with Δ_{jets} being the more sensitive of the two. However, the distribution of $\Delta_{\mathrm{jets}}^{n}$ was used in the template fit to extract f_{DPS} because the normalization makes it less dependent on the jet energy calibration. This lower dependency leads to smaller systematic uncertainties. The Δ_{jets} distribution was used to cross-check the results obtained with $\Delta_{\mathrm{jets}}^{n}$.

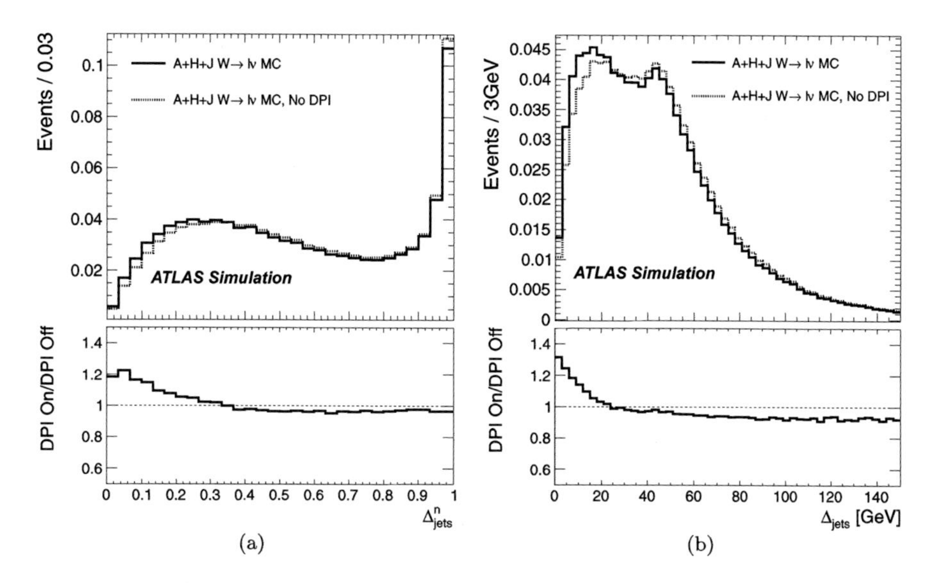

Fig. 4 Detector-level distributions of the (a) Δ^n_{jets} and (b) Δ_{jets} observables in $W \to \ell\nu + 2\text{j}$ events, selected from the A+H+J sample, for the inclusive and SPS samples. The inclusive sample is compared to the SPS template extracted from it, denoted in the legend as "No DPI". The lower panels show the ratio between the inclusive and SPS distributions. Figure taken from the ATLAS publication.[102]

3.6. *DPS-sensitive observables — CMS*

CMS used two DPS-sensitive observables in the template fit to extract f_{DPS}. The first is the same Δ^n_{jets} observable used in the ATLAS measurement. The second is the azimuthal angle between the W boson and the dijet system,

$$\Delta S = \arccos\left(\frac{\mathbf{p}_\text{T}(\mu, E_\text{T}^{\text{miss}}) \cdot \mathbf{p}_\text{T}(\text{jet}_1, \text{jet}_2)}{|\mathbf{p}_\text{T}(\mu, E_\text{T}^{\text{miss}})| \times |\mathbf{p}_\text{T}(\text{jet}_1, \text{jet}_2)|}\right). \tag{15}$$

The parameter $\mathbf{p}_\text{T}(\mu, E_\text{T}^{\text{miss}})$ is the combined transverse momentum of the muon and missing transverse energy in the event. Similarly, $\mathbf{p}_\text{T}(\text{jet}_1, \text{jet}_2)$ is the combined transverse momentum vectors of the two jets. CMS demonstrated that incorporating the W boson kinematics in the observables leads to higher sensitivity to DPS. CMS could include the W boson kinematics thanks to the use of the larger data set with more refined E_T^{miss} reconstruction.

CMS performed a simultaneous fit to the distributions of the two observables using the binned likelihood method. The data distributions were corrected for detector effects before fitting.

3.7. *Measurement strategy validation*

To validate the strategy of the measurement, the experiments estimated f_{DPS} in the simulated $W + 2\text{j}$ sample via a template fit. The estimated fraction, $f^{\text{MC}}_{\text{DPS}}$, was

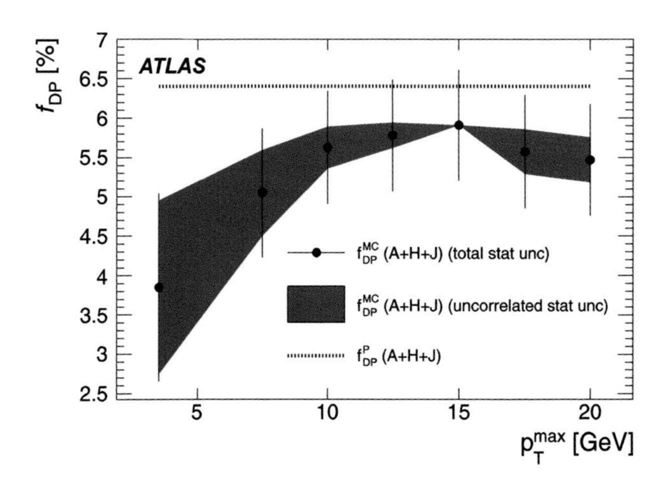

Fig. 5 The extracted fraction of DPS from the $\Delta^{n}_{\mathrm{jets}}$ distribution in the A+H+J sample, $f^{\mathrm{MC}}_{\mathrm{DPS}}$, as a function of the transverse momentum threshold for MPI partons, $p_{\mathrm{T}}^{\mathrm{max}}$. The band illustrates the statistical component of the uncertainty of $f^{\mathrm{MC}}_{\mathrm{DPS}}$, relative to the reference sample with $p_{\mathrm{T}}^{\mathrm{max}} = 15$ GeV. The fraction of DPS obtained directly by counting events with MPI partons fulfilling the DPS signal requirements, $f^{\mathrm{P}}_{\mathrm{DPS}}$, is shown as the dashed line. DPS is denoted as "DP" in the figure. Figure taken from the ATLAS publication.[102]

compared with the fraction obtained directly from the Monte Carlo event record, $f^{\mathrm{P}}_{\mathrm{DPS}}$. Accessing the kinematics of the outgoing partons, $f^{\mathrm{P}}_{\mathrm{DPS}}$ was obtained by counting events with MPI partons which fulfilled the DPS signal requirements.

Making the comparison for various values of the threshold $p_{\mathrm{T}}^{\mathrm{max}}$, the smallest difference between $f^{\mathrm{MC}}_{\mathrm{DPS}}$ and $f^{\mathrm{P}}_{\mathrm{DPS}}$ was seen for $p_{\mathrm{T}}^{\mathrm{max}} = 15$ GeV in the ATLAS measurement (see Fig. 5). Therefore, this value of $p_{\mathrm{T}}^{\mathrm{max}}$ was set as the threshold for MPI partons. The decline in $f^{\mathrm{MC}}_{\mathrm{DPS}}$ values seen in Fig. 5 for $p_{\mathrm{T}}^{\mathrm{max}} > 15$ GeV suggests that MPI partons with ${p_{\mathrm{T}}}^{\mathrm{parton}} > 15$ GeV sometimes result in jets with ${p_{\mathrm{T}}}^{\mathrm{jet}} > 20$ GeV. Seeing that $f^{\mathrm{MC}}_{\mathrm{DPS}}$ was estimated using detector-level distributions, this study also served to demonstrate that in this case the f_{DPS} extracted from uncorrected data distributions is a good approximation to a parton-level measurement.

The same study was repeated in the CMS measurement, but it was observed that for $p_{\mathrm{T}}^{\mathrm{max}} < 12$ GeV, $f^{\mathrm{MC}}_{\mathrm{DPS}}$ does not depend on $p_{\mathrm{T}}^{\mathrm{max}}$. In the case of CMS, the fit was performed using particle-level distributions taken from simulation. The results of the CMS study are shown in Fig. 6. Figure 6 also shows a comparison between the CMS results for $f^{\mathrm{MC}}_{\mathrm{DPS}}$ when performing the fit using each of the observables separately and for a simultaneous fit of both observables. A better agreement between $f^{\mathrm{MC}}_{\mathrm{DPS}}$ and $f^{\mathrm{P}}_{\mathrm{DPS}}$ is observed for the simultaneous fit.

3.8. *Determination of σ_{eff}*

To extract f_{DPS} in data, the template fit was performed to the background-subtracted data distributions. ATLAS performed the fit using the data distribution

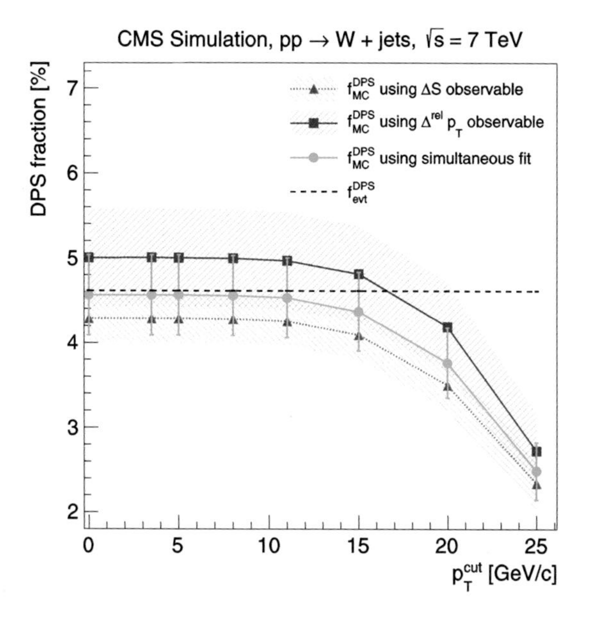

Fig. 6 The extracted value of the DPS fraction in $W + n\mathrm{j}$ events, simulated with MadGraph 5 + Pythia 8, using different background templates obtained by varying the transverse momentum cut off (${p_\mathrm{T}}^{\mathrm{cut}}$) for the second hard interaction. The DPS fractions obtained by performing both simultaneous and individual fits to the $\Delta^{\mathrm{rel}}p_\mathrm{T}$ and ΔS observables are shown. The DPS fraction, $f_{\mathrm{evt}}^{\mathrm{DPS}}$, for the simulated $W + n\mathrm{j}$ events is shown by a dashed black line. The error bars/bands represent the statistical uncertainty added in quadrature to the systematic uncertainty of the DPS template. The ${p_\mathrm{T}}^{\mathrm{cut}}$, $\Delta^{\mathrm{rel}}p_\mathrm{T}$ and $f_{\mathrm{evt}}^{\mathrm{DPS}}$ are equivalent to $p_\mathrm{T}^{\max}$, $\Delta_{\mathrm{jets}}^{n}$ and $f_{\mathrm{DPS}}^{\mathrm{P}}$ in the text, respectively. Figure taken from the CMS publication.[103]

of $\Delta_{\mathrm{jets}}^{n}$, before correcting for detector effects, obtaining

$$f_{\mathrm{DPS}}^{\mathrm{ATLAS}} = 0.08 \pm 0.01\ (\mathrm{stat.}) \pm 0.02\ (\mathrm{syst.}). \tag{16}$$

This value is in agreement with the result of the fit to the Δ_{jets} distribution. The normalized $\Delta_{\mathrm{jets}}^{n}$ and Δ_{jets} distributions in data are compared in Fig. 7 to the sum of the SPS and DPS templates, normalized to $(1 - f_{\mathrm{DPS}}^{\mathrm{ATLAS}})$ and $f_{\mathrm{DPS}}^{\mathrm{ATLAS}}$ respectively. To facilitate future comparisons with MPI models, the distributions were corrected for detector-effects and published as well.

CMS performed a simultaneous fit to the fully-corrected, background-subtracted, data distributions of $\Delta_{\mathrm{jets}}^{n}$ and ΔS using a binned likelihood method. The DPS fraction obtained is

$$f_{\mathrm{DPS}}^{\mathrm{CMS}} = 0.055 \pm 0.002\ (\mathrm{stat.}) \pm 0.014\ (\mathrm{syst.}). \tag{17}$$

The results of the fit are shown in Fig. 8.

It is important to note that the value of f_{DPS} depends on the phase-space of the measurement. Thus, the difference between the f_{DPS} values obtained by the

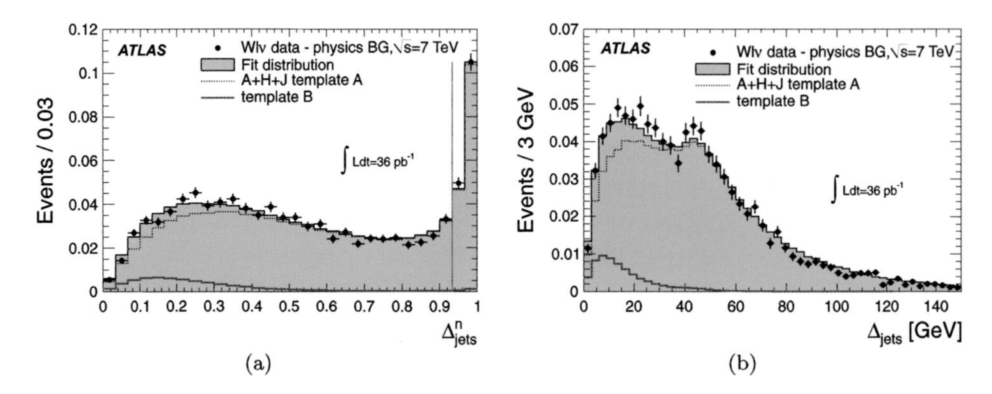

Fig. 7 Distributions of (a) Δ^n_{jets} and (b) Δ_{jets} in the background-subtracted data (dots) compared to the result from the best fit for $f_{\mathrm{DPS}}^{\mathrm{ATLAS}}$. In (a), the bins to the right of the vertical dash-dotted line were excluded from the fit. Data and the fit distribution (green histogram) were normalized to unity, the SPS template (dashed line) to $(1 - f_{\mathrm{DPS}}^{\mathrm{ATLAS}})$ and the DPS template (blue solid line) to $f_{\mathrm{DPS}}^{\mathrm{ATLAS}}$. The SPS and DPS templates are denoted in the legend as "Template A" and "Template B" respectively. Figure taken from the ATLAS publication.[102]

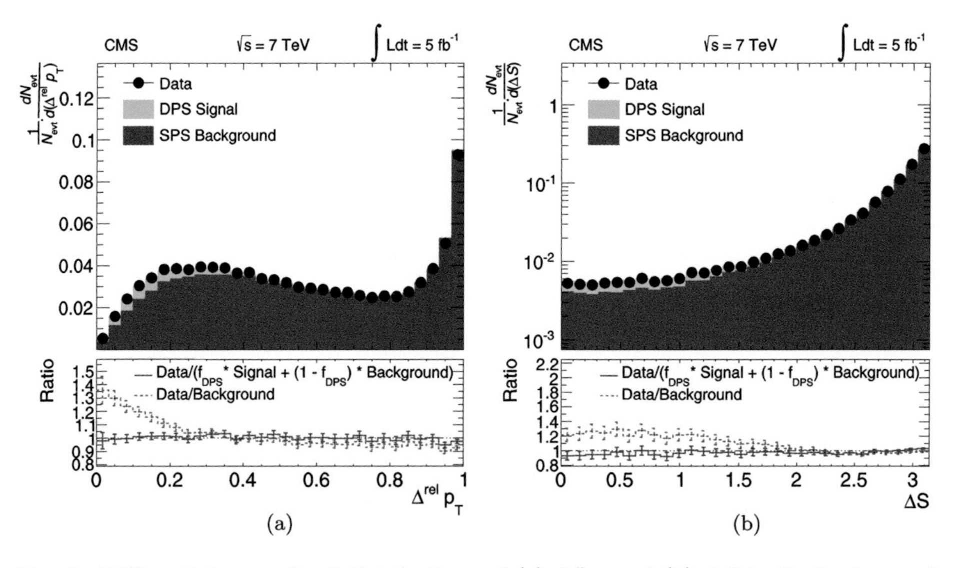

Fig. 8 Differential normalized distributions of (a) Δ^n_{jets} and (b) ΔS in the background-subtracted data, corrected for detector effects, compared to the result from the best fit for $f_{\mathrm{DPS}}^{\mathrm{CMS}}$. The SPS template is normalized to $(1 - f_{\mathrm{DPS}}^{\mathrm{CMS}})$ and the DPS template to $f_{\mathrm{DPS}}^{\mathrm{CMS}}$. The ratio of the data to the fit and background distributions is shown in the bottom panels. The parameter $\Delta^{\mathrm{rel}} p_{\mathrm{T}}$ in (a) is equivalent to Δ^n_{jets}. Figure taken from the CMS publication.[103]

two experiments can be safely ignored. A meaningful comparison between DPS measurements is done via σ_{eff}.

Combining the value of f_{DPS} with measurements of R and the dijet cross-section (see Eqs. (8) and (11)), the experiments determined the following values of σ_{eff}:

$$\sigma_{\text{eff}}^{\text{ATLAS}}(W+2\text{j}) = 15 \pm 3\ (\text{stat.})^{+5}_{-3}\ (\text{syst.})\ \text{mb}, \tag{18}$$

$$\sigma_{\text{eff}}^{\text{CMS}}(W+2\text{j}) = 20.7 \pm 0.8\ (\text{stat.}) \pm 6.6\ (\text{syst.})\ \text{mb}. \tag{19}$$

The leading sources of systematic uncertainty in the ATLAS measurement originated from uncertainties in the influence of pile-up, jet energy calibration and background model dependence, all contributing similarly. The largest source of uncertainty in the CMS measurement is the model dependence of the background template. The σ_{eff} values obtained are consistent with each other within uncertainties. They are also consistent with all of the σ_{eff} measurements performed so far in final states which involve jets.

4. Lower Limits and Preliminary Studies

A number of preliminary DPS studies in final states involving massive vector bosons were performed at the LHC. A short summary of those available at the moment is given below.

4.1. *The W + J/ψ final state*

The ratio of the W + J/ψ cross-section to the inclusive W boson cross-section was measured by ATLAS using data of pp collisions at $\sqrt{s} = 7$ TeV.[193] The DPS contribution to the W + J/ψ final state was estimated to be

$$f_{\text{DPS}}^{\text{ATLAS}}(W+\text{J}/\psi) = \frac{1}{\sigma_{\text{eff}}(W+2\text{j})}\frac{\sigma(W)\sigma(\text{J}/\psi)}{\sigma(W+\text{J}/\psi)} = 0.38^{+0.22}_{-0.20}. \tag{20}$$

The fraction was estimated using the value of σ_{eff} determined by ATLAS in the W + 2j final state, together with the prompt J/ψ cross-section from the ATLAS measurement described in Ref. 194. The uncertainties in Eq. (20) are the combined statistical and systematic uncertainties. This estimate of f_{DPS} is only a rough evaluation of the DPS contribution to the W + J/ψ final state. It was not directly estimated from the W + J/ψ data.

4.2. *The Z + J/ψ final state*

A lower limit on σ_{eff} was extracted in the ATLAS measurement of associated Z + J/ψ production at $\sqrt{s} = 8$ TeV.[195] In the DPS production mechanism of Z + J/ψ, the azimuthal angle between the Z boson and the J/ψ ($\Delta\phi(Z, \text{J}/\psi)$) is expected to be uniform, while in SPS at leading order the angle is expected to be $\Delta\phi(Z, \text{J}/\psi) \sim \pi$. Assuming that all of the events with $\Delta\phi(Z, \text{J}/\psi) < \pi/5$ originate from DPS, a limit

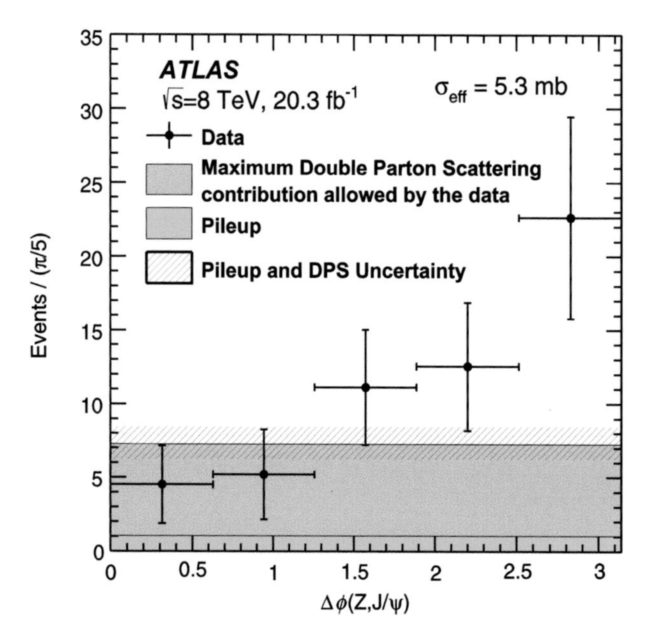

Fig. 9 Distribution of the azimuthal angle between the Z boson and the J/ψ meson for prompt J/ψ production, $\Delta\phi(Z, \mathrm{J}/\psi)$. The estimated pile-up contribution is shown as the cyan band. The maximum DPS contribution allowed by the data is shown as the yellow band. The hashed region shows the DPS and pile-up uncertainties added in quadrature. Figure taken from the ATLAS publication.[195]

on the maximum DPS contribution to the observed $Z+\mathrm{J}/\psi$ signal was obtained (see Fig. 9). The upper limit on f_{DPS} was found to be $f_{\mathrm{DPS}} < 0.29$ at 68% confidence level, corresponding to the lower limit on σ_{eff},

$$\sigma_{\mathrm{eff}}^{\mathrm{ATLAS}}(Z + \mathrm{J}/\psi) > 5 \text{ mb}. \tag{21}$$

4.3. *Same-sign W pair*

A search for events containing a pair of same-sign W bosons produced in a DPS was conducted by CMS using data at $\sqrt{s} = 8$ TeV.[196] The $W \to \mu\nu$ decay channel was used in the study. Various DPS-sensitive observables were combined in a multivariate analysis in the form of boosted decision trees. The available data of 19.75 fb^{-1} were not sufficient to perform a direct measurement of σ_{eff}. However, a lower limit at 95% confidence level was obtained,

$$\sigma_{\mathrm{eff}}^{\mathrm{CMS}}(WW) > 6 \text{ mb}. \tag{22}$$

5. Future Studies

The $W+2$j measurements at $\sqrt{s} = 7$ TeV were important milestones in the ongoing study of DPS. They provided the first evidence of the weak dependence of σ_{eff} on the centre-of-mass energy and the physical processes taking place. Complementing these with measurements at higher energies and in the similar $Z + 2\text{j}$ final state would be highly valuable.

5.1. *Z + 2j final state*

The $Z + 2\text{j}$ final state possesses the same experimental advantages as the $W + 2\text{j}$ final state,

- a clean leptonic tag;
- a secondary process with a large cross-section;
- a clear separation between the products of the two interactions (the vector bosons do not radiate jets);
- no need to take into account multiple combinatorial combinations of the final state partons.

On top of these, the $Z + 2\text{j}$ final state has the advantage that the Z boson can be fully reconstructed from the charged leptons it decays to. These attributes render the $Z + 2\text{j}$ final state a natural candidate for DPS studies at the LHC.

5.2. *Experimental and phenomenological advances*

5.2.1. *Inclusive final states*

Exclusive final states were used in the $W + 2\text{j}$ measurements, a decision motivated by the gain in differentiating power between DPS and SPS (see Section 3.1.1). Both experiments estimated the effect of using an exclusive final state on σ_{eff} is of the order of a few percent. In a recent measurement in the four-jet final state it was shown that combining observables using an artificial neural network allows to perform the measurement in an inclusive final state, while minimizing the effect on the performance.[71] The same approach should be attempted in future $V + 2\text{j}$ measurements. In fact, the $V + 2\text{j}$ case is easier than the four-jet one because the leptonically decaying VB does not radiate extra jets. Moreover, triple parton scattering can also be included in the measurement if more jets are allowed in the final state, $V + 2\text{j} + 2\text{j}$.

5.2.2. *Monte Carlo generators*

The experiments demonstrated that events produced with leading-order generators do not reproduce the kinematic properties of events in the data. A better performance was observed using tree-level matrix-element generators. In current versions of Monte Carlo generators, next-to-leading order calculations are available for some of the final states involving massive vector bosons. Incorporating these in future measurements should result in a more accurate description of the kinematic distributions of the data.

The Pythia 8 and A+H+J generators provide access to the kinematics of MPI partons, an option which proved useful in the ATLAS and CMS measurements when selecting SPS events. Such access to MPI information in more generators would help future DPS measurements to test their model dependency. At the moment, Sherpa[157] allows to produce SPS events by applying an upper p_{T} threshold on MPI partons during event generation. This is similar to the ATLAS method of applying the $p_{\mathrm{T}}^{\max}$ threshold to select SPS events. This feature of Sherpa should be exploited in the future to estimate the systematic uncertainties associated with the SPS models used in measurements.

5.3. *Energy dependence of σ_{eff}*

Other than determining σ_{eff}, the available data allows to measure directly, with greater accuracy, its dependence on centre-of-mass energy. Assuming the energy dependence of σ_{eff} is similar in form to that of the total hadron–hadron cross-section at high energies,[197] $\sigma_{\mathrm{eff}} \sim (\sqrt{s})^{\alpha}$, the following ratio can be defined:

$$r_{\sigma_{\mathrm{eff}}}^{13/7} \equiv \frac{\sigma_{\mathrm{eff}}^{13}}{\sigma_{\mathrm{eff}}^{7}} = \left(\frac{\sqrt{13}}{\sqrt{7}}\right)^{\alpha}. \tag{23}$$

The superscript denotes the energy, in TeV, in which the measurement is performed. The experiments can then perform a measurement of σ_{eff} at the three energies, 7, 8 and 13 TeV, and calculate the multiplication of ratios,

$$R_{\sigma_{\mathrm{eff}}} \equiv r_{\sigma_{\mathrm{eff}}}^{13/7} \cdot r_{\sigma_{\mathrm{eff}}}^{13/8} \equiv \frac{\sigma_{\mathrm{eff}}^{13}}{\sigma_{\mathrm{eff}}^{7}} \cdot \frac{\sigma_{\mathrm{eff}}^{13}}{\sigma_{\mathrm{eff}}^{8}}, \tag{24}$$

$$R_{\sigma_{\mathrm{eff}}} = \left(\frac{\sigma_V^{13} \cdot \sigma_V^{13}}{\sigma_V^{7} \cdot \sigma_V^{8}}\right) \left(\frac{\sigma_{2\mathrm{j}}^{13} \cdot \sigma_{2\mathrm{j}}^{13}}{\sigma_{2\mathrm{j}}^{7} \cdot \sigma_{2\mathrm{j}}^{8}}\right) \left(\frac{\sigma_{\mathrm{DPS}}^{7} \cdot \sigma_{\mathrm{DPS}}^{8}}{\sigma_{\mathrm{DPS}}^{13} \cdot \sigma_{\mathrm{DPS}}^{13}}\right) = \left(\frac{13}{\sqrt{7} \cdot \sqrt{8}}\right)^{\alpha}. \tag{25}$$

If the various cross-sections are measured in the scope of one measurement, it is reasonable to assume that some of the experimental uncertainties would partially cancel in the ratios in Eq. (25).

Consider the $Z+2\mathrm{j}$ final state for the purpose of this discussion. If the same jet definition, calibration and phase-space requirements are used in all cross-section measurements at the three energies, the jet related uncertainties are expected to largely cancel in the ratios. The same is true for the lepton reconstruction uncertainties. In case one set of Monte Carlo generators is used throughout the measurement, theoretical modeling uncertainties should partially cancel as well. These cancellations of uncertainties would result in a smaller uncertainty on α compared to the case of fitting α from separate σ_{eff} measurements.

The biggest difficulty in following this prescription with the available data is to find the lowest common minimum p_{T} threshold for the jets at the three energies. In order to enhance the DPS fraction in the $Z+2\mathrm{j}$ sample, the jet p_{T} threshold should be set as low as possible (see Section 3.1). This does not present a problem for selecting $Z + 2\mathrm{j}$ events, since they are collected using the lepton triggers. However, the measurement of $\sigma_{2\mathrm{j}}$ at each energy requires to select three dijet samples. At high luminosity runs, the jet triggers settings may result in an insufficient amount of low p_{T} dijet events to perform the measurement. Low p_{T} in this case refers to ${p_{\mathrm{T}}}^{\mathrm{jet}} \sim 50$ GeV. Possible ways to overcome this issue are to use dijet events collected by a random trigger or take advantage of low luminosity runs such as the 2015 13 TeV run. Alternatively, the prospect of directly measuring the energy dependence of σ_{eff} may drive the LHC and the experiments to take dedicated low luminosity runs at various energies.

5.4. *Vector boson in association with heavy flavor*

The ATLAS studies of vector boson production in association with heavy flavor mesons showed the potential of DPS studies in such final states. The estimated contribution of DPS to these final states is 3–7 times larger than in final states which involve jets. Moreover, the fully leptonic decay channel of the $V + \mathrm{J}/\psi$ final state provides a very clean signal. This renders the $V + \mathrm{J}/\psi$ channels to be promising ones for future studies.

5.5. *Same-sign W production*

Expanding the CMS DPS study of same-sign W production and determining σ_{eff} may require more data than expected in Run II of the LHC. However, considering that both interactions probe quarks in the protons, extracting lower limits in the WW final state could indicate whether the value of σ_{eff} is different for quark initiated processes. This is an important test of the assumed universality of σ_{eff}.

Chapter 8

Study of Double Parton Scattering Processes with Heavy Quarks

Ivan Belyaev* and Daria Savrina†

Institute for Theoretical and Experimental Physics,
B. Cheremushkinskaya 25, 117218, Moscow, Russia
**Ivan.Belyaev@itep.ru*
†Daria.Savrina@cern.ch

Study of double parton scattering processes (DPS) involving heavy quarks provides the most precise probing of factorization hypothesis for DPS for gluon-mediated processes. The measurements are performed for the different final states, including open and hidden-flavor hadrons and for the different kinematic ranges of incoming gluons.

1. Introduction

The double parton scattering (DPS) processes in the high-energy hadron–hadron interactions attract[88,89,198,199] significant theoretical and experimental interest. While the DPS processes first have been observed thirty years ago by AFS collaboration[162] in the proton–proton collisions at relatively low energy $\sqrt{s} = 63$ GeV, and later studied in the p$\bar{\text{p}}$ collisions at $\sqrt{s} = 1.8$ and 1.96 TeV by CDF[97,153] and D0[98,148,159,200] collaborations, the role of DPS processes becomes much more important[3,67,201] at higher energies, in particular, for pp collisions at $\sqrt{s} = 7, 8$ and 13 TeV at Large Hadron Collider (LHC).

The cross-section of DPS processes can be expressed as

$$\sigma_{\text{DPS}} = \frac{1}{s}\frac{\sigma_A \sigma_B}{\sigma_{\text{eff}}}, \tag{1}$$

where $\sigma_{A,B}$ are the individual cross-sections for processes A and B, $s = 1, 2$ is a symmetry factor and σ_{eff} is an effective cross-section. In the factorization approach the latter is expressed via the integral over the transverse degrees of freedom of the partons in the protons and is a universal process and scale-independent constant.

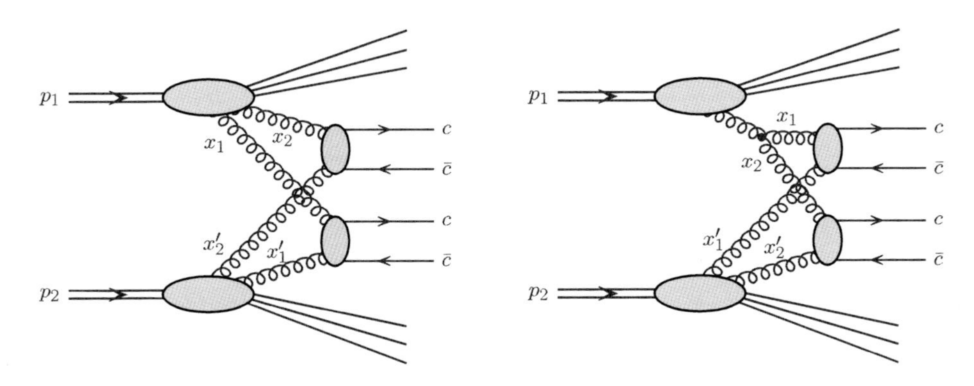

Fig. 1 Diagrams for multi-parton production of $c\bar{c}c\bar{c}$: $2 \otimes 2$ (left) and $1 \otimes 2$ (right) processes.[86] Figure reprinted with permission. Copyright 2014 by the American Physical Society.

The value of σ_{eff} is expected to be of the order of the inelastic cross-section, but currently it cannot be calculated[30,45,202,203] from the basic principles. Significant violations[204,205] of the factorization are expected for some kinematic regions and processes, e.g., for the processes with high-x partons.

In addition to DPS $2 \otimes 2$ process, characterized by the independent scattering of two pairs of partons from the incoming protons, Fig. 1 (left), one also needs to account for the potentially large additional contribution from the $1 \otimes 2$ process, where a parton from one incoming proton splits at some hard scale[9,10,82,85] and creates two partons that further participate in the two independent single-scattering processes, see Fig. 1 (right). This contribution is calculated[90,91,95] in pertubative QCD and provides significant dependency of multi-parton interaction cross-sections on the transverse scale. However for small-x processes with relatively small transverse momenta, this large dependence is expected to stabilize[96] by accounting of soft correlations, turning σ_{eff} to be a constant.

Testing the universality of σ_{eff}, in terms of its (in)dependency on the process, scale and collision energy, allows to shed light on the role of factorization and contribution of $1 \otimes 2$ processes and provides the unique opportunity to probe the parton–parton correlations in the proton. Good understanding of DPS processes is very important for the search of New Physics (NP) effects at LHC, since for certain final states DPS process mimics[206–208] the production of heavy exotic particles. Studies of DPS processes with heavy quarks or quarkonia in the final state are especially important both for QCD tests, mentioned above and for the deep understanding of potentially important background source for NP searches.

For DPS processes producing four heavy quarks in the final state, $Q_1\bar{Q}_1Q_2\bar{Q}_2$, the corresponding elementary subprocesses $gg \to Q_1\bar{Q}_1$ and $gg \to Q_2\bar{Q}_2$ are studied in detail. The production of open-charm hadrons at high-energy hadron collisions has been studied in pp collisions by LHCb collaboration[105,209,210] at $\sqrt{s} = 5$, 7 and 13 TeV, by ATLAS collaboration[211] at $\sqrt{s} = 7$ TeV and by ALICE collaboration[212–214] at $\sqrt{s} = 2.76$ TeV and 7 TeV, and by CDF collaboration[215] in $\mathrm{p\bar{p}}$ collisions at $\sqrt{s} = 1.96$ TeV. The measurements are in reasonable agreement with calculations at the next-to-leading order (NLO) using the generalized mass variable flavor number scheme[216–220] (GMVFNS), POWHEG[221] and fixed order with next-to-leading-log resummation[222–225] (FONLL), see Fig. 2.

Experimentally, the processes with heavy quarks often could be studied in details up to very low transverse momenta, comparable or even smaller than the masses of the involved heavy quarks, especially for the final states with the dimuon decays of the $\mathrm{J/\psi}$ or Υ mesons. It opens the unique possibility to explore the low-p_T region, where DPS contribution[3,67] is not suppressed. Full reconstruction of the final state quarkonia or the open-flavor heavy hadron allow to minimize some large experimental uncertainty, e.g., related to jet reconstuction or jet-energy calibration. Currently practically all (experimental) knowledge of DPS processes in very interesting and important kinematic region $x_1, x_2 \gg x'_1, x'_2$ comes from the measurements involving heavy quarks[78,106,226,227] in the forward region, where studies with e.g., light quarks (jets) and/or the direct photons are experimentally challenging.

2. Studies with Open-Flavor Hadrons

The LHCb experiment studied[78] the associated production of CC and $\mathrm{J/\psi\,C}$ combinations, where C stands for $\mathrm{D^0}$, $\mathrm{D^+}$, $\mathrm{D_s^+}$ or Λ_c^+, in the kinematic region of $2 < y_{\mathrm{J/\psi}}, y_\mathrm{C} < 4.5$, $\mathrm{p^T_{J/\psi}} < 10\,\mathrm{GeV}/c$ and $3 < \mathrm{p^T_C} < 12\,\mathrm{GeV}/c$ using $355 \pm 13\,\mathrm{pb}^{-1}$ of data taken in pp collisions at the centre-of-mass energy of $\sqrt{s} = 7$ TeV. Open-charm hadrons are reconstructed via $\mathrm{D^0 \to K^-\pi^+}$, $\mathrm{D^+ \to K^-\pi^+\pi^+}$, $\mathrm{D_s^+ \to K^-K^+\pi^+}$ and $\Lambda_\mathrm{c}^+ \to \mathrm{pK^-\pi^+}$ modes, while $\mathrm{J/\psi}$ mesons are reconstructed in dimuon final state. Charge-conjugated processes are included. Clear high-statistics signals with significance in excess of five standard deviations have been observed for six CC modes, $\mathrm{D^0D^0}$, $\mathrm{D^0D^+}$, $\mathrm{D^0D_s^+}$, $\mathrm{D^0}\Lambda_\mathrm{c}^+$, $\mathrm{D^+D^+}$ and $\mathrm{D^+D_s^+}$, and for four $\mathrm{J/\psi\,C}$ modes, $\mathrm{J/\psi\,D^0}$, $\mathrm{J/\psi\,D^+}$, $\mathrm{J/\psi\,D_s^+}$ and $\mathrm{J/\psi}\,\Lambda_\mathrm{c}^+$. Large $\mathrm{D^0D^0}$ and $\mathrm{J/\psi\,D^0}$ signals are shown in Fig. 3. The possible backgrounds from hadrons produced in two different pp interactions within the same bunch crossing (pile-up) and contamination from b-hadrons decays were reduced to a negligible level by imposing cuts based on the consistency of the decay chain.

The model-independent cross-sections for double charm production in the fiducial range were calculated using the yields obtained from the fits of two-dimensional distributions. The dominating systematic uncertainties are related to the luminosity determination, to the C-hadrons branching fractions and to the differences between the data and simulated sample used for the reconstruction efficiency determination. The resulting cross-sections for CC and $\mathrm{J/\psi\,C}$ events are shown in the left part of Fig. 4 as well as several SPS theoretical estimations[228–230] of $\mathrm{J/\psi\,C}$ production. The expectations from SPS process are by a factor of about 20 below the

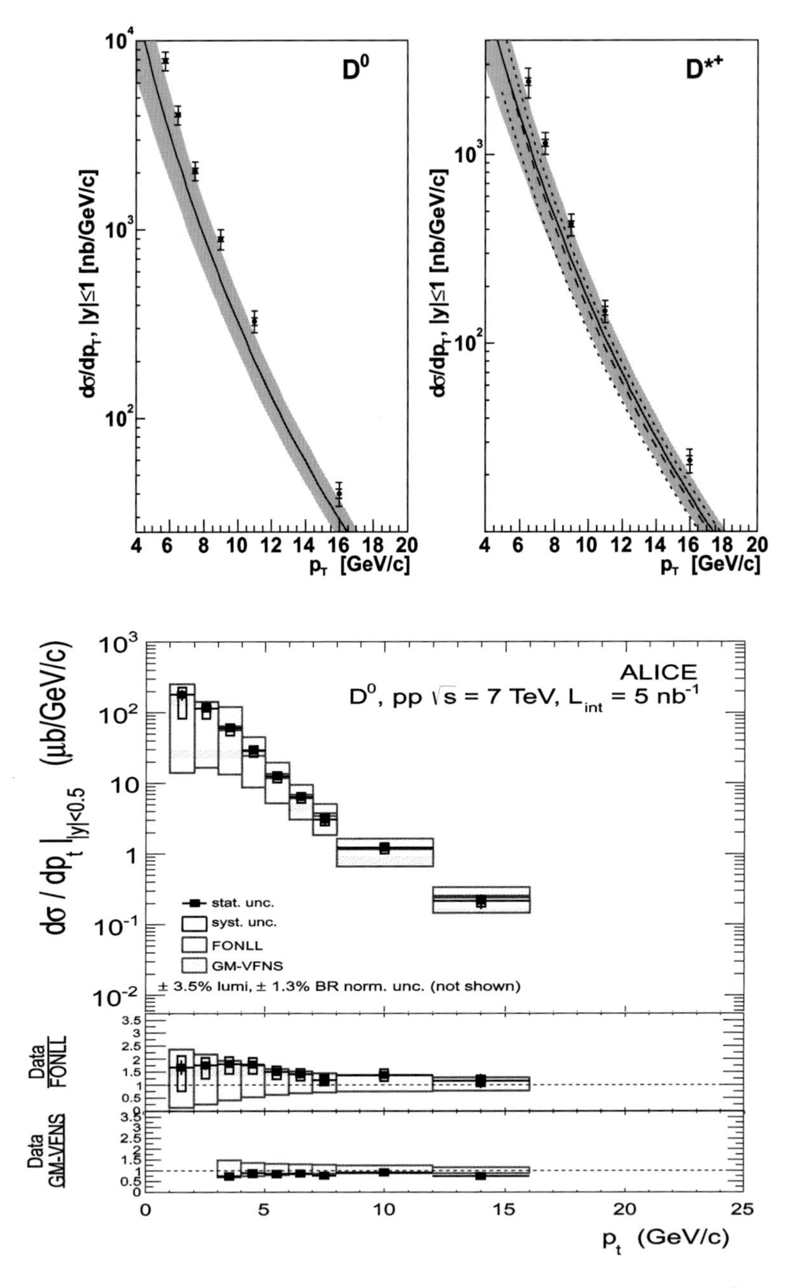

Fig. 2 Top: Inclusive differential production cross-sections for prompt D^0 and D^{*+} mesons[215] in $p\bar{p}$ collisions at $\sqrt{s} = 1.96$ TeV. The solid curves are FONLL theoretical predictions with the uncertainties indicated by the shaded bands. The dashed curve shown with the D^{*+} cross-section corresponds to GMVFNS theoretical prediction; the dotted lines indicate the uncertainty. Figure reprinted with permission. Copyright 2003 by the American Physical Society. Bottom: Inclusive differential production cross-section for prompt D^0 mesons in pp collisions at $\sqrt{s} = 7$ TeV[213] compared with FONLL and GMVFNS theoretical predictions.

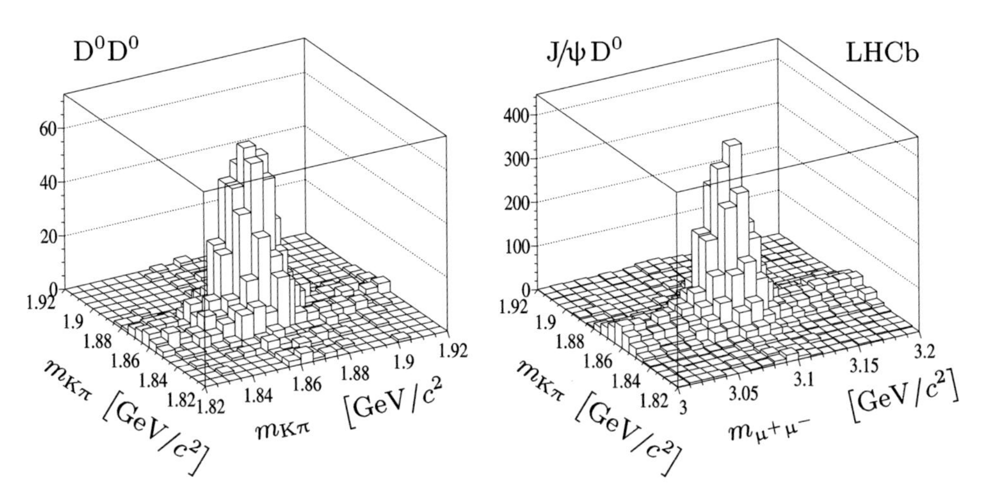

Fig. 3 Mass distributions for (left) D^0D^0 and (right) $J/\psi\, D^0$ signals.

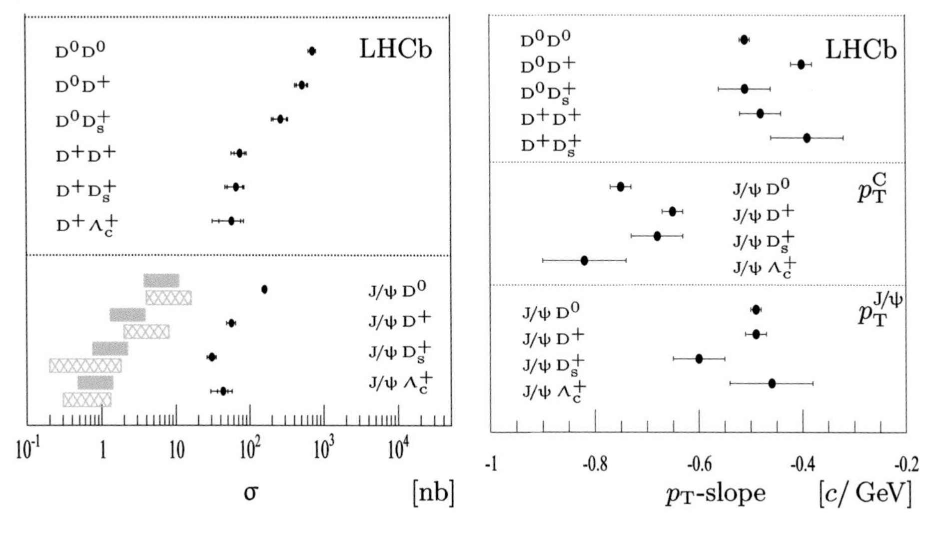

Fig. 4 (left) Measured cross-sections σ_{CC}, $\sigma_{C\overline{C}}$ and $\sigma_{J/\psi\, C}$ (points with error bars). For $J/\psi\, C$ channels, the SPS theory calculations are shown with hatched[228,229] and shaded[230] areas. The inner error bars indicate the statistical uncertainty whilst the outer error bars indicate the sum of the statistical and systematic uncertainties in quadrature. For the $J/\psi\, C$ case the outermost error bars correspond to the total uncertainties including the uncertainties due to the unknown polarization of the prompt J/ψ mesons. (right) The slope parameters for transverse momentum spectra of charm hadrons for CC and $J/\psi\, C$ events.

measured cross-sections. Also the SPS predictions for D^0D^0 are also approximately 30 times smaller[231,232] than the measured production rate. The observed ratio of cross-sections for CC and corresponding $C\overline{C}$ processes, proportional to $\frac{\sigma(c\bar{c}c\bar{c})}{\sigma(c\bar{c})}$, is close to 10%. This value is very large compared with e.g., those measured for $\frac{\sigma_{J/\psi\,J/\psi}}{\sigma_{J/\psi}} = (5.1 \pm 1.0\,(\text{stat.}) \pm 1.1\,(\text{syst.})) \times 10^{-4}$ in the same kinematic region.[226]

The kinematic properties of CC and $J/\psi\,C$ events were studied. To compare the transverse momentum spectra of J/ψ and open-charm hadrons, the spectra were fit to an exponential function. The resulting slope parameters are shown in Fig. 4 (right). The transverse momentum spectra for J/ψ mesons from $J/\psi\,C$ events are significantly harder than those observed in prompt J/ψ meson production[233] while the spectra for open-charm mesons in $J/\psi\,C$ case appear to be very similar to those observed[105] for the prompt charm hadrons. Similar transverse momenta for CC and $C\overline{C}$ events are observed. However these spectra are much harder than those measured[105] for prompt charm events. It indicates large correlations between transverse momenta of two charm hadrons, possibly due to a large role of the gluon splitting process. The rapidity and azimuthal angle distributions, presented in Fig. 5, do not exhibit visible correlations between the two charm hadrons in CC and

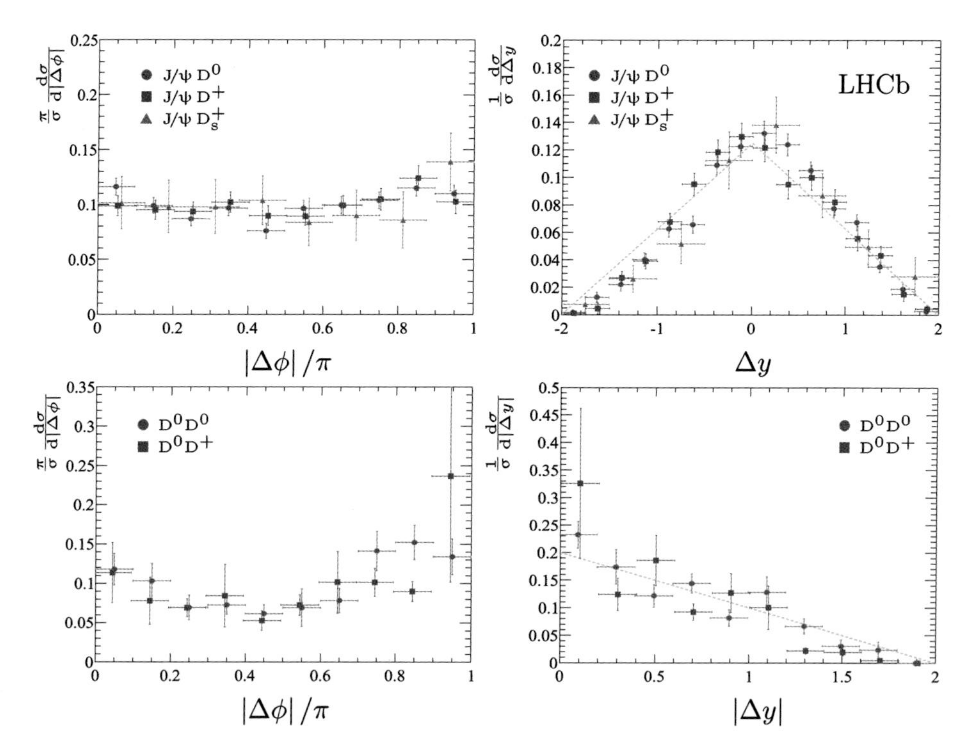

Fig. 5 Distributions of the difference in azimuthal angle (left) and rapidity (right) for $J/\psi\,C$ (top), D^0D^0 and D^0D^+ events (bottom). The dashed line shows the expected distribution for uncorrelated events.

J/ψ C events and are well consistent with uncorrelated production, supporting the dominant role of DPS contribution for CC and J/ψ C production.

Another study exhibiting a good separation power between the SPS and DPS mechanisms performed by the LHCb experiment is the study[106] of the associated production of Υ and an open-charm meson. For such process non-relativistic QCD (NR QCD) color-singlet (CS)[234] and k_T-factorization[229,235–242] predict the ratio $\mathscr{R}_{\Upsilon c\bar{c}} = \frac{\sigma(\Upsilon c\bar{c})}{\sigma(\Upsilon)}$ to be around (0.1–0.6)% in LHCb acceptance, while the DPS predicts this ratio to be of the order of 10%. The measurement was performed with the combined sample of data collected in pp collisions at the centre-of-mass energies of 7 and 8 TeV. Twelve different combinations of $\Upsilon(n\mathrm{S})$ ($n = 1, 2, 3$) and open-charm hadrons, C, were studied in the fiducial volume of $2 < y_\Upsilon, y_\mathrm{C} < 4.5$, $p_\mathrm{T}^\Upsilon < 15$ GeV/c and $1 < p_\mathrm{T}^\mathrm{C} < 20$ GeV/c. In five of them signal significance exceeds five standard deviations: $\Upsilon(1\mathrm{S})\mathrm{D}^0$, $\Upsilon(1\mathrm{S})\mathrm{D}^+$, $\Upsilon(1\mathrm{S})\mathrm{D}_s^+$, $\Upsilon(2\mathrm{S})\mathrm{D}^0$ and $\Upsilon(2\mathrm{S})\mathrm{D}^+$. Two of them with the highest signal yields, $\Upsilon(1\mathrm{S})\mathrm{D}^0$ and $\Upsilon(1\mathrm{S})\mathrm{D}^+$ were used for the production cross-section measurements and DPS studies. The ratios $\mathscr{R}_{\Upsilon c\bar{c}}$ were measured to be $(7.7 \pm 1.0)\%$ and $(8.0 \pm 0.9)\%$ for data sets collected at 7 and 8 TeV, respectively: these significantly exceed SPS calculations, and agrees with DPS expectations.

All differential cross-sections for $\Upsilon(1\mathrm{S})\mathrm{D}$ events nicely agree with expectations from DPS process. Figure 6 (left) shows the distribution of the azimuthal angle difference $\Delta\phi$ for $\Upsilon(1\mathrm{S})\mathrm{D}^0$ events together with both SPS and DPS expectations. The transverse momentum and rapidity distributions of $\Upsilon(1\mathrm{S})$ mesons also agree well with SPS predictions, obtained using k_T-factorization approach, while the shape

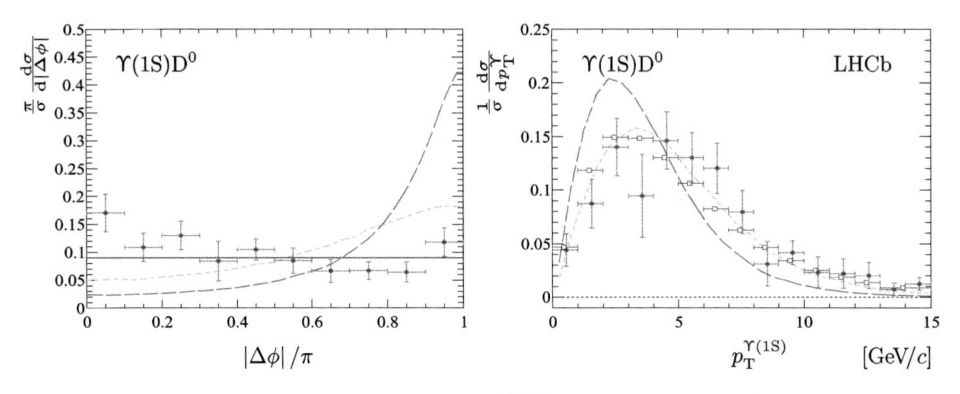

Fig. 6 Distributions for $|\Delta\phi|/\pi$ (left) and $p_\mathrm{T}^{\Upsilon(1\mathrm{S})}$ (right) for $\Upsilon(1\mathrm{S})\mathrm{D}^0$ events (solid red circles). Straight blue line in the $|\Delta\phi|/\pi$ plots shows the result of the fit with a constant function. The SPS predictions for the shapes of $\Delta\phi$ distribution are shown with dashed (orange) and long-dashed (magenta) curves for calculations based on the k_T-factorization and the collinear approximation, respectively. The transverse momentum spectra, derived within the DPS mechanism using the measured production cross-sections[105,243] for Υ and open-charm mesons, are shown with the open (blue) squares. All distributions are normalized to unity.

of the transverse momentum spectra of $\Upsilon(1S)$ mesons, see Fig. 6 (right), disfavors the SPS predictions obtained using the collinear approximation.

3. Double Quarkonia Studies

Unlike the DPS measurements with open-flavor hadrons, that are unique for the LHCb experiment, several experiments contributed to the study of the double quarkonia production. The dimuon decay of quarkonia combines relatively high branching fraction and experimentally favorable signature together with efficient triggering and low background.

First observed by the NA3 experiment in pion–nucleon[244] and proton–nucleon[245] interactions, the J/ψ pair production was later studied both in $p\bar{p}$ collisions at the Tevatron by D0 collaboration and in pp collisions at different energies at the LHC[136,226,227,246] by ATLAS, CMS and LHCb collaborations.

The first measurement in the pp collisions[226] was made by the LHCb experiment. The oppositely charged tracks identified as muons were combined to obtain pairs of J/ψ candidates (J/ψ_1 and J/ψ_2). To determine the signal event yield the invariant mass of the second muon pair was plotted in bins of the first pair invariant mass, see Fig. 7 (left). Fit to this distribution was performed with a function including a signal component for J/ψ and a component for the combinatorial background. Other sources of background were found to be negligible. In total 141 ± 19 J/ψ pairs were found with the signal significance exceeding 6 standard deviations.

The total cross-section in the region $2 < y^{J/\psi} < 4.5$, $p_T^{J/\psi} < 10\,\mathrm{GeV}/c$ was found to be $5.1 \pm 1.0\,(\mathrm{stat.}) \pm 1.1\,(\mathrm{syst.})\,\mathrm{nb}$. This value is not precise enough to distinguish[248,249] between the SPS and DPS contributions. The SPS contributions are calculated to be $4.0 \pm 1.2\,\mathrm{nb}$ and $4.6 \pm 1.1\,\mathrm{nb}$ in the leading-order[247,250,251]

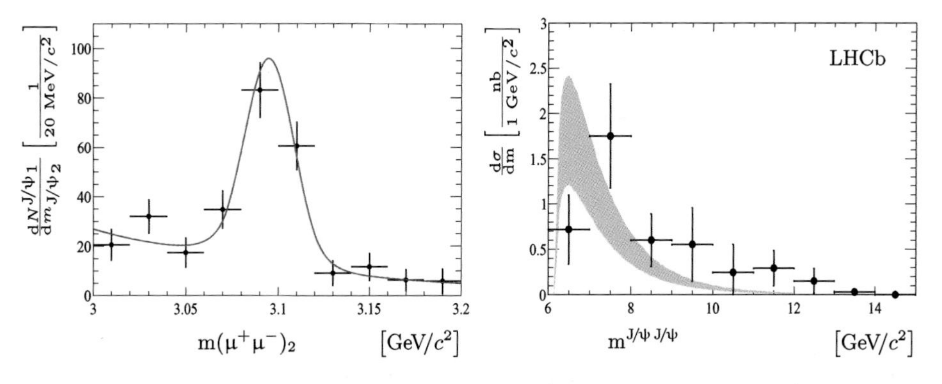

Fig. 7 (left) The fitted yield of $J/\psi \to (\mu^+\mu^-)_1$ in bins of $(\mu^+\mu^-)_2$ mass. (right) Differential production cross-section for J/ψ pairs as a function of mass of the J/ψ J/ψ system. The points correspond to the data. Only statistical uncertainties are included in the error bars. The shaded area corresponds to LO CS predictions.[247]

NR QCD CS approach, and $5.4^{+2.7}_{-1.1}$ nb using complete[251] NLO CS approach. The DPS contribution is expected to be 3.8 ± 1.3 nb. The differential production cross-section as a function of the mass of the J/ψ J/ψ system together with expectations from leading-order (LO) CS calculations is presented in Fig. 7 (right). The statistics did not allow to study the kinematic properties of these events and make firm conclusions.

Another measurement at $\sqrt{s} = 7$ TeV was performed[136] by the CMS collaboration, which included the kinematic regions where the color-octet (CO) models could play a more significant role in the double heavy quarkonium production. After excluding possible sources of background, they observed 446 ± 23 J/ψ pairs produced promptly in the same pp collision. The cross-section in the fiducial region was measured to be $1.49 \pm 0.07\,(\text{stat.}) \pm 0.13\,(\text{syst.})$ nb. The differential cross-sections as a function of the rapidity difference between the two J/ψ mesons, $|\Delta y|$, and $M_{J/\psi\, J/\psi}$, the mass of J/ψ J/ψ system, are shown in Fig. 8. These distributions are very sensitive to the DPS effects[248,249,252,253] and non-vanishing cross-sections for large $|\Delta y|$ and $M_{J/\psi\, J/\psi}$ bins could be considered as a sign of DPS mechanism.

These distributions have been analyzed[253] against incomplete NLO* CS predictions. From this analysis Lansberg and Shao[253] concluded the importance of α_s^5 contributions at medium and large transverse momenta and dominance of DPS mechanism at large $|\Delta y|$ and large $M_{J/\psi\, J/\psi}$. No significant CO contribution was found. The extracted value of σ_{eff} parameter has been found[253] to be $\sigma_{\text{eff}} = 11.0 \pm 2.9$ mb.

A measurement with higher centre-of-mass energy of 8 TeV has been performed by the ATLAS collaboration. In total they observed 1160 ± 70 promptly produced

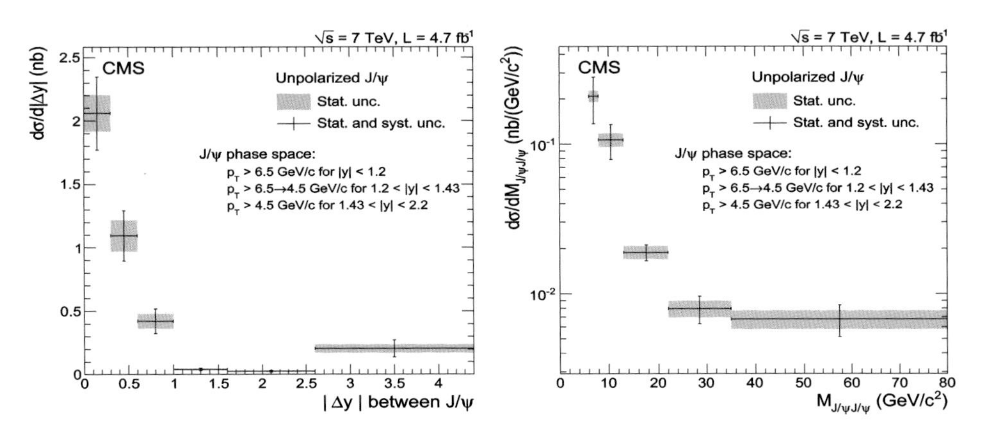

Fig. 8 Differential cross-section for prompt J/ψ pair production as a function of the absolute rapidity difference between J/ψ mesons (left) and the mass of J/ψ J/ψ system (right) The shaded regions represent the statistical uncertainties only, and the error bars represent the statistical and systematic uncertainties added in quadrature.

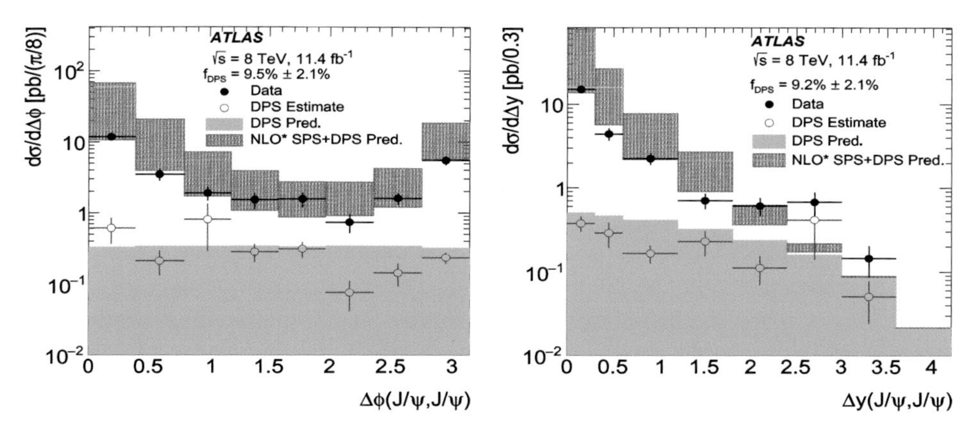

Fig. 9 The DPS and total differential cross-sections as a function of (left) the azimuthal angle between the two J/ψ mesons, and (right) the difference in rapidity between the two J/ψ mesons. Shown are the data as well as the DPS and NLO* SPS predictions. The DPS predictions[254] are normalized to the value of $f_{\rm DPS}$ found in the data and the NLO* SPS predictions are multiplied by a constant feed-down correction factor. The data-driven DPS-weighted distribution and the total data distribution are compared to the DPS theory prediction and the total SPS+DPS prediction.

J/ψ pairs. The total cross-section in the fiducial region was measured to be $160 \pm 12 \pm 14 \pm 2 \pm 3\,\text{pb}$, where the first uncertainty is statistical, the second one is systematic mainly due to the trigger efficiency determination. The third and fourth uncertainties are coming correspondingly from the known $J/\psi \to \mu\mu$ branching fraction uncertainty and the luminosity determination uncertainty. A data-driven model-independent technique was used to separate the DPS and the SPS contributions. The overall differential cross-section and the differential cross-section for DPS contribution are shown in Fig. 9.

The measured DPS-sensitive distributions were compared and found in a good agreement with the DPS predictions,[254] while the SPS distributions have shown some discrepancy with the NLO* predictions[253,255] which however could be explained by feed-down from higher charmonium states. The DPS fraction was found to be $f_{\rm DPS} = (9.2 \pm 2.1\,(\text{stat.}) \pm 0.5\,(\text{syst.}))\%$, resulting in

$$\sigma_{\rm eff} = 6.3 \pm 1.6 \pm 1.0 \pm 0.1 \pm 0.1\,\text{mb},$$

where the first uncertainty is statistical, the second one is systematic, the third one comes from the $J/\psi \to \mu\mu$ branching fraction and the fourth is an uncertainty on the luminosity determination.

One more study of the double J/ψ production was performed by the LHCb collaboration[227] at the centre-of-mass energy of 13 TeV. They observed $(1.05 \pm 0.05) \times 10^3$ signal J/ψ pairs. The total cross-section in the fiducial region was measured to be $15.2 \pm 1.0\,(\text{stat.}) \pm 0.9\,(\text{syst.})\,\text{nb}$. A large statistics of J/ψ pairs allows to study differential cross-sections and compare them with different theory

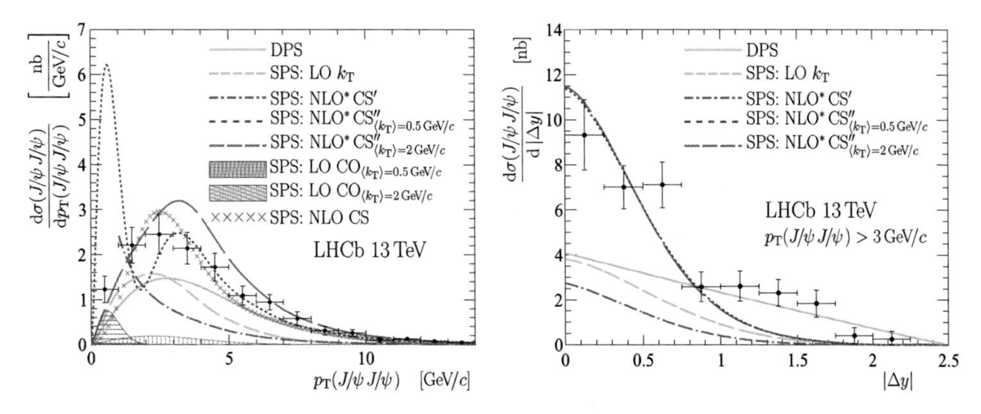

Fig. 10 Comparisons between measurements[227] and theoretical predictions[137,251,253,255–258] for the differential cross-sections as a function of (left) $p_T^{J/\psi\,J/\psi}$ and (right) $|\Delta y|$. The (black) points with error bars represent the measurements.

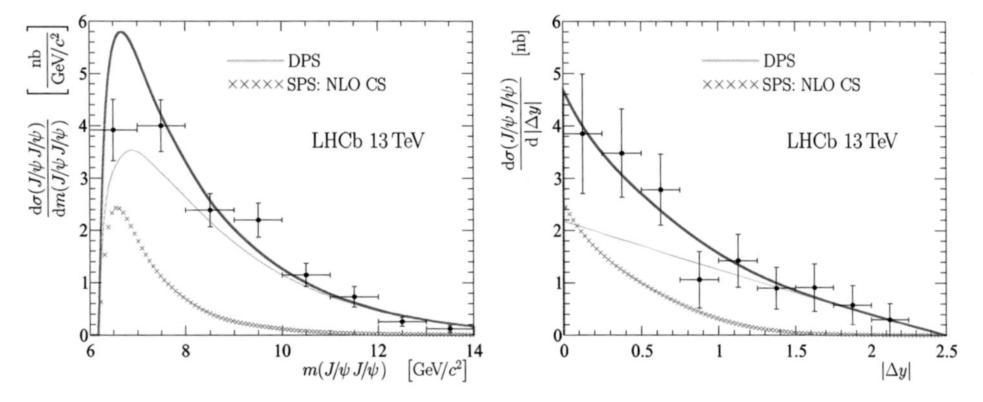

Fig. 11 Result of templated DPS fit for $\frac{d\sigma}{dm}$ (left) and $\frac{d\sigma}{d|\Delta y|}$ (right). The points with error bars represent the data. The total fit result is shown with the thick (red) solid line, the DPS component is shown with the thin (orange) solid line and the SPS component (full NLO CS) is shown with small (red) crosses.

models. Figure 10 shows the normalized differential cross-sections as a function of the transverse momentum of J/ψ J/ψ system and the rapidity difference between two J/ψ mesons.

The obtained cross-section value is interpreted as a sum of the SPS and DPS contributions, which were separated by a study of differential cross-sections. Fit to the data was performed with a function, corresponding to the sum of an SPS and DPS model predictions. The fraction f_{DPS} was treated as free parameters of the fit. Examples of DPS fit to $m(J/\psi\,J/\psi)$ and $|\Delta y|$ distributions using full NLO CS calculations for SPS model are shown in Fig. 11. For all fits the fractions of the DPS contribution were found to be higher than 50%, and for many cases compatible with 100%.

Table 1. Summary of the $\sigma_{\rm eff}$ values (in mb) from DPS fits[227] for different SPS models. The uncertainty is statistical only. The common systematic uncertainty of 12%, accounting for the systematic uncertainty of $\sigma\,(\mathrm{J/\psi\,J/\psi})$ and the total uncertainty for $\sigma(\mathrm{J/\psi})$, is not shown.

		NLO* CS″[137,253,255,257,258]		
Variable	LO $k_{\rm T}$[260]	$\langle k_{\rm T}\rangle = 2\,{\rm GeV}/c$	$\langle k_{\rm T}\rangle = 0.5\,{\rm GeV}/c$	NLO CS[251]
$p_{\rm T}^{\mathrm{J/\psi\,J/\psi}}$	11.3 ± 0.6	10.1 ± 6.5	10.9 ± 1.2	—
$y^{\mathrm{J/\psi\,J/\psi}}$	—	11.9 ± 7.5	10.0 ± 5.0	—
$m^{\mathrm{J/\psi\,J/\psi}}$	10.6 ± 1.1	10.2 ± 1.0		10.4 ± 1.0
$\lvert\Delta y\rvert$	12.5 ± 4.1	12.2 ± 3.7	12.4 ± 3.9	11.2 ± 2.9

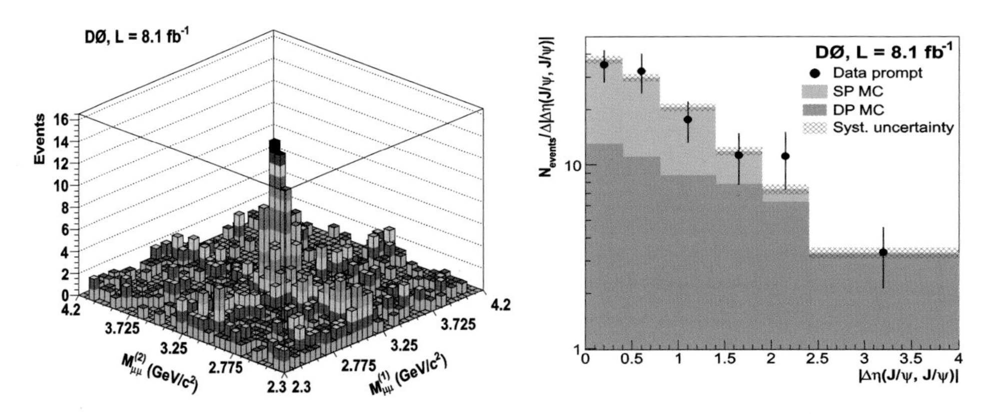

Fig. 12 (left) Dimuon invariant mass distribution in data for two muon pairs. (right) The distribution of the rapidity difference between two J/ψ candidates in data after background subtraction. The distributions for the SPS and DPS templates are shown normalized to their respective fitted fractions. The uncertainty band corresponds to the total systematic uncertainty on the sum of SP and DP events. Figure reprinted with permission from Ref. 159. Copyright 2014 by the American Physical Society.

The obtained values of $\sigma_{\rm DPS}$ and the known[259] J/ψ production cross-section were used to calculate $\sigma_{\rm eff}$. The obtained $\sigma_{\rm eff}$ values are shown in Table 1 depending on the choice of the SPS model used. They turned out to be slightly higher than values measured from the central J/ψ pair production. In spite of the large difference between different SPS models, the values of $\sigma_{\rm eff}$ exhibit only modest model dependency.

The D0 collaboration has for the first time observed the pair J/ψ production in $\mathrm{p\bar{p}}$ collisions at $\sqrt{s} = 1.96$ TeV at Tevatron,[159] see Fig. 12 (left). After the selection 242 pairs of J/ψ mesons were found, about 40% of which were determined to be prompt signal J/ψ pairs. The cross-section in the fiducial region, $\sigma^{\mathrm{J/\psi\,J/\psi}}$, was measured to be $129 \pm 11\,(\mathrm{stat.}) \pm 37\,(\mathrm{syst.})\,\mathrm{fb}$. In order to study the SPS and DPS contributions separately, they were distinguished by fitting the $\lvert\Delta y\rvert$ distribution by a sum of SPS and DPS templates with the corresponding fractions used as the free parameters, see Fig. 12 (right). The DPS fraction f_{DPS} was found to be $(42 \pm 12)\%$, leading to the following values of the cross-sections: $59 \pm 6\,(\text{stat.}) \pm 22\,(\text{syst.})$ fb and $70 \pm 6\,(\text{stat.}) \pm 22\,(\text{syst.})$ fb for SPS and DPS processes, respectively. The corresponding σ_{eff} was measured to be $4.8 \pm 0.5\,(\text{stat.}) \pm 2.5\,(\text{syst.})$ mb.

The measurements of associated productions involving bottomonia resonances were performed by D0 and CMS collaborations. The D0 collaboration observed the associated J/ψ and Υ production[200] using 8 fb^{-1} of data collected at $\sqrt{s} = 1.96$ TeV in $p\bar{p}$ collisions. This combination is expected to be produced predominantly through DPS mechanism with SPS contribution suppressed[137,248] by an additional powers of α_s. It results in the hierarchy of expected SPS production cross-sections $\sigma_{\text{SPS}}^{J/\psi\,\Upsilon} < \sigma_{\text{SPS}}^{\Upsilon\Upsilon} < \sigma_{\text{SPS}}^{J/\psi\,J/\psi}$ that is opposite to more intuitive DPS hierarchy $\sigma_{\text{DPS}}^{\Upsilon\Upsilon} < \sigma_{\text{DPS}}^{J/\psi\,\Upsilon} < \sigma_{\text{DPS}}^{J/\psi\,J/\psi}$. That makes the observation of J/ψ and Υ pair production and the precise measurement of the hierarchy of the production cross-sections very important.

Dimuon invariant mass distribution in data for two muon pairs in the fiducial region of $p_T^{\mu} > 2\,\text{GeV}/c$ and $|\eta^{\mu}| < 2$ is shown in Fig. 13 (left) together with the two-dimensional fit surface. In total $12 \pm 3.8\,(\text{stat.}) \pm 2.8\,(\text{syst.})$ promptly produced J/ψ Υ pairs are observed, corresponding to the significance of 3.2σ.

The distribution of azimuthal angle between J/ψ and Υ mesons is presented in Fig. 13 (right). This distribution showed a good agreement with the dominance of DPS mechanism. It allows to translate the measured production cross-section for J/ψ Υ pairs of $27 \pm 9\,(\text{stat.}) \pm 7\,(\text{syst.})$ fb into the effective cross-section

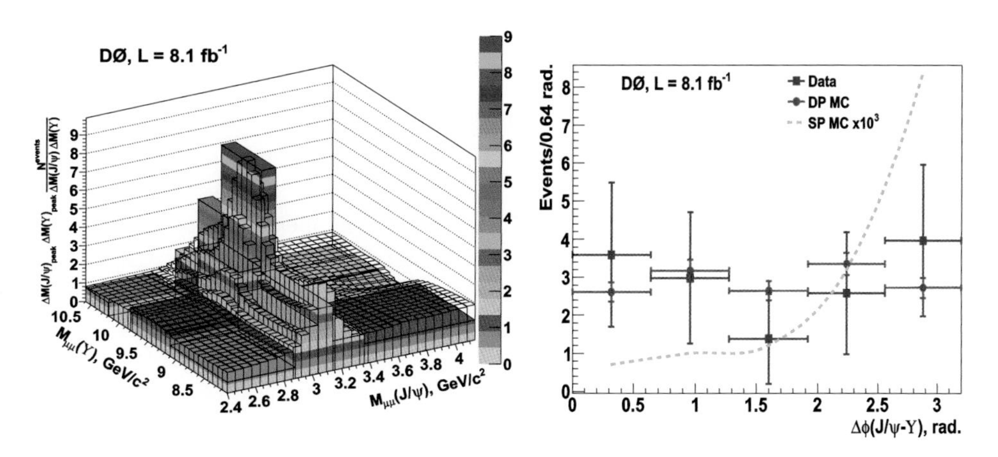

Fig. 13 (left) Dimuon invariant mass distribution in data for two muon pairs together with the two-dimensional fit surface. The scaling factor is applied so that the height of the peak bin is the number of observed events in that bin. (right) The distribution of the azimuthal angle between the J/ψ and Υ candidates in data after background subtraction. Also shown the expectations from DPS and SPS processes in arbitrary units.

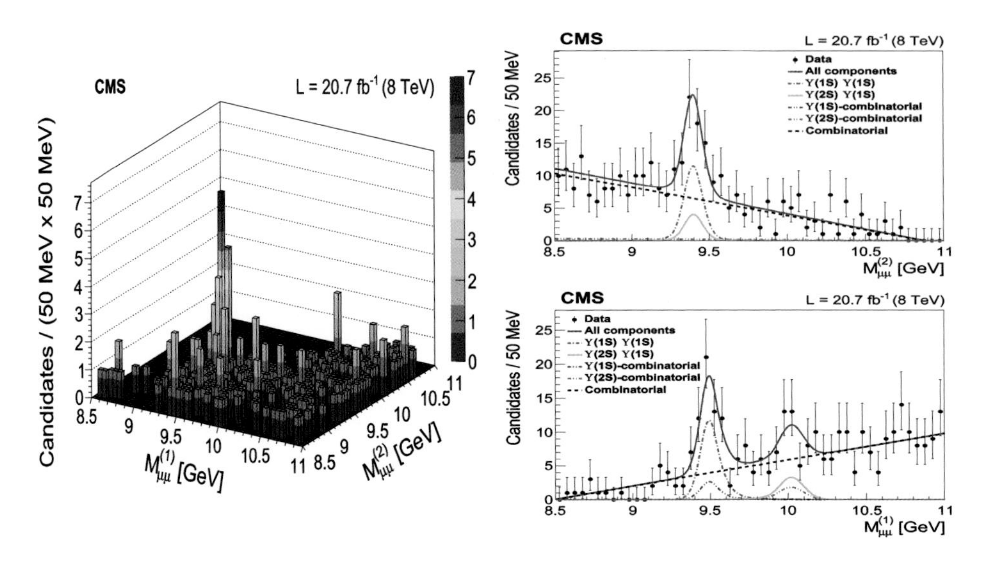

Fig. 14 (left) Dimuon invariant mass distribution in data for two muon pairs. (right) Invariant mass distributions of the lower-mass muon pair (top) and the higher-mass muon pair (bottom). The data are shown by the points. The different curves show the contributions of the various event categories from the fit.

$\sigma_{\text{eff}} = 2.2 \pm 0.7\,(\text{stat.}) \pm 0.9\,(\text{syst.})\,\text{mb}$. The obtained value of σ_{eff} is below the previous measurements involving heavy quarkonium.

The CMS collaboration made a first observation[261] of double bottomonium production using $20.7\,\text{fb}^{-1}$ of data collected in pp collisions at centre-of-mass energy of 8 TeV. The dimuon invariant mass distribution in data for two muon pairs in the fiducial region of $|y^{\Upsilon}| < 2$ is presented in Fig. 14 (left). A signal yield of 38 ± 7 pairs of $\Upsilon(1\text{S})$ mesons is determined from two-dimensional fit. The projections of two-dimensional fit are shown in Fig. 14 (right). The significance of the signal exceeds 5σ.

The production cross-section of $\Upsilon(1\text{S})$ pairs was measured to be $\sigma^{\Upsilon\Upsilon} = 68.8 \pm 12.7 \pm 7.4 \pm 2.8\,\text{pb}$, where the first uncertainty is statistical, the second one is systematic and the third one comes from the known value of $\Upsilon \to \mu^+\mu^-$ branching fraction. The DPS contribution in this channel is expected to be small[262] relative to the SPS one. Using a conservative estimate of $f_{\text{DPS}} = 10\%$, the effective cross-section was estimated to be 6.6 mb. On the other hand the SPS prediction[263] with feed-down from higher states gives a value of $\sigma_{\text{SPS}}^{\Upsilon\Upsilon} = 48\,\text{pb}$ which combined with the $\sigma^{\Upsilon\Upsilon}$ measured in data leads to $f_{\text{DPS}} = 30\%$ and, correspondingly $\sigma_{\text{eff}} = 2.2$ mb. Both estimates are in line with values obtained from other heavy quarkonium measurements. Higher statistics would allow to make a more precise conclusions.

4. Other DPS Measurements Involving Heavy Quarks

Besides the studies described above, there are other processes involving heavy quarks which can tell a lot about charm parton distribution inside the proton, charm production mechanism and the DPS. Good examples of such studies are measurements of an associated production of charm and a W or a Z boson. It should be noted that for these final states one probes the PDFs in the region of relatively high-x partons, where significant violation[204,205] of factorization hypothesis is expected.

The CMS and ATLAS collaborations studied[264,265] an associated production of a W boson with open-charm considering both a case of explicitly produced D meson and a case of jet initiated by the c quark. Each experiment used about $5\,\text{fb}^{-1}$ of data collected in pp collisions at the centre-of-mass energy of $\sqrt{s} = 7$ TeV. The total and differential cross-sections of these processes were measured and found to be consistent between the two experiments and with theoretical predictions.

The LHCb experiment observed[266] the associated production of a Z boson and an open-charm meson (D^0 or D^+) in the forward region using data collected at $\sqrt{s} = 7$ TeV. Seven candidate events for associated production of a Z boson with a D^0 meson and four candidate events for a Z boson with a D^+ meson were observed with a combined significance of 5.1σ. The fiducial cross-sections of these processes were measured, but the lack of statistics did not allow to disentangle the DPS and SPS contributions.

With higher statistics these studies promise to give more interesting and useful information for understanding the heavy-quark production mechanisms and in particular the DPS studies.

5. Summary

Nowadays a study of DPS using heavy quarks provides the most precise determination of the parameter σ_{eff}. Figure 15 summarizes all available measurements of σ_{eff} with heavy quarks.

There are no clear patterns in the measured values of σ_{eff}, however the values of σ_{eff} parameter measured with the double quarkonia final state are a bit lower than the reference value[97] of $\sigma_{\text{eff}} = 14.5 \pm 1.7^{+1.7}_{-2.3}$ mb measured in multi-jet events at the Tevatron. This could be a sign that the spatial region occupied by gluons within the proton is smaller than that occupied by quarks. On the other hand, the measurements[78] with two open-flavor hadrons give values of σ_{eff} larger than the reference value. These values are in very good agreement[95] with calculations. The measurements with quarkonia and open-flavor hadrons are well consistent with the reference value of σ_{eff} and all other measurements[71,199] of σ_{eff} using multi-jet, di-jet + W and γ + 3-jets processes. It should be noted that for many measurements

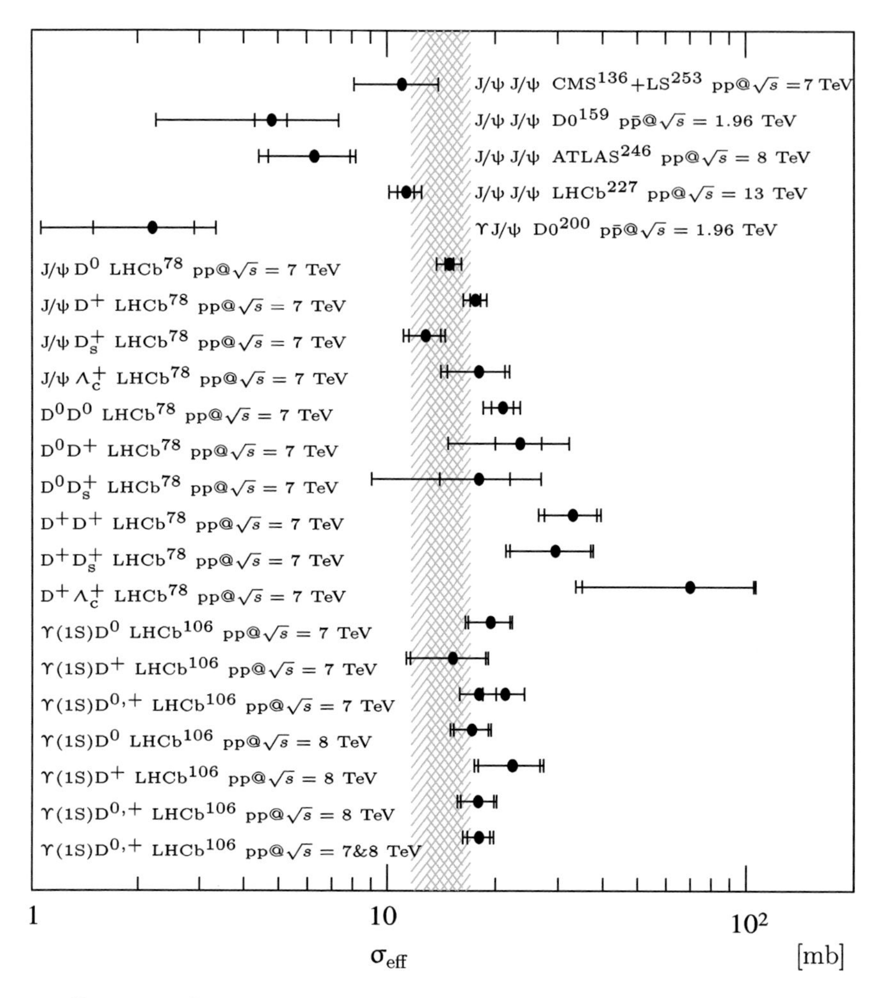

Fig. 15 Summary of $\sigma_{\rm eff}$ measurements with heavy quarks. The inner error bars indicate the statistical uncertainty whilst the outer error bars indicate the sum of statistical and systematic uncertainties in quadrature. The hatched area shows the reference value[97] of $\sigma_{\rm eff} = 14.5 \pm 1.7^{+1.7}_{-2.3}$ mb measured in multi-jet events at the Tevatron.

the uncertainties are large, and formally only one measurement[200] is not consistent with the reference value of $\sigma_{\rm eff}$. Better precision is needed to conclude on universality of $\sigma_{\rm eff}$. For large part of measurements, especially for production cross-section of $\Upsilon\Upsilon$ and $J/\psi\Upsilon$ pairs, the precision is limited by the statistics. Currently none of the LHC experiments used the full Run-I data sample for DPS studies. Extending these analyzes to full Run-I and subsequent larger Run-II data sets will allow significant improvement of the precision of $\sigma_{\rm eff}$ measurements and it will allow a definite conclusion on the universality of $\sigma_{\rm eff}$. Also the analysis of larger data sets will allow to study the DPS processes with open-beauty hadrons in the final state. A huge statistics of events with pairs of open-charm hadrons at LHCb could allow the precise measurement of σ_{eff} separately in the different kinematic regions, providing the ultimate test for the universality of σ_{eff}. It is worth to mention that for the precise measurements of DPS processes in pp collisions at $\sqrt{s} = 13$ TeV one will need to account for the contribution from the triple parton scattering process.[267]

Chapter 9

Double, Triple, and n-Parton Scatterings in High-Energy Proton and Nuclear Collisions

David d'Enterria* and Alexander Snigirev†

*CERN, EP Department, 1211 Geneva, Switzerland

†Skobeltsyn Institute of Nuclear Physics,
Lomonosov Moscow State University, 119991, Moscow, Russia

The framework to compute the cross-sections for the production of particles with high mass and/or large transverse momentum in double (DPS), triple (TPS), and in general n-parton scatterings from the corresponding single–parton ($\sigma_{\rm SPS}$) values in high-energy proton–proton, proton–nucleus, and nucleus–nucleus collisions is reviewed. The basic parameter of the factorized n-parton scattering ansatz is an effective cross-section $\sigma_{\rm eff}$ encoding all unknowns about the underlying generalized n-parton distribution in the proton (nucleon). In its simplest and most economical form, the $\sigma_{\rm eff}$ parameter can be derived from the transverse parton profile of the colliding protons and/or nucleus, using a Glauber approach. Numerical examples for the cross-sections and yields expected for the concurrent DPS or TPS production of heavy quarks, quarkonia, and/or gauge bosons in proton and nuclear collisions at LHC and Future Circular Collider (FCC) energies are provided. The obtained cross-sections are based on perturbative QCD predictions for $\sigma_{\rm SPS}$ at next-to-leading-order (NLO) or next-to-NLO (NNLO) accuracy including, when needed, nuclear modifications of the corresponding parton densities.

1. Introduction

The extended nature of hadrons and their growing parton densities when probed at increasingly higher collision energies make it possible to simultaneously produce multiple particles with large transverse momentum and/or mass ($\sqrt{p_{\rm T}^2 + m^2} \gtrsim 2$ GeV) in independent multiparton interactions (MPI)[35] in proton–(anti)proton (pp, p$\bar{\rm p}$),[88,89,198,199,268] as well as in proton–nucleus (pA),[80,93,127,269–278] and nucleus (AA)[276,277,279] collisions. Double, triple, and in general n-parton scatterings depend chiefly on the transverse overlap of the matter densities of the colliding hadrons and provide valuable information on (i) the badly known three-dimensional (3D) profile of the partons inside the nucleon, (ii) the unknown energy evolution of the parton density as a function of impact parameter (b), and (iii) the role of multiparton (space, momentum, flavor, color,...) correlations in the hadronic wave functions. A good understanding of n-parton scattering (NPS) is not only useful to improve our knowledge of the 3D parton structure of the proton, but it is also of relevance for a realistic characterization of backgrounds in searches of new physics in rare final states with multiple heavy particles.

The interest in MPI has increased in the last years not only as a primary source of particle production at hadron colliders,[280] but also due to their role[281] in the "collective" partonic behavior observed in "central" pp collisions, bearing close similarities to that measured in heavy-ion collisions.[282,283] As a matter of fact, the larger transverse parton density in a nucleus (with A nucleons) compared to that of a proton significantly enhances double (DPS) and triple (TPS) parton scattering cross-sections coming from interactions where the colliding partons belong to the same or different nucleons of the nucleus (nuclei), providing thereby additional information on the underlying multiparton dynamics.

Many final states involving the concurrent production of, e.g., heavy quarks (c, b), quarkonia (J/ψ, ϒ), jets, and gauge bosons (γ, W, Z) have been measured and found consistent with DPS at the Tevatron (see, e.g., early results from CDF[97,284] and more recent ones from D0[148,174]) as well as at the LHC (see, e.g., the latest results from ATLAS,[71,246] CMS[103,152] and LHCb[106]). The TPS processes, although not observed so far, have visible cross-sections for charm and bottom in pp[43] and pA[278] collisions at LHC and future circular (FCC)[285,286] energies. The present writeup reviews and extends our past work on DPS and TPS in high-energy pp, pA and AA collisions,[43,275–279] expanding the basic factorized formalism to generic NPS processes, and presenting realistic cross-section estimates for the double and triple parton production of heavy quarks, quarkonia, and/or gauge bosons in proton and nuclear collisions at LHC and FCC.

2. n-Parton Scattering Cross-Sections in Hadron–Hadron Collisions

In a generic hadronic collision, the inclusive cross-section to produce n hard particles in n independent hard parton scatterings, $hh' \to a_1 \dots a_n$, can be written as a convolution of generalized n-parton distribution functions (PDF) and elementary partonic cross-sections summed over all involved partons,

$$\sigma^{\rm NPS}_{hh'\to a_1\dots a_n} = \left(\frac{m}{n!}\right) \sum_{i_1,\dots,i_n,i'_1,\dots,i'_n} \int \Gamma_h^{i_1\dots i_n}(x_1,\dots,x_n;\mathbf{b_1},\dots,\mathbf{b_n};Q_1^2,\dots,Q_n^2)$$
$$\times\, \hat\sigma_{a_1}^{i_1 i'_1}(x_1,x'_1,Q_1^2)\,\dots\,\hat\sigma_{a_n}^{i_n i'_n}(x_n,x'_n,Q_n^2)$$

$$\times\,\Gamma_{h'}^{i'_1\ldots i'_n}(x'_1,\ldots,x'_n;\mathbf{b_1}-\mathbf{b},\ldots,\mathbf{b_n}-\mathbf{b};Q_1^2,\ldots,Q_n^2)$$
$$\times\,dx_1\ldots dx_n\,dx'_1,\ldots,dx'_n\,d^2b_1,\ldots,d^2b_n\,d^2b. \quad (1)$$

Here, $\Gamma_h^{i_1,\ldots,i_n}(x_1,\ldots,x_n;\mathbf{b_1},\ldots,\mathbf{b_n};Q_1^2,\ldots,Q_n^2)$ are n-parton generalized distribution functions, depending on the momentum fractions $x_1,\ldots,x_n$, and energy scales $Q_1,\ldots,Q_n$, at transverse positions $\mathbf{b_1},\ldots,\mathbf{b_n}$ of the $i_1,\ldots,i_n$ partons, producing final-state particles $a_1,\ldots,a_n$ with subprocess cross-sections $\hat{\sigma}_{a_1}^{i_1i'_1},\ldots,\hat{\sigma}_{a_n}^{i_ni'_n}$. The combinatorial $(m/n!)$ prefactor takes into account the different cases of (indistinguishable or not) final states. For a set of identical particles (i.e., when $a_1=\cdots=a_n$) we have $m=1$, whereas $m=2,3,6,\ldots$ for final states with an increasing number of different particles produced. In the particular cases of interest here, we have

- DPS: $m=1$ if $a_1=a_2$; and $m=2$ if $a_1\neq a_2$.
- TPS: $m=1$ if $a_1=a_2=a_3$; $m=3$ if $a_1=a_2$, or $a_1=a_3$, or $a_2=a_3$; and $m=6$ if $a_1\neq a_2\neq a_3$.

The n-parton distribution function $\Gamma_h^{i_1,\ldots,i_n}(x_1,\ldots,x_n;\mathbf{b_1},\ldots,\mathbf{b_n};Q_1^2,\ldots,Q_n^2)$ theoretically encodes all the 3D parton structure information of the hadron of relevance to compute the NPS cross-sections, including the density of partons in the transverse plane and any intrinsic partonic correlations in kinematical and/or quantum-numbers spaces. Since $\Gamma_h^{i_1\ldots i_n}$ is potentially a very complicated object, one often resorts to simplified alternatives to compute NPS cross-sections based on simpler quantities. As a matter of fact, without any loss of generality, any n-parton cross-section can be always expressed in a more economical and phenomenologically useful form in terms of single parton scattering (SPS) inclusive cross-sections, theoretically calculable in perturbative quantum chromodynamics (pQCD) approaches through collinear factorization[287] as a function of "standard" (longitudinal) PDF, $D_h^i(x,Q^2)$, at a given order of accuracy in the QCD coupling expansion (next-to-next-to-leading order, NNLO, being the current state-of-the-art for most calculations):

$$\sigma_{hh'\to a}^{\mathrm{SPS}}=\sum_{i_1,i_2}\int D_h^{i_1}(x_1;Q_1^2)\,\hat{\sigma}_a^{i_1i_2}(x_1,x'_1)\,D_{h'}^{i_2}(x'_1;Q_1^2)dx_1dx'_1. \quad (2)$$

More precisely, any n-parton cross-section can be expressed as the nth-product of the corresponding SPS cross-sections for the production of each single final-state particle, normalized by the $(n-1)$th power of an effective cross-section,

$$\sigma_{hh'\to a_1\ldots a_n}^{\mathrm{NPS}}=\left(\frac{m}{n!}\right)\frac{\sigma_{hh'\to a_1}^{\mathrm{SPS}}\cdots\sigma_{hh'\to a_n}^{\mathrm{SPS}}}{\sigma_{\mathrm{eff,NPS}}^{n-1}}, \quad (3)$$

where $\sigma_{\mathrm{eff,NPS}}$ encodes all the unknowns related to the underlying generalized PDF. Equation (3) encapsulates the intuitive result that the probability to produce n particles in a given inelastic hadron–hadron collision should be proportional to the n-product of probabilities to independently produce each one of them, normalized by the $(n-1)$th power of an effective cross-section to guarantee the proper units of the final result (3).[a]

The value of $\sigma_{\text{eff,NPS}}$ in Eq. (3) can be theoretically estimated making a few common approximations. First, the n-PDF are commonly assumed to be factorizable in terms of longitudinal and transverse components, i.e.,

$$\begin{aligned}&\Gamma_h^{i_1..i_n}(x_1,\ldots,x_n;\mathbf{b_1},\ldots,\mathbf{b_n};Q_1^2,\ldots,Q_n^2)\\&\quad = D_h^{i_1..i_n}(x_1,\ldots,x_n;Q_1^2,\ldots,Q_n^2)\cdot f(\mathbf{b_1})\cdots f(\mathbf{b_n}),\end{aligned} \tag{4}$$

where $f(\mathbf{b_1})$ describes the transverse parton density of the hadron, often considered a universal function for all types of partons, from which the corresponding hadron–hadron overlap function can be derived:

$$T(\mathbf{b}) = \int f(\mathbf{b_1})f(\mathbf{b_1}-\mathbf{b})d^2b_1, \tag{5}$$

with the fixed normalization $\int T(\mathbf{b})d^2b = 1$. Making the further assumption that the longitudinal components reduce to the product of independent single PDF,

$$D_h^{i_1,\ldots,i_n}(x_1,\ldots,x_n;Q_1^2,\ldots,Q_n^2) = D_h^{i_1}(x_1;Q_1^2)\cdots D_h^{i_n}(x_n;Q_n^2), \tag{6}$$

the effective NPS cross-section bears a simple geometric interpretation in terms of powers of the inverse of the integral of the hadron–hadron overlap function over all impact parameters,

$$\sigma_{\text{eff,NPS}} = \left\{\int d^2b\, T^n(\mathbf{b})\right\}^{-1/(n-1)}. \tag{7}$$

3. Double and Triple Parton Scattering Cross-Sections in Hadron–Hadron Collisions

The generalized expression (1) for the case of double parton scattering cross-sections in hadron–hadron collisions, $hh' \to a_1a_2$, reads

$$\begin{aligned}\sigma_{hh'\to a_1a_2}^{\text{DPS}} = \left(\frac{m}{2}\right)\sum_{i,j,k,l}\int &\Gamma_h^{ij}(x_1,x_2;\mathbf{b_1},\mathbf{b_2};Q_1^2,Q_2^2)\\ &\times\hat{\sigma}_{a_1}^{ik}(x_1,x_1',Q_1^2)\cdot\hat{\sigma}_{a_2}^{jl}(x_2,x_2',Q_2^2)\end{aligned}$$

[a]Indeed, in the simplest DPS case, the probability to produce particles a, b in a pp collision is $P_{\text{pp}\to ab} = P_{\text{pp}\to a}\cdot P_{\text{pp}\to b} = \frac{\sigma_{\text{pp}\to a}}{\sigma_{\text{pp}}^{\text{inel}}}\cdot\frac{\sigma_{\text{pp}\to b}}{\sigma_{\text{pp}}^{\text{inel}}}$, which implies $\sigma_{\text{pp}\to a,b} = \frac{\sigma_{\text{pp}\to a}\cdot\sigma_{\text{pp}\to b}}{\sigma_{\text{eff}}}$, with $\sigma_{\text{eff}} \approx \sigma_{\text{pp}}^{\text{inel}}$. In reality, the measured value of $\sigma_{\text{eff}} \approx 15$ mb is a factor of 2–3 lower (i.e., the DPS probability is 2–3 times *larger*) than the naive $\sigma_{\text{eff}} \approx \sigma_{\text{pp}}^{\text{inel}}$ expectation for typical "hard" (minijet) inelastic pp partonic cross-sections $\sigma_{\text{pp}}^{\text{inel}} \approx$ 30–50 mb. This is so because the independent-scattering assumption does not hold as the probability to produce a second particle is higher in low-impact-parameter (large transverse overlap) pp events where a first partonic scattering has already taken place.

$$\times \Gamma_{h'}^{kl}(x'_1, x'_2; \mathbf{b_1} - \mathbf{b}, \mathbf{b_2} - \mathbf{b}; Q_1^2, Q_2^2)$$
$$\times\, dx_1 dx_2 dx'_1 dx'_2 d^2b_1 d^2b_2 d^2b. \tag{8}$$

Applying the "master" equations (3) and (7) for $n = 2$, one can express this cross-section as a double product of independent single inclusive cross-sections

$$\sigma^{\rm DPS}_{hh'\to a_1a_2} = \left(\frac{m}{2}\right) \frac{\sigma^{\rm SPS}_{hh'\to a_1} \cdot \sigma^{\rm SPS}_{hh'\to a_2}}{\sigma_{\rm eff,DPS}}, \tag{9}$$

where the effective DPS cross-section (7) that normalizes the double SPS product is

$$\sigma_{\rm eff,DPS} = \left[\int d^2b\, T^2(\mathbf{b})\right]^{-1}. \tag{10}$$

Similarly, the generic expression (1) for the TPS cross-section for the process $hh' \to a_1a_2a_3$ reads[288]

$$\sigma^{\rm TPS}_{hh'\to a_1a_2a_3} = \left(\frac{m}{3!}\right) \sum_{i,j,k,l,m,n} \int \Gamma_h^{ijk}(x_1, x_2, x_3; \mathbf{b_1}, \mathbf{b_2}, \mathbf{b_3}; Q_1^2, Q_2^2, Q_3^2)$$
$$\times\, \hat{\sigma}^{il}_{a_1}(x_1, x'_1, Q_1^2) \cdot \hat{\sigma}^{jm}_{a_2}(x_2, x'_2, Q_2^2) \cdot \hat{\sigma}^{kn}_{a_3}(x_3, x'_3, Q_3^2)$$
$$\times\, \Gamma_{h'}^{lmn}(x'_1, x'_2, x'_3; \mathbf{b_1} - \mathbf{b}, \mathbf{b_2} - \mathbf{b}, \mathbf{b_3} - \mathbf{b}; Q_1^2, Q_2^2, Q_3^2)$$
$$\times\, dx_1 dx_2 dx_3 dx'_1 dx'_2 dx'_3 d^2b_1 d^2b_2 d^2b_3 d^2b, \tag{11}$$

which can be reduced to a triple product of independent single inclusive cross-sections

$$\sigma^{\rm TPS}_{hh'\to a_1a_2a_3} = \left(\frac{m}{3!}\right) \frac{\sigma^{\rm SPS}_{hh'\to a_1} \cdot \sigma^{\rm SPS}_{hh'\to a_2} \cdot \sigma^{\rm SPS}_{hh'\to a_3}}{\sigma^2_{\rm eff,TPS}}, \tag{12}$$

normalized by the *square* of an effective TPS cross-section (7), that amounts to[43]

$$\sigma^2_{\rm eff,TPS} = \left[\int d^2b\, T^3(\mathbf{b})\right]^{-1}. \tag{13}$$

One can estimate the values of the effective DPS (10) and TPS (13) cross-sections via Eq. (5) for different transverse parton profiles of the colliding hadrons, such as those typically implemented in the modern pp Monte Carlo (MC) event generators Pythia 8,[192] and Herwig.[110] In Pythia 8, the pp overlap function as a function of impact parameter is often parameterized in the form:

$$T(\mathbf{b}) = \frac{m}{2\pi r_p^2\, \Gamma(2/m)} \exp\left[-(b/r_p)^m\right], \tag{14}$$

normalized to one, $\int T(\mathbf{b}) d^2b = 1$, where r_p is the characteristic "radius" of the proton, Γ is the gamma function, and the exponent m depends on the MC "tune" obtained from fits to the measured underlying-event activity and various DPS cross-sections in pp collisions.[152] It varies between a pure Gaussian ($m = 2$) to a more peaked exponential-like ($m = 0.7, 1$) distribution. From the corresponding integrals of the square and cube of $T(\mathbf{b})$, we obtain:

$$\sigma_{\text{eff,DPS}} = \left(\int d^2b \, T^2(\mathbf{b}) \right)^{-1} = 2\pi r_p^2 \frac{2^{2/m}\Gamma(2/m)}{m}, \quad \text{and} \tag{15}$$

$$\sigma_{\text{eff,TPS}} = \left(\int d^2b \, T^3(\mathbf{b}) \right)^{-1/2} = 2\pi r_p^2 \frac{3^{1/m}\Gamma(2/m)}{m}. \tag{16}$$

From Eq. (15), in order to reproduce the experimental $\sigma_{\text{eff,DPS}} \simeq 15 \pm 5$ mb value extracted in multiple DPS measurements at Tevatron[97,148,174,284] and LHC,[71,88,103,106,152,246] the characteristic proton "radius" parameter amounts to $r_p \simeq 0.11 \pm 0.02, 0.24 \pm 0.04, 0.49 \pm 0.08$ fm for exponents $m = 0.7, 1, 2$ as defined in Eq. (14). The values of $\sigma_{\text{eff,DPS}}$ and $\sigma_{\text{eff,TPS}}$, Eqs. (15)–(16), are of course closely related: $\sigma_{\text{eff,TPS}} = (3/4)^{1/m} \cdot \sigma_{\text{eff,DPS}}$. Such a relationship is independent of the exact numerical value of the proton "size" r_p, but depends on the overall shape of its transverse profile characterized by the exponent m. For typical Pythia 8, $m = 0.7, 1, 2$ exponents tuned from experimental data,[152] one obtains $\sigma_{\text{eff,TPS}} = [0.66, 0.75, 0.87] \times \sigma_{\text{eff,DPS}}$ respectively.

The Herwig event generator uses an alternative parameterization of the proton profile described by the dipole fit of the two-gluon form factor in the momentum representation[82]

$$F_{2g}(\mathbf{q}) = 1/(q^2/m_g^2 + 1)^2, \tag{17}$$

where the gluon mass m_g parameter characterizes the transverse momentum q distribution of the proton, and the transverse density is obtained from its Fourier-transform: $f(\mathbf{b}) = \int e^{-i\mathbf{b}\cdot\mathbf{q}} F_{2g}(\mathbf{q}) \frac{d^2q}{(2\pi)^2}$. The corresponding DPS (10) and TPS (13) effective cross-sections read:[110]

$$\sigma_{\text{eff,DPS}} = \left[\int F_{2g}^4(q) \frac{d^2q}{(2\pi)^2} \right]^{-1} = \frac{28\pi}{m_g^2}, \tag{18}$$

and[43]

$$\sigma_{\text{eff,TPS}} = \left[\int (2\pi)^2 \delta(\mathbf{q_1} + \mathbf{q_2} + \mathbf{q_3}) F_{2g}(\mathbf{q_1}) F_{2g}(\mathbf{q_2}) F_{2g}(\mathbf{q_3}) \right.$$
$$\left. \times \, F_{2g}(-\mathbf{q_1}) F_{2g}(-\mathbf{q_2}) F_{2g}(-\mathbf{q_3}) \frac{d^2q_1}{(2\pi)^2} \frac{d^2q_2}{(2\pi)^2} \frac{d^2q_3}{(2\pi)^2} \right]^{-1/2}.$$

Numerically integrating the latter and combining it with (18), we obtain $\sigma_{\text{eff,TPS}} = 0.83 \times \sigma_{\text{eff,DPS}}$, which is quite close to the value derived for the Gaussian pp overlap function in Pythia 8. In order to reproduce the experimentally measured $\sigma_{\text{eff,DPS}} \simeq 15 \pm 5$ mb values, the characteristic proton "size" for this parameterization amounts to $r_g = 1/m_g \simeq 0.13 \pm 0.02$ fm.

Despite the wide range of proton transverse parton densities and associated effective radius parameters considered, we find that the $\sigma_{\text{eff,TPS}} \lesssim \sigma_{\text{eff,DPS}}$ result is robust with respect to the underlying parton profile. As a matter of fact, from the average and standard deviation of all typical parton transverse distributions studied in Ref. 43, the following relationship between double and triple scattering effective cross-sections can be derived:

$$\sigma_{\mathrm{eff,TPS}} = k \times \sigma_{\mathrm{eff,DPS}}, \quad \text{with } k = 0.82 \pm 0.11. \tag{19}$$

Thus, from the typical $\sigma_{\mathrm{eff,DPS}} \simeq 15 \pm 5$ mb value extracted from a wide range of DPS measurements at Tevatron and LHC, the following numerical effective TPS cross-section is finally obtained:

$$\sigma_{\mathrm{eff,TPS}} = 12.5 \pm 4.5 \text{ mb}. \tag{20}$$

3.1. *TPS cross-sections in pp collisions: Numerical examples*

Many theoretical and experimental studies exist that have extracted $\sigma_{\mathrm{eff,DPS}}$ from computed and/or measured DPS cross-sections for a large variety of final states in pp collisions (see other chapters in the DPS part of this book). In this subsection, we focus therefore on the TPS case for which we presented the first-ever estimates in Ref. 43. The experimental observation of triple parton scatterings in pp collisions requires perturbatively-calculable processes with SPS cross-sections not much smaller than $\mathcal{O}(1\ \mu\mathrm{b})$ since, otherwise, the corresponding TPS cross-sections (which go as the cube of the SPS values) are extremely reduced. Indeed, according to Eq. (12) with the data-driven estimate (20), a triple hard process pp $\to a\,a\,a$, with SPS cross-sections $\sigma^{\mathrm{SPS}}_{\mathrm{pp}\to a} \approx 1\ \mu$b, has a very small cross-section $\sigma^{\mathrm{TPS}}_{\mathrm{pp}\to a\,a\,a} \approx 1$ fb. Evidence for TPS appears thereby challenging already without accounting for additional reducing factors arising from decay branching ratios, and experimental acceptances and reconstruction inefficiencies, of the produced particles. Promising processes to probe TPS, with not too small pQCD cross-sections, are inclusive charm (pp $\to \mathrm{c\bar{c}} + X$), and bottom (pp $\to \mathrm{b\bar{b}} + X$), whose cross-sections are dominated by gluon–gluon fusion ($gg \to \mathrm{Q\bar{Q}}$) at small x, for which one can expect a non-negligible contributions of DPS[231,289,290] and TPS[43,267,278] to their total inclusive production (Fig. 1).

The TPS heavy-quark cross-sections can be computed with Eq. (12) for $m = 1$, i.e., $\sigma^{\mathrm{TPS}}_{\mathrm{pp}\to\mathrm{Q\bar{Q}}} = (\sigma^{\mathrm{SPS}}_{\mathrm{pp}\to\mathrm{Q\bar{Q}}})^3/(6\,\sigma^2_{\mathrm{eff,TPS}})$ with $\sigma_{\mathrm{eff,TPS}}$ given by (20), and $\sigma^{\mathrm{SPS}}_{\mathrm{pp}\to\mathrm{Q\bar{Q}}}$ calculated via Eq. (2) at NNLO accuracy using a modified version[291] of the `Top++` (v2.0) code,[292] with $N_f = 3, 4$ light flavors, heavy-quark pole masses set at $m_{c,b} = 1.67, 4.66$ GeV, default renormalization and factorization scales set at $\mu_R = \mu_F = 2\,m_{c,b}$, and using the ABMP16 proton PDF.[293] Such NNLO calculations increase the total SPS heavy-quark cross-sections by up to 20% at LHC energies compared to the corresponding NLO results,[294,295] reaching a better agreement with the experimental data, and featuring much reduced scale uncertainties ($\pm 50\%, \pm 15\%$ for $\mathrm{c\bar{c}}, \mathrm{b\bar{b}}$).[291] Figure 2 shows the resulting total SPS and TPS cross-sections for charm and bottom production over $\sqrt{s} = 35$ GeV–100 TeV, and Table 1

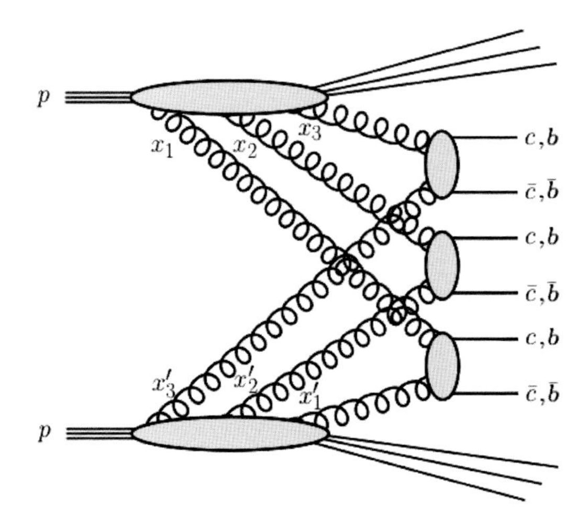

Fig. 1 Schematic diagram for the leading order contribution to triple charm ($c\bar{c}$) and bottom ($b\bar{b}$) pair production via gluon fusion, in TPS processes in pp collisions.

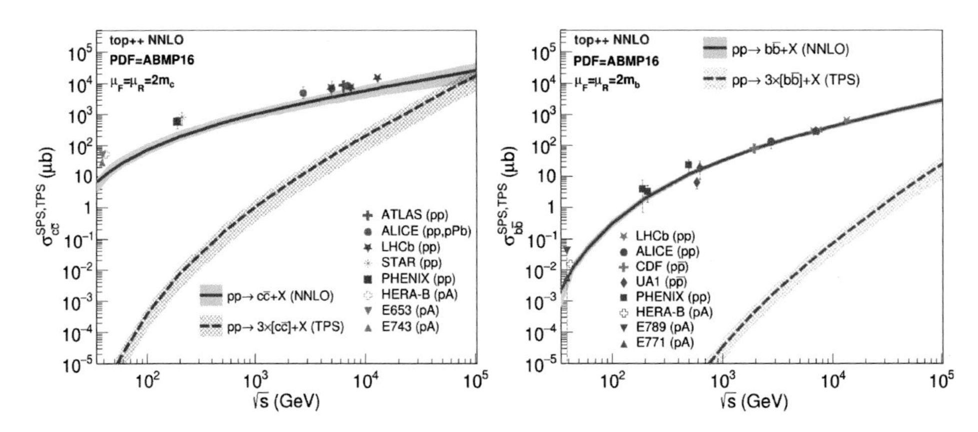

Fig. 2 Total charm (left) and bottom (right) cross-sections in pp collisions as a function of c.m. energy, in single parton (solid line) and triple parton (dashed line) parton scatterings. Bands around curves indicate scale, PDF (and $\sigma_{\text{eff,TPS}}$, in the case of σ^{TPS}) uncertainties added in quadrature. The symbols are experimental data collected in Ref. 291.

lists the results with associated uncertainties for the nominal pp c.m. energies at LHC and FCC. The PDF uncertainties are obtained from the corresponding 28 eigenvalues of the ABMP16 set. The dominant uncertainty comes from the theoretical scales dependence, which is estimated by modifying μ_R and μ_F within a factor of two.

Figure 2 shows that the TPS cross-sections rise fast with $\sqrt{s}$, as the cube of the corresponding SPS cross-sections. Triple-$c\bar{c}$ production from three independent parton scatterings amounts to 5% of the inclusive charm yields at the LHC

Table 1. Cross-sections for charm and bottom production in SPS (NNLO) and TPS processes in pp collisions at LHC and FCC energies. The quoted uncertainties include scales (sc), PDF, and total (quadratic, including $\sigma_{\rm eff,TPS}$) values.

Final state	$\sqrt{s} = 14$ TeV	$\sqrt{s} = 100$ TeV
$\sigma^{\rm SPS}_{c\bar{c}+X}$	$7.1 \pm 3.5_{\rm sc} \pm 0.3_{\rm PDF}$ mb	$25.0 \pm 16.0_{\rm sc} \pm 1.3_{\rm PDF}$ mb
$\sigma^{\rm TPS}_{c\bar{c}\,c\bar{c}\,c\bar{c}+X}$	$0.39 \pm 0.28_{\rm tot}$ mb	$16.7 \pm 11.8_{\rm tot}$ mb
$\sigma^{\rm SPS}_{b\bar{b}+X}$	$0.56 \pm 0.09_{\rm sc} \pm 0.01_{\rm PDF}$ mb	$2.8 \pm 0.6_{\rm sc} \pm 0.1_{\rm PDF}$ mb
$\sigma^{\rm TPS}_{b\bar{b}\,b\bar{b}\,b\bar{b}+X}$	$0.19 \pm 0.12_{\rm tot}$ μb	$24 \pm 17_{\rm tot}$ μb

($\sqrt{s} = 14$ TeV) and to more than half of the total charm cross-section at the FCC. Since the total pp inelastic cross-section at $\sqrt{s} = 100$ TeV is $\sigma_{\rm pp} \simeq 105$ mb,[296] charm-anticharm triplets are expected to be produced in ~15% of the pp collisions at these energies. Triple-$b\bar{b}$ cross-sections remain quite small and reach only about 1% of the inclusive bottom cross-section at FCC (100 TeV). These results indicate that TPS is experimentally observable in triple heavy-quark pair final states at the LHC and FCC. The possibility of detecting triple charm-meson production in pp collisions at the LHC has been discussed in more detail in Ref. 267.

4. Double and Triple Parton Scattering Cross-Sections in Proton–Nucleus Collisions

In proton–nucleus collisions, the parton flux is enhanced by the number A of nucleons in the nucleus and the SPS cross-section is simply expected to be that of proton–proton collisions or, more exactly, that of proton–nucleon collisions (pN, with $N = p, n$ being *bound* protons and neutrons with their appropriate relative fractions in the nucleus) taking into (anti)shadowing modifications of the nuclear PDF,[297] scaled by the factor A, i.e.,[298]

$$\sigma^{\rm SPS}_{\rm pA\to a} = \sigma^{\rm SPS}_{\rm pN\to a} \int d^2b\, T_{\rm pA}(\mathbf{b}) = A \cdot \sigma^{\rm SPS}_{\rm pN\to a}. \tag{21}$$

Here, $T_{\rm pA}(\mathbf{r})$ is the standard nuclear thickness function, analogous to Eq. (5) for the pp case, as a function of the impact parameter $\mathbf{r}$ between the colliding proton and nucleus, given by an integral of the nuclear density function $\rho_A(\mathbf{r})$ over the longitudinal direction

$$T_{\rm pA}(\mathbf{r}) = \int \rho_A\big(\sqrt{r^2+z^2}\big)\, dz, \text{ normalized to } \int T_{\rm pA}(\mathbf{r})\, d^2r = A, \tag{22}$$

which can be easily computed using (simplified) analytical nuclear profiles, and/or employing realistic Fermi–Dirac (aka. Woods–Saxon) nuclear spatial densities determined in elastic eA measurements,[299] via an MC Glauber model.[298]

The most naive assumption is to consider that the NPS cross-sections in pA collisions can be obtained by simply A-scaling the corresponding pp NPS values, as done via Eq. (21) for the SPS cross-sections. We show next that DPS and TPS cross-sections in proton–nucleus collisions can be significantly enhanced, with extra $A^{4/3}$ (for DPS and TPS) and $A^{5/3}$ (for TPS alone) terms complementing the A-scaling, due to additional multiple scattering probabilities among partons from different nucleons.

4.1. *DPS cross-sections in pA collisions*

The larger transverse parton density in nuclei compared to protons results in enhanced DPS cross-sections, pA→ ab, coming from interactions where the two partons of the nucleus belong to (1) the same nucleon, and (2) two different nucleons[80,93,269–275] as shown in Fig. 3. Namely,

$$\sigma_{\rm pA}^{\rm DPS} = \sigma_{\rm pA}^{\rm DPS,1} + \sigma_{\rm pA}^{\rm DPS,2}, \tag{23}$$

where

(1) the first term is just the A-scaled DPS cross-section in pN collisions:

$$\sigma_{\rm pA\to \mathit{ab}}^{\rm DPS,1} = A \cdot \sigma_{\rm pN\to \mathit{ab}}^{\rm DPS}, \tag{24}$$

(2) the second contribution, from parton interactions from two different nucleons, depends on the square of $T_{\rm pA}$,

$$\sigma_{\rm pA\to \mathit{ab}}^{\rm DPS,2} = \sigma_{\rm pN\to \mathit{ab}}^{\rm DPS} \cdot \sigma_{\rm eff,DPS} \cdot F_{\rm pA}, \tag{25}$$

$$\text{with } F_{\rm pA} = \frac{A-1}{A}\int T_{\rm pA}^2(\mathbf{r})\, d^2r = (A-1)/A \cdot T_{\rm AA}(0), \tag{26}$$

where the $(A-1)/A$ factor accounts for the difference between the number of nucleon pairs and the number of *different* nucleon pairs, and $T_{\rm AA}(0)$ is the nuclear overlap function at $b = 0$ for the corresponding AA collision. In the

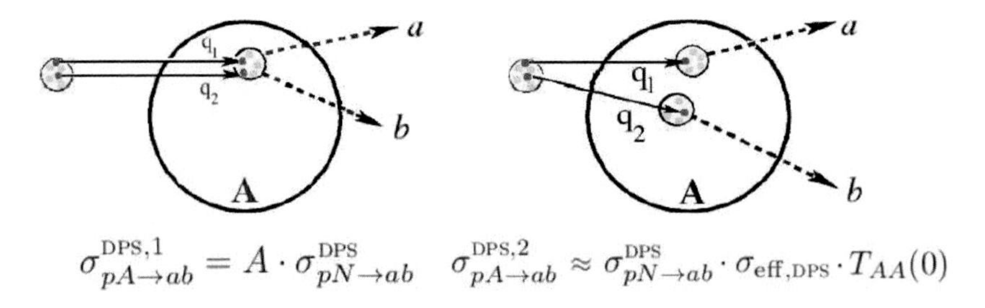

$$\sigma_{pA\to ab}^{\rm DPS,1} = A \cdot \sigma_{pN\to ab}^{\rm DPS} \qquad \sigma_{pA\to ab}^{\rm DPS,2} \approx \sigma_{pN\to ab}^{\rm DPS} \cdot \sigma_{\rm eff,DPS} \cdot T_{AA}(0)$$

Fig. 3 Schematic diagrams of DPS contributions in pA collisions where the two colliding partons belong to the same (left) or a different (right) pair of nucleons in the nucleus. The corresponding cross-sections are described in the text.

simplest approximation of a spherical nucleus with uniform nucleon density with radius $R_A \propto A^{1/3}$, the factor (26) can be written as

$$F_{pA} = \frac{9A(A-1)}{8\pi R_A^2} \approx \frac{A^{4/3}}{14\pi}\ [\text{mb}^{-1}], \tag{27}$$

where the second approximate equality (valid for large A) indicates the corresponding dependence on the A mass-number alone. For Pb, with $A = 208$ and $R_A \approx 7$ fm ≈ 22 mb$^{1/2}$, one obtains $F_{pA} \approx 31.5$ mb^{-1}, in good agreement with the more accurate result, $F_{pA} = 30.25\,\text{mb}^{-1}$, computed with a Glauber MC[298] using the standard Woods–Saxon spatial density of the lead nucleus (radius $R_A = 6.36\,\text{fm}$, and surface thickness $a = 0.54\,\text{fm}$).[299]

The sum of (24) and (25) yields the inclusive cross-section for the DPS production of particles a and b in a pA collision:

$$\sigma^{\text{DPS}}_{\text{pA}\to ab} = A \cdot \sigma^{\text{DPS}}_{\text{pN}\to ab}\,[1 + \sigma_{\text{eff,DPS}}\,F_{\text{pA}}/A] \tag{28}$$

$$\approx A \cdot \sigma^{\text{DPS}}_{\text{pN}\to ab}\left[1 + \frac{\sigma_{\text{eff,DPS}}}{14[\text{mb}]\pi}A^{1/3}\right], \tag{29}$$

which is enhanced by the factor in parentheses compared to the A-scaled DPS cross-section in pN collisions. Given the experimental $\sigma_{\text{eff,DPS}} \approx 15$ mb value, the pp-to-pA DPS enhancement factor can be further numerically simplified as $[1{+}A^{1/3}/\pi]$, which goes from ~1.4 for small to ~3 for large nuclei. Namely, the relative weight of the two DPS terms of Eq. (29) goes from $\sigma^{\text{DPS},1}_{\text{pA}\to ab} : \sigma^{\text{DPS},2}_{\text{pA}\to ab} = 0.7 : 0.3$ (small A) to $0.33 : 0.66$ (large A). Thus, e.g., in the case of p–Pb collisions, 33% of the DPS yields come from partonic interactions within just one nucleon of the Pb nucleus, whereas 66% of them involve parton scatterings from two different Pb nucleons.

The final factorized DPS formula in proton–nucleus collisions can be written as a function of the elementary proton–nucleon single parton cross-sections as

$$\sigma^{\text{DPS}}_{\text{pA}\to ab} = \left(\frac{m}{2}\right)\frac{\sigma^{\text{SPS}}_{\text{pN}\to a}\cdot\sigma^{\text{SPS}}_{\text{pN}\to b}}{\sigma_{\text{eff,DPS,pA}}}, \tag{30}$$

where the effective DPS pA cross-section in the denominator depends on the effective cross-section measured in pp, and on a pure geometric quantity (F_{pA}) that is directly derivable from the well-known nuclear transverse profile, namely

$$\sigma_{\text{eff,DPS,pA}} = \frac{\sigma_{\text{eff,DPS}}}{A + \sigma_{\text{eff,DPS}}\,F_{\text{pA}}} \approx \frac{\sigma_{\text{eff,DPS}}}{A + \sigma_{\text{eff,DPS}}\,T_{\text{AA}}(0)} \approx \frac{\sigma_{\text{eff,DPS}}}{A + A^{4/3}/\pi}. \tag{31}$$

For a Pb nucleus (with $A = 208$, and $F_{\text{pA}} = 30.25$ mb^{-1}) and taking $\sigma_{\text{eff,DPS}} = 15 \pm 5$ mb, one obtains $\sigma_{\text{eff,DPS,pA}} = 22.5 \pm 2.3$ μb. The overall increase of DPS cross-sections in pA compared to pp collisions is $\sigma_{\text{eff,DPS}}/\sigma_{\text{eff,DPS,pA}} \approx [A + A^{4/3}/\pi]$ which, in the case of p–Pb implies a factor of ~600 relative to pp (ignoring nuclear PDF effects here), i.e., a factor of $[1{+}A^{1/3}/\pi] \approx 3$ higher than the naive expectation assuming the same A-scaling of the single parton cross-sections, Eq. (21). One can thus exploit such large expected DPS signals over the SPS backgrounds in proton–nucleus collisions to study double parton scatterings in detail and, in particular, to extract the value of $\sigma_{\rm eff,DPS}$ independently of measurements in pp collisions — given that the parameter $F_{\rm pA}$ in Eq. (31) depends on the comparatively better-known transverse density of nuclei.

DPS cross-sections in pA collisions: Numerical examples

One of the "cleanest" channels to study DPS in pp collisions is same-sign WW production[175] as it features precisely-known pQCD SPS cross-sections, a clean experimental final state with two like-sign leptons plus missing transverse momentum from the undetected neutrinos, and small non-DPS backgrounds.[b] The DPS cross-section in p–Pb for same-sign WW production was first estimated in Ref. 275, computing the SPS $W^{\pm}$ cross-sections ($\sigma^{\rm SPS}_{\rm pN\to W}$) with MCFM (v.6.2)[300,301] at NLO accuracy with CT10 proton[302] and EPS09 nuclear[303] PDF, and setting default renormalization and factorization theoretical scales to $\mu = \mu_R = \mu_F = m_W$. The background $W^{\pm}W^{\pm}$jj cross-sections are computed with MCFM for the QCD part (formally at LO, but setting $\mu_R = \mu_F = 150$ GeV to effectively account for missing higher order corrections), and with VBFNLO (v.2.6)[304,305] for the electroweak contributions with theoretical scales set to the momentum transfer of the exchanged boson, $\mu^2 = t_{W,Z}$. In p–Pb at 8.8 TeV, the EPS09 nuclear PDF modifies the total W^+ (W^-) production cross-section by about -7% ($+15\%$) compared to that obtained using the free proton CT10 PDF.[306]

We extend here the results of Ref. 275, using Eq. (30) with $m = 1$ and $\sigma_{\rm eff,DPS,pA} = 22.5 \pm 2.3$ μb, and including FCC p–Pb energies ($\sqrt{s_{_{\rm NN}}} = 63$ TeV). The resulting cross-sections are listed in Table 2. The uncertainties of the SPS NLO single-W cross-sections amount to about $\pm 10\%$ by adding in quadrature those from the EPS09 PDF eigenvector sets (the proton PDF uncertainties are much lower in the relevant x, Q^2 regions) and from the theoretical scales (obtained by independently varying μ_R and μ_F within a factor of two). The QCD $W^{\pm}W^{\pm}$jj cross-sections uncertainties are those from the full-NLO calculations,[307] whereas those of the VBF cross-sections are much smaller as they do not involve any gluons in the initial state. The DPS cross-section uncertainties are dominated by a propagated $\pm 30\%$ uncertainty from $\sigma_{\rm eff,DPS}$.

Figure 4 shows the computed total cross-sections for all W processes considered over the c.m. energy $\sqrt{s_{_{\rm NN}}} = 2$–65 TeV range. At the nominal LHC p–Pb c.m. energy of 8.8 TeV, the same-sign WW DPS cross-section is $\sigma^{\rm DPS}_{\rm p-Pb\to WW} \approx 150$ pb (thick curve), larger than the sum of SPS backgrounds, $\sigma^{\rm SPS}_{\rm p-Pb\to WWjj}$ (lowest dashed curve) obtained adding the QCD and electroweak cross-sections for the production

[b]The lowest order at which two same-sign W bosons can be produced is accompanied with two jets ($W^{\pm}W^{\pm}$jj), $q\,q \to W^{\pm}W^{\pm}\,q'\,q'$ with $q = u, c, \ldots$ and $q' = d, s, \ldots$ whose leading contributions are $\mathcal{O}(\alpha_s^2\alpha_w^2)$ for the mixed QCD-electroweak diagrams, and $\mathcal{O}(\alpha_w^4)$ for the pure vector-boson fusion (VBF) processes, where α_w is the electroweak coupling.

Table 2. Cross-sections for the production of single-W, and same-sign W pairs in p–Pb collisions at LHC and FCC c.m. energies, computed at NLO with MCFM and VBFNLO for the processes quoted. The last column lists the same-sign DPS cross-sections (sum of positive and negative W pairs) obtained with Eq. (30) for $\sigma_{\text{eff,DPS,pA}} = 22.5 \pm 2.3\ \mu$b.

p–Pb	W^+, W^- NLO (μb)	W^+W^+jj (QCD), (VBF) NLO (pb)	$W^\pm W^\pm$ (DPS) (pb)
5.0 TeV	6.85 ± 0.68, 5.88 ± 0.59	12.1 ± 1.2, 12.4 ± 0.6	44 ± 13
8.8 TeV	12.6 ± 1.3, 11.1 ± 1.1	40.4 ± 4.0, 51.8 ± 2.0	152 ± 45
63 TeV	83.4 ± 8.4, 77.9 ± 7.8	$166. \pm 17.$, $2150. \pm 220.$	$6700. \pm 2000.$

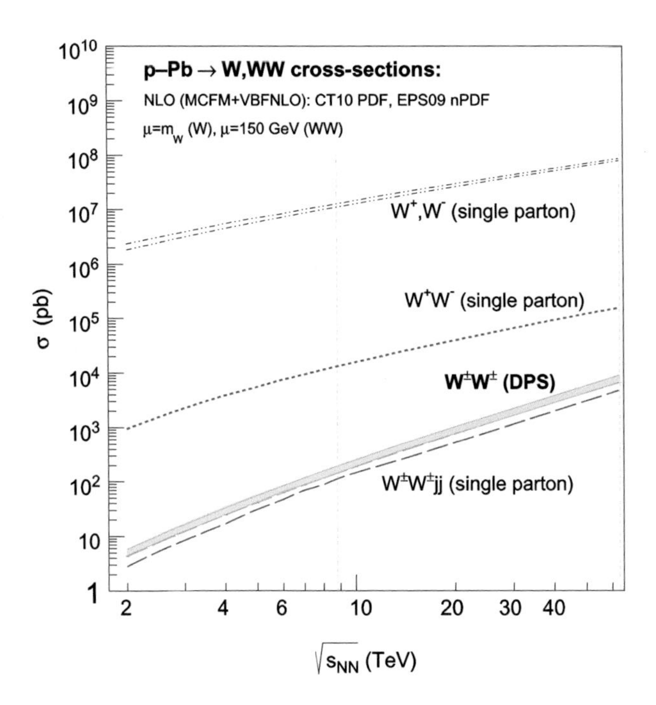

Fig. 4 Cross-sections as a function of c.m. energy for single-W, and W-pair (both opposite-sign and same-sign) production from SPS and from DPS in p–Pb collisions. Dotted vertical lines indicate the nominal 8.8 and 63 TeV p–Pb energies at LHC and FCC.

of W^+W^+ (W^-W^-) plus 2 jets. In the fully-leptonic final state ($W^\pm W^\pm \to \ell\nu\,\ell'\nu'$, with $\ell = e^\pm, \mu^\pm$) and accounting for decay branching ratios and standard ATLAS/CMS acceptance and reconstruction cuts ($|y^\ell| < 2.5$, $p_T^\ell > 15$ GeV), one expects up to 10 DPS same-sign WW events in $\mathscr{L}_{\text{int}} = 2$ pb^{-1} integrated luminosity.[275] At FCC energies ($\sqrt{s_{_{\rm NN}}} = 63$ TeV), the ssWW DPS cross-section is more than twice larger than the ssWW(jj) SPS one. With $\mathscr{L}_{\text{int}} \approx 30$ pb^{-1}, and

Table 3. Production cross-sections at $\sqrt{s_{\rm NN}}$ = 8.8 TeV for SPS quarkonia and electroweak bosons in pN collisions, and for DPS double-J/ψ, J/ψ + Υ, J/ψ+W, J/ψ+Z, double-Υ, Υ+W, Υ+Z, and same-sign WW, in p–Pb. DPS cross-sections are obtained via Eq. (30) for $\sigma_{\rm eff,DPS,pA}$ = 22.5 μb (uncertainties, not quoted, are of the order of 30%), and the corresponding yields, after dilepton decays and acceptance+efficiency losses (note that the J/ψ yields are *per unit of rapidity* at mid- and forward-y, see text), are given for the nominal 1 pb^{-1} integrated luminosity.

p–Pb (8.8 TeV)	J/ψ + J/ψ	J/ψ + Υ	J/ψ+W	J/ψ+Z
$\sigma^{\rm SPS}_{{\rm pN}\to a}, \sigma^{\rm SPS}_{{\rm pN}\to b}$	45 μb (×2)	45 μb, 2.6 μb	45 μb, 60 nb	45 μb, 35 nb
$\sigma^{\rm DPS}_{\rm p-Pb}$	45 μb	5.2 μb	120 nb	70 nb
$N^{\rm DPS}_{\rm p-Pb}$ (1 pb^{-1})	~65	~60	~15	~3
	Υ + Υ	Υ+W	Υ+Z	ss WW
$\sigma^{\rm SPS}_{{\rm pN}\to a}, \sigma^{\rm SPS}_{{\rm pN}\to b}$	2.6 μb (×2)	2.6 μb, 60 nb	2.6 μb, 35 nb	60 nb (×2)
$\sigma^{\rm DPS}_{\rm p-Pb}$	150 nb	7 nb	4 nb	150 pb
$N^{\rm DPS}_{\rm p-Pb}$ (1 pb^{-1})	~15	~8	~1.5	~4

a factor twice larger rapidity coverage,[286] one expects $\mathcal{O}(10^4)$ ssWW pairs from DPS processes. Same-sign WW production in p–Pb collisions constitutes thereby a promising channel to measure $\sigma_{\rm eff,DPS}$, independently of the standard pp-based extractions of this quantity.

Table 3 collects the estimated DPS cross-sections for the combined production of quarkonia (J/ψ, Υ) and/or electroweak bosons (W, Z) in p–Pb collisions at the nominal LHC energy of $\sqrt{s_{\rm NN}}$ = 8.8 TeV. The quoted SPS pN cross-sections have been obtained at NLO accuracy with the color evaporation model (CEM)[308] for quarkonia (see details in Section 5.1), and with MCFM for the electroweak bosons, using CT10 proton and EPS09 nuclear PDF. The DPS cross-sections are estimated via Eq. (30) with $\sigma_{\rm eff,DPS,pA}$ = 22.5 μb, and the visible DPS yields are quoted for $\mathcal{L}_{\rm int}$ = 1 pb^{-1} integrated luminosities, taking into account the branching fractions BR(J/ψ, Υ,W,Z) = 6%, 2.5%, 11%, 3.4% per dilepton decay; plus simplified acceptance and efficiency losses: $\mathcal{A} \times \mathcal{E}$(J/ψ) ≈ 0.01 (over 1-unit of rapidity at $|y| = 0$, and $|y| = 2$), and $\mathcal{A} \times \mathcal{E}(\Upsilon; {\rm W,Z}) \approx 0.2;\ 0.5$ (over $|y| < 2.5$). All listed processes are in principle observable in the LHC proton-lead runs, whereas rarer DPS processes like W+Z and Z+Z have much lower cross-sections and require much higher luminosities and/or c.m. energies such as those reachable at the FCC.

4.2. *TPS cross-sections in pA collisions*

Similarly to the DPS case, the proton–nucleus TPS cross-section for the pA → *abc* process, is obtained from the sum of three contributions:

$$\sigma^{\rm TPS}_{\rm pA} = \sigma^{\rm TPS,1}_{\rm pA} + \sigma^{\rm TPS,2}_{\rm pA} + \sigma^{\rm TPS,3}_{\rm pA}, \text{ with} \tag{32}$$

(1) A cross-section, scaling like Eq. (21) for the SPS case, corresponding to the TPS value in pN collisions scaled by A, namely:

$$\sigma^{\text{TPS},1}_{\text{pA}\to abc} = A \cdot \sigma^{\text{TPS}}_{\text{pN}\to abc}. \tag{33}$$

(2) A second contribution, involving interactions of partons from two different nucleons in the nucleus, depending on the square of T_{pA},

$$\sigma^{\text{TPS},2}_{\text{pA}\to abc} = \sigma^{\text{TPS}}_{\text{pN}\to abc} \cdot 3\,\frac{\sigma^2_{\text{eff,TPS}}}{\sigma_{\text{eff,DPS}}}\,F_{\text{pA}}, \tag{34}$$

with F_{pA} given by Eq. (26).

(3) A third term, involving interactions among partons from three different nucleons, depending on the cube of T_{pA},

$$\sigma^{\text{TPS},3}_{\text{pA}\to abc} = \sigma^{\text{TPS}}_{\text{pN}\to abc} \cdot \sigma^2_{\text{eff,TPS}} \cdot C_{\text{pA}}, \quad \text{with} \tag{35}$$

$$C_{\text{pA}} = \frac{(A-1)(A-2)}{A^2} \int d^2b\, T^3_{\text{pA}}(\mathbf{b}), \tag{36}$$

with the $(A-1)(A-2)/A^2$ factor introduced to take into account the difference between the total number of nucleon TPS and that of *different* nucleon TPS. By using a hard-sphere approximation for a nucleus of radius $R_{\text{A}} \propto A^{1/3}$, the C_{pA} factor can be analytically calculated as

$$C_{\text{pA}} = \frac{27}{4}\frac{A\,(A-1)\,(A-2)}{5\pi^2 R_{\text{A}}^4} \approx \frac{A^{5/3}}{160\,\pi^2}\,[\text{mb}^{-2}], \tag{37}$$

where the last approximate equality holds for large A. For a Pb nucleus ($A = 208$, $R_{\text{A}} = 22$ mb$^{1/2}$) this factor amounts to $C_{\text{pA}} \approx 5.1$ mb^{-2}, in agreement with the $C_{\text{pA}} = 4.75$ mb^{-2} numerically obtained through a Glauber MC with a realistic Woods-Saxon Pb profile.

The inclusive TPS cross-section for the independent production of three particles a, b, and c in pA collisions is obtained from the sum of the three terms (33), (34), and (35):

$$\sigma^{\text{TPS}}_{\text{pA}\to abc} = A\,\sigma^{\text{TPS}}_{\text{pN}\to\text{abc}}\left[1 + 3\,\frac{\sigma^2_{\text{eff,TPS}}}{\sigma_{\text{eff,DPS}}}\frac{F_{\text{pA}}}{A} + \sigma^2_{\text{eff,TPS}}\frac{C_{\text{pA}}}{A}\right] \tag{38}$$

$$\approx A\,\sigma^{\text{TPS}}_{\text{pN}\to\text{abc}}\left[1 + \frac{\sigma^2_{\text{eff,TPS}}}{\sigma_{\text{eff,DPS}}}\frac{3\,A^{1/3}}{14[\text{mb}]\pi} + \sigma^2_{\text{eff,TPS}}\frac{A^{2/3}}{160[\text{mb}^2]\pi^2}\right], \tag{39}$$

where the last approximation holds for large A, and can be written as a function of $\sigma_{\text{eff,TPS}}$ and A alone making use of Eq. (19):

$$\sigma^{\text{TPS}}_{\text{pA}\to abc} \approx A\,\sigma^{\text{TPS}}_{\text{pN}\to\text{abc}}\left[1 + \sigma_{\text{eff,TPS}}\frac{A^{1/3}}{5.7[\text{mb}]\pi} + \sigma^2_{\text{eff,TPS}}\frac{A^{2/3}}{160[\text{mb}^2]\pi^2}\right]. \tag{40}$$

The TPS cross-section in pA collisions is enhanced by the factor in parentheses in Eqs. (38)–(40) compared to the corresponding one in pN collisions scaled by A. The final formula for TPS in proton–nucleus reads

$$\sigma^{\rm TPS}_{{\rm pA}\to abc} = \left(\frac{m}{6}\right) \frac{\sigma^{\rm SPS}_{{\rm pN}\to a} \cdot \sigma^{\rm SPS}_{{\rm pN}\to b} \cdot \sigma^{\rm SPS}_{{\rm pN}\to c}}{\sigma^2_{\rm eff,TPS,pA}}, \tag{41}$$

where the effective TPS pA cross-section in the denominator depends on the effective TPS cross-section measured in pp, and on purely geometric quantities $(F_{\rm pA}, C_{\rm pA})$ directly derivable from the well-known nuclear profiles,[278]

$$\sigma_{\rm eff,TPS,pA} = \left[\frac{A}{\sigma^2_{\rm eff,TPS}} + \frac{3\,F_{\rm pA}[{\rm mb}^{-1}]}{\sigma_{\rm eff,DPS}} + C_{\rm pA}[{\rm mb}^{-2}]\right]^{-1/2}, \tag{42}$$

which can be numerically approximated as a function of the number A of nucleons in the nucleus (for A large) alone, as follows:

$$\sigma_{\rm eff,TPS,pA} \approx \left[\frac{A}{\sigma^2_{\rm eff,TPS}} + \frac{A^{4/3}}{5.7[{\rm mb}]\,\pi\,\sigma_{\rm eff,TPS}} + \frac{A^{5/3}}{160[{\rm mb}^2]\,\pi^2}\right]^{-1/2}. \tag{43}$$

For a Pb nucleus ($A = 208$, $F_{\rm pA} = 30.25$ mb^{-1}, and $C_{\rm pA} = 4.75$ mb^{-2}) and taking $\sigma_{\rm eff,TPS} = 12.5 \pm 4.5$ mb, the effective TPS cross-section amounts to $\sigma_{\rm eff,TPS,pA} = 0.29 \pm 0.04$ mb. Thus, for p–Pb the relative importance of the three TPS terms of Eq. (38) is $\sigma^{\rm TPS,1}_{{\rm pA}\to abc} : \sigma^{\rm TPS,2}_{{\rm pA}\to abc} : \sigma^{\rm TPS,3}_{{\rm pA}\to abc} = 1 : 4.54 : 3.56$. Namely, in p–Pb collisions, 10% of the TPS yields come from partonic interactions within just one nucleon of the lead nucleus, 50% involve scatterings within two nucleons, and 40% come from partonic interactions in three different Pb nucleons. The sum of the three contributions in Eq. (38), ignoring differences between pN and pp collisions, indicates that the TPS cross-sections in p–Pb are about nine times larger than the naive expectation based on A-scaling of the corresponding pN TPS cross-sections, Eq. (33). One can thus exploit the large expected TPS signals in proton–nucleus collisions to extract the $\sigma_{\rm eff,TPS}$ parameter, and thereby $\sigma_{\rm eff,DPS}$ via Eq. (19), independently of TPS measurements in pp collisions—given that the $F_{\rm pA}$ and $C_{\rm pA}$ parameters in Eq. (38) depend on the comparatively better known transverse density of nuclei.

TPS cross-sections in pA collisions: Numerical examples

As a concrete numerical example in Ref. 278 we have computed the TPS cross-sections for charm ($c\bar{c}$) and bottom ($b\bar{b}$) production, following the motivation for the similar measurement in pp collisions (Section 3.1), over a wide range of c.m. energies, $\sqrt{s_{\rm NN}} \approx$ 5–500 TeV, of relevance for collider (LHC and FCC) and ultra-high-energy cosmic rays physics. The TPS heavy-quark cross-sections are computed via Eq. (41) for $m = 1$, i.e., $\sigma^{\rm TPS}_{{\rm pA}\to c\bar{c},b\bar{b}} = (\sigma^{\rm SPS}_{{\rm pN}\to c\bar{c},b\bar{b}})^3/(6\,\sigma^2_{\rm eff,TPS,pA})$ with the effective TPS cross-sections given by Eq. (42): $\sigma_{\rm eff,TPS,pA} = 0.29 \pm 0.04$ mb for p–Pb, and

$\sigma_{\text{eff,TPS,pA}} = 2.2 \pm 0.4$ mb for p–air collisions.[c] The SPS cross-sections, $\sigma^{\text{SPS}}_{\text{pN}\to c\bar{c},b\bar{b}}$, are calculated at NNLO via Eq. (2) with `Top++` (v.2.0) with the same setup as described in Section 3.1, using the ABMP16 proton and EPS09 nuclear PDF. In the p–Pb case, the inclusion of EPS09 nuclear shadowing reduces moderately the total charm and bottom cross-sections in pN compared to pp collisions, by about 10% (15%) and 5% (10%) at the LHC (FCC). At $\sqrt{s_{\text{NN}}} = 5.02$ TeV, our prediction ($\sigma^{\text{SPS,NNLO}}_{\text{p-Pb}\to c\bar{c}} = 650 \pm 290_{\text{sc}} \pm 60_{\text{PDF}}$ mb) agrees well with the ALICE total D-meson measurement[309] extrapolated using FONLL[294] to a total charm cross-section of $\sigma^{\text{ALICE}}_{\text{p-Pb}\to c\bar{c}} = 640 \pm 60_{\text{stat}}\, {}^{+60}_{-110}\big|_{\text{syst}}$ mb (data point in the top-left panel of Fig. 5).

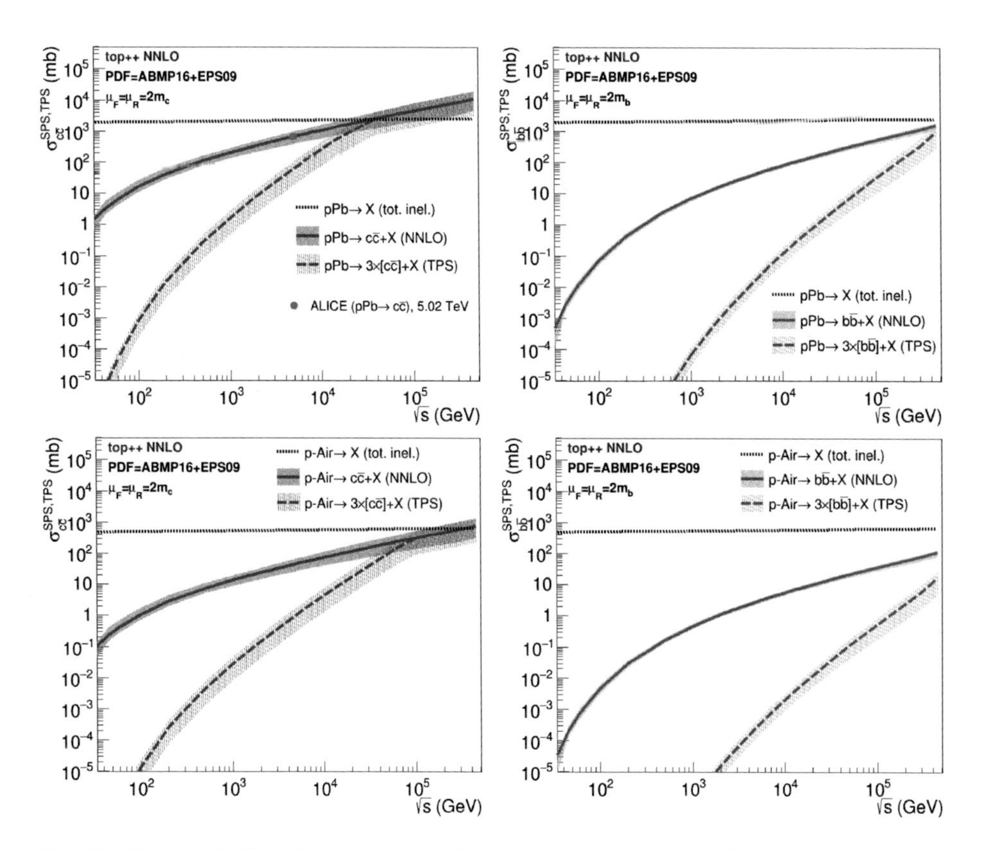

Fig. 5 Charm (left) and bottom (right) cross-sections in p–Pb (top panels) and p–air (bottom panels) collisions as a function of c.m. energy, in single parton (solid band) and triple parton (dashed band) scatterings, compared to the total inelastic pA cross-sections (dotted line in all panels). Bands around curves indicate scale, PDF (and $\sigma_{\text{eff,TPS}}$, in the TPS case) uncertainties added in quadrature. The p–Pb $\to c\bar{c} + X$ charm data point on the top-left plot has been derived from the ALICE D-meson data.[309]

[c] Using $A = 14.3$ for a 78:21% mixture of ^{14}N:^{16}O, with $F_{\text{pA}} = 0.51$ mb^{-1}, and $C_{\text{pA}} = 0.016$ mb^{-2} obtained via a Glauber MC.[298]

Table 4. Cross-sections for inclusive inelastic, and for SPS and TPS charm and bottom production in p–Pb (at LHC and FCC energies) and p–air (at GZK-cutoff c.m. energies) collisions. For the SPS and TPS cross-sections the quoted values include scales, PDF, and total (quadratically added, including $\sigma_{\rm eff,TPS}$) uncertainties.

Process	p–Pb (8.8 TeV)	p–Pb (63 TeV)	p–air (430 TeV)
$\sigma^{\rm inel}_{\rm pA}$	2.2 ± 0.4 b	2.4 ± 0.4 b	0.61 ± 0.10 b
$\sigma^{\rm SPS}_{{\rm c\bar{c}}+X}$	$0.96 \pm 0.45_{\rm sc} \pm 0.10_{\rm PDF}$ b	$3.4 \pm 1.9_{\rm sc} \pm 0.4_{\rm PDF}$ b	$0.75 \pm 0.5_{\rm sc} \pm 0.1_{\rm PDF}$ b
$\sigma^{\rm TPS}_{{\rm c\bar{c}\,c\bar{c}\,c\bar{c}}+X}$	$200 \pm 140_{\rm tot}$ mb	$8.7 \pm 6.2_{\rm tot}$ b	$5.0 \pm 3.6_{\rm tot}$ b
$\sigma^{\rm SPS}_{{\rm b\bar{b}}+X}$	$72 \pm 12_{\rm sc} \pm 5_{\rm PDF}$ mb	$370 \pm 75_{\rm sc} \pm 30_{\rm PDF}$ mb	$110 \pm 25_{\rm sc} \pm 5_{\rm PDF}$ mb
$\sigma^{\rm TPS}_{{\rm b\bar{b}\,b\bar{b}\,b\bar{b}}+X}$	$0.084 \pm 0.045_{\rm tot}$ μb	$11 \pm 7_{\rm tot}$ μb	$17 \pm 11_{\rm tot}$ μb

Since the TPS p–Pb cross-section go as the cube of $\sigma^{\rm SPS}_{{\rm pN}\to{\rm Q\bar{Q}}}$, the impact of shadowing is amplified and leads to 15–35% depletions of the TPS cross-sections compared to results obtained with the free proton PDF.

Table 4 collects the total inelastic and the heavy-quarks cross-sections at $\sqrt{s_{_{\rm NN}}}$ = 8.8 TeV and 63 TeV in p–Pb collisions, and at $\sqrt{s_{_{\rm NN}}}$ = 430 TeV in p–air collisions. The latter c.m. energy corresponds to the so-called "GZK cutoff"[310,311] reached in collisions of $\mathcal{O}(10^{20}\,{\rm eV})$ proton cosmic-rays, with N and O nuclei at rest in the upper atmosphere. The PDF uncertainties include those from the proton and nucleus in quadrature, as obtained from the corresponding $28 \oplus 30$ eigenvalues of the ABMP16 $\oplus$ EPS09 sets. The dominant uncertainty is linked to the theoretical scale choice, estimated by modifying μ_R and μ_F within a factor of two. At the LHC, the large SPS $\rm c\bar{c}$ cross-section ($\sim$1 b) results in triple-$\rm c\bar{c}$ cross-sections from independent parton scatterings amounting to about 20% of the inclusive charm yields. Since the total inelastic p–Pb cross-sections is $\sigma^{\rm inel}_{\rm p-Pb} \approx 2.2$ b, charm TPS takes place in about 10% of the p–Pb events at 8.8 TeV. At the FCC, the theoretical TPS charm cross-section even overcomes the inclusive charm one. Such a result indicates that quadruple, quintuple,... parton–parton scatterings are expected to produce extra $\rm c\bar{c}$ pairs with non-negligible probability. Recall that inclusive cross-sections can be related with the factorial moments of the multiplicity distribution and that the SPS, DPS, TPS... cross-sections are not bounded by the inelastic cross-section, and also the DPS, TPS... cross-sections are not bounded by the SPS one, as is discussed in Chapter 3. The huge TPS $\rm c\bar{c}$ cross-sections in p–Pb at $\sqrt{s_{_{\rm NN}}}$ = 63 TeV, will make triple-J/ψ production, with $\sigma({\rm J/\psi\,J/\psi\,J/\psi}+X) \approx 1$ mb, observable. Triple-$\rm b\bar{b}$ cross-sections remain comparatively small, in the 0.1 mb range, at the LHC but reach $\sim$10 mb (i.e., 3% of the total inclusive bottom cross-section) at the FCC.

Figure 5 plots the cross-sections over $\sqrt{s_{_{\rm NN}}} \approx$ 40 GeV–500 TeV for SPS (solid bands), TPS (dashed bands) for charm (left) and bottom (right) production, and total inelastic (dotted curve) in p–Pb (top panels) and p–air (bottom panels) collisions. Whenever the central value of the theoretical TPS cross-section overcomes the inclusive charm cross-section, indicative of multiple (beyond three) $c\bar{c}$-pair production, we equalize it to the latter. At $\sqrt{s_{_{\rm NN}}} \approx 25$ TeV, the total charm and inelastic p–Pb cross-sections are equal implying that, above this c.m. energy, the average number of charm pairs produced in p–Pb collisions is larger than 1. In the $b\bar{b}$ case, such a situation only occurs at much higher c.m. energies, above 500 TeV. For p–air collisions at the GZK cutoff, the cross-section for inclusive as well as TPS charm production equal the total inelastic cross-section ($\sigma^{\rm inel}_{\rm pAir} \approx 0.61$ b) indicating that the average number of $c\bar{c}$-pairs produced in p–air collisions is larger than 1. In the $b\bar{b}$ case, about 20% of the p–air collisions produce bottom hadrons, but only about 4% of them have TPS production. These results emphasize the numerical importance of TPS processes in proton–nucleus collisions at colliders, and their relevance for hadronic MC models commonly used for the simulation of ultrarelativistic cosmic-ray interactions with the atmosphere[312] which, so far, do not include any heavy-quark production.

5. Double and Triple Parton Scattering Cross-Sections in Nucleus–Nucleus Collisions

In nucleus–nucleus collisions, the parton flux is enhanced by A nucleons in each nucleus, and the SPS cross-section is simply expected to be that of NN collisions, taking into account (anti)shadowing effects in the nuclear PDF, scaled by the factor A^2, i.e.,[298]

$$\sigma^{\rm SPS}_{\rm AA\to a} = \int T_{\rm AA}(\mathbf{b})d^2b = A^2 \cdot \sigma^{\rm SPS}_{\rm NN\to a}, \tag{44}$$

where $T_{\rm AA}(\mathbf{b})$ the standard nuclear overlap function, normalized to A^2,

$$T_{\rm AA}(\mathbf{b}) = \int T_{\rm pA}(\mathbf{b_1})T_{\rm pA}(\mathbf{b_1}-\mathbf{b})d^2b_1d^2b, \tag{45}$$

with $T_{\rm pA}(\mathbf{b})$ being the nuclear thickness function at impact parameter $\mathbf{b}$, Eq. (22), connecting the centres of the colliding nucleus in the transverse plane. In the next two subsections, we present the estimates for DPS and TPS cross-sections in AA collisions from the corresponding SPS values.

5.1. *DPS cross-sections in AA collisions*

The DPS cross-section in AA is the sum of three terms, corresponding to the diagrams of Fig. 6,

$$\sigma^{\rm DPS}_{\rm AA} = \sigma^{\rm DPS,1}_{\rm AA} + \sigma^{\rm DPS,2}_{\rm AA} + \sigma^{\rm DPS,3}_{\rm AA}, \text{ where} \tag{46}$$

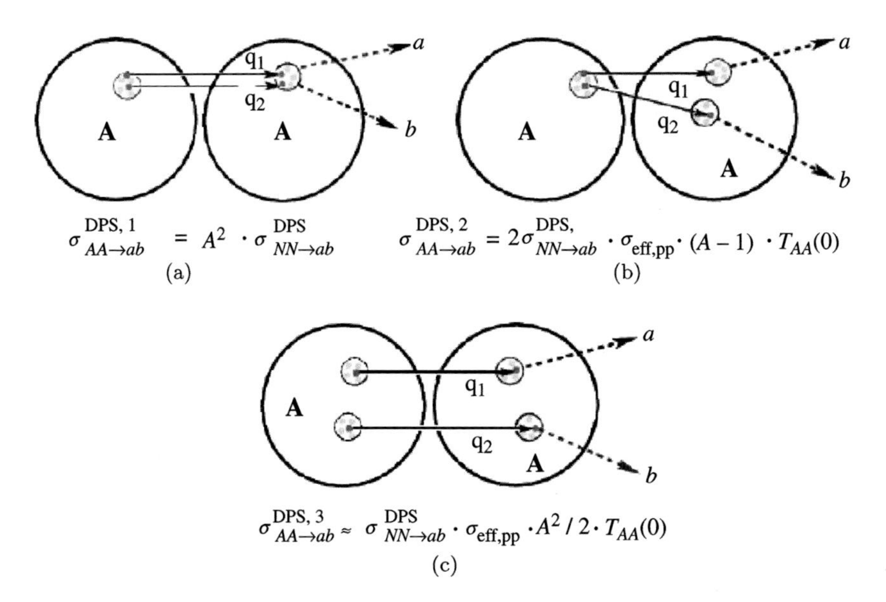

Fig. 6 Schematic diagrams contributing to DPS cross-sections in AA collisions: The two colliding partons belong to the same pair of nucleons (a), partons from one nucleon in one nucleus collide with partons from two different nucleons in the other nucleus (b), and the two colliding partons belong to two different nucleons from both nuclei (c).

(1) The first term, similarly to the SPS cross-sections (44), is just the DPS cross-section in NN collisions scaled by A^2,

$$\sigma^{\rm DPS,1}_{\rm AA\to \mathit{ab}} = A^2 \cdot \sigma^{\rm DPS}_{\rm NN\to \mathit{ab}}. \tag{47}$$

(2) The second term accounts for interactions of partons from one nucleon in one nucleus with partons from two different nucleons in the other nucleus,

$$\sigma^{\rm DPS,2}_{\rm AA\to \mathit{ab}} = 2\sigma^{\rm DPS}_{\rm NN\to \mathit{ab}} \cdot \sigma_{\rm eff,DPS} \cdot A \cdot F_{\rm pA}, \tag{48}$$

with $F_{\rm pA} \approx T_{\rm AA}(0)$ given by Eq. (26).

(3) The third contribution from interactions of partons from two different nucleons in one nucleus with partons from two different nucleons in the other nucleus, reads

$$\sigma^{\rm DPS,3}_{\rm AA\to \mathit{ab}} = \sigma^{\rm DPS}_{\rm NN\to \mathit{ab}} \cdot \sigma_{\rm eff,DPS} \cdot T_{3,\rm AA} \tag{49}$$

with,

$$T_{3,\rm AA} = \left(\frac{A-1}{A}\right)^2 \int T_{\rm pA}(\mathbf{b_1}) T_{\rm pA}(\mathbf{b_2}) T_{\rm pA}(\mathbf{b_1}-\mathbf{b}) T_{\rm pA}(\mathbf{b_2}-\mathbf{b}) d^2b_1 d^2b_2 d^2b \tag{50}$$

$$= \left(\frac{A-1}{A}\right)^2 \int d^2r\, T^2_{\rm AA}(\mathbf{r}) \approx \frac{A^2}{2} \cdot T_{\rm AA}(0), \tag{51}$$

where the latter integral of the nuclear overlap function squared does not depend much on the precise shape of the transverse parton density in the nucleus, amounting to $A^2/1.94 \cdot T_{\rm AA}(0)$ for a hard-sphere and $A^2/2 \cdot T_{\rm AA}(0)$ for a Gaussian profile. The factor $((A-1)/A)^2$ takes into account the difference between the number of nucleon pairs and the number of *different* nucleon pairs.

Adding (47)–(49), the inclusive cross-section of a DPS process with two hard parton subprocesses a and b in AA collisions can be written as

$$\sigma^{\rm DPS}_{\rm AA\to ab} = A^2\, \sigma^{\rm DPS}_{\rm NN\to ab}\left[1+\frac{2}{A}\sigma_{\rm eff,DPS}\, F_{\rm pA} + \frac{(A-1)^2}{A^2}\sigma_{\rm eff,DPS}\int d^2r T^2_{\rm AA}(\mathbf{r})\right] \tag{52}$$

$$\approx A^2\, \sigma^{\rm DPS}_{\rm NN\to ab}\left[1+\frac{2}{A}\,\sigma_{\rm eff,DPS}\, T_{\rm AA}(0) + \frac{1}{2}\,\sigma_{\rm eff,DPS}\, T_{\rm AA}(0)\right] \tag{53}$$

$$\approx A^2\, \sigma^{\rm DPS}_{\rm NN\to ab}\left[1+\frac{\sigma_{\rm eff,DPS}}{7[\rm mb]\,\pi}\, A^{1/3} + \frac{\sigma_{\rm eff,DPS}}{28[\rm mb]\,\pi}\, A^{4/3}\right], \tag{54}$$

where the last approximation, showing the A-dependence of the DPS cross-sections, applies for large nuclei. The factor in parentheses in Eqs. (52)–(54) indicates the enhancement in DPS cross-sections in AA compared to the corresponding A^2-scaled values in nucleon–nucleon collisions, Eq. (47), which amounts to ~27 (for small $A = 40$) or ~215 (for large $A = 208$). The overall mass-number scaling of DPS cross-sections in AA compared to pp collisions is given by a $(A^2 + k\, A^{7/3} + w\, A^{10/3})$ factor with $k, w \approx 0.7, 0.2$, which is clearly numerically dominated by the $A^{10/3}$ term. The final DPS cross-section "pocket formula" in heavy-ion collisions can be written as

$$\sigma^{\rm DPS}_{\rm AA\to ab} = \left(\frac{m}{2}\right)\frac{\sigma^{\rm SPS}_{\rm NN\to a}\cdot \sigma^{\rm SPS}_{\rm NN\to b}}{\sigma_{\rm eff,DPS,AA}}, \tag{55}$$

with the effective AA normalization cross-section amounting to

$$\sigma_{\rm eff,DPS,AA} \approx \frac{1}{A^2\left[\sigma^{-1}_{\rm eff,DPS} + \frac{2}{A}\,{\rm T}_{\rm AA}(0) + \frac{1}{2}\,{\rm T}_{\rm AA}(0)\right]}. \tag{56}$$

For a value of $\sigma_{\rm eff,DPS} \approx 15$ mb and for nuclei with mass numbers $A = 40$–240, we find that the relative weights of the three components contributing to DPS scattering in AA collisions are $1:2.3:23$ (for $A = 40$) and $1:4:200$ (for $A = 208$). Namely, only 13% (for ^{40}Ca+^{40}Ca) or 2.5% (for ^{208}Pb+^{208}Pb) of the DPS yields in AA collisions come from the first two diagrams of Fig. 6 involving partons from one single nucleon. Clearly, the "pure" DPS contributions arising from partonic collisions within a single nucleon (first and second terms of Eq. (54)) are much smaller than the last term from double particle production coming from two independent *nucleon–nucleon* collisions. The DPS cross-sections in AA are practically unaffected by the value of $\sigma_{\rm eff,DPS}$, but dominated instead by double parton interactions from *different nucleons* in both nuclei. In the case of ^{208}Pb–^{208}Pb collisions, the numerical value of Eq. (56) is $\sigma_{\rm eff,DPS,AA} = 1.5 \pm 0.1$ nb, with uncertainties dominated by those of the Glauber MC determination of $T_{\rm AA}(0)$. Whereas the single parton cross-sections in Pb–Pb collisions, Eq. (44), are enhanced by a factor of $A^2 \simeq 4 \cdot 10^4$ compared to that in pp collisions, the corresponding double parton cross-sections are enhanced by a much higher factor of $\sigma_{\rm eff,DPS}/\sigma_{\rm eff,DPS,AA} \propto 0.2\,A^{10/3} \simeq 10^7$.

Centrality dependence of DPS cross-sections in AA collisions

The DPS cross-sections discussed above are for "minimum bias" AA collisions without any selection in reaction centrality. The cross-sections for single and double-parton scattering within an impact-parameter interval $[\mathrm{b}_1,\mathrm{b}_2]$, corresponding to a given centrality percentile $\mathrm{f}_{\%}$ of the total AA cross-section $\sigma_{\rm AA}^{\rm inel}$, with average nuclear overlap function $\langle T_{\rm AA}[b_1,b_2]\rangle$ read (for large A, so that $A-1\approx A$):

$$\sigma^{\rm SPS}_{{\rm AA}[b_1,b_2]\to a} = A^2\,\sigma^{\rm SPS}_{{\rm NN}\to a}\,f_1[b_1,b_2] = \sigma^{\rm SPS}_{{\rm NN}\to a}\cdot \mathrm{f}_{\%}\,\sigma^{\rm inel}_{\rm AA}\cdot\langle \mathrm{T}_{\rm AA}[\mathrm{b}_1,\mathrm{b}_2]\rangle, \tag{57}$$

$$\begin{aligned}\sigma^{\rm DPS}_{{\rm AA}[b_1,b_2]\to ab} &= A^2\,\sigma^{\rm DPS}_{{\rm NN}\to ab}\,f_1[b_1,b_2]\\ &\quad\times\left[1+\frac{2\sigma_{\rm eff,DPS}}{A}T_{\rm AA}(0)\frac{f_2[b_1,b_2]}{f_1[b_1,b_2]}+\sigma_{\rm eff,DPS}T_{\rm AA}(0)\frac{f_3[b_1,b_2]}{f_1[b_1,b_2]}\right],\end{aligned} \tag{58}$$

where the latter has been obtained integrating Eq. (52) over $b_1 < b < b_2$, and where the three dimensionless and appropriately-normalized fractions f_1, f_2, and f_3 are:

$$f_1[b_1,b_2] = \frac{2\pi}{A^2}\int_{b_1}^{b_2} bdb\,T_{\rm AA}(b) = \frac{\mathrm{f}_{\%}\,\sigma^{\rm inel}_{\rm AA}}{A^2}\,\langle T_{\rm AA}[b_1,b_2]\rangle,$$

$$f_2[b_1,b_2] = \frac{2\pi}{A\,T_{\rm AA}(0)}\int_{b_1}^{b_2} bdb\int d^2b_1\,T_{\rm pA}(\mathbf{b_1})T_{\rm pA}(\mathbf{b_1}-\mathbf{b})T_{\rm pA}(\mathbf{b_1}-\mathbf{b}),$$

$$f_3[b_1,b_2] = \frac{2\pi}{A^2\,T_{\rm AA}(0)}\int_{b_1}^{b_2} bdb\,T^2_{\rm AA}(b).$$

The integrals f_2 and f_3 can be evaluated[313] for small enough centrality bins around a given impact parameter b. The dominant f_3/f_1 contribution in Eq. (58) is simply given by the ratio $\langle T_{\rm AA}[b_1,b_2]\rangle/T_{\rm AA}(0)$ which is practically insensitive (except for very peripheral collisions) to the precise shape of the nuclear density profile. The second centrality-dependent DPS term, f_2/f_1, cannot be expressed in a simple form in terms of $T_{\rm AA}(b)$, but it is of order unity for the most central collisions, $f_2/f_1 = 4/3$, and $16/15$ for Gaussian and hard-sphere profiles respectively, and it is suppressed in comparison with the third leading term by an extra factor $\sim$2/A. Finally, for not very-peripheral collisions ($\mathrm{f}_{\%} \lesssim$ 0–65%), the DPS cross-section in a (thin) impact-parameter $[\mathrm{b}_1,\mathrm{b}_2]$ range can be approximated by

$$\sigma^{\rm DPS}_{{\rm AA}\to ab}[b_1,b_2] \approx \sigma^{\rm DPS}_{{\rm NN}\to ab}\cdot\sigma_{\rm eff,DPS}\cdot \mathrm{f}_{\%}\,\sigma^{\rm inel}_{\rm AA}\cdot\langle \mathrm{T}_{\rm AA}[\mathrm{b}_1,\mathrm{b}_2]\rangle^2 \tag{59}$$

$$= \left(\frac{m}{2}\right)\sigma^{\rm SPS}_{{\rm NN}\to a}\cdot\sigma^{\rm SPS}_{{\rm NN}\to b}\cdot \mathrm{f}_{\%}\,\sigma^{\rm inel}_{\rm AA}\cdot\langle \mathrm{T}_{\rm AA}[\mathrm{b}_1,\mathrm{b}_2]\rangle^2. \tag{60}$$

Dividing this last expression by Eq. (57), one finally obtains the corresponding ratio of double- to single-parton scattering cross-sections as a function of impact parameter[d]:

$$(\sigma^{\rm DPS}_{\rm AA\to ab}/\sigma^{\rm SPS}_{\rm AA\to a})[b_1,b_2] \approx \binom{m}{2}\,\sigma^{\rm SPS}_{\rm NN\to b}\cdot\langle T_{\rm AA}[b_1,b_2]\rangle. \quad (61)$$

DPS cross-sections in AA collisions: Numerical examples

Quarkonia has been historically considered a sensitive probe of the quark–gluon–plasma (QGP) formed in heavy-ion collisions,[314] and thereby their production channels need to be theoretically and experimentally well understood in pp, pA and AA collisions.[315] Double-quarkonium (J/ψ J/ψ, ϒϒ) production is a typical channel for DPS studies in pp, given their large cross-sections and relatively well-understood double-SPS backgrounds.[249,251,253] In Ref. 279, the DPS cross-section for double-J/ψ production in Pb–Pb collisions has been estimated via Eq. (55) with $m = 1$, $\sigma_{\rm eff,DPS,AA} = 1.5 \pm 0.1$ nb, and prompt-J/ψ SPS cross-section computed at NLO via CEM[308] with the CT10 proton and the EPS09 nuclear PDF, and theoretical scales $\mu_R = \mu_F = 1.5\,m_c$ for a c-quark mass $m_c = 1.27$ GeV. The EPS09 nuclear modification factors result in a reduction of 20–35% of the J/ψ cross-sections compared to those calculated using the free proton PDFs.

Figure 7 shows the $\sqrt{s}$-dependence of single-J/ψ in pp, NN and Pb–Pb collisions (top panel), and of double-J/ψ cross-sections in Pb–Pb, as well as the fraction of J/ψ events with double-J/ψ produced via DPS (bottom panel). Our theoretical setup with CT10 (anti)proton PDF alone agrees well with the experimental pp, p$\bar{\rm p}$ data[233,316–320] extrapolated to full phase space[279] (squares in Fig. 7). At the nominal Pb–Pb energy of 5.5 TeV, the single prompt-J/ψ cross-sections is ∼1 b, and ∼20% of such collisions are accompanied by the production of a second J/ψ from a double parton interaction. Accounting for dilepton decays, acceptance and efficiency, which reduce the yields by a factor of $\sim 3{\cdot}10^{-7}$ in the ATLAS/CMS (central) and ALICE (forward) rapidities, the visible cross-section is $d\sigma^{\rm DPS}_{\rm J/\psi\,J/\psi}/dy|_{y=0,2} \approx 60$ nb, i.e., about 250 double-J/ψ events per unit-rapidity (both at central and forward y) are expected in the four combinations of dielectron and dimuon channels for a $\mathscr{L}_{\rm int} = 1$ nb^{-1} integrated luminosity (assuming no net in-medium J/ψ suppression or enhancement).

Following Eq. (61), the probability of J/ψ J/ψ DPS production increases rapidly with decreasing impact parameter and ∼35% of the most central Pb–Pb → J/ψ + X collisions have a second J/ψ produced in the final state (Fig. 8). These results show quantitatively the large probability for double-production of J/ψ mesons in high-energy nucleus–nucleus collisions. Thus, the observation of a J/ψ pair in a given Pb–Pb event should not be (blindly) interpreted as, e.g., indicative of J/ψ

[d]Such analytical expression neglects the first and second terms of Eq. (58). In the $f_{\%} \approx 65$–100% centrality percentile, the second term would add about 20% more DPS cross-sections, and for very peripheral collisions ($f_{\%} \approx 85$–100%, where $\langle T_{\rm AA}[b_1,b_2]\rangle$ is of order or less than $1/\sigma_{\rm eff,DPS}$) the contributions from the first term are also non-negligible.

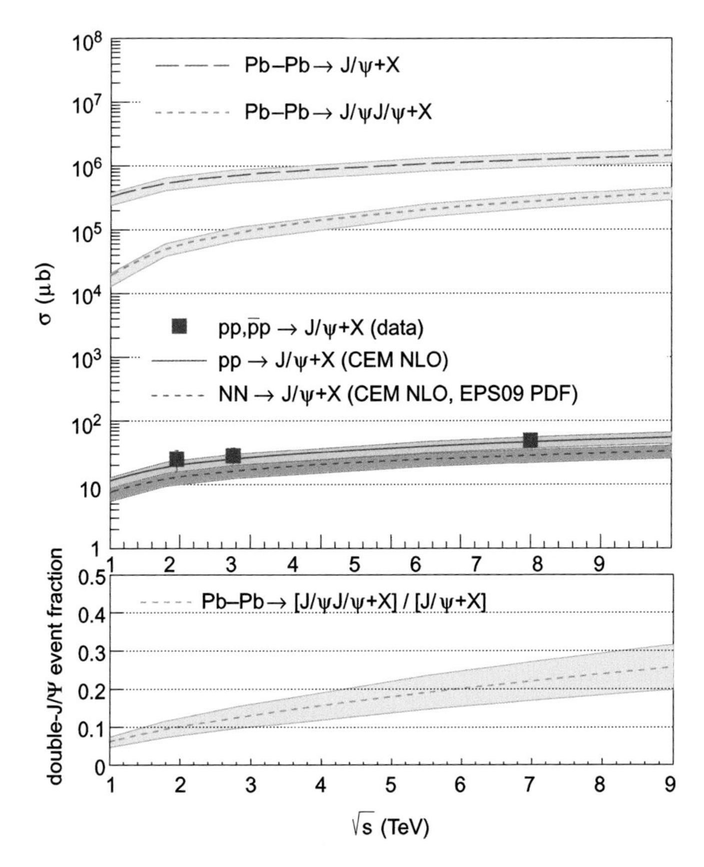

Fig. 7 Top: Production cross-sections as a function of c.m. energy for prompt-J/ψ in pp, NN, and Pb–Pb, and for DPS J/ψ J/ψ in Pb–Pb collisions. Bottom: Fraction of J/ψ events in Pb–Pb collisions with a pair of J/ψ mesons produced, as a function of $\sqrt{s_{\text{NN}}}$. Bands show the nuclear PDF and scales uncertainties in quadrature. [Figures from Ref. 279.]

production via $c\bar{c}$ regeneration in the QGP,[321] since DPS constitute an important fraction of the inclusive J/ψ yield, with or without final-state dense medium effects.

Table 5 collects the DPS cross-sections for the (pair) production of quarkonia (J/ψ, ϒ) and/or electroweak bosons (W, Z) in Pb–Pb collisions at the nominal LHC energy of 5.5 TeV, obtained via Eq. (55) with $\sigma_{\text{eff,DPS,AA}} = 1.5$ nb. The visible DPS yields for $\mathscr{L}_{\text{int}} = \text{nb}^{-1}$ are quoted taking into account BR(J/ψ, ϒ,W,Z) = 6%, 2.5%, 11%, 3.4% per dilepton decay; plus simplified acceptance and efficiency losses: $\mathscr{A} \times \mathscr{E}(\text{J}/\psi) \approx 0.01$ (over 1-unit of rapidity at $|y| = 0$, and $|y| = 2$), and $\mathscr{A} \times \mathscr{E}(\Upsilon;\text{W,Z}) \approx 0.2;\ 0.5$ (over $|y| < 2.5$). All listed processes are in principle observable in the LHC heavy-ion runs, whereas rarer DPS processes like W+Z and Z+Z have much lower visible cross-sections and would require much higher luminosities and/or c.m. energies such as those reachable at the FCC.

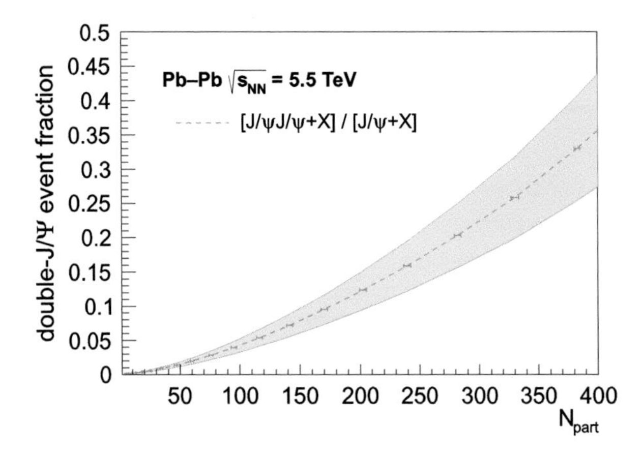

Fig. 8 Fraction of J/ψ events in Pb–Pb collisions at 5.5 TeV where a J/ψ-pair is produced from DPS as a function of the reaction centrality (given by N_{part}), as per Eq. (61). The band shows the EPS09 PDF plus scale uncertainties. [Figure from Ref. 279.]

Table 5. Production cross-sections at $\sqrt{s_{NN}}$ = 5.5 TeV for SPS quarkonia and electroweak bosons in NN collisions, and for DPS double-J/ψ, J/ψ + Υ, J/ψ+W, J/ψ+Z, double-Υ, Υ+W, Υ+Z, and same-sign WW, in Pb–Pb. DPS cross-sections are obtained via Eq. (55) for $\sigma_{\rm eff,DPS,AA}$ = 1.5 nb (uncertainties, not quoted, are of the order of 30%), and the corresponding yields, after dilepton decays and acceptance+efficiency losses (note that the J/ψ yields are *per unit of rapidity* at mid- and forward-y, see text), are given for the nominal 1 nb^{-1} integrated luminosity.

Pb–Pb (5.5 TeV)	J/ψ + J/ψ	J/ψ + Υ	J/ψ +W	J/ψ +Z
$\sigma^{\rm SPS}_{{\rm NN}\to a}, \sigma^{\rm SPS}_{{\rm NN}\to b}$	25 μb (×2)	25 μb, 1.7 μb	25 μb, 30 nb	25 μb, 20 nb
$\sigma^{\rm DPS}_{\rm Pb-Pb}$	210 mb	28 mb	500 μb	330 μb
$N^{\rm DPS}_{\rm Pb-Pb}$ (1 nb^{-1})	~250	~340	~65	~14
	Υ + Υ	Υ+W	Υ+Z	ss WW
$\sigma^{\rm SPS}_{{\rm NN}\to a}, \sigma^{\rm SPS}_{{\rm NN}\to b}$	1.7 μb (×2)	1.7 μb, 30 nb	1.7 μb, 20 nb	30 nb (×2)
$\sigma^{\rm DPS}_{\rm Pb-Pb}$	960 μb	34 μb	23 μb	630 nb
$N^{\rm DPS}_{\rm Pb-Pb}$ (1 nb^{-1})	~95	~35	~8	~15

5.2. *TPS cross-sections in AA collisions*

For completeness, we estimate here the expected scaling of TPS cross-sections in nucleus–nucleus compared to proton–proton collisions. Following our discussion for pA in Section 4.2, the TPS cross-section in AA collisions results from the sum of nine terms, schematically represented in Fig. 9, generated by three independent

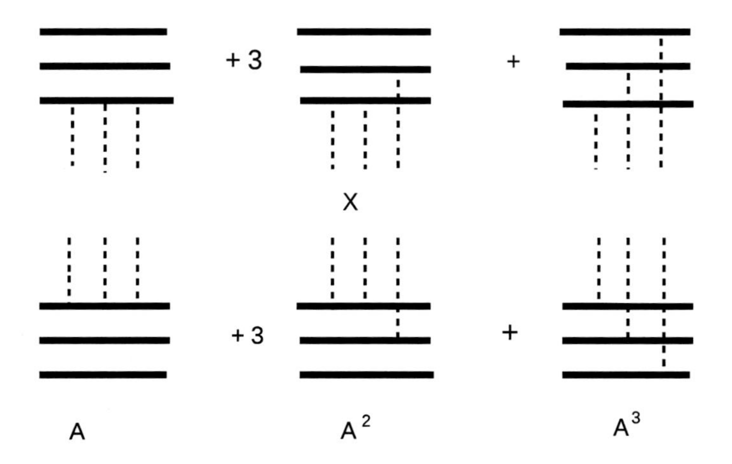

Fig. 9 Schematic diagrams contributing to TPS cross-sections in AA collisions.

structures appearing in triple parton scatterings in pA:

$$\begin{aligned}\sigma^{\text{TPS}}_{\text{AA}\to abc} \propto\; & A\cdot A + 3A\cdot A^2 + A\cdot A^3 \\ & + 3A^2\cdot A + 9A^2\cdot A^2 + 3A^2\cdot A^3 \\ & + A^3\cdot A + 3A^3\cdot A^2 + A^3\cdot A^3.\end{aligned} \tag{62}$$

These nine terms have different prefactors that can be expressed as a function of the nuclear thickness function, and the effective TPS and DPS cross-sections, as done previously for the simpler pA case, see, e.g., Eq. (38). For instance, the first $A\cdot A$ term is just the TPS cross-section in NN collisions scaled by A^2:

$$\sigma^{\text{TPS},1}_{\text{AA}\to abc} = A^2\cdot\sigma^{\text{TPS}}_{\text{NN}\to abc}, \tag{63}$$

whereas the last $A^3\cdot A^3$ contribution arises from interactions of partons from three different nucleons in one nucleus with partons from three different nucleons in the other nucleus (i.e., they result from triple *nucleon–nucleon* scatterings):

$$\sigma^{\text{TPS},9}_{\text{AA}\to abc} = \sigma^{\text{TPS}}_{\text{NN}\to abc}\cdot\sigma^2_{\text{eff,TPS}}\cdot T_{9,\text{AA}}\,,\ \text{with}\ T_{9,\text{AA}} = \int d^2r T^3_{\text{AA}}(\mathbf{r}). \tag{64}$$

In this latter expression, for simplicity, we omitted the $[(A-1)(A-2)/A^2]^2$ factor needed to account for the difference between the total number of nucleon triplets and that of different nucleon triplets. The ratio

$$\sigma^{\text{TPS},1}_{\text{AA}\to abc}/\sigma^{\text{TPS},9}_{\text{AA}\to abc} \approx [2/\sigma_{\text{eff,DPS}}T_{\text{AA}}(0)]^2 \tag{65}$$

shows that the "pure" TPS contributions arising from partonic collisions within a single nucleon (which scale as A^2) are negligible compared to triple particle production coming from three independent nucleon–nucleon collisions which scale as $A^6(r_p/R_{\text{A}})^4 \propto A^{14/3}$. In the Pb–Pb case, the relative weights of these two "limiting"

TPS contributions are 1 : 40 000, to be compared with 1 : 200 for the similar DPS weights. The many other intermediate terms of Eq. (62) correspond to the various "mixed" parton–nucleon contributions, which can be also written in analytical form in this approach but, however, are suppressed by additional powers of A compared to the dominant nucleon–nucleon triple scattering.

Thus, as found in the DPS case, TPS processes in AA collisions are not so useful to derive $\sigma_{\text{eff,DPS}}$ or $\sigma_{\text{eff,TPS}}$ and thereby study the intranucleon partonic structure as in pp or pA collisions. The estimates presented here demonstrate that double- and triple- (hard) nucleon–nucleon scatterings represent a significant fraction of the inelastic hard AA cross-section, and the standard Glauber MC provides a simper approach to compute their occurrence in a given heavy-ion collision.

6. Summary

Multiparton interactions are a major contributor to particle production in proton and nuclear collisions at high centre-of-mass energies. The possibility to concurrently produce multiple particles with large transverse momentum and/or mass in independent parton–parton scatterings in a given proton (nucleon) collision increases with $\sqrt{s}$, and provides valuable information on the badly-known 3D partonic profile of hadrons, on the unknown energy evolution of the parton density as a function of impact parameter b, and on the role of partonic spatial, momentum, flavor, color,... correlations in the hadronic wave functions.

We have reviewed the factorized framework that allows one to compute the cross-sections for the simultaneous perturbative production of particles in double (DPS), triple (TPS), and in general n-parton (NPS) scatterings, from the corresponding single-parton scattering (SPS) cross-sections in proton–proton, proton–nucleus, and nucleus–nucleus collisions. The basic parameter of the factorized ansatz is an effective cross-section parameter, σ_{eff}, encoding all unknowns about the underlying generalized n-parton distribution function in the proton (nucleon). In the simplest and most phenomenologically-useful approach, we have shown that σ_{eff} bears a simple geometric interpretation in terms of powers of the inverse of the integral of the hadron–hadron overlap function over all impact parameters. Simple recursive expressions can thereby be derived to compute the NPS cross-section from the nth product of the SPS ones, normalized by $(n-1)$th power of σ_{eff}. In the case of pp collisions, a particularly simple and robust relationship between the effective DPS and TPS cross-sections, $\sigma_{\text{eff,DPS}} = (0.82 \pm 0.11) \times \sigma_{\text{eff,TPS}}$, has been extracted from an exhaustive analysis of typical parton transverse distributions of the proton, including those commonly used in Monte Carlo hadronic generators such as Pythia 8 and Herwig.

In proton–nucleus and nucleus–nucleus collisions, the parton flux is augmented by the number A and A^2, respectively, of nucleons in the nucleus (nuclei). The larger nuclear transverse parton density compared to that of protons, results in enhanced probability for NPS processes, coming from interactions where the colliding partons belong to the same nucleon, and/or to two or more different nucleons. Whereas the standard SPS cross-sections scale with the mass-number A in pA relative to pp collisions, we have found that the DPS and TPS cross-sections are further enhanced by factors of order $(A + (1/\pi)\,A^{4/3})$ and $(A + (2/\pi)\,A^{4/3} + (1/\pi^2)\,A^{5/3})$ respectively. In the case of p–Pb collisions, this implies enhancement factors of $\sim$600 (for DPS) and of $\sim$1900 (for TPS) with respect to the corresponding SPS cross-sections in pp collisions. The relative roles of intra- and inter-nucleon parton contributions to DPS and TPS cross-sections in pA collisions have been also derived. In p–Pb, 1/3 of the DPS yields come from partonic interactions within just one nucleon of the Pb nucleus, whereas 2/3 involve scatterings from partons of two Pb nucleons; whereas for the TPS yields, 10% of them come from partonic interactions within one nucleon, 50% involve scatterings within two nucleons, and 40% come from partonic interactions in three different Pb nucleons. In proton–nucleus collisions, one can thereby exploit the large expected DPS and TPS signals over the SPS backgrounds to study DPS and TPS in detail and, in particular, to extract the value of the key $\sigma_{\rm eff,DPS}$ parameter independently of measurements in pp collisions, given that the corresponding NPS yields in pA depend on the comparatively better-known nuclear transverse density profile.

For heavy ions, the A^2-scaling of proton–proton SPS cross-sections becomes $\propto (A^2 + (2/\pi)\,A^{7/3} + 1/(2\pi)\,A^{10/3})$ for DPS cross-sections, and includes much larger powers of A (up to $A^{14/3}$) for TPS processes. In the Pb–Pb case, these translate into many orders-of-magnitude enhancements (e.g., the DPS cross-sections are $\sim 10^7$ larger than the corresponding SPS pp ones). In addition, the MPI probability is significantly enhanced for increasingly central collisions: the impact-parameter dependence of DPS cross-sections is basically proportional to the AA nuclear overlap function at a given b. The huge DPS and TPS cross-sections expected in AA collisions are, however, clearly dominated by scatterings among partons of *different* nucleons, rather than by partons belonging to the same proton or neutron. For nuclei with mass numbers $A = 40$–240, the relative weights of the three components contributing to DPS scattering in AA collisions are $1 : 2.3 : 23$ (for $A = 40$) and $1 : 4 : 200$ (for $A = 208$). Namely, only 13% (for ^{40}Ca+^{40}Ca) or 2.5% (for ^{208}Pb+^{208}Pb) of the DPS yields in AA collisions come from diagrams involving partons from one single nucleon. Clearly, the "pure" DPS contributions involving partonic collisions within a nucleon are much smaller than those issuing from two independent *nucleon–nucleon* collisions. In the TPS case, the relative weights of the two extreme contributions (three parton collisions within two single nucleons versus those from three different nucleon–nucleon collisions) are $1 : 40\,000$ for Pb–Pb. The NPS cross-sections in AA are practically unaffected by the value of $\sigma_{\rm eff,DPS}$ and, although DPS and TPS processes account for a significant fraction of the inelastic hard AA cross-section, they are not as useful as those in pp or pA collisions to study the partonic structure of the proton (nucleon).

Numerical examples for the cross-sections and visible yields expected for the concurrent DPS and TPS production of heavy quarks, quarkonia, and/or gauge bosons in proton and nuclear collisions at LHC, FCC, and at ultra-high cosmic-ray energies have been provided. The obtained DPS and TPS cross-sections are based on perturbative QCD predictions for the corresponding single inclusive processes at NLO or NNLO accuracy including, when needed, nuclear modifications of the corresponding parton densities. Processes such as double-J/ψ, J/ψ Υ, J/ψ W, J/ψ Z, double-Υ, Υ W, Υ Z, and same-sign W W production have large cross-sections and visible event rates for the nominal LHC and FCC luminosities. The study of such processes in proton–nucleus collisions provides an independent means to extract the effective $\sigma_{\rm eff,DPS}$ parameter characterising the transverse parton distribution in the nucleon. In addition, we have shown that double-J/ψ and double-Υ final states have to be explicitly taken into account in any event-by-event analysis of quarkonia production in heavy-ion collisions. The TPS processes, although not observed so far, have visible cross-sections for charm and bottom in pp and pA collisions at LHC and FCC energies. At the highest c.m. energies reached in collisions of cosmic rays with the nuclei in the upper atmosphere, the TPS cross-section for triple charm-pair production equals the total p–air inelastic cross-section, indicating that the average number of $c\bar{c}$-pairs produced in multiple partonic interactions is above unity. The results presented here emphasize the importance of having a good understanding of the NPS dynamics in hadronic collisions at current and future colliders, both as genuine probes of QCD phenomena and as backgrounds for searches of new physics in rare final states with multiple heavy particles, and their relevance in our comprehension of ultrarelativistic cosmic-ray interactions with the atmosphere.

Part II

Soft MPI: Phenomenology and Description in MC Generators

Chapter 10

The Development of MPI Modeling in Pythia

Torbjörn Sjöstrand

Department of Astronomy and Theoretical Physics, Lund University, Sölvegatan 14A, SE-223 62 Lund, Sweden

Many of the basic ideas in multiparton interaction (MPI) phenomenology were first developed in the context of the Pythia event generator, and MPIs have been central in its modeling of both minimum-bias (MB) and underlying-event physics in one unified framework. This chapter traces the evolution towards an increasingly sophisticated description of MPIs in Pythia, including topics such as the ordering of MPIs, the regularization of the divergent QCD cross-section, the impact-parameter picture, color reconnection, multiparton PDFs and beam remnants, interleaved and intertwined evolution, and diffraction.

1. Introduction

The Pythia event generator[322] was initially created to explore the physics of color flow in hadronic collisions, in analogy with how the Lund string model[323] had successfully predicted string effects in e^+e^- annihilation.[324,325] Initially only $2 \to 2$ partonic (q, g, γ) processes were implemented, with color flow connecting the scattered partons to the beam remnants, followed by string fragmentation using Jetset.[326] At the 1984 Snowmass workshop on the SSC, when I first got directly involved in the physics of high-energy hadron colliders, it was obvious that this approach was too primitive to be of relevance. During the autumn I implemented initial- and final-state radiation (ISR and FSR),[327] with the expectation that this further activity would give event topologies more comparable with Sp$\bar{p}$S data. In terms of jet phenomenology it did, but underlying events were still much less active than in data.

The natural explanation, in my opinion, was that the composite nature of the proton would lead to several parton–parton interactions, giving more activity. Thus in the spring of 1985, I developed a first multiparton interaction (MPI) model, still primitive but offering a significantly improved description of data, convincing me that MPIs was the way to go. Not everybody approved; the first writeup[328] was not accepted for publication. In 1986 studies resumed, and several further key aspects were introduced.[35] In its basic ideology, this formalism has remained, even if the details have been improved and extended many times over the years.

This evolution will be described in the following, and in the process, an overview will be given of all the components of the current framework. While Pythia-centred, external sources of inspiration (in a positive or negative sense) will be mentioned, with emphasis on the early days, when the basic ideas were formulated. Much more information can be obtained from the companion articles of this book, about other models and generators, and about all the experimental studies that have been undertaken over the years. Notably, no experimental plots are shown, since relevant ones are already reproduced elsewhere, see Chapters 6–8 and 11–16, often compared with Pythia and other generators.

2. Early Data and Models

In the eighties, the S$p\overline{p}$S was providing new data on hadronic collisions, at an order of magnitude higher CM energy than previously available, from 200 to 900 GeV. It came to change our understanding of hadronic collisions. Some observations are of special interest for the following:

- The width $\sigma(n_{\mathrm{ch}})$ of the charged multiplicity distribution is increasing with energy such that $\sigma(n_{\mathrm{ch}})/\langle n_{\mathrm{ch}}\rangle$ stays roughly constant,[329,330] "KNO scaling",[331] actually even slowly getting broader. A close-to Poissonian process, in longitudinal phase space or in the fragmentation of a single straight string, instead would predict a $1/\sqrt{\langle n_{\mathrm{ch}}\rangle}$ narrowing.
- Multiplicity fluctuations show long-range "forward–backward" correlations,[332] defined by

$$b_{\mathrm{FB}}(\Delta\eta) = \frac{\langle n_{\mathrm{F}} n_{\mathrm{B}}\rangle - \langle n_{\mathrm{F}}\rangle^2}{\langle n_{\mathrm{F}}^2\rangle - \langle n_{\mathrm{F}}\rangle^2} , \tag{1}$$

  where n_{F} and n_{B} is the (charged) multiplicity in two symmetrically located unit-width pseudorapidity bins, separated by a central variable-width $\Delta\eta$ gap. Again this is not expected in Poissonian processes.
- The average transverse momentum $\langle p_{\perp}\rangle$ increases with increasing charged multiplicity.[333,334] This is opposite to the behavior at lower energies, where energy–momentum conservation effects dominate, with a crossover at the highest ISR energies.[335]
- A non-negligible fraction of the total cross-section is associated with minijet production,[336,337] increasing from $\sim$5% at 200 GeV to $\sim$15% at 900 GeV. Here UA1 defined a minijet as a region $\Delta R = \sqrt{(\Delta\eta)^2 + (\Delta\varphi)^2} \leq 1$ with $\sum E_{\perp} > 5$ GeV.

- The increase of the total $p\overline{p}$ cross-section $\sigma_{\mathrm{tot}}(s)$ rather well matches that of the minijet one $\sigma_{\mathrm{minijet}}(s)$, i.e., $\sigma_{\mathrm{tot}}(s) - \sigma_{\mathrm{minijet}}(s)$ is almost constant.[336,337]
- Events with a minijet have a rather flat $\langle p_{\perp} \rangle (n_{\mathrm{ch}})$, while ones without show a strong rise, starting from a lower level.[336]
- The fraction of events having several minijets is non-negligible. (Rates up to 5 are quoted from workshop presentations in Ref. 35, but apparently never published.)
- Events containing a hard jet also have an above-average level of particle production well away from the jet core,[338] the "pedestal effect". Of note is that the pedestal increases rapidly up to $E_{\perp \mathrm{jet}} \sim 10$ GeV, and then flattens out, even dropping slightly.[337]
- Also the jet profiles are affected by this extra source of activity.
- By contrast, there were no early studies on double parton scattering (DPS) at the S$p\overline{p}$S. The first observation instead came from AFS at ISR,[162] in a study of pairwise balancing jets in four-jet events, but it did not convince everybody.

On the theoretical side, the basic idea of MPI existed,[6,30,31,161,339–341] see also Chapter 3. These first studies almost exclusively considered DPS, without a vision of an arbitrary number of scatterings. Studies often only included scattering of valence quarks, since the large-x region was needed to access "large" jet $p_{\perp}$ scales. Therefore DPS/MPI was only expected to correspond to a tiny fraction of the total cross-section. If needed, a $p_{\perp \mathrm{min}}$ cutoff would be introduced at a sufficiently high value to make it so.

For soft physics, the pomeron language was predominant, notably in its dual topological unitarization (DTU) formulation, both to describe total cross-sections and event topologies.[38,342–352] In it a cut pomeron corresponds to two multiperipheral chains, or strings in Lund language, stretched directly between the two beam remnants after the collision. In most of the earlier phenomenological studies only one cut pomeron was used, but extensions to multiple pomerons were introduced for S$p\overline{p}$S applications. Then the number of cut pomerons can vary freely, e.g., according to a Poissonian. Uncut pomerons, i.e., virtual corrections, ensure unitarity. This approach was quite successful in describing aspects of the data such as the charged multiplicity distribution and forward–backward correlations.

In contrast to the unitarization approach, the good match between the rise of $\sigma_{\mathrm{tot}}(s)$ and $\sigma_{\mathrm{minijet}}(s)$ led to speculations that $\sigma_{\mathrm{tot}}(s)$ (or at least its inelastic component) could be written as an incoherent sum $\sigma_{\mathrm{tot}}(s) = \sigma_{\mathrm{soft}} + \sigma_{\mathrm{minijet}}(s)$.[353–355]

At the time, there appears to have been little "middle ground" between the hard MPI, the soft multi-pomeron and the UA1-minijet ways of approaching physics.

On the generator side, ISAJET[356] was state of the art. It described one hard interaction with its showers, and then added an underlying event (UE) based on the pomeron approach. The UE was intended to reproduce minimum-bias (MB) event properties at the hard-interaction-reduced collision energy. Since it was based on independent fragmentation the two components could be easily decoupled. Other generators for hard interactions[357,358] had more primitive UE descriptions, and the one generator for MB[359] did not include hard interactions. In addition (unpublished) longitudinal phase-space models tuned to inclusive data were used within the experimental collaborations, ultimately refined into the UA5 generator.[360]

3. The First Pythia Model

Against this backdrop, the key new idea of the first Pythia model[328] was to reinterpret the multi-pomeron picture in terms of multiple perturbative QCD interactions. Thus there would no longer be the need for separate descriptions of MB and UE physics. A hard-process event would just be the high-$p_\perp$ tail of the MB class, and a soft-process event just one where the hardest jet was too soft to detect as such. MPIs come out as an unavoidable consequence, not only as a tiny tail of hard DPS events, but as representing the bulk of the inelastic non-diffractive cross-section σ_{nd} at higher energies.

By contrast, no importance could be attached to the 5 GeV UA1 minijet cutoff scale or to the seemingly simple relationship between $\sigma_{\mathrm{tot}}(s)$ and $\sigma_{\mathrm{minijet}}(s)$ that it led to. On the contrary, MPIs had to extend much lower in $p_\perp$ in order to give enough varying activity to describe, e.g., the approximate KNO scaling. Here jet universality was assumed, i.e., that the underlying fragmentation mechanism was the same string as described e^+e^- data so well, only applied to a more complicated partonic state.

In its technical implementation, the starting point of the model is the differential perturbative QCD $2 \to 2$ cross-section

$$\frac{d\sigma}{dp_\perp^2} = \sum_{i,j,k} \iiint f_i(x_1, Q^2)\, f_j(x_2, Q^2)\, \frac{d\hat{\sigma}_{ij}^k}{d\hat{t}}\, \delta\left(p_\perp^2 - \frac{\hat{t}\hat{u}}{\hat{s}}\right) dx_1\, dx_2\, d\hat{t}, \tag{2}$$

with $Q^2 = p_\perp^2$ as factorization and renormalization scale. The corresponding integrated cross-section depends on the chosen $p_{\perp\mathrm{min}}$ scale:

$$\sigma_{\mathrm{int}}(p_{\perp\mathrm{min}}) = \int_{p_{\perp\mathrm{min}}^2}^{s/4} \frac{d\sigma}{dp_\perp^2} dp_\perp^2, \tag{3}$$

see Fig. 1.

Diffractive events presumably give a small fraction of the perturbative jet activity, and elastic none, so the simple model sets out to describe only inelastic non-diffractive events, with an approximately known σ_{nd}. It is thus concluded that the average such event ought to contain

$$\langle n_{\mathrm{MPI}}(p_{\perp\mathrm{min}}) \rangle = \frac{\sigma_{\mathrm{int}}(p_{\perp\mathrm{min}})}{\sigma_{\mathrm{nd}}} \tag{4}$$

hard interactions. An average above unity corresponds to more than one such subcollision per event, which is allowed by the multiparton structure of the incoming

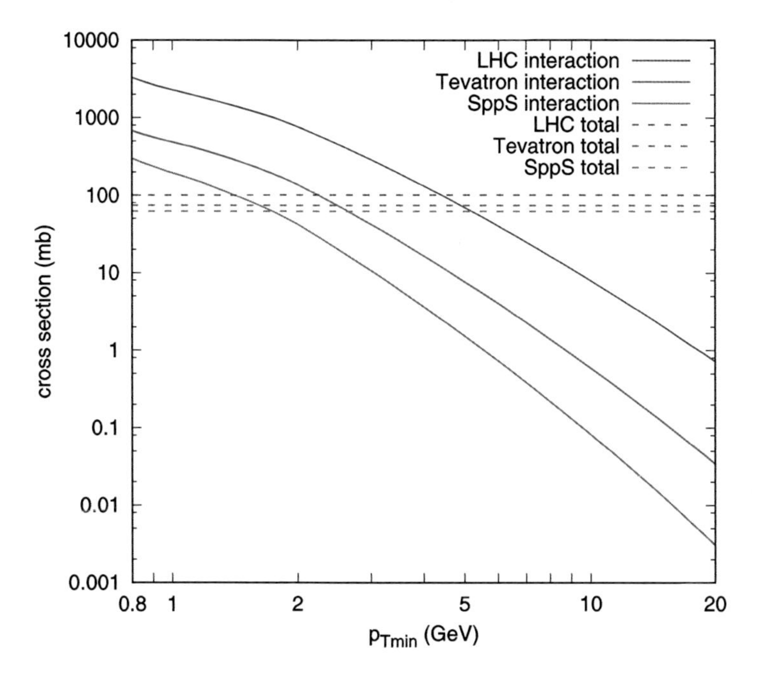

Fig. 1 The integrated interaction cross-section $\sigma_{\rm int}(p_{\perp\rm min})$ for the Sp$\bar{p}$S at 630 GeV, Tevatron at 1.96 TeV and LHC at 13 TeV. For comparison the total cross-section $\sigma_{\rm tot}$ at the respective energy is indicated by a horizontal line, with the non-diffractive part $\sigma_{\rm nd}$ at order 60% of this. Results have been obtained with the Pythia 8.223 default values, including the NNPDF2.3 QCD + QED LO PDF set with $\alpha_{\rm s}(M_Z) = 0.130$.[361]

hadrons. If the interactions were to occur independently of each other, $n_{\rm MPI}(p_{\perp\rm min})$ would be distributed according to a Poissonian. But such an approach would be flawed, e.g., sometimes using up more energy for collisions than is available.

The solution to this problem was inspired by the parton-shower paradigm. The generation of consecutive MPIs is formulated as an evolution downwards in $p_\perp$, resulting in a sequence of n interactions with $\sqrt{s}/2 > p_{\perp 1} > p_{\perp 2} > \cdots > p_{\perp n} > p_{\perp\rm min}$. The probability distribution for $p_{\perp 1}$ becomes

$$\frac{d\mathscr{P}}{dp_{\perp 1}} = \frac{1}{\sigma_{\rm nd}} \frac{d\sigma}{dp_{\perp 1}} \exp\left(- \int_{p_{\perp 1}}^{\sqrt{s}/2} \frac{1}{\sigma_{\rm nd}} \frac{d\sigma}{dp'_\perp} dp'_\perp \right). \tag{5}$$

Here, the naive probability is corrected by an exponential factor expressing that there must not be any interaction in the range between $\sqrt{s}/2$ and $p_{\perp 1}$ for $p_{\perp 1}$ to be the hardest interaction. The procedure can be iterated, to give

$$\frac{d\mathscr{P}}{dp_{\perp i}} = \frac{1}{\sigma_{\rm nd}} \frac{d\sigma}{dp_{\perp i}} \exp\left(- \int_{p_{\perp i}}^{p_{\perp i-1}} \frac{1}{\sigma_{\rm nd}} \frac{d\sigma}{dp'_\perp} dp'_\perp \right). \tag{6}$$

The exponential factors resemble Sudakov form factors of parton showers,[362,363] or uncut pomerons for that matter, and fills the same function of ensuring probabilities bounded by unity. Summing up the probability for a scattering at a given $p_{\perp}$ scale to happen at any step of the generation chain gives back $(1/\sigma_{\text{nd}})\, d\sigma/dp_{\perp}$, and the number of interactions above any $p_{\perp}$ is a Poissonian with an average of $\sigma_{\text{int}}(p_{\perp})/\sigma_{\text{nd}}$, as it should. The downwards evolution in $p_{\perp}$ is routinely handled by using the veto algorithm,[158] like for showers.

The similarities with showers should not be overemphasized, however. While the shower $p_{\perp}$ scale has some approximate relationship to an evolution in time, this is not so for MPIs. Rather, when the two Lorentz-contracted hadron "pancakes" collide, the MPIs can be viewed as occurring simultaneously in different parts of the overlap region. What is instead gained is a way to handle the parton distribution functions (PDFs) of several partons in the same hadron, at the very least to conserve overall energy and momentum. Specifically, it is for the hardest MPI that conventional PDFs have been tuned and tested, so we had better respect that. For subsequent MPIs no PDF data exist, so some adjustments are acceptable. In this first implementation only rescaled PDFs

$$f(x_i', Q^2) \quad \text{with } x_i' = \frac{x_i}{1 - \sum_{j=1}^{i-1} x_j} < 1 \tag{7}$$

are used for the ith interaction. This rescaling suppresses the tail towards events with many MPIs, so the n_{MPI} distribution becomes narrower than Poissonian.

To complement the model, a number of further details of the simulation had to be specified, often intended as temporary solution.

- There is a finite probability that no MPIs at all are generated above $p_{\perp\text{min}}$. For this set of events, small but not negligible, an infinitely soft gluon exchange is assumed, leading to two strings stretched directly between the beam remnants.
- Only the hardest interaction is allowed to be any combination of incoming and outgoing flavors, weighted according to Eq. (2). For subsequent ones kinematics is chosen the same way, with modified PDFs, but afterwards the process is always set up to be of the $gg \to gg$ type. The reason is to avoid complicated beam-remnant structures.
- The color flow of the hardest interaction is described by the original Pythia algorithm.[322] Two extreme scenarios for the color flow of the non-hardest MPIs were compared. In the simplest one, each such gives rise to a double string stretched between the two outgoing gluons of the MPI, disconnected from the rest of the event.
- By the choices above, where only the hardest interaction affects the flavor and color of the beam remnant, a limited number of remnant types can be obtained. If a valence quark is kicked out, a diquark is left behind. If a gluon, the leftover color octet state of a proton can be split into a quark and a diquark that attach to

two separate strings. If the two remnants then share the longitudinal momentum evenly, it maximizes the particle production. This gives too few low-multiplicity events, and also leaves less room for MPIs to build up a high-multiplicity tail, assuming that the average is kept fixed. Therefore a probability distribution is used wherein the quark often obtains much less momentum than the diquark. Finally, if a sea (anti)quark is kicked out, the remnant can be split into a single hadron plus a quark or diquark.

- Only the hardest interaction is dressed up with showers, whereas the subsequent ones are not. Again the reason is beam-remnant issues, but one excuse is that most non-hardest MPIs appear at low $p_\perp$ scales, where little further radiation should be allowed.

The key free parameter of this framework is $p_{\perp\mathrm{min}}$. The lower it is chosen, the higher the average number of MPIs, cf. Eq. (3), and thus the higher the average charged multiplicity $\langle n_{\mathrm{ch}} \rangle$. To agree with 540 GeV UA5 data[329] a value of $p_{\perp\mathrm{min}} = 1.6$ GeV was required. The dependence of $\langle n_{\mathrm{ch}} \rangle$ on $p_{\perp\mathrm{min}}$ is quite strong so, if everything else is kept fixed, the $p_{\perp\mathrm{min}}$ uncertainty is of order ± 0.1 GeV. The most agreeable aspect, however, is that with $p_{\perp\mathrm{min}}$ tuned, the shape of the n_{ch} distribution is reasonably well described. Without MPIs the shape is Poissonian-like, also when a single hard interaction is allowed. With MPIs instead an approximate KNO scaling behavior is obtained, driven by the n_{MPI} distribution. (Which, even if also a Poissonian, obeys $\langle n_{\mathrm{MPI}} \rangle \ll \langle n_{\mathrm{ch}} \rangle$, meaning much larger relative fluctuations $\sigma / \langle n \rangle$.) By the same mechanism also strong forward–backward correlations are obtained, where before these were tiny. That is, the n_{MPI} is a kind of global quantum number of an event, that affects whether particle production is high or low over the whole rapidity range. With some damping, since not all strings are stretched equally far out to the beam remnants.

In part this is nothing new; the number of cut pomerons in soft models fills a similar function for both n_{MPI} distributions and forward–backward correlations. What is new is that an application of perturbation theory, in combination with string fragmentation, can give a reasonable description also of MB physics. This unifies hard and soft physics at colliders, as being part of the same framework. It also introduces a new cutoff scale in QCD, with a value different from other scales, such as the proton mass.

It was clear from the onset that the model was incomplete in its details, and the listed open questions for the model well matches the problems that have later been studied.

- What is the correct behavior of $d\sigma / dp_\perp^2$ at small $p_\perp$? A sharp cutoff, below which cross-sections vanish, is not plausible.
- How to remove (or, if not, interpret) the class of events with no MPIs, currently represented by a $p_\perp = 0$ interaction?

- How to introduce an impact-parameter picture, giving more activity for central collisions and less for peripheral? This is needed to give an a bit wider n_{ch} distribution. Also, for UA1 jets the MPI formalism as it stands at this stage only gives about a quarter of the observed pedestal effect.
- How to achieve a better description of multiparton PDFs, that also consistently includes e.g., flavor conservation and correlations?
- Where does the baryon number go if several valence quarks are kicked out from a proton?
- How does the color singlet nature of the incoming beams translate into color correlations between the different MPIs?
- What is the structure and role of beam remnants?
- By confinement and the uncertainty relation the incoming partons must have some random non-perturbative transverse motion. How should such "primordial $k_{\perp}$" effects be included? These then have to be compensated in the remnants, and furthermore the remnant parts may have relative $k_{\perp}$ values of their own.
- How should parton-shower effects be combined consistently between the systems? The flavor, color and beam-remnant issues reappear here.
- How important is ISR evolution wherein a parton branches into two that participate in two separate interactions?
- How important is rescattering, i.e., when one parton can scatter consecutively from two or more partons from the other hadron?
- How do diffractive topologies contribute to the picture? Typical experimental "minimum bias" triggers catch a fraction of these events, which have different properties from the non-diffractive ones. The low-multiplicity end of the n_{ch} distribution was left unexplained in the studies, with the motivation that it is dominated by diffraction.
- How do the results scale with collision energy? With a fixed $p_{\perp\mathrm{min}}$ scale it was possible to reproduce the $\langle n_{\mathrm{ch}} \rangle$ evolution from fixed-target to 900 GeV, and this was the basis for extrapolations.

4. Smooth Damping and Impact-Parameter Dependence

For the first published MPI article,[35] the original framework was extended to address some of the most pressing shortcomings above. (The older approach was also retained as a simpler alternative. Unfortunately the new approach led to longer computer generation times, which was a real issue at the time.)

The sharp cutoff $p_{\perp\mathrm{min}}$ is replaced by a smooth turnoff at a scale $p_{\perp 0}$. To be specific, the cross-section of Eq. (2) is multiplied by a damping factor

$$\left(\frac{\alpha_{\mathrm{s}}(p_{\perp 0}^2 + p_{\perp}^2)}{\alpha_{\mathrm{s}}(p_{\perp}^2)} \, \frac{p_{\perp}^2}{p_{\perp 0}^2 + p_{\perp}^2} \right)^2 . \tag{8}$$

Since the QCD $2 \to 2$ processes are dominated by t-channel gluon exchange, which behaves like $1/\hat{t}^2 \sim 1/p_\perp^4$, this means that

$$\frac{d\sigma}{dp_\perp^2} \sim \frac{\alpha_s^2(p_\perp^2)}{p_\perp^4} \to \frac{\alpha_s^2(p_{\perp 0}^2 + p_\perp^2)}{(p_{\perp 0}^2 + p_\perp^2)^2}, \tag{9}$$

which is finite in the limit $p_\perp \to 0$. This behavior can be viewed as a consequence of color screening: in the $p_\perp \to 0$ limit a hypothetical exchanged gluon would not resolve individual partons but only (attempt to) couple to the vanishing net color charge of the hadron. Technically the damping factor is multiplying the $d\hat{\sigma}/d\hat{t}$ expressions, but it could equally well have been imposed (half each) on the PDFs instead, since neither can be trusted for $p_\perp \to 0$.

In this modified framework all interactions are associated with a $p_\perp > 0$ scale, and at least one interaction must occur when two hadrons pass by for there to be an event at all. Thus we require $\sigma_{\mathrm{int}}(0) > \sigma_{\mathrm{nd}}$, where the σ_{int} integration, Eq. (3), now includes the damping factor. A tune to $\langle n_{\mathrm{ch}} \rangle$ at 540 GeV gives $p_{\perp 0} \approx 2.0$ GeV, i.e., of the same order as the sharp $p_{\perp\mathrm{min}}$ cutoff. The two would have been even closer, had not factorization and renormalization scales here been multiplied by 0.075, as suggested at the time to obtain an approximate NLO jet cross-section.[364] Below Sp$\bar{p}$S energies a fixed $p_{\perp 0}$ gives too small a $\sigma_{\mathrm{int}}(0)$, so in this form the model is primarily intended for high-energy collider physics.

The other main change was to introduce a dependence on the impact parameter b of the collision process. To do this, a spherically symmetric matter distribution $\rho(\mathbf{x})\, d^3x = \rho(r)\, d^3x$ is assumed. In a collision process the overlap of the two hadrons is then given by

$$\begin{aligned} \widetilde{\mathcal{O}}(b) &= \iint d^3x\, dt\, \rho_{\mathrm{boosted}}\left(x - \frac{b}{2}, y, z - vt\right) \rho_{\mathrm{boosted}}\left(x + \frac{b}{2}, y, z + vt\right) \\ &\propto \iint d^3x\, dt\, \rho(x, y, z)\, \rho(x, y, z - \sqrt{b^2 + t^2}), \end{aligned} \tag{10}$$

where the second line is obtained by suitable scale changes.

A few different ρ distributions were studied, see Fig. 3 of Ref. 365. Using Gaussians is especially convenient, since the convolution then becomes trivial. A simple Gaussian was the starting point, but we found it did not give a good enough description of the data. Instead the preferred choice was a double Gaussian

$$\rho(r) = (1 - \beta)\, \frac{1}{a_1^3} \exp\left(-\frac{r^2}{a_1^2}\right) + \beta\, \frac{1}{a_2^3} \exp\left(-\frac{r^2}{a_2^2}\right). \tag{11}$$

This corresponds to a distribution with a small core region, of radius a_2 and containing a fraction β of the total hadronic matter, embedded in a larger hadron of radius a_1. The choice of a not-so-smooth shape was largely inspired by the "hot spot" ideas popular at the time.[235,366] The starting point is that, as a consequence of parton cascading, partons may tend to cluster in a few small regions, typically associated with the three valence quarks.

More convoluted ansätze could have been considered, but having two free parameters, β and a_2/a_1, was sufficient to give the necessary flexibility.

It is now assumed that the interaction rate, to first approximation, is proportional to the overlap

$$\langle \widetilde{n}_{\mathrm{MPI}}(b) \rangle = k\, \widetilde{\mathcal{O}}(b). \tag{12}$$

For each given b the number of interactions is assumed distributed according to a Poissonian, at least before energy–momentum conservation issues are considered. Zero interactions means that the hadrons pass each other without interacting. The $\widetilde{n}_{\mathrm{MPI}}(b) \geq 1$ interaction probability therefore is

$$\mathscr{P}_{\mathrm{int}}(b) = 1 - \exp\left(-\langle \widetilde{n}_{\mathrm{MPI}}(b) \rangle\right) = 1 - \exp(-k\, \widetilde{\mathcal{O}}(b)). \tag{13}$$

We notice that $k\widetilde{\mathcal{O}}(b)$ is essentially the same as the eikonal $\Omega(s,b) = 2\,\mathrm{Im}\chi(s,b)$ of optical models,[367–370] but split into one piece $\widetilde{\mathcal{O}}(b)$ that is purely geometrical and one $k = k(s)$ that carries the information on the parton–parton interaction cross-section. Furthermore, this is (so far) only a model for non-diffractive events, so does not attempt to relate the eikonal to total or diffractive cross-sections.

Simple algebra shows that the average number of interactions in events, i.e., hadronic passes with $n_{\mathrm{MPI}} \geq 1$, is given by

$$\langle n \rangle = \frac{\int k\, \widetilde{\mathcal{O}}(b)\, d^2 b}{\int \mathscr{P}_{\mathrm{int}}(b)\, d^2 b} = \frac{\sigma_{\mathrm{int}}(0)}{\sigma_{\mathrm{nd}}}, \tag{14}$$

which fixes the absolute value of k (numerically).

For event generation, Eq. (5) generalizes to

$$\frac{d\mathscr{P}}{d^2 b\, dp_{\perp 1}} = \frac{\widetilde{\mathcal{O}}(b)}{\langle \widetilde{\mathcal{O}} \rangle} \frac{1}{\sigma_{\mathrm{nd}}} \frac{d\sigma}{dp_{\perp 1}} \exp\left(- \frac{\widetilde{\mathcal{O}}(b)}{\langle \widetilde{\mathcal{O}} \rangle} \int_{p_{\perp 1}}^{\sqrt{s}/2} \frac{1}{\sigma_{\mathrm{nd}}} \frac{d\sigma}{dp'_{\perp}} dp'_{\perp} \right), \tag{15}$$

with the definition

$$\langle \widetilde{\mathcal{O}} \rangle = \frac{\int \widetilde{\mathcal{O}}(b) d^2 b}{\int \mathscr{P}_{\mathrm{int}}(b) d^2 b}. \tag{16}$$

Hence $\widetilde{\mathcal{O}}(b)/\langle \widetilde{\mathcal{O}} \rangle$ represents the enhancement at small b and depletion at large b. The simultaneous selection of $p_{\perp 1}$ and b is somewhat more tricky. In practice different schemes are used, depending on context.

- For MB events Eq. (15) can be integrated oven $p_{\perp 1}$ to give $\mathscr{P}_{\mathrm{int}}(b)$ of Eq. (13). To pick such a b it is useful to note that $\mathscr{P}_{\mathrm{int}}(b) < \min(1, k\, \widetilde{\mathcal{O}}(b))$ and split the b range accordingly. Once b is fixed the selection of $p_{\perp 1}$ can be done as for Eq. (5), only with an extra fix $\widetilde{\mathcal{O}}(b)/\langle \widetilde{\mathcal{O}} \rangle$, both in the prefactor and in the exponential.

If $p_{\perp 1} = 0$ is reached in the downwards evolution without an interaction having been found, which happens with probability $\exp(-k\,\widetilde{\mathcal{O}}(b))$, then the generation is restarted at the maximum scale $\sqrt{s}/2$.

- For a very hard process the exponential of Eq. (15) is very close to unity and can be dropped. Then the selection of b and $p_{\perp 1}$ decouples and can be done separately. Here, $d\sigma/dp_{\perp 1}$ can represent any hard process, not only QCD jets, and $p_{\perp 1}$ any set of relevant kinematic variables.
- For a medium-hard process one can begin as in the hard case, and then use the exponential as an acceptance probability. If the hard-process kinematics is considered fixed then only a new b value is chosen in case of rejection. Note that it is always the QCD cross-section that enters in the exponential. (Or, to be proper, the sum of all possible reactions, but that is completely dominated by QCD.) For non-QCD processes the $p_{\perp 1}$ scale in the exponential has to be associated with some suitable hardness scale, like the mass for the production of a resonance.

Once the hardest interaction is chosen, the generation of subsequent ones proceeds by a logical extension of Eq. (6) to

$$\frac{d\mathcal{P}}{dp_{\perp i}} = \frac{\widetilde{\mathcal{O}}(b)}{\langle\widetilde{\mathcal{O}}\rangle}\frac{1}{\sigma_{\mathrm{nd}}}\frac{d\sigma}{dp_{\perp i}}\exp\left(-\frac{\widetilde{\mathcal{O}}(b)}{\langle\widetilde{\mathcal{O}}\rangle}\int_{p_{\perp i}}^{p_{\perp i-1}}\frac{1}{\sigma_{\mathrm{nd}}}\frac{d\sigma}{dp'_{\perp}}\,dp'_{\perp}\right). \tag{17}$$

There is one subtlety to note about ordering, however. QCD interactions have to be ordered in $p_{\perp}$ for the formalism to reproduce the correct inclusive cross-section. This applies for the MB generation, which gives an arbitrary $p_{\perp 1}$, and also in a sample of hard jets above some large $p_{\perp\mathrm{min}}$ scale. But it does not hold for non-QCD hard processes. For Z^0 production, say, which is not part of the normal MPI machinery, the second MPI (counting the Z^0 as the first) can go all the way up to the kinematic limit in $p_{\perp}$ without involving any double-counting, with $p_{\perp}$-ordering only kicking in for the third MPI.

With MPIs stretching down to $p_{\perp} = 0$, the need arises to evaluate PDFs below their lower limit Q_0 scale, typically 1 – 2 GeV. To first approximation this is done by freezing them below Q_0, but some attempts were made to enhance the relative importance of valence quarks for $Q \to 0$, since this is what one should expect to happen.

As before, color drawing for all MPIs except the hardest one is handled in a primitive manner. Given that the kinematics of an interaction has been chosen with the full cross-section, the final state is picked among three possibilities, Fig. 2, by default with equal probability.

(a) Assume the collision to have produced a gg pair and stretch two string pieces directly between them, giving a closed gluon loop.
(b) Assume the collision to have produced a $q\overline{q}$ pair and stretch a string directly between them.

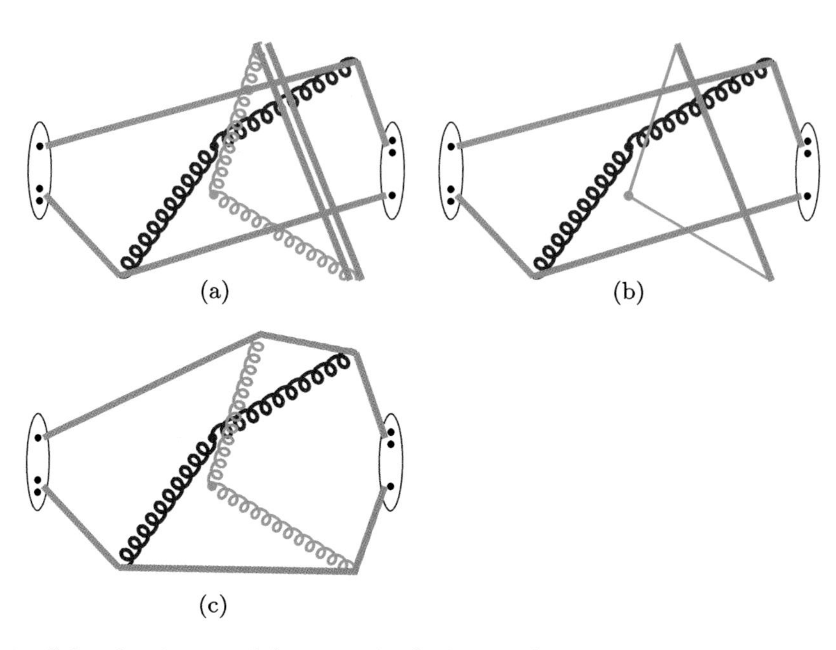

Fig. 2 Color drawing possibilities in the final state for the simple model. Thick blue gluons denote outgoing partons from the primary interaction, thin green gluons or quark lines the partons of a further MPI, black ovals the beam remnants with valence quarks, and orange thick lines the color strings stretched between the partons. While the primary interaction and its connection to the beam remnants is handled according to the color flow of the matrix elements, in the $N_C \to \infty$ limit,[371] the further MPIs give a mix of behaviors (a), (b) and (c), as described in the text. Note that the figure is not to scale; e.g., that the strings have a transverse width of hadronic size.

(c) Assume the collision to have produced a gg pair, but insert them separately on an already existing string in such a way so as to minimize the increase of string length λ.[372] Here,

$$\lambda \approx \sum_{i=0}^{n} \ln\left(1 + \frac{m_{i,i+1}^2}{m_0^2}\right), \quad m_{i,i+1}^2 = (\epsilon_i p_i + \epsilon_{i+1} p_{i+1})^2, \tag{18}$$

for a string $q_0 g_1 g_2 \cdots g_n \overline{q}_{n+1}$, where $\epsilon_q = 1$ but $\epsilon_g = 1/2$ because a gluon momentum is shared between two string pieces it is connected to.

Neither of these three follow naturally from any color flow rules, such as t-channel gluon exchange. Rather the first two represent the simplest way to decouple different interaction systems from each other, not having to trace colors back through the beam remnants. If MPIs are such separated systems, and thus on the average each gives the same $\langle p_\perp \rangle$, then an essentially flat $\langle p_\perp \rangle(n_{\mathrm{ch}})$ would be expected, since the study of the n_{ch} distribution tells us that higher n_{ch} values is a consequence of more MPIs rather than of harder jets. To obtain a rising $\langle p_\perp \rangle(n_{\mathrm{ch}})$ it is therefore essential to have a mechanism to connect the different MPI subsystems in color, not only at random but specifically so as to reduce the total string length of the event, more and more the more MPIs there are. Each further MPI on the average then contributes less n_{ch} than the previous, while still the same (semi)hard $p_{\perp}$ kick is to be shared between the hadrons, thus inducing the rising trend. This is precisely what the third and last component is intended to do. It is the first large-scale application of color reconnection (CR) ideas, previous applications having been for more specific channels such as $B \to J/\psi$ decays.[373–375]

A very simple model for diffraction was also added, wherein the diffractive mass M is selected according to dM^2/M^2 and is represented by a single string stretched between a diquark in the forward direction and a quark in the backward one.

With these changes to the original model it is possible to obtain a quite reasonable description of essentially all the key experimental data outlined in Section 2. Above all, the model offered physics explanations for the behavior observed in data.

- For the charged multiplicity distribution, improvements in the high-multiplicity tail originate from the introduction of an impact-parameter picture, whereas the addition of diffraction improves the low-multiplicity one. To describe the energy dependence, where $\sigma(n_{\mathrm{ch}})/\langle n_{\mathrm{ch}}\rangle$ is slightly increasing with energy, the impact-parameter dependence is crucial, since the $\sigma(n_{\mathrm{MPI}})/\langle n_{\mathrm{MPI}}\rangle$ then does not fall, which it otherwise would when $\langle n_{\mathrm{MPI}}\rangle$ increases with energy. Also forward-backward correlations now are even stronger, reflecting the broader n_{MPI} distribution, and actually even somewhat above data. A number of other minimum-bias distributions look fine, like the $\mathrm{d}n_{\mathrm{ch}}/\mathrm{d}\eta$ spectrum, inclusively as well as split into multiplicity bins, except for the lowest one, which is dominated by diffraction.
- The $\langle p_{\perp}\rangle(n_{\mathrm{ch}})$ distribution is well described, both inclusively and split into samples with our without minijets. As already mentioned, the new CR mechanism here plays a key role to get the correct rising trend in the inclusive case, and to counteract the drop otherwise expected in the jet sample. Not only the slope but also the absolute value of $\langle p_{\perp}\rangle$ is well reproduced, without any need to modify the fragmentation $p_{\perp}$ width tuned to e^+e^- data. This is one of the key observations that lend credence to the jet universality concept.
- The UA1 minijet studies are rather well reproduced. Notably the default double Gaussian is needed to obtain the observed fraction with several minijets. The simpler alternatives with a single Gaussian, no b dependence, or no MPIs at all fared increasingly worse, even with α_{s} tuned to give the same average number of minijets.
- The pedestal effect, i.e., how the underlying event activity first rises with the trigger jet/cluster $E_{\perp}$, is well described, and explained. The rise is caused by a shift in the composition of events, from one dominated by fairly peripheral

collisions to one strongly biased towards central ones. In the model there is a limit for how far this biasing can go: the exponential in Eq. (15)) can be neglected once $p_{\perp 1} \simeq E_\perp$ is so large that $\sigma_{\text{int}}(p_{\perp 1}) \ll \sigma_{\text{nd}}$. This happens at around 10 GeV, explaining the origin of that scale. The probability distribution is then given by $\widetilde{\mathcal{O}}(b)\, d^2b$ independently of the $p_{\perp 1}$ value. The double Gaussian is required to obtain the correct pedestal height, whereas a single Gaussian undershoots. A slight drop of the pedestal height for $E_\perp > 25$ GeV can be attributed to a shift from mainly gg interactions to mainly $q\overline{q}$ ones.

In summary, most if not all of MB and UE physics at collider energies is explained and reasonably well described once the basic MPI framework has been complemented by a smooth turnoff of the cross-section for $p_\perp \to 0$, a requirement to have at least one MPI to get an event, an impact-parameter dependence, and a color reconnection mechanism.

5. Interlude

While the S$p\overline{p}$S had paved the way for a new view on hadronic collisions, the Tevatron rather contributed to cement this picture. KNO distributions kept on getting broader,[376] forward-backward correlations got stronger,[377] and $\langle p_\perp \rangle (n_{\text{ch}})$ showed the same rising trend,[378,379] to give some examples. The Tevatron emphasis was on hard physics, however, and it took many years to go beyond the S$p\overline{p}$S MB and UE studies. Notable is the CDF study on the production of $\gamma + 3$ jets,[97] which came to be the first generally accepted proof of the existence of DPS. Studies of the pedestal effect eventually also became quite sophisticated,[128,380–382] providing differential information on activity in towards, away and transverse regions in azimuth relative to the trigger, including a Z^0 trigger. All of these observations were in qualitative agreement with Pythia predictions. An improved quantitative agreement was obtained in a succession of tunes,[383,384] see also Chapter 11, like the much-used Tune A.

Even if agreement may not have been perfect, there was no obvious pattern of disagreement between S$p\overline{p}$S/Tevatron data and the Pythia model. Therefore it could routinely be used for experimental studies and for extrapolations to LHC and SSC energies. But it also meant that further MPI development was slow in the period 1988–2003, with only some relevant points, as follows.

More up-to-date formulae for total, elastic and diffractive cross-sections were implemented,[385] starting from the $\sigma_{\text{tot}}(s)$ parametrizations of Donnachie and Landshoff (DL).[197] They are based on an effective pomeron description, with parameters adjusted to describe existing data and also give a reasonable extrapolation to high energies. They worked well through the Tevatron era, but overestimated the diffractive rate for LHC and have since been modified.

The $p_{\perp 0}$ parameter went through several iterations, as new PDF sets appeared on the market and became defaults in Pythia. Notably HERA data showed that there is a non-negligible rise in the small-x region, even for small Q^2 scales, whereas pre-HERA PDFs had tended to enforce a flat $xf(x, Q_0^2)$ at small x. This implies that the all-$p_\perp$ integrated QCD cross-section rises much faster with s than assumed before, and thereby generates a faster rising $\langle n_{\mathrm{ch}} \rangle(s)$. The need for an s dependence of $p_{\perp 0}$, which previously had been marginal, now became obvious. Initially a logarithmic s dependence was used. Later on a power-like form was introduced, such as

$$p_{\perp 0}(s) = (2.1 \text{ GeV}) \left(\frac{s}{1 \text{ TeV}^2} \right)^{\epsilon} \tag{19}$$

with $\epsilon = 0.08$, inspired by the DL ansatz $\sigma_{\mathrm{tot}}(s) \simeq s^{\epsilon}$, which also qualitatively matches well with a HERA $xf(x, Q_0^2) \simeq x^{-\epsilon}$ behavior.

In an attempt to understand the behavior of $p_{\perp 0}(s)$, a simple toy study was performed.[386] As we already argued, the origin of a $p_{\perp 0}$ damping scale in the first place is that the proton is a color singlet, which means that individual parton color charges are screened. A very naive estimate is that the screening distance should be the inverse of the proton size, $p_{\perp 0} \approx \hbar / r_p \approx 0.3$ fm. But this assumes that the proton only consists of very few partons, such that the typical distance between two partons is r_p. In reality we expect the evolution of PDFs, especially at small x, to lead to a much higher density. Therefore the typical color screening distance — how far away you need to go to find partons with opposing color charges — to be much smaller than r_p. In order to test this, we built a model for the transverse structure of the proton as follows. Start out from a picture with three valence quarks and two "valence gluons" that share the full momentum of the proton at a scale $Q_0 \approx 0.5$ GeV, based on the GRV PDF approach,[387] distributed across a transverse proton disc, and with net vanishing color. They are then evolved upwards in Q^2, to create ISR cascades. A technical problem is that the $x \to 0$ singularity would generate infinitely many partons. Therefore branchings are only allowed if both daughters have an $x > x_{\mathrm{min}} \simeq p_{\perp 0}/\sqrt{s}$. Colors are preserved in branchings, and daughters can drift a random amount in transverse space of order $\hbar / Q$ if produced at a scale Q. A damping factor can then be defined by

$$\frac{\left| \sum_k q_k \, e^{i \mathbf{r}_k \mathbf{p}_\perp} \right|^2}{\sum_k |q_k|^2}, \tag{20}$$

where $\mathbf{p}_\perp$ represents a gluon plane wave probing the proton, consisting of partons with color charge q_k located at $\mathbf{r}_k$. This approach indeed gives results consistent with a damping at scales around 2 GeV, varying with s about as outlined above, but it contains too many uncertainties to be used for any absolute predictions.

The MPI framework was extended to γp[388] and $\gamma\gamma$[389] collisions. It there was applied to the vector meson dominance (VMD) part of the photon wave function, where the γ fluctuates into a virtual meson, predominantly a ρ^0. The same framework as for $pp/p\overline{p}$ collisions can then be recycled, with modest modifications.

To finish this section, a few words on theory and on MPI modeling in some other Tevatron-era (and beyond) generators.

Generally, MPI ideas were gradually becoming more accepted. An interesting (partial) alternative was raised by the CCFM equations,[390,391] which interpolate between the DGLAP[392–394] and BFKL[171,172] ones. Already BFKL allows $p_\perp$-unordered evolution chains, and with CCFM such a behavior can be extended to higher $p_\perp$ scales. As illustrated in the LDC model,[395] this can then give what looks like several (semi)hard interactions within one single chain.

ISAJET remained in use, even if slowly losing ground, with an essentially unchanged description of the underlying event.

When Herwig was extended to hadronic collisions,[396] it used an UE model/parametrization based on the UA5 MB generator,[360] which is purely soft physics. MPIs were never made part of the Fortran Herwig core code. Instead the UA5-based default could be replaced by the separate Jimmy[36] add-on. Its basic ideas resemble the ones in Pythia, but with several significant differences. The impact-parameter profile is given by the electromagnetic form factor, and at each given b the number of MPIs (in addition to the hard process itself) is given by a Poissonian with an average proportional to the convolution of two form factors. These MPIs are unordered in $p_\perp$, and all use unmodified PDFs. Instead interactions that break energy–momentum conservation are rejected. To handle beam remnants, it is assumed that each ISR shower initiator, except the first, is a gluon; if not an additional ISR branching is made to ensure this. Thereby it is possible to chain each MPI to the next in color, such that the remnant flavor structure is related only to the first interaction. This handling allows all MPIs to be associated with ISR and FSR, unlike Pythia at the time. Note that Jimmy was intended for UE studies, and that Herwig + Jimmy did not offer an MB option. Such a framework was developed[397] but the code for it was never made public. Only with the C++ version[398] did MPIs become a fully integrated part of the core code, for UE and MB.[37,399]

Another (later) multipurpose generator entrant is Sherpa,[157] which so far has based itself on the Pythia MPI framework, but a new separate MPI model is under development.[400]

Many generators geared towards MB physics also adapted semihard MPI ideas. Notably, generators based on eikonalization procedures typically already had contributions for soft and diffractive MPI-style physics, and could add a further contribution for hard MPIs. This means that a non-diffractive event can contain variable numbers of soft $p_\perp = 0$ and hard $p_\perp > p_{\perp\mathrm{min}}$ MPIs. Typically a Gaussian b dependence is used, not necessarily with the same width for all contributions. Main examples of such programs are DTUjet,[401,402] PhoJet,[403,404] DPMjet,[405] Sibyll,[406,407] EPOS,[408,409] see also Chapter 19, and QGSjet.[410,411] It would carry too far to go into the details of these programs. Some of them are in use at the LHC, and describe MB data quite successfully. They are not only used for pp collisions but often also for pA and AA, and for cosmic-ray cascades in the atmosphere.

6. Multiparton PDFs and Beam Remnants

In 2004 the Pythia MPI model was significantly upgraded,[365] specifically to allow a more realistic description of multiparton PDFs and beam remnants. Then ISR and FSR could also be included for each MPI, not only the hardest one.

To extend the PDF framework, it is assumed that quark distributions can be split into one valence and one sea part. In cases where this is not explicit in the PDF parametrizations, it is assumed that the sea is flavor–antiflavor symmetric, so that one can write, e.g.,

$$u(x, Q^2) = u_{\mathrm{val}}(x, Q^2) + u_{\mathrm{sea}}(x, Q^2) = u_{\mathrm{val}}(x, Q^2) + \overline{u}(x, Q^2). \tag{21}$$

The parametrized $u(x, Q^2)$ and $\overline{u}(x, Q^2)$ distributions can then be used to find the relative probability for a kicked-out u quark to be either valence or sea.

For valence quarks two effects should be considered. One is the reduction in content by previous MPIs: if a u valence quark has been kicked out of a proton then only one remains, and if two then none remain. In addition the constraint from momentum conservation should be included, as already introduced in Eq. (7). Together this gives

$$u_{i,\mathrm{val}}(x, Q^2) = \frac{N_{u,\mathrm{val,remain}}}{N_{u,\mathrm{val,original}}} \frac{1}{X} u_{\mathrm{val}}\left(\frac{x}{X}, Q^2\right) \quad \text{with} \quad X = 1 - \sum_{j=1}^{i-1} x_j, \tag{22}$$

for the u quark in the ith MPI, and similarly for the d. The $1/X$ prefactor ensures that the u_i integrates to the remaining number of valence quarks. The momentum sum is also preserved, except for the downwards rescaling for each kicked-out valence quark. The latter is compensated by an appropriate scaling up of the gluon and sea PDFs.

When a sea quark (or antiquark) q_{sea} is kicked out of a hadron, it must leave behind a corresponding antisea parton in the beam remnant, by flavor conservation, which can then participate in another interaction. We can call this a companion antiquark, $\overline{q}_{\mathrm{cmp}}$. In the perturbative approximation the pair comes from a gluon branching $g \to q_{\mathrm{sea}} + \overline{q}_{\mathrm{cmp}}$. This branching often would not be in the perturbative regime, but we choose to make a perturbative ansatz, and also to neglect subsequent perturbative evolution of the q_{cmp} distribution. Even if approximate, this procedure should catch the key feature that a sea quark and its companion should not be expected too far apart in x (or, better, in $\ln x$). Given a selected x_{sea}, the distribution in $x = x_{\mathrm{cmp}} = y - x_{\mathrm{sea}}$ then is

$$\begin{aligned} q_{\mathrm{cmp}}(x; x_{\mathrm{sea}}) &= C \int_0^1 g(y)\, P_{g \to q_{\mathrm{sea}} \overline{q}_{\mathrm{cmp}}}(z)\, \delta(x_{\mathrm{sea}} - zy)\, dz \\ &= C\, \frac{g(x_{\mathrm{sea}} + x)}{x_{\mathrm{sea}} + x}\, P_{g \to q_{\mathrm{sea}} \overline{q}_{\mathrm{cmp}}}\left(\frac{x_{\mathrm{sea}}}{x_{\mathrm{sea}} + x}\right). \end{aligned} \tag{23}$$

Here, $P_{g\to q\bar{q}}(z)$ is the standard DGLAP branching kernel, $g(y)$ an approximate gluon PDF, and C gives an overall normalization of the companion distribution to unity. Furthermore an X rescaling is necessary as for valence quarks. The addition of a companion quark does break the momentum sum rule, this times upwards, and so is compensated by a scaling down of the gluon and sea PDFs.

In summary, in the downwards evolution, the kinematic limit is respected by a rescaling of x, as before. In addition the number of remaining valence quarks and new companion quarks is properly normalized. Finally, the momentum sum is preserved by a scaling of gluon and (non-companion) sea quarks. All of these scalings should not be interpreted as a physical change of the beam hadron, but merely as reflecting an increasing knowledge of its contents, akin to conditional probabilities.

At the end of the MPI + ISR generation sequence, a set of initiator partons have been taken out of the beam, i.e., partons that initiate the ISR chains that stretch in to the hard interactions. The beam remnant contains a number of leftover valence and companion quarks that carry the relevant flavor quantum numbers, plus gluons and sea to make up the total momentum. The latter are not book-kept explicitly, except for the rare case when the remnant contains no valence or companion quarks at all, and where a gluon is needed to carry the leftover momentum.

When the initiators are taken out of the incoming beam particle they are assumed to have a primordial $k_\perp$. Naively this would be expected to be of the order of the Fermi motion inside the proton, i.e., a few hundred MeV. In order to describe the low-$p_\perp$ tail of the Z^0 spectrum a rather higher value of the order of 2 GeV seems to be required. This suggests imperfections in the modeling of ISR at small scales, specifically how and at what $p_\perp$ scales it should be stopped. Also note that ISR dilutes the $k_\perp$ at the hard interaction by a factor $x_{\mathrm{hard}}/x_{\mathrm{init}}$, i.e., by the fraction of the initiator longitudinal momentum that reaches the hard interaction. More ISR means a higher x_{init} and thus more dilution, counteracting the $p_\perp$ gained by the ISR itself.

Given such considerations, a Gaussian distribution is used for the primordial $k_\perp$, with a width that depends on the scale Q ($p_\perp$) of each MPI, increasing smoothly from 0.36 GeV (= the string hadronization $p_\perp$) at small Q to 2 GeV at large. There is also a check that the $k_\perp$ does not become too big for a low-mass system. The beam remnants are each given a $k_\perp$ at the lower scale, but in addition they collectively have to take the recoil to ensure that the net $p_\perp$ vanishes among the initiators and remnants.

The beam remnants also share leftover energy and longitudinal momentum. This is done by an ansatz of specific x spectra for valence quarks, valence diquarks, and companion quarks. The x values determine the relative fractions partons take of the lightcone momentum $E \pm p_z$, with + (−) for the beam moving in the $+z$ ($-z$) direction. It is not possible to fully conserve four-momentum inside each remnant + initiators system individually, however. Actually, by their relative motion the beam remnants together obtain a spectrum of invariant masses stretching well above the proton mass. Instead overall energy and momentum is preserved by longitudinal boosts of the two remnant subsystems, which effectively corresponds to a shuffling of four-momentum between the two sides. It is possible to generate too big remnant masses, but usually this can be fixed by a reselection of remnant x values.

What remains to consider is how the colors of partons are connected with each other to give the string pieces that eventually hadronize. This was one of the key stumbling blocks in the original model, especially what to do if several valence quarks are kicked out of a proton, Fig. 3. The main new tool at our disposal at this point is an implementation of junction fragmentation.[412] A junction is a vertex at which three string pieces come together, in a Y-shaped topology, and with each string stretching out to a quark, in the simplest case. The net baryon number then gets to be associated with the junction: given enough energy each string piece can break by the production of new $q\overline{q}$ (or $qq\overline{q}\overline{q}$) pairs, splitting off mesons (or baryon–antibaryon pairs), leaving the innermost q on each string to form a baryon together. An antijunction carries a negative baryon number, since the three strings in this case stretch out to antiquarks.

The rest frame of the junction is obtained in a symmetric configuration, where the opening angle between any pair of outgoing string ends is $2\pi/3 = 120°$. This also defines the approximate rest frame of the central baryon. In cases where only one quark is kicked out of an incoming proton, the remaining two quarks in the beam remnant have a tiny opening angle in the collision rest frame, meaning the junction is strongly boosted in the forward direction, along with the two quarks, and these can then together be treated as a single unresolved diquark. If two valence quarks are kicked out, however, the junction can end up far away from the beam

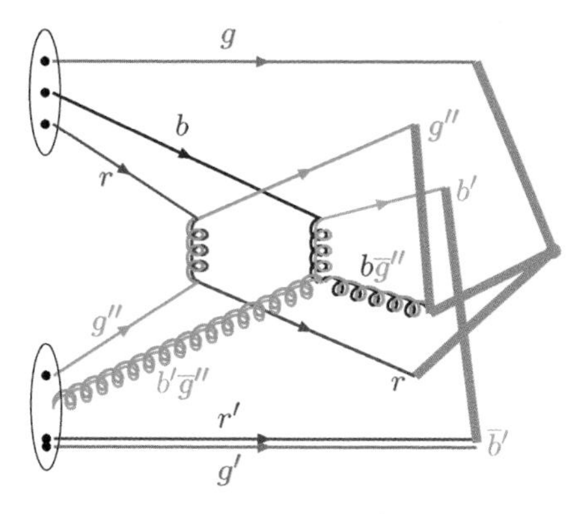

Fig. 3 Example of an event where two valence quarks are kicked out from a proton, giving a junction topology. A possible color flow is indicated, where primed colors are distinguishable from unprimed ones in the $N_C \to \infty$ limit. The remnant diquark is bookkept as a unit with $r' + g' = \overline{b}'$. Thick orange lines indicate strings stretched between outgoing partons, with the junction placed rightmost to avoid clutter.

remnant itself. Note that a junction is normally not associated with the original quarks after a collision, owing to color exchange.

A major complication is that the three strings may be stretched via various intermediate gluons out to the (new) endpoint quarks, and then the string motion and fragmentation becomes far more complex. It is such general issues that had taken time to resolve, at least approximately. Also systems containing a junction and an antijunction connected to each other need to be described.

Colors can be traced within each MPI individually, both through the hard interaction and the related ISR and FSR cascades, in the $N_C \to \infty$ limit.[371] If this limit is taken seriously, however, the beam remnants have to compensate the colors of all initiator partons, which means that they build up a high color charge, which has to be carried by an unrealistically large number of remnant partons. It is more plausible, although a bit extreme in the other direction, to assume that the color taken by one initiator is the anticolor of another one. It is such a strategy that allows us to work with the minimal number of beam remnants that preserves net flavor (or a single gluon if all valence quarks are kicked out). A sea quark initiator can be associated with its companion antiquark, be that another initiator or a remnant parton, and together be traced back to an imagined gluon that can be attached as above.

Thus only the valence colors remain. A proton can be described as a quark plus a diquark if none or one valence quark is kicked out, else as three quarks in a junction topology. It is along these original color lines that the gluon initiators are attached one by one. Three main alternatives are implemented for the order of these attachments, from completely random to ones that favor smaller string lengths λ. (These connections can give a gluon the same color as anticolor, which clearly is unphysical. Such color associations are rejected and others tried.) Not even in the latter case does $\langle p_{\perp} \rangle(n_{\mathrm{ch}})$ rise as fast as observed in the data, however, which suggests that a mere arrangement of colors in the initial state is not enough. A mechanism is also needed for CR in the final state, as already used in the earlier models.

Again the λ measure is used to pick such reconnections: a string piece ij stretched between partons i and j and another mn between m and n can reconnect to in and mj if $\lambda_{in} + \lambda_{mj} < \lambda_{ij} + \lambda_{mn}$. A free strength parameter is introduced to regulate the fraction of pairs that are being tested this way. With this further mechanism at hand it now again becomes possible to describe $\langle p_{\perp} \rangle(n_{\mathrm{ch}})$ data approximately.

7. Interleaved Evolution

In models up until now, MPIs have been considered one by one. Once an MPI has been picked, the ISR and FSR associated with it has been generated before moving on to the next. This ordering is not trivial, since both the MPI and ISR mechanisms need to take momentum from the beam remnants, and therefore are in direct competition. If instead all MPIs had been generated first, and all ISR added only afterwards, the number of MPIs would have been higher and the amount of ISR less.

Time ordering does not give any clear guidance what is the correct procedure. We have in mind a picture where all MPIs happen simultaneously at the collision moment, while ISR stretches backwards in time from it, and FSR forwards. But we have no clean way of separating the hard interactions themselves from the virtual ISR cascades that "already" exist in the colliding hadrons.

Instead we choose the same guiding principle as we did when we originally decided to consider MPIs ordered in $p_\perp$: it is most important to get the hardest part of the story "right", and then one has to live with an increasing level of approximation for the softer steps. With the introduction of $p_\perp$-ordered showers[107] it became possible to choose $p_\perp$ as common evolution scale. Initially only MPI and ISR were interleaved, with FSR left to the end. This caught the important momentum competition between MPI and ISR, so was the big step. When Pythia 8 was written[192] full MPI/IRS/FSR interleaving[114] was default from the beginning. Going straight for the latter formulation, the scheme is characterized by one master formula

$$\frac{d\mathscr{P}}{dp_\perp} = \left(\frac{d\mathscr{P}_{\mathrm{MPI}}}{dp_\perp} + \sum \frac{d\mathscr{P}_{\mathrm{ISR}}}{dp_\perp} + \sum \frac{d\mathscr{P}_{\mathrm{FSR}}}{dp_\perp} \right) \\ \times \exp\left(- \int_{p_\perp}^{p_{\perp\mathrm{max}}} \left(\frac{d\mathscr{P}_{\mathrm{MPI}}}{dp'_\perp} + \sum \frac{d\mathscr{P}_{\mathrm{ISR}}}{dp'_\perp} + \sum \frac{d\mathscr{P}_{\mathrm{FSR}}}{dp'_\perp} \right) dp'_\perp \right) \quad (24)$$

that probabilistically determines what the next step will be. Here the ISR sum runs over all incoming partons, two per already produced MPI, the FSR sum runs over all outgoing partons, and $p_{\perp\mathrm{max}}$ is the $p_\perp$ of the previous step. Starting from the hardest interaction, Eq. (24) can be used repeatedly to construct a complete parton-level event.

Since each of the three terms contains a lot of complexity, with matrix elements, splitting kernels and PDFs in various combinations, it would seem quite challenging to pick a $p_\perp$ according to Eq. (24). Fortunately the "winner-takes-it-all" trick (which is exact[413]) comes to the rescue. In it you select a $p_{\perp\mathrm{MPI}}$ value as if the other terms did not exist in the equation, and correspondingly a $p_{\perp\mathrm{ISR}}$ and a $p_{\perp\mathrm{FSR}}$. Then the one of the three that is largest decides what is to come next. Inside the ISR and FSR sums one can repeat the same trick, i.e., only consider one term at a time and decide which term gives the highest $p_\perp$.

The multiparton PDFs introduced in Section 6 play a key role, to help select a new MPI or perform ISR on an already existing one. Note that momentum and flavor should not be deducted for the current MPI itself when doing ISR. To exemplify, if the valence d quark has been kicked out of a proton in a given MPI, then there are no such d's left for other MPIs, neither in ISR nor in MPI steps, but for the given MPI a valence d at higher x is still available as a potential mother to the current d.

In summary, $p_\perp$ fills the function of a kind of factorization scale, where the perturbative structure above it has been resolved, while the one below it is only given an effective description, e.g., in terms of multiparton PDFs. A decreasing $p_\perp$ scale should then be viewed as an evolution towards increasing resolution; given that the event has a particular structure when viewed at some $p_\perp$ scale, how might that picture change when the resolution cutoff is reduced by some infinitesimal $\mathrm{d}p_\perp$? That is, let the "harder" features of the event set the pattern to which "softer" features have to adapt.

8. Intertwined Evolution

The above interleaving introduces a strong but indirect connection between different MPIs, in that each parton still has a unique association with exactly one MPI and its associated ISR and FSR. But this is likely not the full story; there are several ways in which the different MPIs may be much closer intertwined,[3,5,9,13] see also Chapter 2. The complexity then is significantly increased, and none of these further mechanisms are included by default in Pythia, but some have been studied and partly implemented.

The first possibility is joined interactions (JI),[107] Fig. 4(a). In it two partons participating in two separate MPIs may turn out to have a common ancestor when the backwards ISR evolution traces their prehistory. The joined interactions are well-known in the context of the forwards evolution of multiparton densities.[15–17,204,414] It can approximately be turned into a backwards evolution probability for a branching $a \to bc$

$$d\mathscr{P}_{bc}(x_b, x_c, Q^2) \simeq \frac{dQ^2}{Q^2}\frac{\alpha_\mathrm{s}}{2\pi}\frac{x_a f_a(x_a, Q^2)}{x_b f_b(x_b, Q^2)\, x_c f_c(x_c, Q^2)}\, z(1-z) P_{a\to bc}(z), \qquad (25)$$

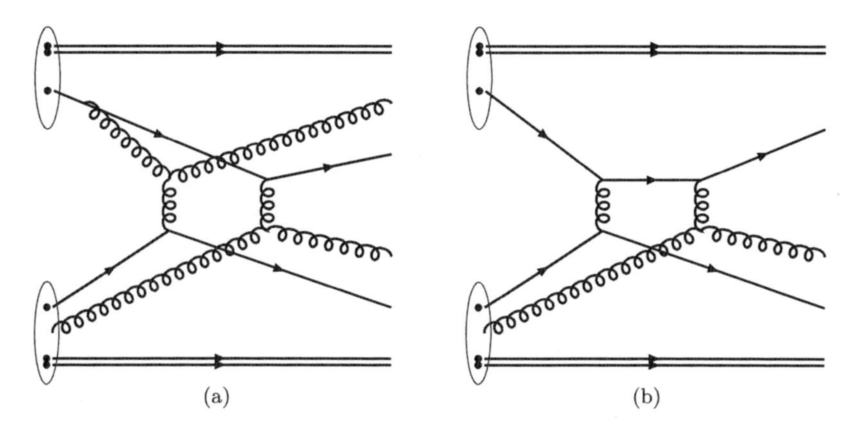

Fig. 4 (a) Joined interactions. (b) Rescattering.

with $x_a = x_b + x_c$ and $z = x_b/(x_b + x_c)$. The main approximation is that the two-parton differential distribution has been factorized as $f_{bc}^{(2)}(x_b, x_c, Q^2) \simeq f_b(x_b, Q^2)\, f_c(x_c, Q^2)$ to put the equation in terms of more familiar quantities.

Just like for the other processes considered, a form factor is given by integration over the relevant Q^2 range and exponentiation. Associating $Q \simeq p_\perp$, as is already done for normal ISR, Eq. (25) can be turned into a $\sum d\mathscr{P}_{\mathrm{JI}}/dp_\perp$ term that can go into Eq. (24) along with the other three. The JI sum runs over all pairs of initiator partons with allowable flavor combinations, separately for the two incoming hadrons. A gluon line can always be joined with a quark or another gluon one, and a sea quark and its companion can be joined into a gluon. The parton densities are defined in the same spirit as before, e.g., $f_b(x_b, p_\perp^2)$ and $f_c(x_c, p_\perp^2)$ are squeezed to be smaller than X, where X is reduced from unity by the momentum carried away by all but the own interaction, and for $f_a(x_a, p_\perp^2)$ by all but the b and c interactions.

Technical complications arise when the kinematics of JI branchings has to be reconstructed, notably in transverse momentum, and the code to overcome these was never written. The reason is that already the evolution itself showed that JI effect are small. Most events do not contain any JI at all above the ISR cutoff scale, and in those that do the JI tends to occur at a low $p_\perp$ value. There are two reasons for this. One is numerics: the number of parton pairs that can be joined increases as more MPIs have already been generated. The other is the PDF behavior: for all but the smallest $Q^2 = p_\perp^2$ scales the huge number of small-x gluons and sea quarks dominate, and it is only close to the lower evolution cutoff that the few valence quarks and high-x gluons play an increasingly important role in the ISR backwards evolution.

The second intertwining possibility is rescattering, i.e., that a parton from one incoming hadron consecutively scatters against two or more partons from the other hadron, Fig. 4(b). The simplest case, $3 \to 3$, i.e., one rescattering, has been studied.[6,30,340] The conclusion is that it should be less important than two separated $2 \to 2$ processes: $3 \to 3$ and $2 \times (2 \to 2)$ contain the same number of vertices and propagators, but the latter wins by involving one parton density more. The exception could be large $p_\perp$ and x values, but there $2 \to n, n \geq 3$ QCD radiation anyway is expected to be the dominant source of multijet events.

For rescattering, a detailed implementation is available as an option in Pythia,[413] as follows. Previously we have described how partons are taken out of the conventional PDFs after each MPI (and ISR), such that less and less of the original momentum remains. If we now should allow a rescattering then a scattered parton has to be put back into the PDF, but now as a δ function. It can be viewed as a quantum mechanical measurement of the wave function of the incoming hadron, where the original "squared wave function" $f(x, Q^2)$ in part collapses by the measurement process of one of the partons in the hadron. That is, one degree of freedom has now been fixed, while the remaining ones are still undetermined.

A hadron can therefore be characterized by a new PDF

$$f(x, Q^2) \to f_{\mathrm{rescaled}}(x, Q^2) + \sum_i \delta(x - x_i) = f_{\mathrm{u}}(x, Q^2) + f_\delta(x, Q^2), \tag{26}$$

where f_{u} represents the unscattered part of the hadron and f_δ the scattered one. The scattered partons have the same x values as originally picked, in the approximation that small-angle t-channel gluon exchange dominates, but more generally there will be shifts. The sum over delta functions runs over all partons that are available to rescatter, including outgoing states from hard or MPI processes and partons from ISR or FSR branchings. All the partons of this disturbed hadron can scatter, and so there is the possibility for an already extracted parton to scatter again.

With the PDF written in this way, the original MPI probability in Eq. (24) can now be generalized to include the effects of rescattering

$$\frac{d\mathcal{P}_{\mathrm{MPI}}}{dp_\perp} \to \frac{d\mathcal{P}_{\mathrm{uu}}}{dp_\perp} + \frac{d\mathcal{P}_{\mathrm{u}\delta}}{dp_\perp} + \frac{d\mathcal{P}_{\delta\mathrm{u}}}{dp_\perp} + \frac{d\mathcal{P}_{\delta\delta}}{dp_\perp}, \tag{27}$$

where the uu component represents the original MPI probability, the uδ and δu components a single rescattering and the $\delta\delta$ component a double rescattering. In this way, rescattering interactions are included in the common $p_\perp$ evolution of MPI, ISR and FSR.

Again the evolution in $p_\perp$ should be viewed as a resolution ordering. In a time-ordered sense, a parton could scatter at a high $p_\perp$ scale and rescatter at a lower one, or the other way around, with comparable probabilities. To simplify the already quite complicated machinery, it is chosen to set up the generation as if the rescattering occurs both at a lower $p_\perp$ and a later time than the "original" scattering, while retaining the full rescattering rate.

In simple low-angle scatterings there is a unique assignment of scattered partons to either of the two colliding hadrons A and B, but for the general case this is not so simple. Different prescriptions have been tried, including the most extreme where all scattered partons can scatter again against partons from either beam. It turns out that differences are small for single rescatter. The scheme dependence is bigger for double rescattering, but this process is anyway significantly suppressed relative to the single rescatters.

The real problem, however, is how to handle kinematics when ISR, FSR and primordial $k_\perp$ are added to rescattering systems. To propagate recoils between systems that are partly intertwined but also partly separate requires what is the most non-trivial code in the whole Pythia MPI framework, and it would carry too far to describe it here.

More important are the results. At the LHC we predict that of the order of half of all MB events contains at least one rescattering, and for events with hard processes the fraction is even larger. Double rescattering is rare, however, and can be neglected. Evaluating the consequences of rescattering is still challenging, since it is a secondary effect within the MPI machinery. The precise amount of MPI has to be tuned to data, e.g., by varying the $p_{\perp 0}$ turn-off parameter, so the introduction of rescattering to a large extent can be compensated by a slight decrease in the amount of "normal" $2 \to 2$ MPIs. Worse still, most MPIs are soft to begin with, and rescattering then introduces a second scale, even softer than the "original" scattering. Thus, rescattering is typically associated with particle production at the lower limit of what can be reliably detected. There are some effects of the Cronin type[415] in hadronic $p_{\perp}$ spectra, i.e., a shift towards higher $p_{\perp}$, but too small to offer a convincing signal.

If one zooms in on the tail of events at larger scales, the rate of (semi-)hard three-jet events from rescattering is of the same order in α_s as four-jets from DPS, but the background from ISR and FSR starts one order earlier for three-jets. Furthermore, DPS has some obvious characteristics to distinguish it from $2 \to 4$ radiation topologies: pairwise balanced jets with an isotropic relative azimuthal separation. No corresponding unique kinematic features are expected for $3 \to 3$ rescattering vs. $2 \to 3$ ISR/FSR, and we have not found any either. Hopefully smarter people will one day find the right observable.

The third and most dramatic intertwining possibility is that the perturbative cascades grip into each other. An example is the "swing" mechanism, whereby two dipoles in the initial state can reconnect colors, which is a key aspect of the DIPSY generator,[416,417] see also Chapter 17. Currently there is no mechanism of this kind in Pythia.

9. An x-Dependent Impact-Parameter Profile

As described in Section 4, a double Gaussian impact-parameter profile was the initial choice, and remained the default for many years. With the introduction of full ISR/FSR in all MPIs, Section 6, a single Gaussian is almost giving enough fluctuations. The "hot spots", that the double Gaussian was introduced to represent, now are obtained in part by the ISR/FSR cascades associated with each MPI, with somewhat fewer separate MPIs needed for the same overall activity. Put another way, the varying ISR/FSR activity introduces another mechanism for fluctuations, while a smaller $\langle n_{\mathrm{MPI}} \rangle$ means that the scaled width $\sigma_{\mathrm{MPI}}(b)/\langle n_{\mathrm{MPI}}(b) \rangle$ goes up, both reducing the need for broader-than-Gaussian profiles.

Given the reduced sensitivity to non-Gaussian profiles, a one-parameter alternative was therefore introduced

$$\widetilde{\mathcal{O}}(b) = \exp(-b^d), \tag{28}$$

where $d < 2$ gives more fluctuations than a Gaussian and $d > 2$ less. Do note that the expression is for the overlap, not for the individual hadrons, for which no simple analytical form is available.

One aspect, however, is that we have assumed the transverse b profile to be decoupled from the longitudinal x one. This is not the expected behavior, because low-x partons should diffuse out in b during the evolution down from higher-x ones,[116] see also Chapter 5. Additionally, if $b = 0$ is defined as the centre of energy of a hadron, then by definition a parton with $x \to 1$ also implies $b \to 0$.

Such diffusion was already studied in Section 5. In the later study,[66] described here, there was no attempt to trace the evolution of cascades in x. Rather it is assumed that the b distribution of partons at any x can be described by a simple Gaussian, but with an x-dependent width:

$$\rho(r,x) \propto \frac{1}{a^3(x)} \exp\left(-\frac{r^2}{a^2(x)}\right) \quad \text{with} \quad a(x) = a_0 \left(1 + a_1 \ln \frac{1}{x}\right), \tag{29}$$

with a_0 and a_1 to be determined. The overlap is then given by

$$\widetilde{\mathcal{O}}(b, x_1, x_2) = \frac{1}{\pi} \frac{1}{a^2(x_1) + a^2(x_2)} \exp\left(-\frac{b^2}{a^2(x_1) + a^2(x_2)}\right). \tag{30}$$

In principle one could argue that also a third length scale should be included, related to the transverse distance the exchanged propagator particle, normally a gluon, could travel. This distance should be made dependent on the $p_\perp$ scale of the interaction. For simplicity this further complication is not considered but, on the other hand, a finite effective radius is allowed also for $x \to 1$. The $x \to 1$ limit is not much probed in MB, since the bulk of MPIs occur at small x, but can be relevant for UE studies, e.g., for the production of new particles at high mass scales.

The two parameters a_0 and a_1 could be tuned at each CM energy. One combination of them is fixed so as to reproduce σ_{nd}. If the wider profile of low-x partons is related to the growth of $\sigma_{\mathrm{nd}}(s)$, then a_1 can be constrained by the requirement that a_0 should be independent of energy. This is reasonably well fulfilled for $a_1 = 0.15$, which is therefore taken as default in this scenario.

The generation of events is more complicated with an x-dependent overlap, but largely involves the same basic principles. Again the b value of an event is selected in conjunction with the kinematics of the hardest interaction, and is thereafter fixed. The acceptance of subsequent MPIs is proportional to $\widetilde{\mathcal{O}}(b, x_1, x_2)$, where x_1 and x_2 are the values for the MPI under consideration.

In overall MB and UE phenomenology, the scenario with $a_1 = 0.15$ ends up roughly halfway between the single and double Gaussian ones. It depends on the process under consideration, however. A 10 GeV γ^* would give an MPI activity close to the single Gaussian, while a 1 TeV Z' would have markedly higher MPI activity, since larger x values would be accessed in the latter process.

Unfortunately, in the experimental tunes to MB and UE data that were made to this model variant,[479] no convincing evidence were found for an $a_1 > 0$. A dedicated study of how the UE varies as a function of the mass of $\mu^+\mu^-$ pairs could provide the definitive test.

10. Diffraction

The nature of diffraction is not obvious, and has been addressed from different points of view over the years, see Chapter 14. From the perspective of this paper the story begins with the Ingelman–Schlein model,[418] wherein it is assumed that the exchanged pomeron $\mathbb{P}$ can be viewed as a hadronic state, and that therefore the creation of a diffractive system can be described as a hadron–hadron collision. This implies that the $\mathbb{P}$ has PDFs, which enables hard processes to occur, as was confirmed by the observation of jet production in diffractive systems.[419] The PomPyt program[420] combined the $\mathbb{P}$ flux inside the proton with the $\mathbb{P}$ PDFs, both largely determined from HERA data, and used Pythia to produce complete hadronic final states.

Originally Pythia itself had a more primitive description, based on a purely longitudinal structure of the diffractive system, for single diffraction either stretched between a kicked-out valence quark and a diquark remnant, or stretched in a hairpin configuration with a kicked-out gluon connected to both a quark and a diquark in the remnant. This was sufficient for low-mass diffraction, but clearly not for high-mass one. Therefore a complete MPI machinery was implemented.[114,421] The diffractive mass is selected as a first step. Thereafter $\mathbb{P}$ and ordinary p PDFs are used to generate an ordinary sequence of MPIs.

There are some catches, however. One is that $\mathbb{P}$ PDFs often are not normalized to unit momentum, in part based on theoretical arguments but mainly because what hard processes probe is the convolution of $\mathbb{P}$ flux and PDFs, not each individually. For the handling of momentum conservation issues in a generator, however, it is essential to let the MPI machinery have access to properly rescaled PDFs, as in Section 6, and for that an implicit normalization to unity is used. Another is that the MPI machinery also requires a normalization to an effective $\sigma_{\mathbb{P}p}$, to replace the σ_{nd} used for normal non-diffractive events, starting from Eq. (4). Again this is not directly measurable, so a default value of 10 mb has been chosen to give $\mathbb{P}p$ properties, such as $\langle n_{\mathrm{ch}} \rangle$, comparable with $pp/p\overline{p}$ at the same mass.

The diffractive machinery generates a mass spectrum that stretches down to ~ 1.3 GeV, but obviously it is not possible to apply an MPI philosophy that low. Therefore, below 10 GeV, the original longitudinal description is recovered. At low masses the kicked-out valence quark scenario is assumed to dominate, but then fall off in favour of a kicked-out gluon. Above 10 GeV there is a transition region, wherein the MPI description gradually takes over from the longitudinal one.

In its details several further deliberations and parameters are involved. There is also a somewhat separate MBR model[422] as an option.

Diffraction raises more MPI questions. In the collision of two incoming protons, one $\mathbb{P}$ exchange could imply a diffractive topology with a rapidity gap, but other simultaneously occurring MPIs would fill in that gap and produce an ordinary MB event. A spectacular example is Higgs production by gauge-boson fusion,

$W^+W^- \to H^0$ and $Z^0Z^0 \to H^0$, where the naive process should result in a large central gap only populated by the Higgs decay products, since no color exchange is involved. Including MPIs, this gap largely fills up,[423] although a fraction of the events should contain no further MPIs,[424] a fraction denoted as the Rapidity Gap Survival Probability. Such a picture has been given credence by the observation of "factorization breaking" between HERA and the Tevatron: the $\mathbb{P}$ flux and PDFs determined at HERA predict about an order of magnitude more QCD jet production than observed at the Tevatron.[425]

In a recent study[426] a dynamical description of such factorization breaking is implemented, as a function of the hard-process kinematics, to predict the resulting event structure for hard diffraction in hadronic collisions. This is done in three steps. Firstly, given a hard process which has been selected based on inclusive PDFs, the fraction of a PDF that should be associated with diffraction is calculated, obtained as a convolution of the $\mathbb{P}$ flux and its PDFs. Secondly, the full MPI framework of Pythia, including also ISR and FSR effects, is applied to find the fraction of events without any further MPIs. Those events that survive these two steps define the diffractive event fraction, while the rest remain as regular non-diffractive events. Thirdly, diffractive events may still have MPIs within the $\mathbb{P}p$ subsystem, and therefore the full hadron–hadron underlying-event generation machinery is repeated for this subsystem. The non-diffractive events are kept as they are in this step.

For typical processes, such as QCD jets or Z^0 production, the PDF step gives $\sim$10% of the events to be of diffractive nature. The requirement of having no further MPIs gives an approximate factor of 10 further suppression, so that only around 1% of the processes show up as diffraction. There numbers are in overall agreement with the experimental ones, but the availability of a complete implementation should allow more detailed tests. Unfortunately there is still a non-negligible uncertainty, both for the model itself and for the external input, such as $\mathbb{P}$ PDFs.

11. Color Reconnection

Color reconnection is a research field on its own, although tightly connected to MPIs, and has been reviewed elsewhere.[427,428] Here only some brief notes follow, again with emphasis on Pythia aspects.

As mentioned in Section 4, CR was essential to obtain a rising $\langle p_\perp \rangle(n_{\mathrm{ch}})$, and that has remained a constant argument over the years, still valid today: separate MPIs must be color-connected in such a way that topologies with a reduced λ measure, Eq. (18), are favored.

LEP 2 offered a good opportunity to search for CR effects. Specifically, in a process $e^+e^- \to W^+W^- \to q_1\overline{q}_2q_3\overline{q}_4$, CR could lead to the formation of alternative "flipped" singlets $q_1\overline{q}_4$ and $q_3\overline{q}_2$, and correspondingly for more complicated parton/string topologies. Such CR would be suppressed at the perturbative level, since it would force some W propagators off the mass shell. This suppression would not apply in the soft region. Two main models were developed in the Pythia context.[429] Strings are viewed as elongated bags in scenario I, and reconnection is proportional to the space–time overlap of these bags. In scenario II, strings are instead imagined as vortex lines, and two cores need to cross each other for a reconnection to occur. In either case it is additionally possible to allow only reconnections that reduce λ, scenarios I$'$ and II$'$. Based on a combination of results from all four LEP collaborations, the no-CR null hypothesis is excluded at 99.5% CL.[430] Within scenario I the best description is obtained for $\sim$50% of the 189 GeV W^+W^- events being reconnected, in qualitative agreement with the Pythia predictions.

Unfortunately it is more difficult to formulate a similarly detailed model of the space–time picture of hadronic collisions, and this has not been done for Pythia. (In part such pictures are presented e.g., in Dipsy and Epos.) Instead simpler scenarios for reducing the λ have been used. In total Pythia 6 came to contain twelve models, many of them involving annealing strategies.

The current Pythia 8[113] initially only contained one model, which still is the default. In it two MPIs can be merged with a probability $\mathscr{P} = r^2 p_{\perp 0}^2/(r^2 p_{\perp 0}^2 + p_{\perp \mathrm{lower}}^2)$, where r is a free parameter, $p_{\perp 0}$ is the standard dampening scale of MPIs, and $p_{\perp \mathrm{lower}}$ is the scale of the lower-$p_\perp$ MPI. Each gluon of the latter MPI is put where it increases λ the least for the higher-$p_\perp$ MPI. The procedure is applied iteratively, so for any MPI the probability of being reconnected is $\mathscr{P}_{\mathrm{tot}} = 1 - (1 - \mathscr{P})^{n_>}$, where $n_>$ is the number of MPIs with higher $p_\perp$.

A new QCD-based CR model[431] implements a further range of reconnection possibilities, notably allowing the creation of junctions by the fusion of two or three strings. The relative rate for different topologies is given by SU(3) color rules in combination with a minimization of the λ measure. The many junctions leads to an enhanced baryon production, although partly compensated by a shift towards strings with masses too low for baryon production.

A specific application of CR is for t, Z^0 and $W^\pm$ decays. With widths around 2 GeV, i.e., $c\tau \approx 0.1$ GeV, their decays happen after other hard perturbative activity (ISR/FSR/MPI) but still inside the hadronization color fields, thereby allowing CR with the rest of the event. It was already for the Tevatron noted that this is a non-negligible source of uncertainty in top mass determinations,[432] and for similar LHC studies several new CR models were implemented in Pythia 8.[433] These fall in two classes: either the t and W decay products undergo CR on equal footing with the rest of the event, or their decays and CR are considered after the rest of the partonic event has had a chance to reconnect. The latter scenario allows more flexibility, to explore also extreme possibilities.

12. LHC Lessons

LHC has been productive in presenting data of relevance for MB/UE/MPI/DPS studies, and there is no possibility to cover even a fraction of it here. The following therefore is a very subjective selection.

To begin with, some words on tunes, see also Chapters 11 and 13. Generators contain a large number of free parameters, by necessity, that attempt to parametrize our ignorance. Many of these are correlated, so cannot be determined separately. A tune is then the outcome of an effort to determine a set of key parameters simultaneously. This is an activity that generator authors do at a basic level all the time, and occasionally as a more concerted effort, e.g., Refs. 114, 434–436. It is also an activity that experimental collaborations undertake, given the direct access to data and the needs of their communities. With no claims of completeness, the Pythia 6 code contains settings for more than 100 different tunes, e.g.,[383,384,437,471,477] whereof about half precede the LHC startup, and the Pythia 8 one for 34 so far, e.g., see Refs. 152, 479, 481, 438. Many of these are minor variations around a common theme. Some tunes have been made by hand whereas others use automated procedures such as PROFESSOR.[439] The access to validated RIVET analysis routines[440] have played an increasingly important role. Data comparisons for many tunes are available in MCPLOTS.[441]

Whereas theorists aim for the best overall description, experimentalists often produce separate MB and UE tunes. With less data to fit, it is possible to obtain a better description for the intended applications. So far it has not been settled how much of the differences in MB and UE parameters represent true shortcomings of the model and how much is a consequence of the fitting process itself.

Several Pythia 6 tunes served as a basis for predictions prior to the LHC startup. Early 7 TeV data showed that most of them undershot the level of activity by some amount, with one being close, but none above. Fortunately a modest change of the energy dependence of the $p_{\perp 0}$ is enough to bring up the activity to a reasonable level, and further extrapolations to 13 TeV have worked better.

Generally speaking, Pythia has been able to explain most phenomenology observed at LHC so far, at least qualitatively, and often also quantitatively. Nevertheless, a significant number of observations do not look as nice. A common theme for many of them is that high-multiplicity pp events have properties similar to those observed in heavy-ion AA collisions. Examples are[442]

- High-multiplicity events have a higher fraction of heavier particles, notably multi-strange baryons,[443] whereas the composition stays rather constant in Pythia.
- The $\langle p_{\perp} \rangle$ is larger for heavier particles[444] (also observed at RHIC[445]), more so than Pythia predicts.
- The charged particle $p_{\perp}$ spectrum is underestimated at low $p_{\perp}$ scales,[446–448] e.g., leading to problems in simultaneously fitting MB data analyzed with $p_{\perp} > 0.1$ GeV and $p_{\perp} > 0.5$ GeV. The deficit is mainly associated with too little $\pi^{\pm}$ production.[449]
- The Λ/K spectrum ratio has a characteristic peak at around 2.5 GeV,[450] which is not at all reproduced.

- The observation of a ridge in two-particle correlations, stretching out in rapidity on both sides of a jet peak, especially for high-multiplicity events.[282,283,451] Correlation functions also points to azimuthal flow, similarly to AA observations.

An alternative model has been formulated[442] to explore at least some of these discrepancies, with three main deviations from the standard framework. First, the standard Gaussian $p_\perp$ spectrum for primary hadron production is replaced by an exponential one, $\exp(-m_{\perp\mathrm{had}}/T)$. It is loosely inspired by thermodynamics, with T an effective temperature, and is intended to enhance the pion rate at small $p_\perp$, while increasing $\langle p_\perp \rangle$ for heavier particles. Second, it is assumed that the normal string tension, alternatively the T above, is increased in regions of phase space where strings are close-packed, which typically is caused by a high MPI activity. This is intended to change both particle composition and $p_\perp$ spectra. Third, a simple model for hadronic rescattering is introduced, whereby hadrons tend to obtain more equal velocities, i.e., larger $\langle p_\perp \rangle$ for heavier particles.

Unfortunately, effects are not as large as one might hope. Specifically, most pions come from decays of heavier hadrons, and so the mechanisms intended to give less $p_\perp$ to pions and more to kaons and protons are counteracted. Nevertheless the thermodynamical model is able to provide significantly improved descriptions of observables such as the $p_\perp$ spectrum of charged hadrons, the average transverse momentum as a function of the hadron mass, and the enhanced production of strange hadrons at large multiplicities.

Even more successful are the Dipsy and Epos generators, however. In Dipsy dense string packing is assumed to lead to the formation of color ropes,[417] wherein an increased string tension favours the production of heavier hadrons and larger $p_\perp$ values, and a shoving mechanism can induce ridge and related phenomena.[452] In Epos the central region of pp collisions can form a quark–gluon plasma, which also allows strangeness enhancements, and strings in the outer regions can again be shoved by the central pressure, to give ridges.[409]

These new data, and the models they favor, may have consequences for the way we think about MPIs. Having MPIs as the origin of QGP formation in AA is an old idea,[453] but now it might even apply to pp. One could also note that the rising trend of $\langle p_\perp \rangle (n_{\mathrm{ch}})$, once the key reason to introduce CR, now partly might be attributed to collective effects.

13. Current State of the Pythia MPI Machinery

The MPI machinery implemented in Pythia has evolved over the years, as we have seen. It is therefore useful to make a quick rundown of the current state, and also mention a few odds and ends that were not yet covered.

The starting point is to define a $d\sigma/dp_\perp^2$, Eq. (2), that decides which processes can occur in an MPI, and then also occurs in the Sudakov-like factor. Originally it only contained QCD $2 \to 2$ processes, but now also includes $2 \to 2$ with photons in the final state, or mediated by an s-channel γ^*, or by t-channel $\gamma^*/Z^0/W^\pm$ exchange, or charmonium and bottomonium production via color singlet and octet channels. This combined cross-section is then regularized in the $p_\perp \to 0$ limit by the damping factor in Eq. (8).

There are two main options to begin the generation. One is if a hard process already has been selected, with a generic hardness scale, e.g., the factorization scale, which we for the sake of bookkeeping equate with $p_{\perp 1}$. Then the impact parameter b can be selected according to one of five possibilities: no b dependence, single Gaussian, double Gaussian, $\exp(-b^p)$ overlap and x-dependent Gaussian. The other main option is, e.g., for inclusive non-diffractive production, where $p_{\perp 1}$ and b have to be selected correlated according to Eq. (15). (Also with an explicit x_1, x_2 dependence for the x-dependent Gaussian.)

The downwards evolution in $p_\perp$ can then proceed, Eq. (17). For a preselected hard process there is an ambiguity, however, whether to allow $p_{\perp 2}$ to be restricted by $p_{\perp 1}$ or go all the way up to the kinematic limit. The former is required for QCD jets, to avoid double-counting, while the latter would be sensible for more exotic processes, where there is no such risk. By default Pythia will make this decision based on some simple rules, but it is also possible to choose strategy explicitly.

The downwards evolution of MPIs is interleaved in $p_\perp$ with ISR and FSR, Eq. (24). Optionally one may also include rescattering in the MPI framework, Eq. (27), but this comes at a cost, so is not recommended for normal usage. Joined interactions are not implemented in the Pythia 8 code, and there is also no swing mechanism.

Modified multiparton PDFs during the evolution follow the strategy of Section 6. The same section also describes the related handling of beam remnants. The subsequent color reconnection stage allows for several different models, with many options. String fragmentation and decays is added at the end of the generation chain, with junction topologies playing a key role for the preservation of the incoming baryon numbers.

Diffractive topologies are included in a picture where the pomeron is given a hadronic substructure, implying that $\mathbb{P}p$ and $\mathbb{P}\mathbb{P}$ subcollisions should be handled with a full MPI machinery of its own, at least for higher diffractive masses.

In total, essentially all the questions raised about the original model, end of Section 3, have since been studied, and tentative answers have been implemented in the existing code.

The generation of DPS is implicit in the MPI machinery. Whereas the first interaction can always be selected hard, normally the second one would tend to be soft. There is a special machinery in Pythia to generate two hard interactions. To understand how it operates, start from the Poissonian distribution $\mathscr{P}_n = \langle n \rangle^n \exp(-\langle n \rangle)/n!$. If $\langle n \rangle \ll 1$, as it should be for a hard process, the exponential can be neglected and $\mathscr{P}_2 = \langle n \rangle^2/2$. Now imagine two processes a and b with cross-sections σ_a and σ_b, meaning that they are produced with rates $\langle n_a \rangle = \sigma_a/\sigma_{\mathrm{nd}}$ and $\langle n_b \rangle = \sigma_b/\sigma_{\mathrm{nd}}$ inside the inelastic event class. Then the cross section for having two such MPIs is

$$\sigma_2 = \sigma_{\mathrm{nd}}\,\mathscr{P}_2 = \sigma_{\mathrm{nd}}\,\frac{(\langle n_a \rangle + \langle n_b \rangle)^2}{2!} = \frac{\sigma_a^2 + 2\sigma_a\sigma_b + \sigma_b^2}{2\,\sigma_{\mathrm{nd}}}. \tag{31}$$

The above equation neglects the impact-parameter dependence. A hard collision implies a smaller average b than for MB events, as we have discussed before, and thus an enhanced rate for the second collision. Poissonian statistics applies for a fixed b, but when averaging over all b an enhancement factor $\mathscr{E}$ is generated. Conventionally such effects are included by replacing the final σ_{nd} in Eq. (31) by an σ_{eff}. Unintuitively a lower σ_{eff} corresponds to a higher $\mathscr{E}$. It can be written as[42]

$$\mathscr{E} = \frac{\sigma_{\mathrm{nd}}}{\sigma_{\mathrm{eff}}} = \frac{\int \widetilde{\mathscr{O}}^2(b)\,d^2b \times \int \mathscr{P}_{\mathrm{int}}(b)\,d^2b}{\left(\int \widetilde{\mathscr{O}}(b)\,d^2b\right)^2}. \tag{32}$$

Thus, $\mathscr{E}$ depends on the shape of $\widetilde{\mathscr{O}}(b)$, with a distribution more spiked at $b = 0$ giving a larger $\mathscr{E}$. But $\mathscr{E}$ also depends on the CM energy and the $p_{\perp 0}$ scale, which enter via $\mathscr{P}_{\mathrm{int}}(b)$.

There is also a dynamical depletion factor related to PDF weights. With the two hard processes initially generated independently of each other, flavor and momentum constraints are not taken into account. Therefore, afterwards, PDFs are re-evaluated as if either interaction were the first one, giving modified PDFs for the second one, as described in Section 6. The average PDF weight change of the two orderings is used as extra event weight, leading to some configurations being rejected.

The program allows the two hard interactions to be selected in partly overlapping channels, and/or (with some warnings) phase space regions. Assume e.g., that process 1 can be either a or c and process 2 either b or c. Then an extension of Eq. (31) tells us the numerator should be

$$2\sigma_{1a}\sigma_{2b} + 2\sigma_{1a}\sigma_{2c} + 2\sigma_{1c}\sigma_{2b} + \sigma_{1c}\sigma_{2c} = 2(\sigma_{1a} + \sigma_{1c})(\sigma_{2b} + \sigma_{2c}) - \sigma_{1c}\sigma_{2c}. \tag{33}$$

To obtain the correct answer the prescription is thus to generate process 1 according to $\sigma_a + \sigma_c$ and process 2 according to $\sigma_b + \sigma_c$, but to throw half of those events where both 1 and 2 were picked to be process c.

14. Summary and Outlook

In this chapter we have traced the evolution of MPI ideas in Pythia over more than 30 years. The emphasis has been on the early developments, since that set the stage for what has come later, and more generally on the theoretical ideas and concepts that have been explored over the years. It is intended to offer a complementary view to other chapters in this book, and so we have avoided detailed comparisons with data, and also not gone into details of other models.

By and large, the original MPI ideas have stood the test of time; they are still at the core of the current Pythia framework. The further work that has been done since is much more extensive than the original one, but often suffers from diminishing returns, i.e., that major upgrades only moderately improve the general agreement with data. Nevertheless, it is important to explore as many aspects of MPIs as possible. They do encode important information on the borderline between perturbative and non-peturbative physics, and we should become better at decoding this information.

The Pythia development described here has been very much influenced by perceived experimental needs, and inspired by theoretical ideas, but has been decoupled from detailed theoretical calculations. The reason is obvious: already DPS offers a formidable challenge, enough to keep theorists busy, and so useful results for truly *MPIs* are rare. While it is interesting to understand two-parton PDFs better, say, in Pythia we need to be able to address twenty-parton PDFs, and nothing less will do.

In this spirit it is important to recall that, even though studies of DPS is an important way to explore MPI, it is not the only one. An example on the to-do list is the lumpiness of particle production in general, e.g., as probed by the minijet rate for different jet clustering R and $p_{\perp\mathrm{min}}$ parameter values, down to the $p_{\perp 0} \approx 2.5$ GeV scale (at LHC energies). And it should not be forgotten that (for some people, like me) the most convincing — and earliest — evidence of MPIs is the broadness of multiplicity distributions. The intermittency[454,455] interpretations of multiplicity fluctuations at the $\mathrm{Sp\bar{p}S}$[456,457] may have been over-enthusiastic at times, but more parts of the same experimental program could be carried out at the LHC and yield useful results.

Nevertheless, the LHC has provided new impetus to the whole MB/UE field of studies, by the observation of new and unexpected phenomena. Both the ridge effect and the enhancement of multi-strange baryons in high-multiplicity events, to take the two most spectacular examples, remain to be fully understood. These observations do not invalidate the MPI concept. On the contrary, plausible explanations start out from a MPI picture and add some kind of collective behavior among the MPIs. Color ropes or other ways to obtain an increased string tension is one example, the formation of a quark-gluon-plasma-like (multiparton!) state another.

Clearly much work lies ahead of us to fully understand what has already been observed, and hopefully also many further surprises will come along to stimulate us further.

Acknowledgments

Thanks to all collaborators on MPI physics through the years, and to Jonathan Gaunt and Peter Skands for helpful comments on the draft manuscript. This project has received funding in part by the Swedish Research Council, contracts number 621-2013-4287 and 2016-05996, in part from the European Research Council (ERC) under the European Union's Horizon 2020 research and innovation programme (grant agreement No 668679), and in part through the European Union Marie Curie Initial Training Networks MCnetITN PITN-GA-2012-315877 and MCnetITN3 722104.

Chapter 11

Measurement of the Observables Sensitive to Underlying Event

Sunil Bansal*, Rick Field† and Deepak Kar‡

Panjab University, Chandigarh, India
†*University of Florida, Gainesville, FL 32611, USA*
‡*School of Physics, University of the Witwatersrand, Johannesburg, Wits 2050, South Africa*

Underlying event is defined to be everything except the hard scatter in hadronic collisions, which must be modeled phenomenologically in Monte Carlo (MC) event generators. This chapter gives a historical overview of measurement of observables sensitive to underlying event from Tevatron and LHC experiments, and simultaneously discusses the evolution of MC tunes based on those measurements.

1. Introduction

1.1. *Definition of underlying event*

To perform precise measurements or to search for signatures of new physics phenomena at hadron colliders, it is important to have a good understanding not only of the high energy particles produced in the collisions, but of the whole interaction between the protons (or protons and antiprotons, as may be the case). The study of jet events at the UA1 experiment at CERN's SppS collider observed the presence of a jet pedestal,[458] i.e., energy deposited outside the core of the jet. This can be attributed to the interactions from the partons not participating in the hardest scattering,[a] and came to be known as the underlying event (UE)

[a]By soft, low transverse momenta (denoted by $p_{\rm T}$) transfer from initial to final state is indicated, additionally with very few or no particles produced with significant transverse momentum, $p_{\rm T}$. In contrast, interactions involving the creation of at least one particle with appreciable $p_{\rm T}$ is termed hard scattering.

following the first CDF measurement to probe properties of charged-particles inside and outside of charged-particle jets.[459]

The UE is technically defined as everything (experimentally tracks and calorimeter energy deposit) in a hadronic collision event which is not coming from the primary[b] hard scattering process. However, experimentally it is impossible to uniquely separate out the different components of a collision on an event-by-event basis. The initial paper from CDF points this out: "Of course, from a certain point of view there is no such thing as an *underlying event* in a proton–antiproton collision. There is only an *event* and one cannot say where a given particle in the event originated".

Therefore, the experimental analyses are focused on exploiting the topology of the event, and measuring activity which is least affected by the particles originating in the hard scattering process. The common strategy is to perform a measurement of observables sensitive to the properties of the underlying event in a well defined fiducial phase space, and comparing that to the predictions of Monte Carlo (MC) event generator models. The MC models simulate the collision events, therefore it is critical that the modelling of the underlying event, along with everything else are as realistic as possible. The measurements help to constraint the modeling of UE in the generators.

1.2. *Importance of UE*

Underlying event activity is an unavoidable background to collider observables for most measurements and searches. In fact it can be imagined as forming a pedestal of activity associated with high transverse momentum objects coming from the hard scattering. However, because of event-by-event fluctuations, the pedestal is not constant, so the UE activity cannot be precisely subtracted. However, there exists techniques using jet area[460] to remove part of the UE.

Although the fraction of activity/energy deposited due to UE may not appear significant in high-p_T processes, they nevertheless have non-negligible effects. In precision measurements, where the accuracy is limited only by systematic uncertainty coming from generator modeling, better constraining the UE improves the precision of the results. Two of the most obvious examples are measurements of top-quark and W-boson masses. A previous study[432] showed the effect of different tunes (which are different plausible models of UE and other soft-QCD processes in MC generators, discussed in more detail in Section 2) on final measured value and uncertainty. The recent ATLAS W-boson mass measurement[461] also has the largest uncertainty coming from MC modeling, although not all of it is due to UE.

[b]The importance of this will become clearer later, when the concept of pile-up will be discussed.

There are also more subtle effects. A useful tool for suppressing mis-reconstructed particles and other backgrounds in experimental analyses is enforcing some isolation criteria, whereby a cone of a certain radius around the target particle is required to contain energy deposited below some threshold value. If the underlying event is not properly modeled by the MC simulation sample used to determine the isolation criteria, then final result will be biased. A similar situation exists for MC-based jet energy scale calibrations, where mis-modeling of UE may lead to overestimation of uncertainties.

Currently *boosted* topologies are being used more and more for searches, where decay products of a heavy hadronically-decaying particle are collected in a large-radius jet. Then the internal (sub)structure of this large-radius jet is analyzed in order to distinguish large-radius jets coming from signal to that coming from light quark or gluon backgrounds. These large-radius jets are contaminated by soft radiation, which obscures the radiation pattern from main decay chain, and part of this soft radiation comes from underlying event. The mass of the jet is one of the most sensitive observable here, which is affected strongly by UE. Many grooming techniques have been developed[462] which discard the soft radiation, but these techniques are again mostly developed and tested on MC simulated samples.

1.3. *Connection to other soft processes*

While distributions sensitive to underlying event contribute to the modeling of soft-QCD processes in MC generators, there are other ways to construct distributions sensitive to soft-QCD effects. Minimum-bias (MB) is a term which is often used interchangeably with underlying event, and although the origin of pile-up is different, it is also often used in a similar context.

- **Minimum-bias (MB):** MB (or min-bias) is a generic term which refers to events that are selected with a loose trigger[c] that accepts a large fraction of the inelastic cross section. The specific definition always depends on the particular experiment. Most experiments require at least one charged-particle detected in some forward detector.[d] In principle they contain all types of interactions proportionally to their natural production rate, and consequently it is characterized by having very few high-p_{T} objects, since their production cross section is much smaller.
- **Pile-up (PU):** At hadron colliders, bunches (of more than 10^{11} for LHC) of protons (or antiprotons) are circulated along the ring, and brought to the

[c]It is impossible to save every event to data storage due to high event rate, so based on preliminary information from tracking, calorimetry, and muon systems, only interactions that satisfy some specific preselected criteria are selected and recorded.

[d]Forward is defined in terms of pseudo-rapidity. The pseudorapidity is given by $\eta = -\ln\tan(\theta/2)$, where the polar angle θ is measured with respect to the beam axis. In ATLAS and CMS experiments, $|\eta| > 2.5$ is considered forward.

collision at the centre of the detectors (termed bunch-crossing). The detector resolution is not often enough to distinguish between separate collisions (usually the ones just before and after), or even multiple collisions in a single bunch crossing. This contamination is termed pile-up, with the former termed out-of-time pile-up, and the latter in-time pile-up. Pile-up is parametrized in terms of the parameter μ, which denotes the average number of additional proton-proton collisions per bunch crossing. A measure of the in-time pile-up is given by the parameter N_{vtx}, which is the average number of reconstructed primary vertices in the event. Pile-up interactions are mostly similar to isotropic MB interactions, however high-p_{T} objects can be produced in a small fraction of events.

2. MC Models

As mentioned in earlier, the MC event generators are used to obtain predictions for what happens in collisions. This section focuses on the aspects of the MC generators relevant to modeling of UE, as shown in Fig. 1 (see also Chap. 10). The full event consists not only the hard partons emitted from the collisions, which can be calculated by perturbation theory in some order (of perturbative expansion in α_s, the strong coupling constant in QCD), but also of multiple softer gluon emissions and their subsequent splitting into quark anti-quark pairs, which cannot be calculated. This is referred to as parton shower (PS) approach. Additionally the soft processes (or semi-hard, the distinction is often subjective) from rest of the proton debris cannot be calculated either. These are phenomenologically modeled in MC event generators, by making use of most available data from the experiments.

The following processes, along with PS, is used to model the full event structure:

- Initial- and Final-State Radiation (I/FSR): A parton can emit further partons both before and after the QCD interaction vertex giving rise to initial and final state parton showers, respectively.

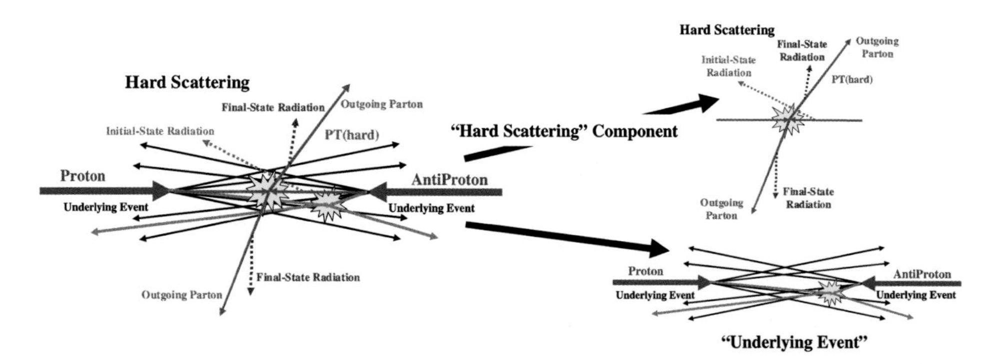

Fig. 1 Illustration of a QCD MC model simulation of a proton–antiproton collision. Figure reprinted with permission from Ref. 492. Copyright 2010 by the American Physical Society.

- Multiple parton interactions (MPI): additional *semi-hard* 2-to-2 parton–parton scatterings in the same event. Double parton interaction (DPI) is a subset of MPI.
- Beam–beam remnants (BBR): particles that come from the partons not participating in hard scatter. QCD evolution of color connections between the hard scattering and the beam–proton remnants are an important component of BBR modeling.
- Fragmentation and hadronization: since free partons can not exist in nature, they need to form color neutral hadrons. This is performed by splitting partons into other partons (termed fragmentation) such that they can combine to form color neutral hadrons (termed hadronization).

There are many parameters controlling all these and other aspects of the program. Each program also have some specific modeling features, resulting in parameters specific to that program. In Table 1, some of the most commonly referred parameters are listed. Since Pythia 8[192] program has been tuned most extensively, this table is strongly influenced by Pythia 8 terminology. However, the specific Pythia 8 parameter names have not been mentioned to keep the discussion sufficiently general.

A tune for a particular generator is a specific set of values (of some or all) of these parameters, usually arrived by trying to describe a set of experimental distributions. For a complete description of the tuning methodology, the reader is pointed to Chap. 13. Table 2 lists some of the commonly used tunes. There are other models, like Sherpa,[157] based on PS approach, and PHOJET,[403] EPOS[463] or QGSJETII[464] which are not, which sometime describe certain MB or UE distributions well, which have not been mentioned.

It also must be noted that there is always a strong interplay between different processes. For example, jets originating from ISR can have sufficiently high-p_T to be considered part of hard scattering, or be soft enough to be part of UE. Additionally, it has been suggested,[96] that soft correlations contribute significantly to UE at low p_T scale. Additionally the two main components of the UE, MPI and BBR necessarily do not vary the same way with collision energy. So a good description of data at Tevatron energy does not automatically lead to a similar agreement at higher LHC energies, everything else being the same.

3. UE Measurements

3.1. *CDF*

The UE (and other soft-QCD) measurements from Tevatron were mostly limited to CDF, as DØ tracking was limited by absence of magnetic field. The first underlying event measurements came out of Run I,[459] and it established many of the techniques that have become standard now. Therefore it will be discussed in some detail.

Table 1. PS generator parameters controlling UE. It has become standard practice to use the mass of the Z-boson, M_Z, which is simply a convenient reference scale large enough to be in the perturbative domain.

Process	Parameter	Effect
Hard process	$\alpha_{\rm strong}$ value at scale $Q^2 = M_Z^2$	More/less activity from hard process for higher/lower value
MPI	$\alpha_{\rm strong}$ value at scale $Q^2 = M_Z^2$	More/less activity from MPI for higher/lower value
MPI	$p_{\rm T}$ cutoff scale	Larger/smaller values result in less/more MPI
MPI	Energy extrapolation exponent	Determines the $p_{\rm T}$ cutoff scale as a function of c.m energy
MPI	Impact parameter/inverse hadron radius dependence	Smaller/larger values indicate more/less overlap between colliding hadrons, hence more/less MPI
BBR	Primordial/intrinsic $k_{\rm T}$	Parton intrinsic transverse momentum assigned due to uncertainty principle, important for describing vector boson $p_{\rm T}$ peak
BBR	Strength of color reconnection	Higher/lower value indicates more/less chances of two systems merging
BBR	Length of color strings	Longer string length gives more multiplicity per interaction, resulting in more high $p_{\rm T}$ particles
ISR	$\alpha_{\rm strong}$ value at scale $Q^2 = M_Z^2$	More/less activity from ISR for higher/lower value
ISR	$\alpha_{\rm strong}$ order	Order at which ISR strong coupling constant runs
ISR	$p_{\rm T}$ cutoff scale	Larger/smaller values result in less/more ISR
ISR	Setting of maximum shower evolution scale	Depends on the matching to hard process, multiplicative/damping factors can be applied
FSR	$\alpha_{\rm strong}$ value at scale $Q^2 = M_Z^2$	More/less activity from FSR for higher/lower value
FSR	$\alpha_{\rm strong}$ order	Order at which FSR strong coupling constant runs
FSR	$p_{\rm T}$ cutoff scale	Larger/smaller values result in less/more FSR
FSR	Setting of maximum shower evolution scale	Depends on the matching to hard process, multiplicative/damping factors can be applied

Table 2. Different MC model tunes used over the years, focusing on parton shower generators.

Generator	Tune (PDF)	Comment
(a) Pre-LHC		
Pythia 6[158,327,465,466]	CDF A[467] (CTEQ5L[468])	CDF Run I UE, *first* (MPI) *tune*
Pythia 6	CDF AW[467] (CTEQ5L)	Tune A + CDF Z-p_T, tuned ISR and intrinsic k_T parameters
Pythia 6	CDF DW[467] (CTEQ5L)	DW+ DØ dijet decorrelation, considered stable Tevatron era tune
Pythia 6	CDF DWT[467] (CTEQ5L)	Same as DW, changed MPI energy scaling
Pythia 6	CDF D6[467] (CTEQ6L1[469])	Same as DW, changed PDF
Pythia 6	ATLAS Rome/DC2[470] (CTEQ6L1)	Same as DW, but DWT energy scaling
Pythia 6	ATLAS MC08/CSC[470] (CTEQ6L1)	CDF Run I and Run II UE
Pythia 6	ATLAS MC09/MC09c[471] (MRST LO*[472])	CDF Run I UE, Run II MB, updated fragmentation function
Pythia 6	Perugia0[473] (CTEQ5L)	UA5 and CDF MB, CDF and DØ Z p_T, targeted for LHC
Pythia 6	Perugia2010[434] (CTEQ5L)	Same as Perugia0, updated fragmentation/shower parameters
Herwig[474] + Jimmy[36]	CDF[383] (CTEQ5L)	CDF Run II UE
Herwig + Jimmy	ATLAS MC08[470] (CTEQ6L1)	CDF Run II UE
Herwig + Jimmy	ATLAS MC09[471] (MRST LO*)	Changed PDF
(b) LHC-era		
Pythia 6	AMBT1[475] (MRST LO*)	*First tune using LHC data,* ATLAS MB
Pythia 6	AMBT2[476] (MRST LO**[472])	CDF and ATLAS MB, ATLAS Jjet shapes and dijet $\Delta\phi$
Pythia 6	AUET2[476] (MRST LO**)	ATLAS UE, ATLAS jet shapes and dijet $\Delta\phi$
Pythia 6	AMBT2B[477] (Several)	Same as AMBT2, bug fixed[e]

(*Continued*)

[e]A bug was discovered in A**T2 tunes, where shower α_S values was not properly propagated from PDF to Pythia 6.

Table 2. (*Continued*)

Generator	Tune (PDF)	Comment
Pythia 6	AUET2B[477] (Several)	Same as AUET2, bug fixed
Pythia 6	Perugia2012[434] (CTEQ6L1)	Perugia10 + LHC MB, identified particle spectra from ALICE and STAR
Pythia 6	Z1[191] (CTEQ5L)	ATLAS MB + CMS UE
Pythia 6	Z2[467] (CTEQ6L1)	Same as Z1, changed PDF
Pythia 6	Z2*[467] (CTEQ6L1)	Same as Z1, tuned MPI energy extrapolation
Pythia 6	CUETP6S1[152] (Several)	CMS UE at different energies
Herwig + Jimmy	AUET1[478] (Several)	ATLAS UE
Herwig + Jimmy	AUET2[476] (Several)	More weight on higher energy ATLAS UE
Pythia 8	A2[479] (MSTW2008LO[480])	ATLAS MB and sum E_T
Pythia 8	A14[481] (NNPDF2.3LO[482])	LHC Run 1, MPI and shower tune, UE/high p_T focused
Pythia 8	A3[483] (NNPDF2.3LO)	Run 1 and Run 2 MB, inelastic cross-section
Pythia 8	4C[114] (CTEQ6L1)	LHC Run 1 data
Pythia 8	4Cx[66] (CTEQ6L1)	4C-like but x-dependent MPI matter profile
Pythia 8	Monash[436] (NNPDF2.3LO)	Update of 4C with more Run 1 MB+UE, updated hadronisation parameters
Pythia 8	CUETP8S1/M1[152] (several/NNPDF23LO)	CMS UE at different energies
Pythia 8	CDPSTP8S1/2-Wj[152] (CTEQ6L1)	DPI in W+dijet
Pythia 8	CDPSTP8S1/2-4j[152] (CTEQ6L1)	DPI in 4-jet
Herwig++[109,484]	LHC-MU900-1	First generation LHC MB and UE tunes
	LHC-UE7-1, LHC-UE-EE-*-1[485] (several)	Tunes with EE are valid for all collision energies, rest are for fixed energies
Herwig++	LHC-UE7-2, MU900-2[485]	Fixed strange hadron production and the b-fragmentation function from before
Herwig++	UE-EE-2, UE-EE-3, UE-EE-4[485] (several)	Updates in MPI +CR parameters, SCR: new CR model
	UE-EE-SCR (CTEQ6L1)[485]	

(*Continued*)

Table 2. (*Continued*)

Generator	Tune (PDF)	Comment
Herwig++	UE-EE-5[485] (CTEQ6L1)	Model update
Herwig++	CUETHppS1[152] (CTEQ6L1)	CMS UE at different energies
Herwig 7[398]	UE-MMHT[110] (MMHT2014 LO[486])	Model update, UE+DPI

Notes: For Herwig, the MPI is added by the Jimmy module, and hence the latter is tuned. It must be emphasized, that although the last column tries to describe the most important *focus* of the tune, it is hardly complete. Each tune builds up on previous tunes, and non-trivial correlation between different parameters controlling different processes implies more changes than that can be fitted in such a table.

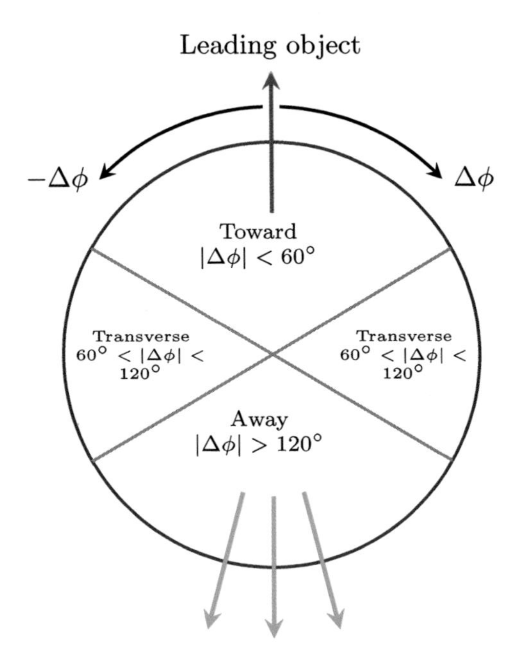

Fig. 2 Definition of UE regions as a function of the azimuthal angle with respect to the chosen leading object.

Since underlying event by definition requires an identified hard scatter, events with a high $p_{\rm T}$ object is usually chosen. Then the event activity can be measured by (charged-particle) tracks or by calorimeter energy clusters.[f] In this measurement at centre-of-mass energy of $\sqrt{s} = 1.8$ TeV, the leading (i.e., the highest $p_{\rm T}$) track jet is taken as the direction of dominant energy flow in the hard scattering, and the charged-tracks are used to determine the activity. Then as illustrated in Fig. 2, in

[f]For measurements that are corrected for detector effects, i.e., unfolded back to particle level, these will be simply referred to as charged or neutral particles later.

each event, the azimuthal angular difference between tracks and the leading object (track jet) $|\Delta\phi| = |\phi - \phi_{\text{leading}}|$, is used to assign each track to one of these regions (a concept first used in Ref. 380):

- $|\Delta\phi| < 60°$, the *toward* region,
- $60° < |\Delta\phi| < 120°$, the *transverse* regions, and
- $|\Delta\phi| > 120°$, the *away* region.

While the underlying event populates the whole region, these azimuthal regions have different sensitivity to the UE, which is exploited to extract the UE activity. The transverse region is sensitive to the underlying event, since it is by construction perpendicular to the direction of the hard scatter and hence it is expected to have a lower level of activity from the hard scattering process compared to the away region, which in most cases would contain the objects which will balance the leading object.

Then the average charged-particle multiplicity and scalar p_{T} sum in the transverse regions are measured, and shown in Fig. 3. The min-bias (i.e., minimum-bias) and JET20 respectively denote the trigger used to select the events, and it is interesting to note that the transition between the triggers in both the distributions are rather continuous.

Some features of the distributions are worth noting, since these will be encountered repeatedly.

- The y-axis contains a measure of activity, and in transverse region the activity is most sensitive to the underlying event. The activity is measured as a function of p_{T} of the leading object, which can be taken as the measure of Q^2 scale of the event. These distributions are so-called profiles, where the average value in each bin is plotted at the bin centre.
- The activity (be it the multiplicity or p_{T} sum) initially shows a sharp rise, then saturates at a characteristic plateau height. This observation is consistent with the predictions by theoretical models,[487] according to which MPI increases with increase of the event energy scale, as collisions become more central (i.e., proton and antiproton overlap increases) from more peripheral at low p_{T} exchange. The plateau height is reached when MPI no longer increases, as the BBR level is roughly constant.
- (later) Since different azimuthal areas are studied, to facilitate direct comparisons, the densities are constructed by dividing the by the $\Delta\eta\Delta\phi$ areas. For example, the transverse multiplicity density corresponds to the number of charged-particles divided by $\Delta\eta\Delta\phi = (2 * 1.5) * (2\pi * 2/3) = 4\pi$, as the CDF inner tracker extends between $|\eta| < 1.5$ and the angular area of the transverse region is twice of 60°. For brevity, the term density will be omitted in some cases.
- The data is compared with certain MC model predictions, which is helpful in interpreting the results, and improving the models in general.

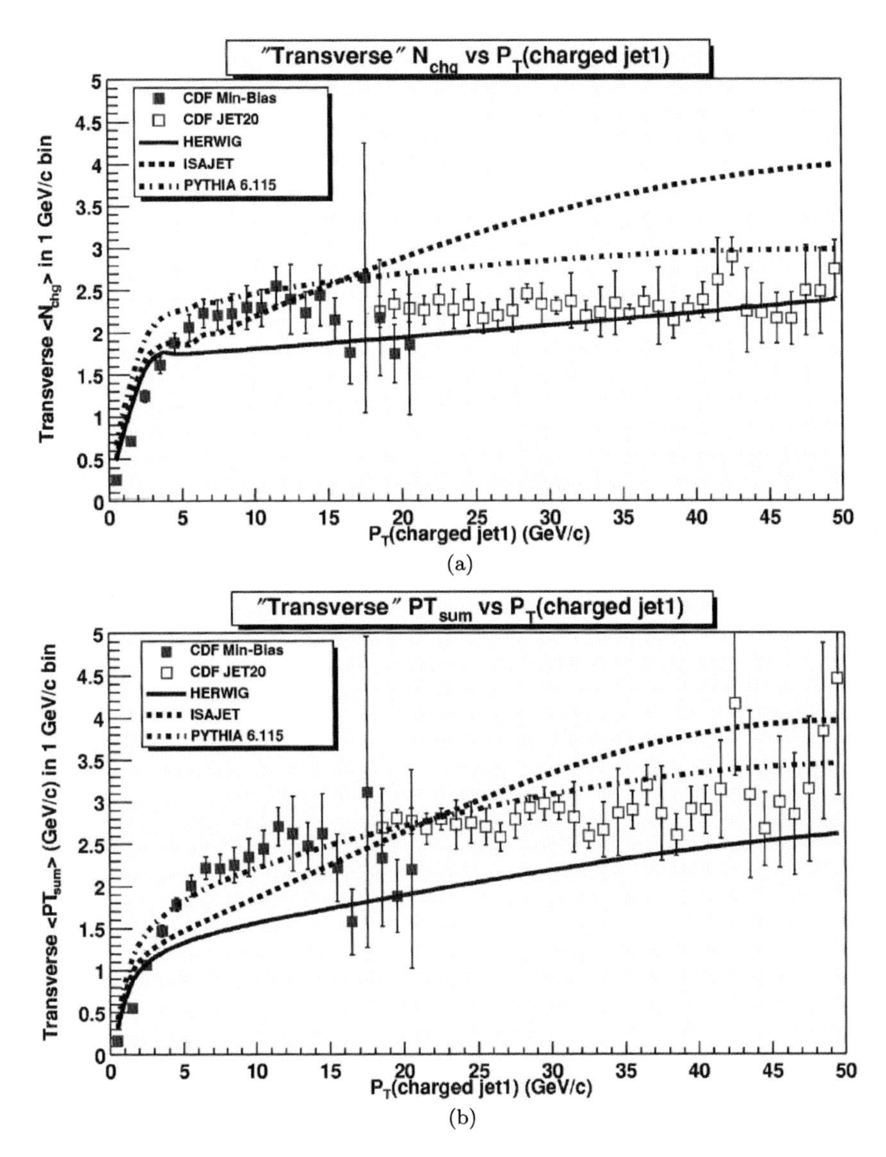

Fig. 3 CDF Run I UE results with charged-particle jet. (a) Charged-particle density as a function of p_T of leading charged-particle jet. (b) Charged-particle scalar sum p_T density as a function of p_T of leading charged-particle jet. Figure reprinted with permission from Ref. 128. Copyright 2002 by the American Physical Society.

In this analysis, the results were compared to Herwig, Pythia 6 and ISAJET[488] (the latter not being discussed here, as it is rather old fashioned) predictions, and none of them described the transverse region data well. Whereas Pythia model had a p_T cutoff scale to regulate perturbative parton–parton scattering cross-section divergence, Herwig and ISAJET did not, resulting in the requirement of a $p_{T(\text{hard})} =$ 3 GeV for these models.

It should be noted that this was before any of the models were actually tuned, so these result provided the impetus for tuning at least MPI part. Additionally the data was not corrected back to particle level, rather tracking inefficiency was applied to MC distributions, which makes this data of limited use.[g]

However, some conclusions can still be drawn. The description of data by Herwig with the angular ordered shower was better than ISAJET, but Pythia performed the best. Only Pythia had a MPI model built in, and in absence of MPI, the charged-particles produced from BBR was too soft for Herwig. This established that MPI are an important ingredient in the modeling of hadron–hadron collisions. This was also seen in parallel in the CDF measurement of charged-particle multiplicity in MB events.[489] The broad multiplicity spectrum, and large multiplicity fluctuation seen in data can only be caused by MPI.

The other CDF Run I UE measurement has come to be known as *Swiss Cheese* result,[490] because it calculated track p_T sum in within two conical regions, which resemble holes in the rectangular η–ϕ projection as in Fig. 4(a). These two cones are constructed at the same η as of the leading jet, but at $\phi = \pm 90°$. The cone with higher (lower) sum p_T are denoted as max and min-cone. The idea is that min-cone will be sensitive to UE, while the max-cone will be dominated by ISR and FSR. The data, shown in Fig. 4. supported that assumption. Also the first tuning of the Pythia 6 MPI parameters was done to better describe this data.

An explicit Swiss cheese topology is also defined, where p_T of all the tracks except those in the two or three highest energy jets are used to calculate the observables.

Subsequently CDF used data at $\sqrt{s} = 600$ GeV and 1.8 TeV[491] to probe the interplay between soft and hard processes in more detail. Charged-particle multiplicity, p_T spectrum, average p_T and event-by-event p_T dispersion was measured by dividing the events into soft and hard categories based on the absence or presence of a cluster[h] of energy 1.1 GeV. This essentially demonstrated the transition between soft MB and semi-hard UE. The p_T distribution at fixed multiplicity in the soft sample was seen to be collision energy invariant. These measurements also showed for the first time that the UE events are indeed more active than generic MB events, roughly by a factor of two, at the same centre-of-mass energy. In Pythia, this difference comes from the fact that there are more MPI in a hard scattering process than in a typical MB collision. By demanding a hard scattering the collision is forced to be more central (i.e., smaller impact parameter), which increases the MPI cross sections.

Most comprehensive CDF UE measurement was performed in Run II at $\sqrt{s} = 1.96$ TeV, which looked at the UE activity Z-boson and inclusive jet events, and compared them to tuned versions of Pythia 6, as well as to each other.[492]

[g] The tracking efficiency numbers were reported, so meaningful comparison can be made at detector level.

[h] Constructed from calorimeters towers or tracks within a cone of radius 0.7.

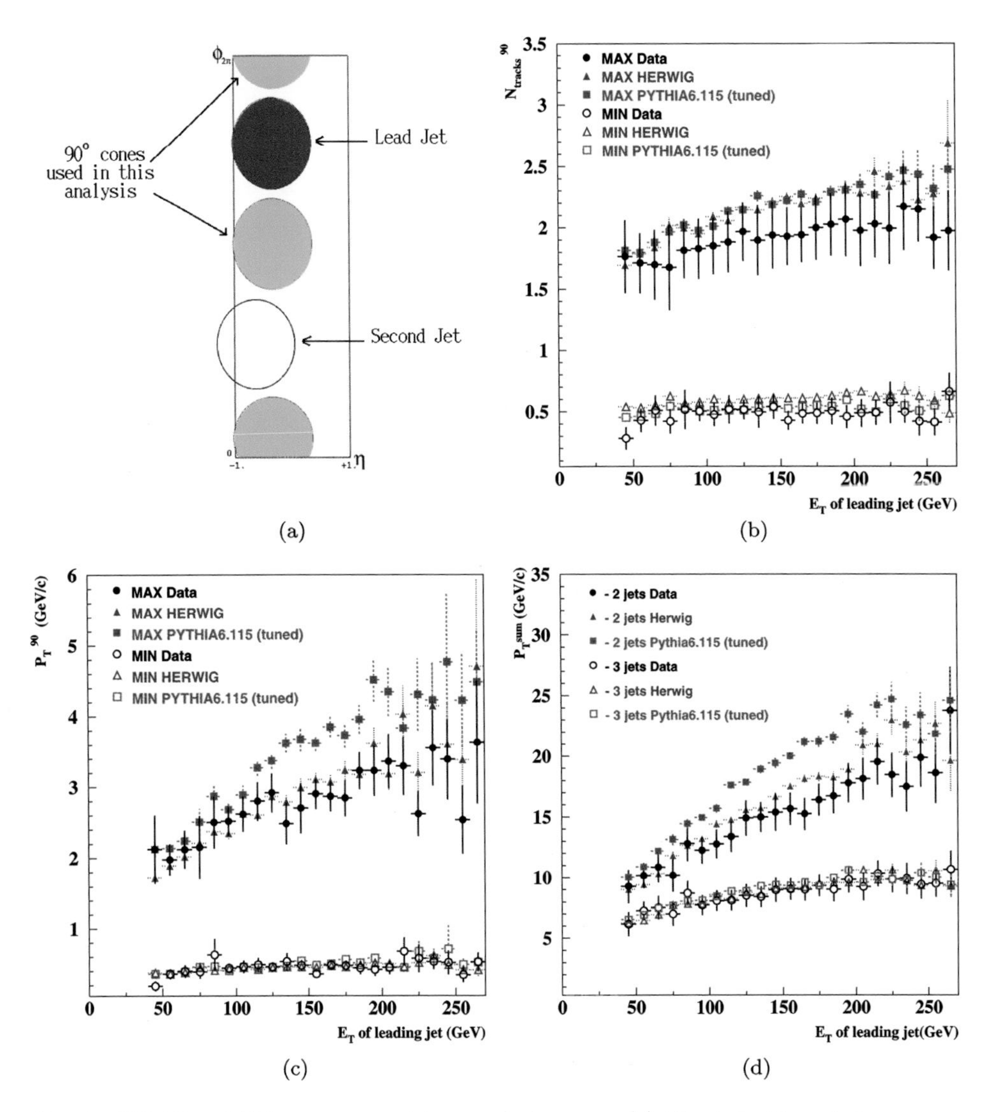

Fig. 4 CDF Run I UE measurement in max/min-cones. (a) The definition of phase space in CDF *swiss cheese* UE measurement. (b) Number of tracks in max/min-cones as a function of E_{T} of the leading jet. (c) Track p_{T} in max/min-cones as a function of E_{T} of the leading jet. (d) Track p_{T} in swiss cheese topology as a function of E_{T} of the leading jet. Figure reprinted with permission from Ref. 381. Copyright 2004 by the American Physical Society.

Z-boson production provides an excellent place to study the UE. Unlike the high-p_{T} jet production, for Z-boson production there is no final-state gluon radiation off the Z-boson.

The concept of max/min-cone, introduced in Run I, was merged with the *trans-max/min* area definition[493,494] in this measurements, and has been used extensively ever since. The max (min) transverse regions are defined on an event-by-event basis containing the larger (smaller) number of charged-particles or to the region containing the larger (smaller) scalar p_T sum of charged-particles. Again densities are formed by dividing by the area of each region.

This is useful in trying to disentangle UE activity further, as in the events with large initial or final-state radiation the trans-max region will usually contain the third jet in high-p_T jet production or the second jet in Z-boson production while both the trans-max and trans-min regions receive contributions from the BBR and soft MPI. Thus, the trans-min region is very sensitive to the BBR. Also the event-by-event difference between trans-max and trans-min can be constructed, which is very sensitive to initial and final-state radiation. Certain results use the *trans-ave* or average transverse to denote the transverse region, and this two terms will be used interchangeably.

The activity in all the regions were separately measured for both leading jet and inclusive Z-boson events. In Fig. 5, in the top row, the comparison

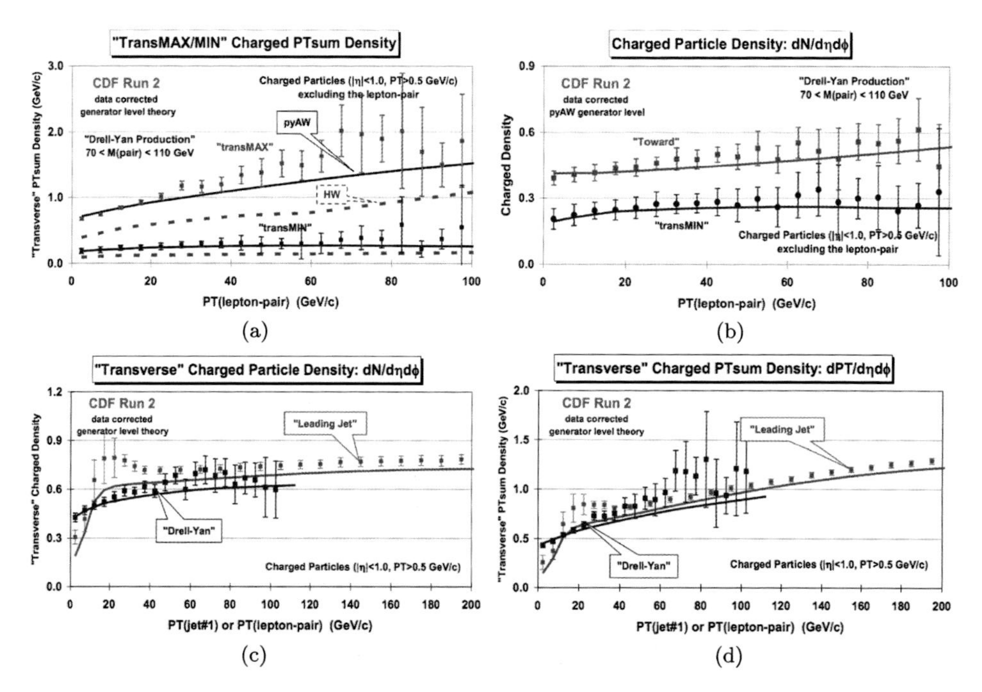

Fig. 5 CDF Run II measurement of UE in Z-boson and leading jet events. (a) Charged-particle sum p_T density as a function of p_T of Z-boson in trans-max/min regions. (b) Charged-particle density as a function of p_T of Z-boson in trans-min and toward regions. (c) Charged-particle density as a function of p_T of Z-boson or leading jet in transverse region. (d) Charged-particle sum p_T density as a function of p_T of Z-boson or leading jet in transverse region. Figure reprinted with permission from Ref. 492. Copyright 2010 by the American Physical Society.

of sum p_T in trans-max and min regions, and the comparison of multiplicity in toward and trans-min regions are shown in Z-boson events. The interesting features are:

- At small leading jet p_T the toward, away, and transverse densities become equal and go to zero as leading jet p_T goes to zero. If the leading jet has no transverse momentum then there can be no particles anywhere. In addition, there are numerous low transverse momentum jets and, hence for leading jet $p_T < 30$ GeV the leading jet is not always the jet resulting from the hard scattering. This produces a bump in the transverse density in the range where the toward, away, and transverse densities become similar in size.
- For Z-boson events the toward and transverse densities are both small and almost equal. The away density is large due to the away-side balancing jet. The toward, away, and transverse densities become equal as the p_T of the lepton pair goes to zero, but unlike the leading jet case the densities do not vanish at lepton pair $p_T = 0$. For Z-boson events with lepton pair $p_T = 0$ the hard scale is set by the lepton-pair mass which is in the region of the Z-boson, whereas in leading jet events the hard scale goes to zero as transverse momentum of the leading jet goes to zero.
- The now familiar plateau-like behavior of UE activity is seen in both the cases.
- The trans-max activity is higher then trans-min. This is because the extra jet contribution in trans-max, so trans-min is more sensitive to UE.
- For Z-boson events, the toward region is also interesting, as unlike jet events, the activity excluding the lepton pair is expected to be sensitive to UE. For high transverse momentum lepton-pair production, particles from initial-state radiation are more likely to populate the overall transverse region than the toward region and hence the densities are slightly larger in the overall transverse region. The most sensitive observables to the UE in Z-boson production are those constructed in the toward and trans-min, since these observables are less likely to receive contributions from ISR and the away-side jet. The toward activity is seen to be higher than in trans-min as well, as it also receives contribution from semi-hard MPI, which trans-min region by construction does not.

The activity was seen to be modeled reasonably well by the Pythia 6 tunes, and were very similar in the two classes of events, as in the bottom row.

Another additional important distribution is the mean p_T of the charged-particles against multiplicity, as it shows the interplay of hard against soft UE, and is also very sensitive to the modeling of color reconnection. In Fig. 6, a gradual increase of mean p_T can be seen with increasing multiplicity. This can be caused either by MPI or hard scattering, as BBR is scale independent. However, the CDF measurement showed that for Herwig (with the MPI being added by Jimmy), the mean p_T increases more rapidly than data. In this case demanding large multiplicity will preferentially select the hard process and lead to a high mean

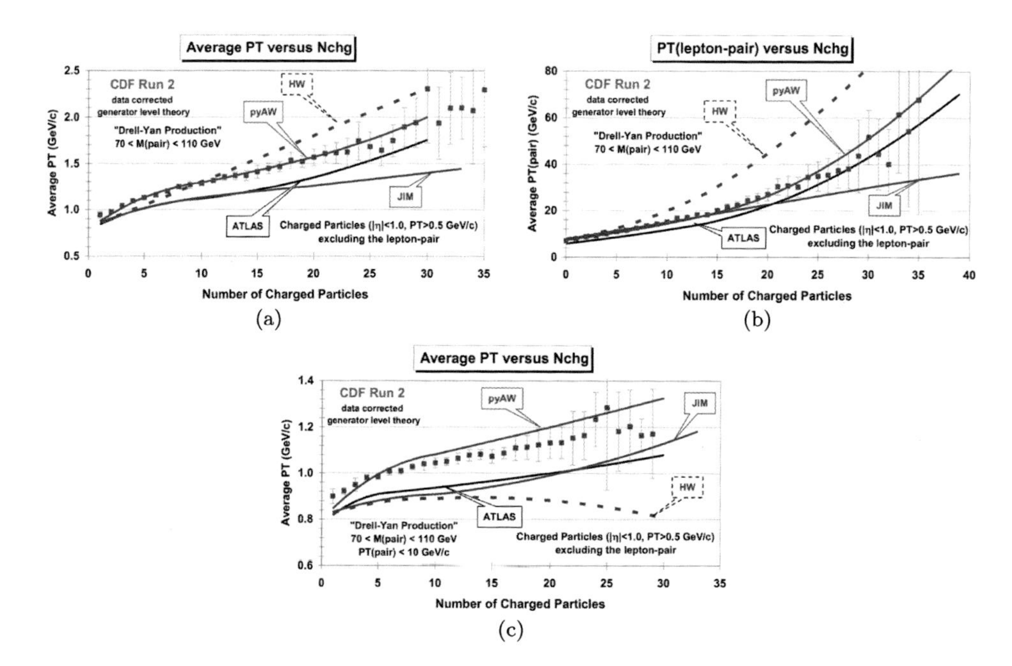

Fig. 6 CDF Run II measurement of UE in Z-boson and leading jet events. (a) Charged-particle mean p_T against multiplicity. (b) Lepton pair p_T against multiplicity. (c) Charged-particle mean p_T against multiplicity for lepton pair $p_T < 10$ GeV. Figure reprinted with permission from Ref. 492. Copyright 2010 by the American Physical Society.

p_T. This can be seen from the rapid increase in the distribution of lepton pair mean p_T against charged-particle multiplicity. However, Pythia models with MPI included describe the data better, indicating that MPI produces large multiplicities that are harder than the beam–beam remnants, but not as hard as the primary 2-to-2 hard scattering. Figure 6 shows the average p_T of charged-particles against the multiplicity for Z-boson events in which the lepton pair $p_T < 10$ GeV. The average p_T still increases as the multiplicity increases, although not as fast. This requirement, which essentially selects events without the away-side jet, means for Herwig without MPI, large multiplicities can come from ISR. Since the particles coming from ISR are usually soft, the average p_T decreases slightly in this case with the increase in multiplicity, in contrast with the data.

However, colored partons in BBR are also affected by color reconnection, modeled by creation and breakup of colored strings between them. Breakup of longer strings lead to production of more charged-particles with smaller p_T, while shorter strings lead to less sharing of same energy between hadrons. Tuning these helps to better model this distribution.

Also event shapes using calorimeter energy deposits were measured[495] in Run II with $\sqrt{s} = 1.96$ TeV. The transverse thrust and thrust minor are sensitive to UE modeling.

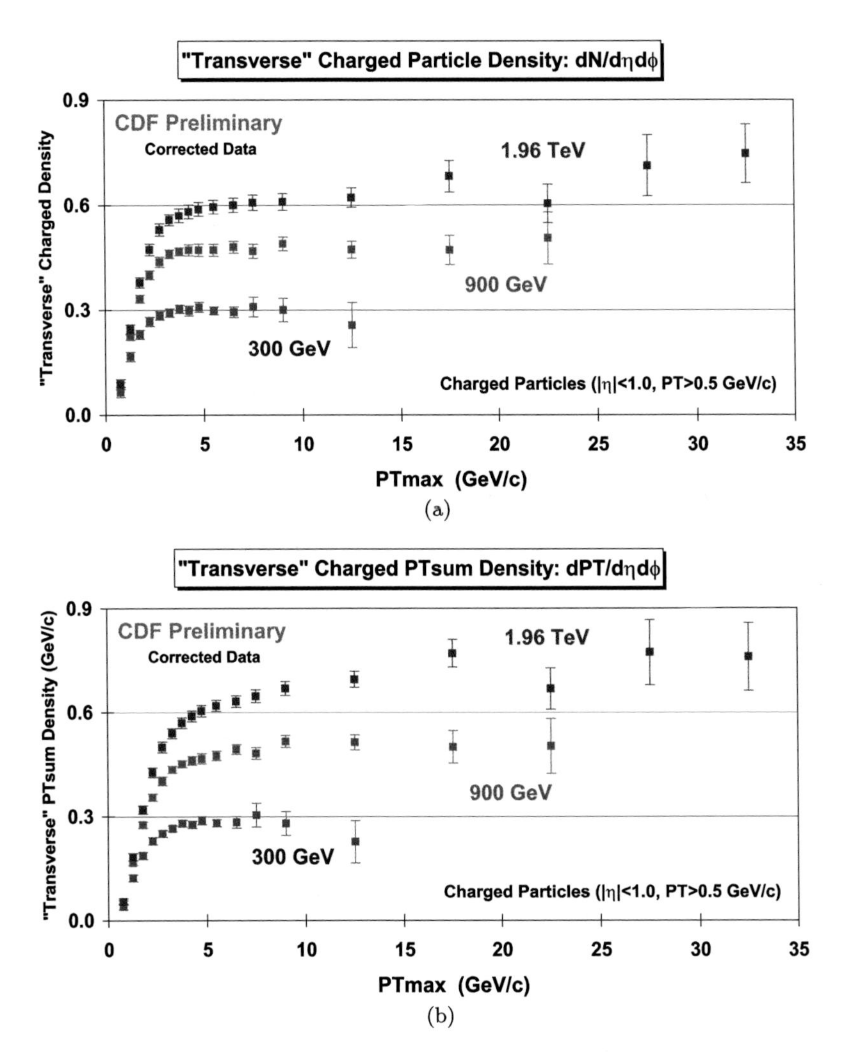

Fig. 7 CDF energy scan UE results. (a) Charged-particle density as a function of p_T of the leading charged-particle in transverse region. (b) Charged-particle sum p_T density as a function of p_T of the leading charged-particle in transverse region.

Subsequently before the eventual shut down of Tevatron, it ran in energy-scan mode, where events were collected at different centre-of-mass collision energies at 300 GeV and 900 GeV. This data can be combined with data at 1.96 TeV allowing for a study of the energy dependence of the UE at three centre-of-mass energies. Furthermore, the 900 GeV data can be directly compared with LHC UE measurements at the same centre-of-mass energy. UE with leading charged-particle measurements were performed,[496] at these energies. The evolution of UE plateau heights, as seen for example in Fig. 7 for multiplicity and sum p_T, are useful inputs to constraining the energy extrapolation of MPI models.

4. UE Measurements at the LHC

4.1. *Predictions from Tevatron*

Before LHC started taking data in 2009, a good way to test of our understanding of the modeling of UE was to make predictions of UE activity at the LHC. The prediction was done for the simplest case, with the leading charged-particle setting the direction and scale of the hard scatter, and using the Tevatron state-of-the-art Pythia 6 DW tune. Figure 8 shows the prediction for charged-particle multiplicity

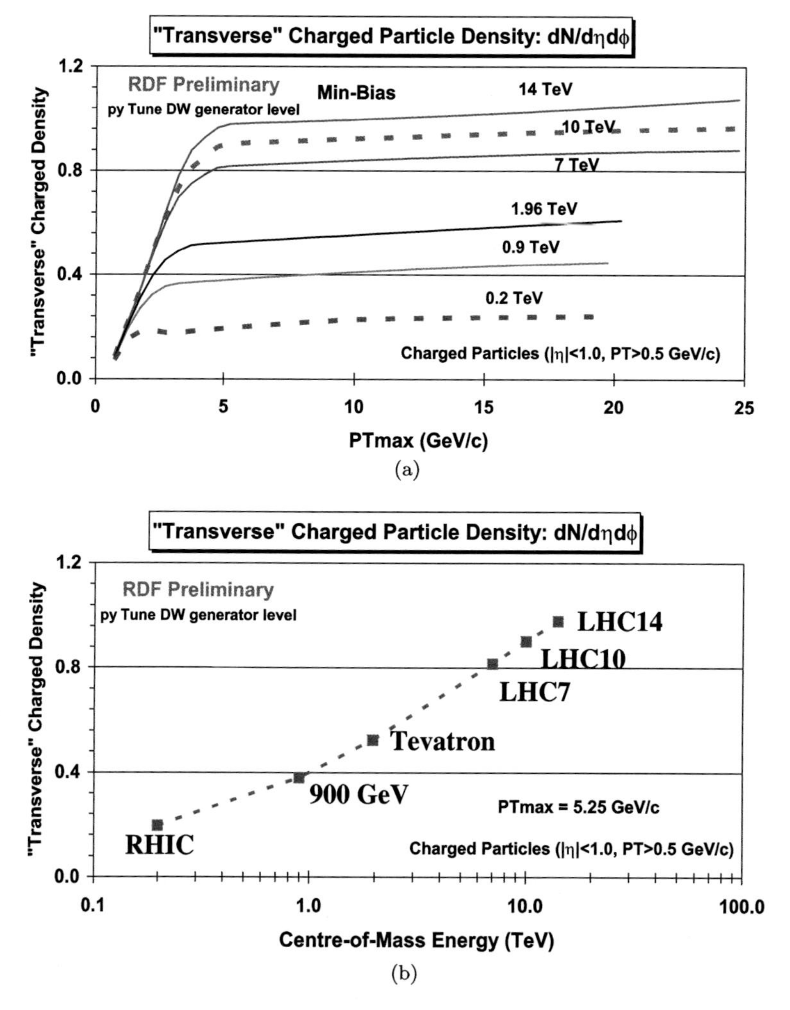

Fig. 8 LHC UE activity prediction using Pythia 6 DW Tune. (a) Transverse region charged-particle density as a function of leading charged-particle at five different centre-of-mass energies. (b) Evolution of the transverse region plateau height (taken at leading charged-particle $p_T = 5.25$ GeV, as indicated by the black dots), at logarithmic scale.

and sum p_T density profiles. As can be seen, the height of the plateau in the overall transverse region does not increase linearly with the centre-of-mass energy. For energies above the Tevatron it increases more like a straight line on a log plot. The UE activity was predicted by Tune DW to increase by about a factor of two in going from 900 GeV to 7 TeV and then to increase by only about 20% in going from 7 TeV to 14 TeV. These predictions proven to be correct, as described at end of Section 4.2.

4.2. *UE measurements using leading charged-particle*

The first ATLAS measurements by taking the leading charged-particle as the leading object at $\sqrt{s}$ = 900 GeV and 7 TeV at the LHC[101,130] showed the inadequacy of Tevatron-era tunes. The first hint of the mis-modeling came from the angular distribution of the charged-particles, where none of the MC models describe the shape or the level of activity in data. The Pythia 6 Tune DW however predicted the factor of 2 increase in UE activity going from 900 GeV to 7 TeV.

Both at 900 GeV and 7 TeV, the data showed more activity than predicted by the pre-LHC era models, as in Fig. 9. While 7 TeV was indeed an unprecedented energy, the discrepancy at 900 GeV was puzzling, in view of the good agreement at 1.96 TeV. The expected difference between the three regions at 7 TeV was seen, and the mean p_T and multiplicity correlation was also seen to not modeled well.

The leading charged-particle underlying event results were the first available results fully corrected to particle level, therefore they were most useful for tune comparison as well as for actual tuning. This, along with early ATLAS min-bias measurements led to tuning of generators, both inside and outside the experimental collaborations.

One of the other distributions that was published in this paper was the standard deviation of charged-particle multiplicity and sum p_T profiles, as in Fig. 10. This for the first time conclusively showed that UE activity is dominated by event-by-event fluctuations, and a flat pedestal activity cannot be subtracted to remove the effect of UE. In parallel to this analysis, another study was performed including both charged and neutral-particles to probe the UE activity, and conclusions were very similar.

Similar measurement was performed by the ALICE experiment at $\sqrt{s} = 900$ GeV and 7 TeV[497] during the start of LHC Run 1 using leading charged-particle. Figure 11 show the UE activity as a function of p_T of leading charged-particle in transverse region. Observables are constructed using charged-particles with $|\eta| < 0.8$ and $p_T > 0.5$ GeV (different from ATLAS and CMS, due to ALICE detector acceptance). Different MC predictions describe the measurements within 20%. UE activity almost doubles as collision increases from 900 GeV to 7 TeV which is consistent with observations by ATLAS and CMS experiments.

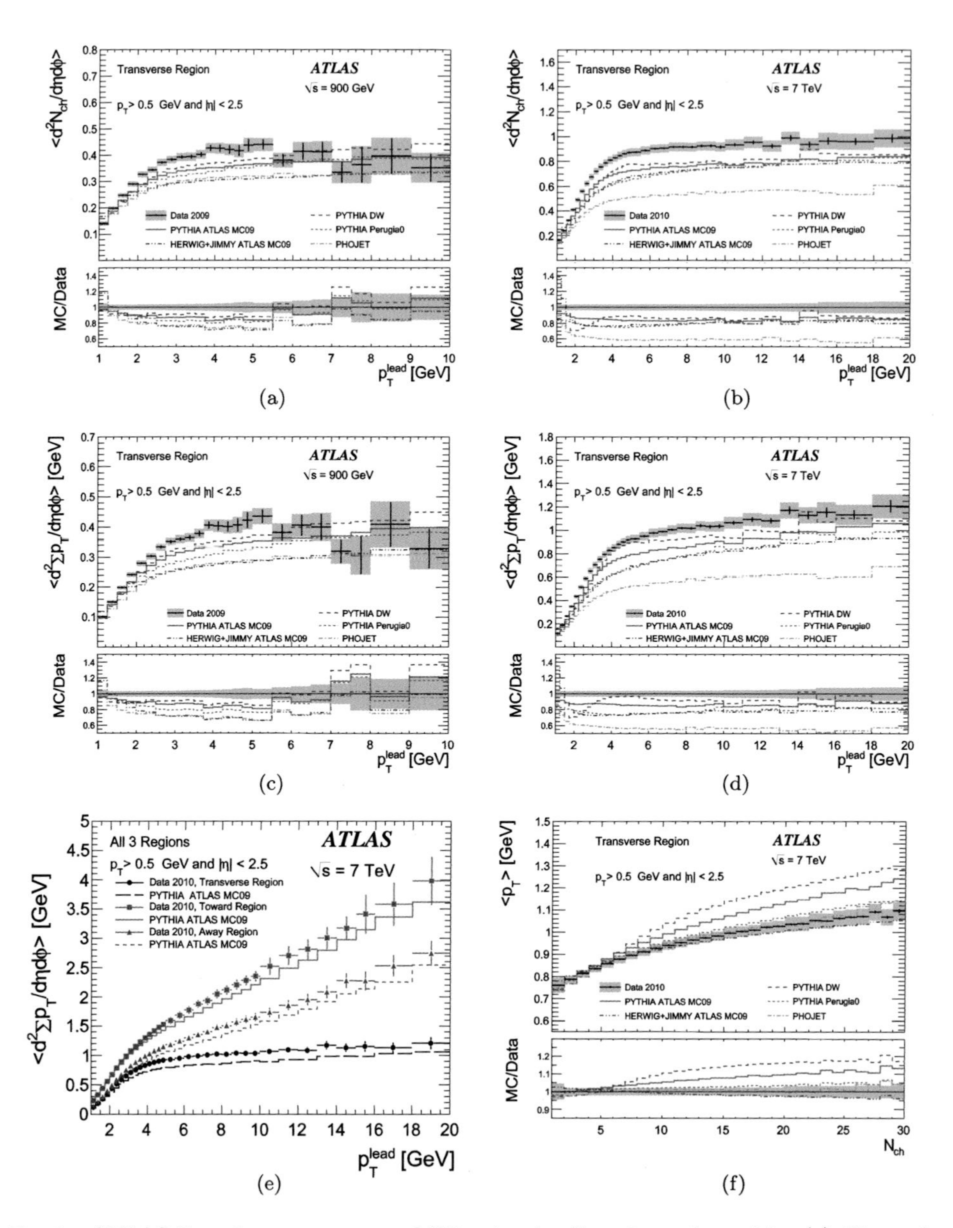

Fig. 9 ATLAS Run 1 measurement of UE using leading charged-particle. (a) Charged-particle multiplicity density as a function of p_{T} of leading charged-particle in transverse region. (b) Charged-particle multiplicity density as a function of p_{T} of leading charged-particle in transverse region. (c) Charged-particle scalar sum p_{T} density as a function of p_{T} of leading charged-particle in transverse region. (d) Charged-particle sum p_{T} as a function of p_{T} of leading charged-particle in transverse region. (e) Charged-particle sum p_{T} as a function of p_{T} of leading charged-particle in transverse, toward and away regions. (f) Charged-particle mean p_{T} as a function of charged-particle multiplicity in transverse region.

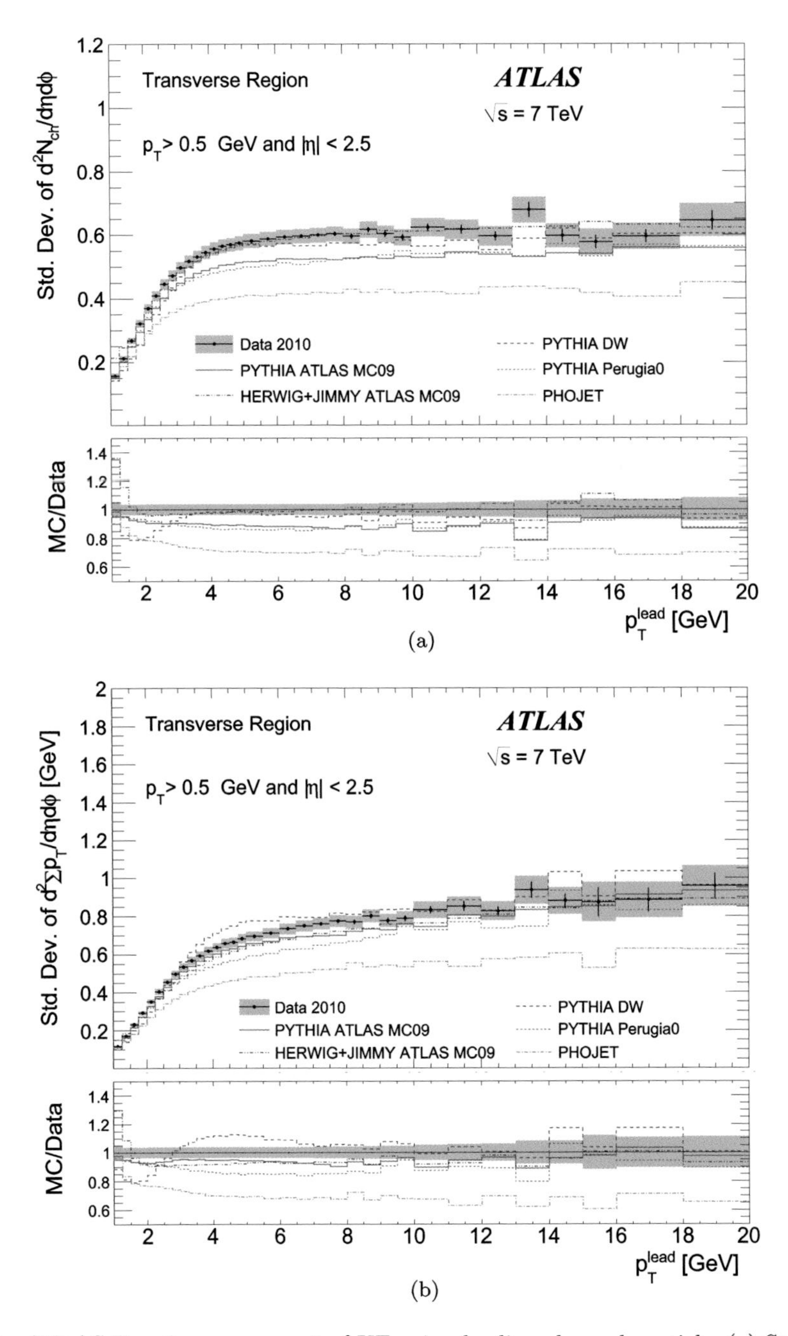

Fig. 10 ATLAS Run 1 measurement of UE using leading charged-particle. (a) Standard deviation of charged-particle multiplicity density as a function of p_{T} of leading charged-particle in transverse region. (b) Standard deviation of charged-particle sum p_{T} as a function of p_{T} of leading charged-particle in transverse region.

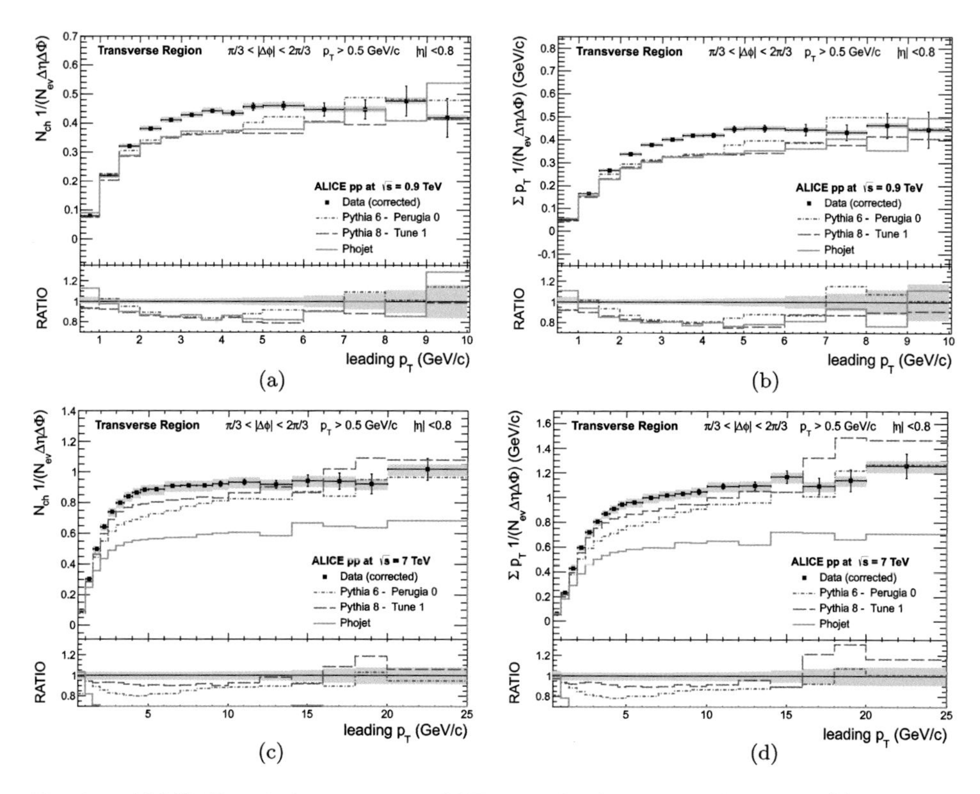

Fig. 11 ALICE Run 1 measurement of UE using leading charged-particle. (a) Charged-particle multiplicity density as a function of p_T of leading charged-particle. (b) Charged-particle sum p_T density as a function of p_T of leading charged-particle. (c) Charged-particle multiplicity density as a function of p_T of leading charged-particle. (d) Charged-particle sum p_T density as a function of p_T of leading charged-particle.

The real test of the tunes derived using Run 1 data was to see how they describe the $\sqrt{s} = 13$ TeV data from Run 2. The ATLAS measurement with the leading charged-particle[498] showed in general reasonable description of data by the tunes being used. Measurements at 13 TeV showed three times increase in UE activity as compared to UE activity at 900 GeV as can be seen from Fig. 12. The angular distribution of charged-particle sum p_T with respect to the leading charged-particle, which was seen to be very badly modeled at the beginning of Run 1, is modeled reasonably well by UE tunes. Similarly for the charged-particle sum p_T profile in the trans-min region, and mean p_T against charged-particle correlations, the agreement is not perfect but not significantly bad. Finally, the evolution UE activity plateau measured at three different centre-of-mass energies in ATLAS is shown.

Finally, the leading charged-particle UE data from CDF at 300 GeV, 900 GeV, and 1.9 TeV can be combined with LHC data at 900 GeV, 7 TeV and 13 TeV allowing for a study of the energy dependence of the UE at five centre-of-mass

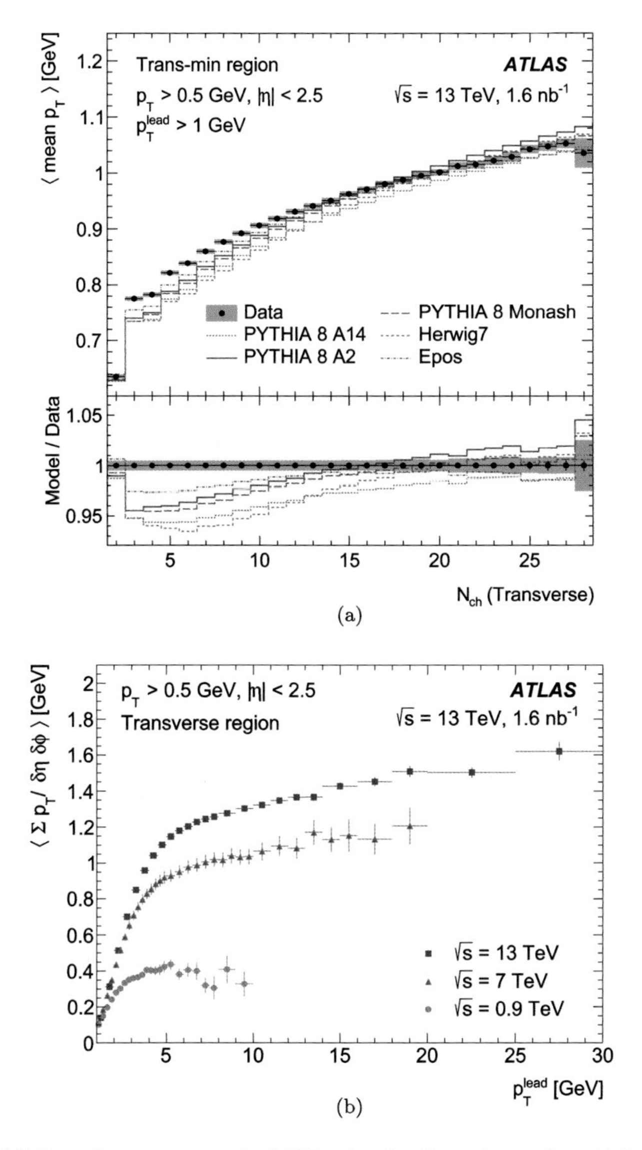

Fig. 12 ATLAS Run 2 measurement of UE using leading charged-particle. (a) Charged-particle mean p_T as a function of charged-particle multiplicity in trans-min region. (b) Charged-particle sum p_T density as a function of p_T of leading charged-particle for three different collision energies.

energies, as shown in the top row of Fig. 13. According to Rick Field, this *is the culmination of his 18 years of UE studies.* This information can be used to test and improve the MPI energy extrapolation of the MC tunes as in the bottom row of Fig. 13, where the right plot shows the evolution of the UE plateau height for charged-particle multiplicity as a function of centre-of-mass energy. Similar

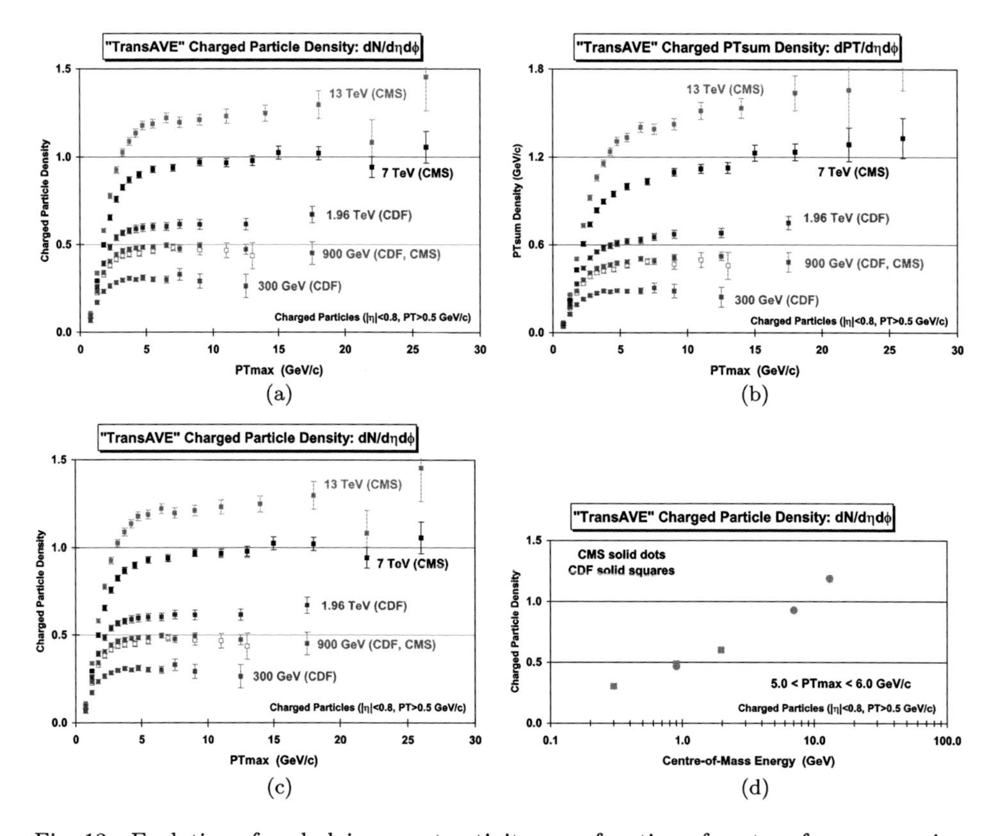

Fig. 13 Evolution of underlying event activity as a function of centre-of-mass energies. (a) Transverse region charged-particle density as a function of leading charged-particle. (b) Transverse region charged-particle sum p_T density as a function of leading charged-particle. (c) Transverse region charged-particle density at five centre-of-mass energies from CDF and CMS. (d) Evolution of the transverse region plateau height (taken at leading charged-particle $p_T = 5$ GeV).

distributions can be constructed for all UE observables in different angular regions, which provides a wealth of information for tuning.

4.3. *UE measurements using leading jet*

UE measurements were also performed using leading jet as a hard object to define the scale and direction. The CMS experiment performed UE measurement using leading charged-particle jet at $\sqrt{s} = 0.9$, 2.76, 7 and 13 TeV.[129,499,500] Experiments tend to perform first measurements with charged-particle jets as they require less detailed calibrations than a calorimeter jets.

Figure 14 shows the UE activity as a function of p_T of leading jet in the transverse region. Charged-particle multiplicity and sum p_T densities increase by more than 300% as centre-of-mass energy rises from 900 GeV to 13 TeV. The

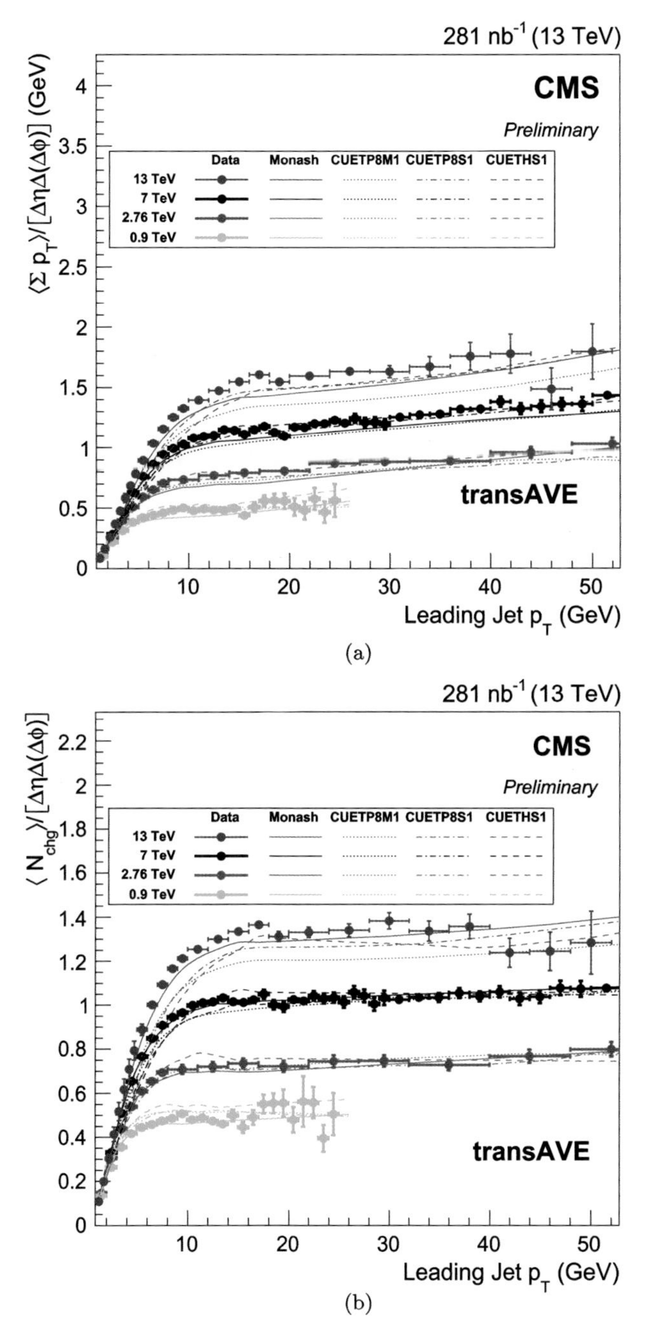

Fig. 14 CMS Run 1 and Run 2 measurement of UE using leading jet. (a) Charged-particle sum p_{T} density as a function of p_{T} of leading jet in trans-ave region. (b) Charged-particle multiplicity density as a function of p_{T} of leading jet in trans-ave region.

jet p_{T} corresponding to the plateau increases from 5–6 GeV to 10–12 GeV with the increase in centre-of-mass energy from 900 GeV to 13 TeV. The Pythia 8 Monash tune is observed to give the best description of the increase as well as plateau region for the CMS measurements whereas other predictions describe various distributions within 10–15%.

As can be seen in Fig. 15, the UE activity in the trans-min region reaches a plateau but the activity in the trans-diff region keeps on increasing slowly with increasing jet p_{T}. This behavior is expected as the trans-diff region has a radiative contribution that increases with jet p_{T}. The activity in the trans-min region increases faster with centre-of-mass energy than the activity in the trans-diff region. This observation leads to the conclusion that MPI has stronger dependence on the centre-of-mass energy than hard scattering process. The comparison with MC predictions show that the radiation contribution is equally well described whereas the particle production from MPI has different level of agreements with various predictions.

The ATLAS experiment also performed UE measurements with leading charged-particle jet.[501] The qualitative features of the measurements are the same as observed by the CMS and UE measurements with leading charged-particle. While the data-MC comparisons did not yield any surprises, an initial enhancement of UE activity with the increase of the radius of charged-particle jets was observed, as shown in Fig. 16(f). This can be understood as a selection bias — only an event with higher overall activity will contain a larger radius jet, so UE activity will be higher as well. However, this is more pronounced at lower jet p_{T} as at higher jet p_{T} events get more active regardless, and the same events will be considered for any radius jet selection.

In the jet events, apart from the inclusive topology where the activity is measured as a function of leading jet, an exclusive dijet topology was also considered by ATLAS.[502] The advantage of this is again to eliminate events with extra jets potentially contaminating the UE activity, and indeed in Fig. 16, the MC models describe the data better in exclusive case than in inclusive, and the activity is also much flatter. In fact the comparison of transverse regions conform that, as the trans-diff (which is sensitive to extra radiation) is rather flat in exclusive case.

4.4. *UE measurements with Z-boson events*

ATLAS and CMS experiments performed UE measurements in Z-boson events. The ATLAS results at 7 TeV[503] included a detailed comparison with UE measurements in dijet events.[502] In both the analyses, apart from the usual profiles, one-dimensional distributions for charged-particle multiplicity and sum p_{T} were looked at, in different Z p_{T} or leading jet p_{T} ranges. These give much more fine-grained information than the average values.

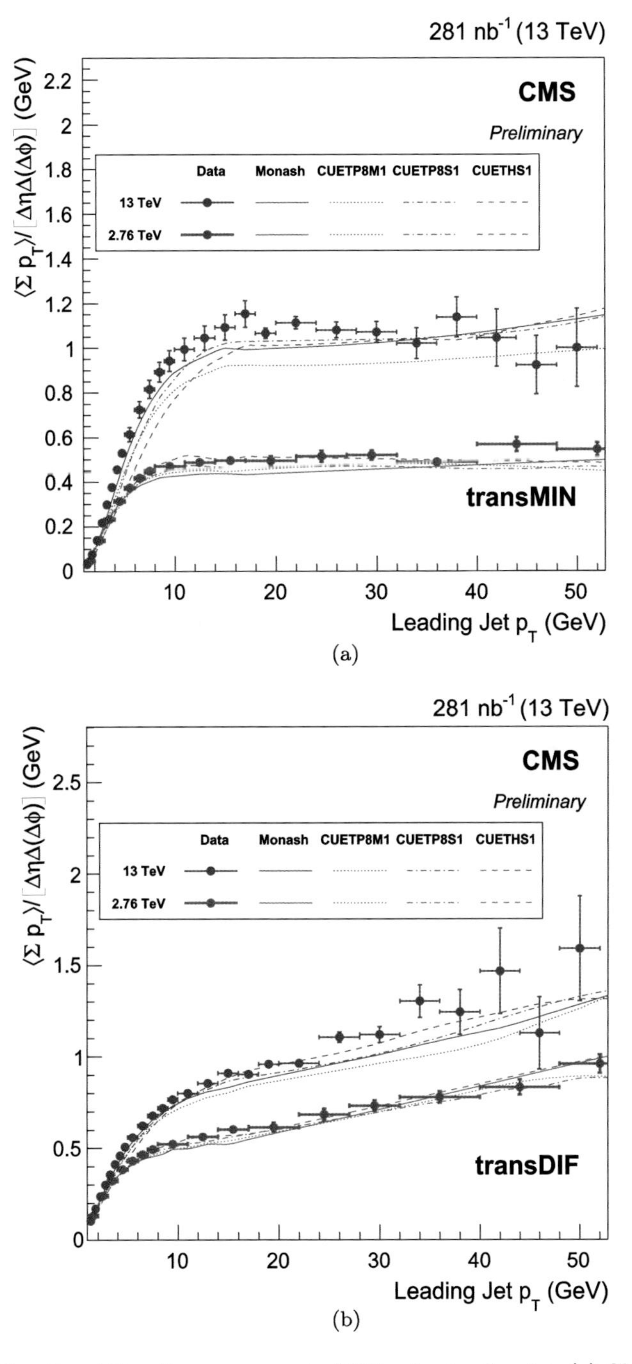

Fig. 15 CMS Run 1 and Run 2 measurement of UE using leading jet. (a) Charged-particle sum p_{T} density as a function of p_{T} of leading jet in trans-min region. (b) Charged-particle multiplicity density as a function of p_{T} of leading jet in trans-diff region.

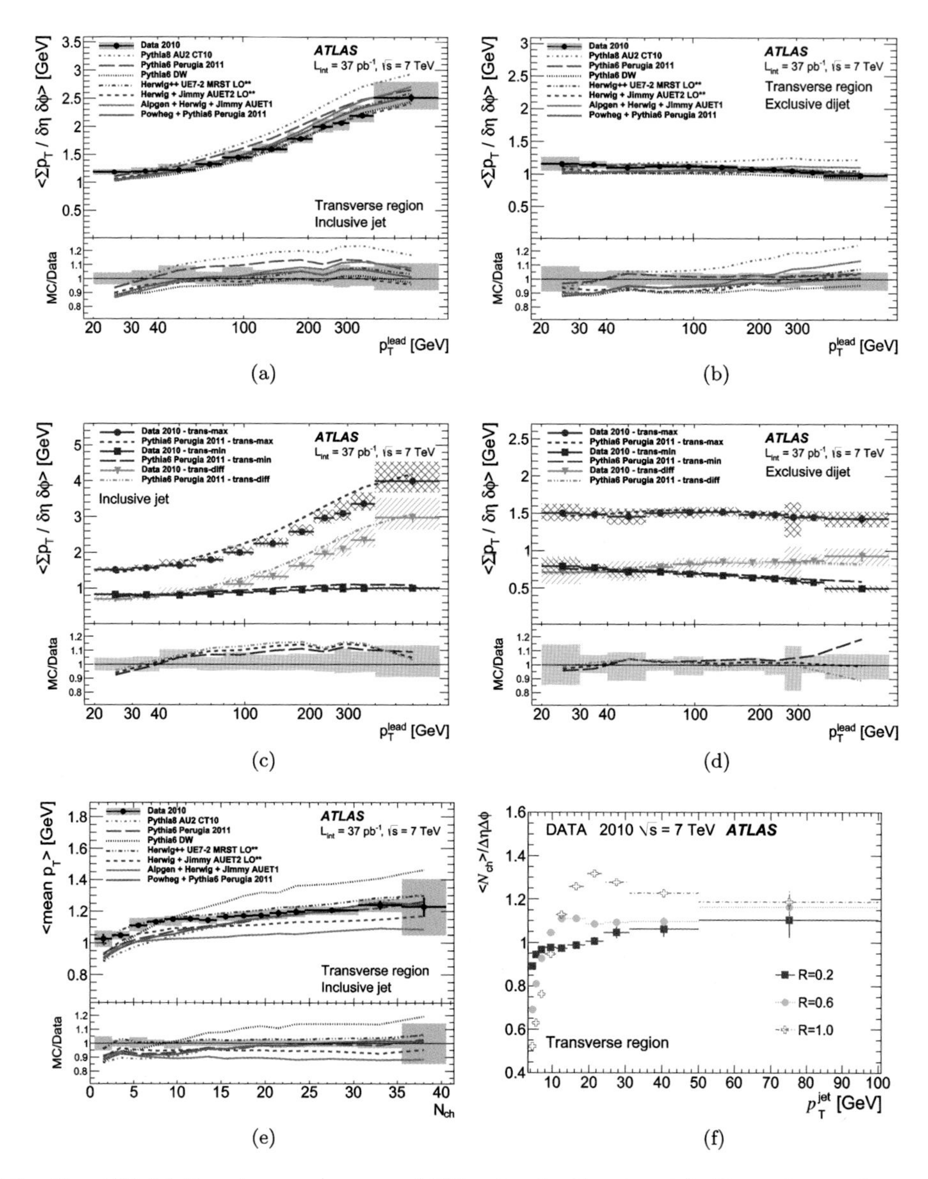

Fig. 16 ATLAS Run 1 measurement of UE using leading jet. (a) Charged-particle sum p_{T} density as a function of p_{T} of leading jet in transverse region for inclusive topology. (b) Charged-particle sum p_{T} density as a function of p_{T} of leading jet in transverse region for exclusive topology. (c) Charged-particle sum p_{T} density as a function of p_{T} of leading jet in trans-max, trans-min and trans-diff regions for inclusive topology. (d) Charged-particle sum p_{T} density as a function of p_{T} of leading jet in trans-max, trans-min and trans-diff regions for exclusive topology. (e) Charged-particle mean p_{T} as a function of charged-particle multiplicity in transverse region for inclusive topology. (f) Charged-particle multiplicity density as a function of p_{T} of leading charged-particle jet in transverse region, for different radius charged-particle jets.

For Z-boson events, apart from the transverse region, the toward region is also sensitive to UE with the leptons removed. In Fig. 17, the charged-multiplicity and sum p_T in toward and trans-min regions show the increase then the plateau-like feature, with none of the models describing the data well over the whole range. Furthermore, the transverse region is divided into trans-max and trans-min, as before. The trans-min region, by construction is expected to be most sensitive the UE activity, and the profile is described better by parton shower models. The toward region activity is slightly lower than transverse, indicating the presence of additional jets in transverse region. The comparison of trans-min, trans-max (and trans-diff) regions indeed show the trans-min activity is flattening out. The distribution of (normalized) sum p_T in three different Z p_T ranges elucidate the effect of additional jets. If the trans-min received no contamination from such jets and was only populated by UE, then sum p_T distribution would have been independent of Z-boson p_T. However, that is the case only till sum p_T of about 10 GeV (accounting for the normalization), then more jet activity with increasing Z p_T contaminates even the trans-min region. Finally, the mean p_T and multiplicity correlation shows the inadequacy of color reconnection models in the generators.

At CDF, the UE activity was seen to be very similar between jet and Z-boson events. In Fig. 18, while the multiplicity profiles seem rather similar (and also the smooth transition from leading charged-particle events to leading jet events is quite remarkable), the sum p_T do not. This can be investigated more by comparing trans-min and trans-max activities in both cases. While for multiplicity, there is no significant difference, for sum p_T, the trans-max in more active in Z-boson events compared to jet events, indicating the overall higher transverse activity in Z-boson events. This was traced back to the fact in certain Z-boson events, the leading jet was seen to have a higher p_T than the Z-boson, and measuring the activity as a function Z-boson p_T and direction introduces a reorientation bias. When such events were removed, the sum p_T was seen to be similar between jet and Z-boson events as well.

The CMS results at $\sqrt{s} = 7$ and 13 TeV,[504,505] included comparison with CDF Run II results at $\sqrt{s} = 1.96$ TeV.[492] Figure 22(a) shows the charged-particle density as a function of p_T of the Z-boson in three usual regions. Figure 22(b) shows the comparison of charged-particle density (in toward region) measured, as a function of $p_T^{\mu\mu}$, at three centre-of-mass energies. It is observed that UE contribution increases by 25–30% as centre-of-mass energy increases from 7 to 13 TeV whereas increase is 60–80% for centre-of-mass energy increase from 1.96 TeV to 7 TeV. Different MC models predict slower rise in UE contribution but agreements improves at higher values of the Z-boson p_T.

Another important observation of UE measurement using Z-boson events is that it helps in partially separating the MPI and additional radiation contributions. Figure 22(a) shows that activities in toward, transverse and away region are similar as p_T of the Z-boson approaches zero. Hence, increase in activities with p_T of

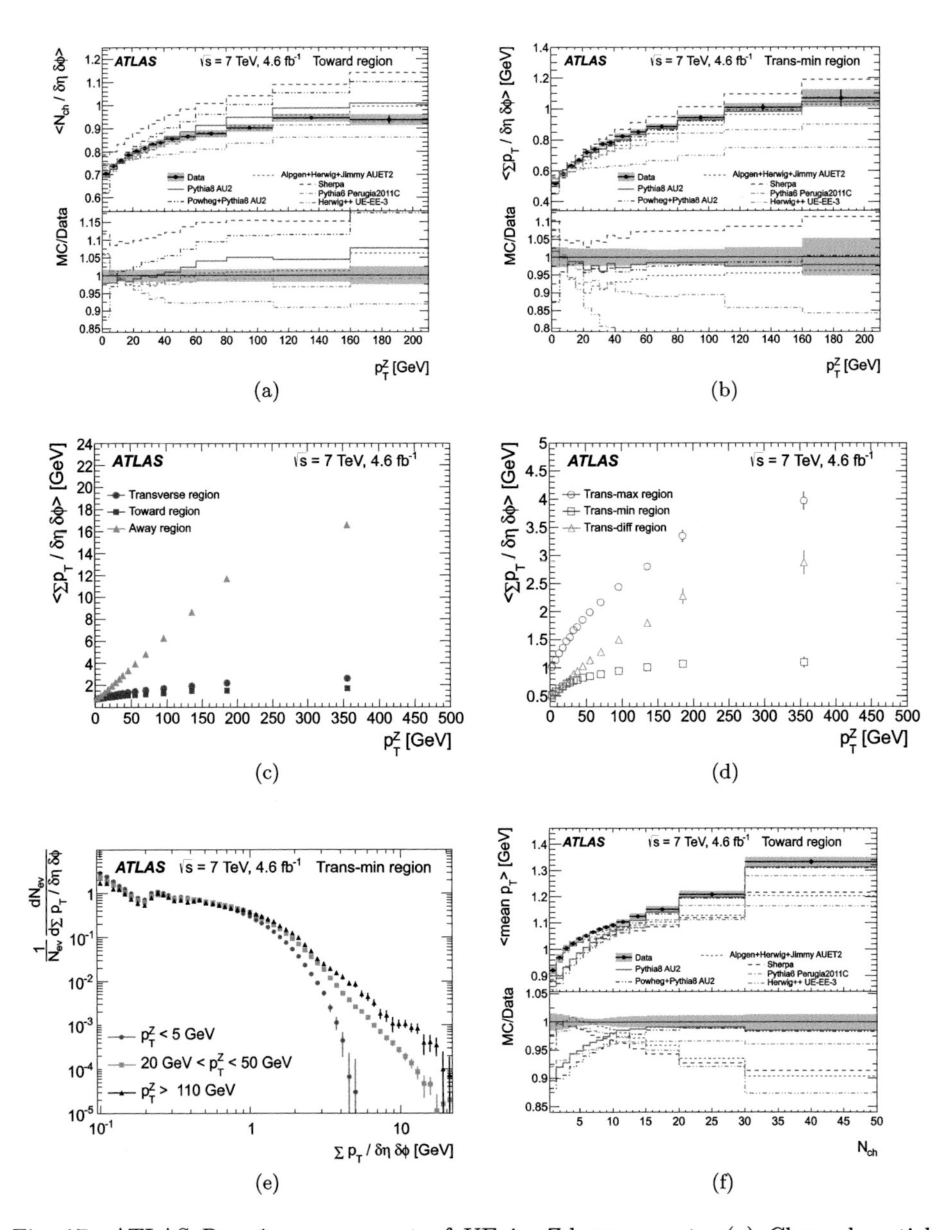

Fig. 17 ATLAS Run 1 measurement of UE in Z-boson events. (a) Charged-particle multiplicity density as a function of p_{T} of Z-boson in toward region. (b) Charged-particle sum p_{T} density as a function of p_{T} of Z-boson in trans-min region. (c) Charged-particle sum p_{T} density as a function of p_{T} of Z-boson in toward, transverse and away regions. (d) Charged-particle sum p_{T} density as a function of p_{T} of Z-boson in trans-max, trans-min and trans-diff regions. (e) Distributions of charged-particle sum p_{T} density in different of Z-boson p_{T} ranges in trans-min region. (f) Charged-particle mean p_{T} as a function of charged-particle multiplicity in toward region.

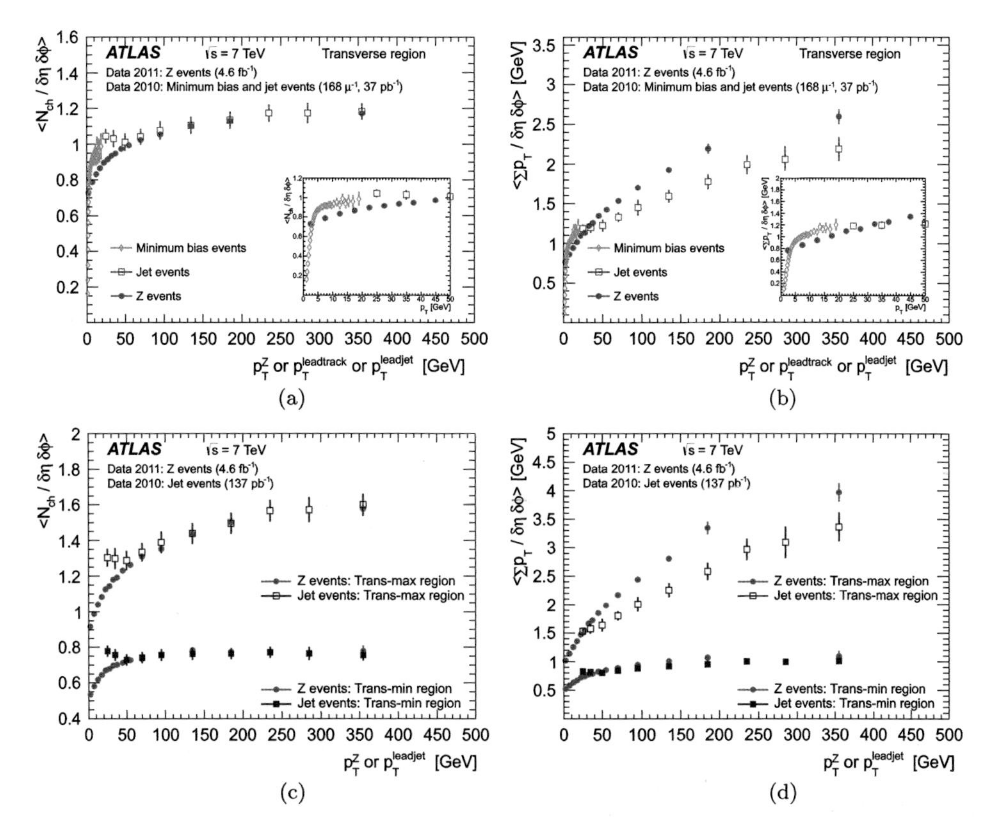

Fig. 18 ATLAS Run 1 comparison in UE activity in Z-boson and leading jet events. (a) Charged-particle multiplicity density as a function of p_{T} of leading jet or Z-boson in transverse region. (b) Charged-particle sum p_{T} density as a function of p_{T} of leading jet or Z-boson in transverse region. (c) Charged-particle multiplicity density as a function of p_{T} of leading jet or Z-boson in trans-max and trans-min regions. (d) Charged-particle sum p_{T} density as a function of p_{T} of leading jet or Z-boson in trans-max and trans-min regions.

the Z-boson is due to increasing radiative contribution. True MPI contribution can be estimated by restricting the boost of such events. CMS observed that the activity is almost flat, after requiring the Z-boson $p_{\mathrm{T}} < 5$ GeV, as a function of the invariant mass of two muons, as shown in Fig. 22(c). By comparing MC predictions with and without MPI, it is clear that the activity is largely (about 80%) due to MPI. The MPI evolution with centre-of-mass energy is further investigated by comparing the activity before and after requiring Z-boson $p_{\mathrm{T}} < 5$ GeV as shown in Fig. 22(d). There is a logarithmic increase in MPI contributions with the centre-of-energy which is qualitatively described by the Powheg events showered by PS generators. Powheg events showered by Pythia 8 describe the energy evolution better than Herwig++.

Most of the UE tunes use MB and UE measurements with jet events, but it is observed that these tunes also nicely describe the UE measurements in Z-boson events as well. This observation corroborates the universality of MPI.

All of these results led to model developments and retuning. One of the main takeaways was that the measurement of pure UE activity in the busy LHC environment is nearly impossible. In Tevatron, the trans-min (and toward for Z-boson events) was not contaminated that much by extra radiation, however at the LHC, even they receive non-negligible contributions. Therefore, it is unrealistic to expect leading order parton shower generators to describe the UE sensitive distributions completely, and it was seen that multiloop or multileg generators (sometimes interfaced with the parton shower ones) described the data better in many cases. This also led to the situation where just tuning the MPI and BBR part is not enough to describe these distributions, rather a full tuning of the shower is necessary, as was done in ATLAS A14 tune of Pythia 8.

4.5. *Other UE sensitive measurements*

Apart from the direct UE measurements, a few other results probed the effect of UE. The ATLAS measurement of transverse energy as a function of pseudorapidity in dijet events, as shown in Fig. 19(a), which is very sensitive to the effect to UE. The profile of transverse thrust is shown in Fig. 19(b) as a function of charged-particle p_T. It shows a similar saturation like behavior as UE activity. Event shape variables have been also measured in Z-boson events, as shown in Fig. 19(c), and for low Z-boson p_T, they are very sensitive to MPI modeling.

To understand the effect of kinematical phase-space, UE measurements were also performed in forward pseudorapidity ($-6.6 < \eta < -5.2$) region by CMS.[506] Ratios of forward energy density ($dE/d\eta$) for the events having a central ($|\eta| < 2$) charged-particle jets and inclusive events were measured. The measurements were performed at $\sqrt{s} = 0.9$, 2.76 and 7 TeV. Figure 20 shows this ratio as a function of the p_T of leading charged-particle jet for different centre-of-mass energies.

At $\sqrt{s} = 7$ TeV, this ratio increases with the p_T of leading jet which is consistent with increase in UE activity in conventional UE measurements. The pre-LHC Pythia 6 tunes fail to describe this measurement whereas the tunes derived with LHC data give better agreement. At $\sqrt{s} = 2.76$ TeV, the increase in ratio is comparatively small which is consistent with the 900 GeV results where ratio decreases below unity. For 900 GeV measurements, the energy density is lower for events with leading jets as compared to inclusive events. This can be understood as a kinematic effect, where higher UE activity accompanying high p_T jet depletes the energy of proton remnants fragmenting within this forward pseudo-rapidity range. These features are reasonably described by different MC predictions.

The measurement of UE activity was also performed using p_T density per unit jet area[8] at detector level[507] at $\sqrt{s} = 0.9$ and 7 TeV. The qualitative features of

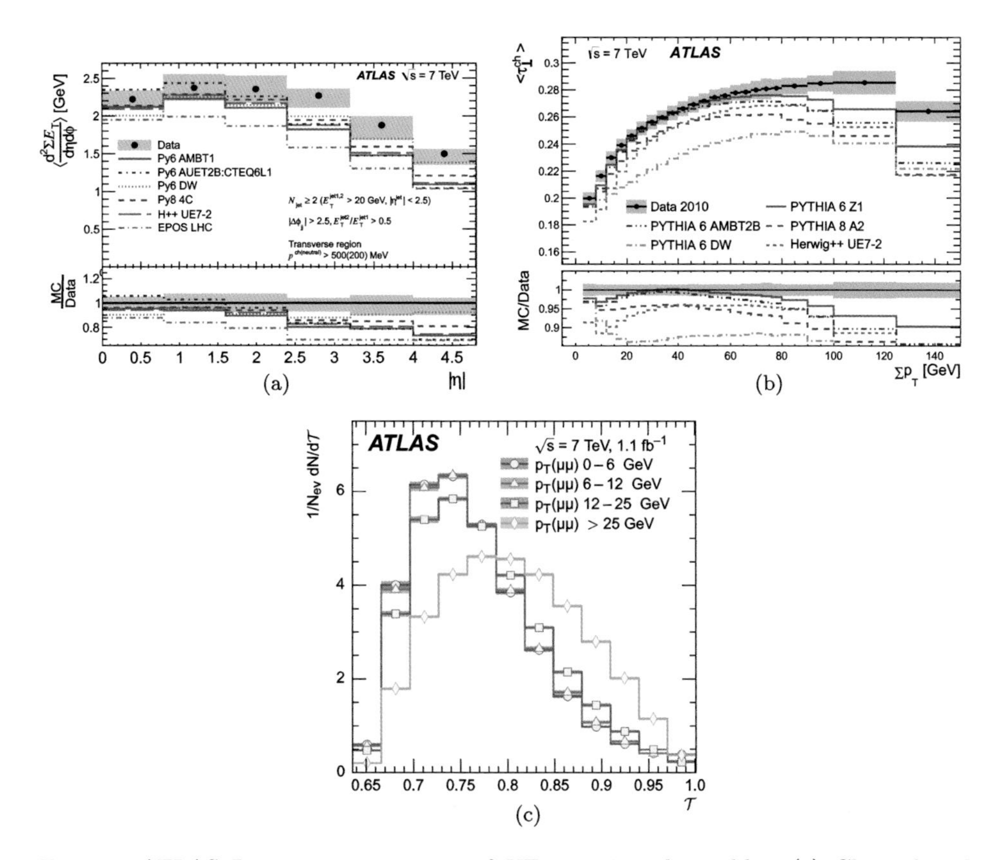

Fig. 19 ATLAS Run 1 measurements of UE sensitive observables. (a) Charged and neutral-particle sum E_{T} density as a function of pseudorapidity. (b) Mean transverse thrust calculated from charged-particles as a function of sum p_{T} of charged-particles. (c) Distributions of transverse thrust calculated from charged-particles in different of Z-boson p_{T} ranges.

the UE measurements were found to be consistent with the conventional methods as shown in Figs. 21(a) and 21(b).

The CMS experiment investigated the UE activity as a function of charged-particle multiplicity at $\sqrt{s} = 7$ TeV.[154] Charged-particles were separated into two classes; those belonging to jets and those belonging to UE. Charged-particles falling within a jet cone of radius 0.5 were considered as *intra-jet particles*, whereas all the remaining particles were considered to be a part of UE. Figure 21(c) shows the mean p_{T} of charged-particles belonging to UE as a function of charged-particle multiplicity for data and different MC predictions. The behavior of UE charged-particles with multiplicity is well described by predictions of Pythia 6 tune Z2* and Pythia 8 tune 4C, whereas Herwig++ predictions fails to describe the measurements. Figure 21(d) shows the average p_{T} of intra-jet charged-particles as a function of charged-particles multiplicity. The average p_{T} of intra-jet charged-particles shows

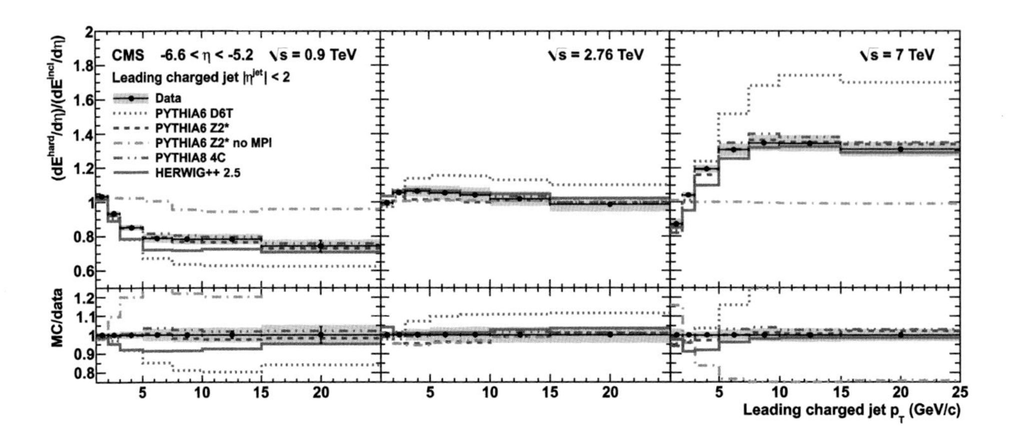

Fig. 20 Ratio of the energy deposited in the pseudorapidity range $-6.6 < \eta < -5.2$ for events with a charged-particle jet with $|\eta^{\rm jet}| < 2$ with respect to the energy in inclusive events, as a function of the jet transverse momentum $p_{\rm T}$ for $\sqrt{s} = 0.9$ (left), 2.76 (middle), and 7 TeV (right). Results are compared to the Pythia and Herwig++ models.

opposite trend as compared to average $p_{\rm T}$ of UE charged-particles. The events with increasing multiplicities are biased towards final-states resulting mostly from jets which fragment into a (increasingly) large number of hadrons. Since the produced hadrons share the energy of the parent parton, a larger amount of them results in overall softer intra-jet and leading-hadron $p_{\rm T}$ spectra. Pythia 6, Pythia 8 and Herwig++ predictions nicely describe the measurements but fail if MPI effects are not included in the simulation.

CMS also performed UE measurements using top–antitop ($t\bar{t}$) quark events at $\sqrt{s} = 7$ and 13 TeV[508,509] at detector level. Here, the $t\bar{t}$ system is used to define the event scale and the reference direction of the hard collision. It is observed that towards and transverse regions are most sensitive to the UE contribution as depicted in Fig. 21(e). Figure 21(f) shows that the average number of charged-particle increases slowly with resultant $p_{\rm T}$ of $t\bar{t}$ pair as expected increase in radiation contributions. MC events generated with Powheg and showered with Pythia 8 (CUETP8M1 tune) and Herwig++ (UE-EE-5C tune) describe measurements well. These UE tunes are derived using inclusive events (dominated by light flavor quarks) and describe UE in $t\bar{t}$, therefore there it can concluded that there is no need to have separate UE tune for the process involving heavy-quarks. This excellent agreement between data and MC predictions further corroborate the universality of UE.

5. UE Measurements at RHIC

The measurement of UE activity also has been performed at the STAR experiment in Relativistic Heavy Ion Collider (RHIC) at proton-proton collisions at $\sqrt{s} = 200$ GeV.[510] Although the measurement is at a low centre-of-mass energy compared

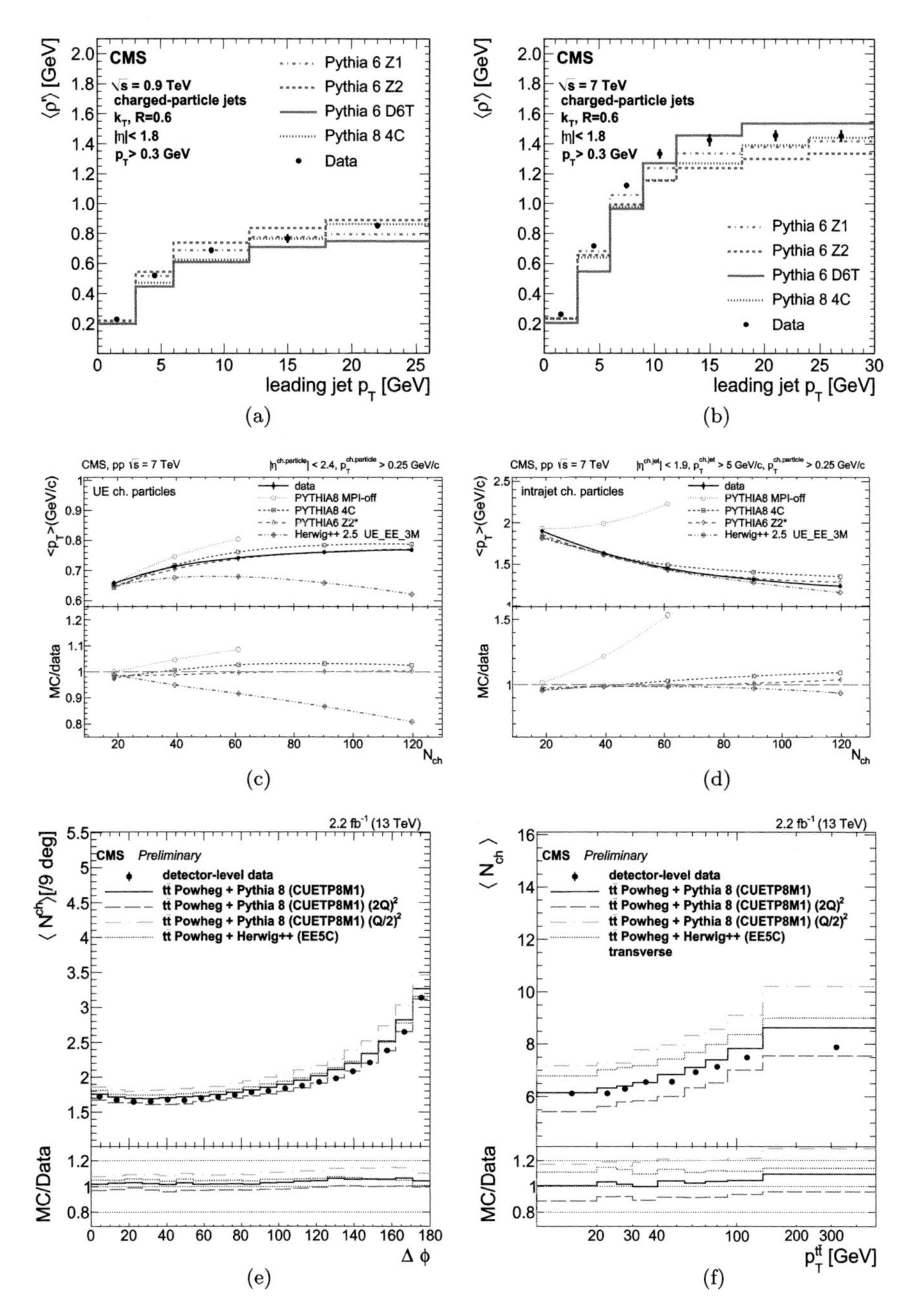

Fig. 21 CMS Run 1 measurements of UE sensitive distribution. (a) Detector level median p_T density per unit jet area as a function of leading jet p_T. (b) Detector level median p_T density per unit jet area as a function of leading jet p_T. (c) Mean p_T of *UE* charged-particles as a function of charged-particle multiplicity. (d) Mean p_T of *intra-jet* charged-particles as a function of charged-particle multiplicity. (e) Average number of charged-particles as a function $\Delta\Phi$ between charged-particles and the resultant azimuthal direction of $t\bar{t}$ system. (f) Average number of charged-particles as a function of the resultant p_T of $t\bar{t}$ system.

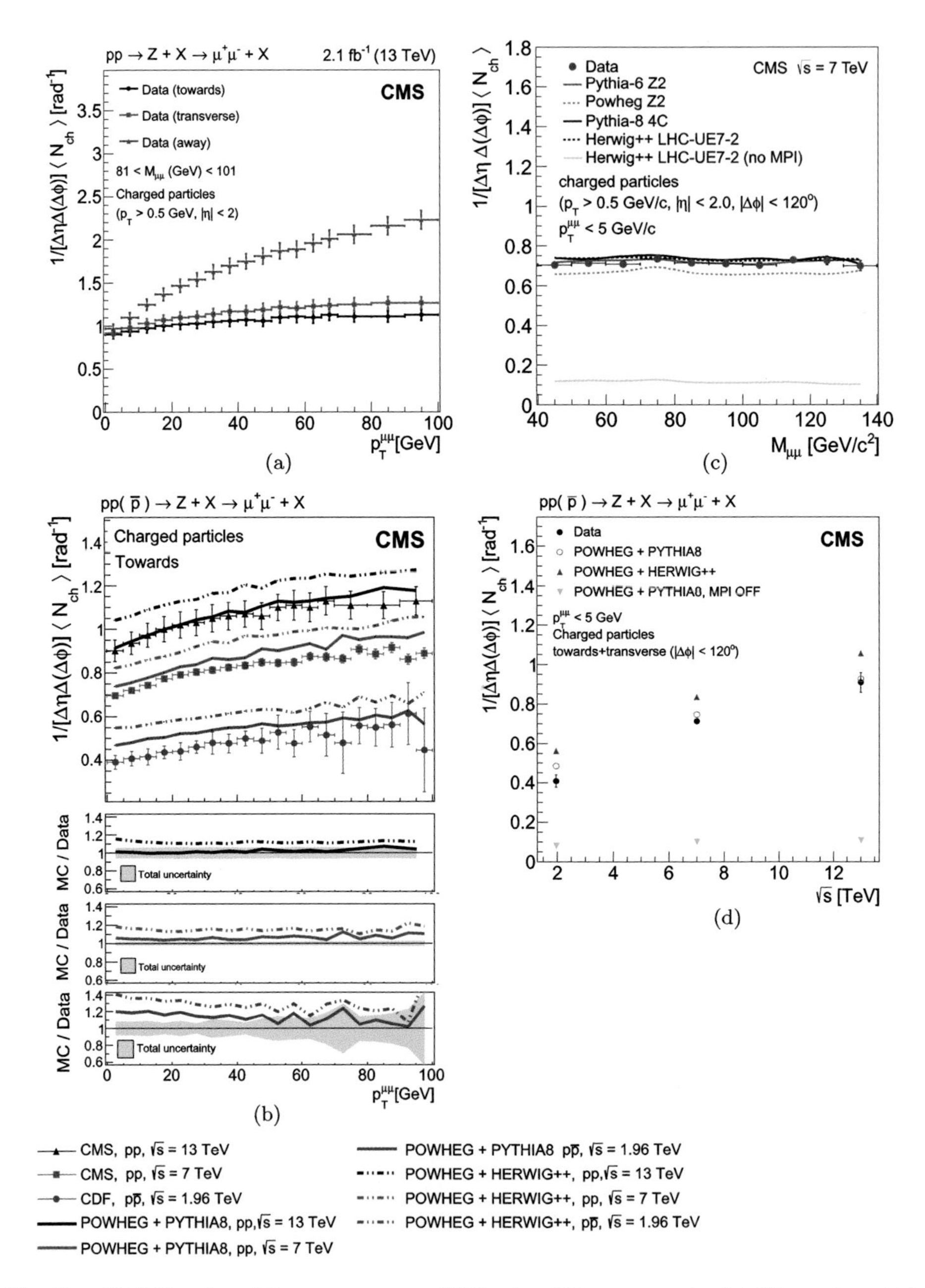

Fig. 22 CMS Run 1 and 2 measurement of UE using Z-boson events. (a) Charged-particle multiplicity density as a function of p_{T} of Z-boson in toward, transverse and away regions. (b) Charged-particle multiplicity density as a function of p_{T} of Z-boson in toward region at different collision energies. (c) Charged-particle multiplicity density as a function of dimuon invariant mass, combined in toward and transverse regions. (d) Charged-particle multiplicity density as a function collision energy, combined in toward and transverse regions, for Z-boson $p_{\mathrm{T}} < 5$ GeV.

to the LHC experiments, it provides a valuable piece of information in the energy evolution of different components of UE, specifically MPI. So far the data has been presented at detector level.

Figure 23 shows the mean p_T of tracks inside the leading jet and in trans-max and trans-min regions constructed with respect to the leading jet, as a function of the leading jet p_T. The activity inside the jets is much higher, and it is fairly flat in both the transverse regions, and MC predictions show reasonable agreement with the data. When exclusive dijet events are selected, no significant difference between trans-max and trans-min activity was observed, which is in stark contrast to the results from CDF and ATLAS. This indicates that only very small amounts of initial/final state radiation are emitted at large angle at RHIC energies.

6. Summary and Future Ideas

The host of measurements from different experiments, spanning centre-of-mass energies from 200 GeV to 13 TeV, using many different observables and topologies, has provided us with an enormous amount of data to understand the behavior of UE, and improve its modeling in the MC generators. In fact to better synchronize the effort between different LHC experiments, to define common observables, common phase space, and common generator model and tunes to compare to, the LHC MB and UE Working Group[511] was formed just before LHC Run 1. This resulted in comparisons of early results from different experiments, which was critical at that point.

The results also led to a massive effort in improving and tuning the MC generators. Table 3 lists the tunes listed in Table 2, but now depicts their status with respect to the most up-to-date results.

While LHC measurements clearly helped in tuning aspects of UE in MC models, newer ways of looking at UE are also being pursued. This is more important than ever, because at LHC energies, UE sensitive observables receives significant contamination from addition jets created in the event. The trans-min region (or the toward in case of Z-boson events), which were considered most sensitive to UE, are also affected substantially by it. This makes disentangling the MPI and shower effects very difficult. Measuring ratios of identified charged-particles has been proposed[512] as a deeper probe of UE mechanism.

In parallel, the focus is shifting to using multi-leg or multi-loop generators in conjunction with shower generators. This mean rather than the large emphasis on tuning the UE modeling in shower generators, the optimization of matched set-ups are more relevant. These set-ups are expected to model the extra jets from the hard matrix element more precisely, and letting shower model the softer activity only. So rather than just tuning the MPI, BBR and CR parameters, shower parameters and matching/merging definition of ME and PS generators are being looked at, and need to be tuned simultaneously as well.

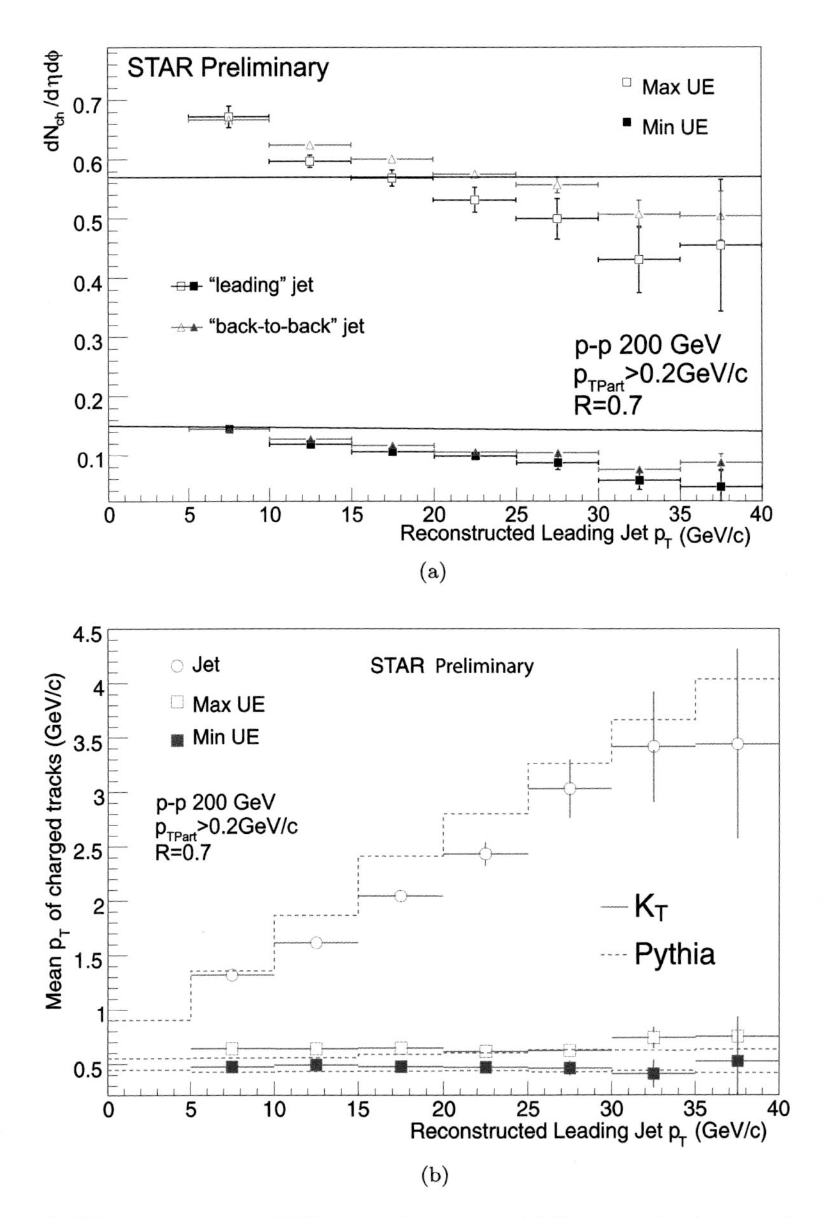

Fig. 23 STAR measurement of UE using jet events. (a) Detector level charged-particle density as a function of p_T of leading jet in inclusive and back-to-back topology. (b) Detector level charged-particle mean p_T as a function of p_T of leading jet in inclusive and back-to-back topology. Reprinted from Ref. 510, Copyright 2011, with permission from Elsevier.

Table 3. Status of MC tunes after comparing with LHC data.

Generator	Tune or tune group	Status
(a) Pre-LHC		
Pythia 6	CDF A, AW, DW, DWT, D6, D6T	Disfavored by LHC data, occasionally DW used as a historical CDF tune
Pythia 6	ATLAS DC2, MC08, MC09/MC09c	Disfavored by LHC data, disagreement with CDF data as well
Pythia 6	Pergia0/10	Superseded by Perugia2010 to update shower, CR model
Herwig + Jimmy	CDF, ATLAS MC08/MC07	Disfavored by LHC data
(b) LHC-era		
Pythia 6	AMBT1, AMBT2, AMBT2B	Disfavored against Pythia 8 MB tunes in ATLAS
Pythia 6	AUET2, AUET2B	Disfavored against Pythia 8 UE tunes in ATLAS
Pythia 6	Perugia2012	Stable LHC Perguia tune
Pythia 6	Z1, Z2, Z2*, CUETP6S1	Z2* and CUETP6S1 are the Run 1 stable CMS tunes, others superseded by it.
Pythia 8	A2, A3	A3 current MB tune in ATLAS, improvement over A2 for inelastic cross section
Pythia 8	A14	Stable UE+Shower tune in ATLAS
Pythia 8	4C, 4Cx, Monash	Monash is the stable author tune, rest deprecated
Pythia 8	CUETP8S1, CUETP8M1	CMS preferred tunes.
Pythia 8	CDPSTP8S*	CMS preferred DPI focused tunes
Herwig + Jimmy	AUET1, AUET2	Deprecated model.
Herwig++	LHC*, UE*	UE-EE-5 is the final tune in the series, recommended by authors
Herwig 7	UE-MMHT	Recommended by authors

Acknowledgments

The authors would like to thank the following people, who have contributed to our understanding of the UE in hadronic collisions and to the improvement of the QCD Monte Carlo models: Darin Acosta, Filippo Ambroglini, Maxim Azarkin, Monika Bansal, Paolo Bartalini, Florian Bechtel, Andy Buckley, You-Hao Chang, Yuan Chao, Sergei Chekanov, Diego Ciangottini, Alberto Cruz, Livio Fano, Craig Group,

Paolo Gunnellini, Richard Haas, Hans Van Haevermaet, Gabriel Hare, Hendrik Hoeth, Xavier Janssen, Hannes Jung, Oldrich Kepka, Kristian Kotov, Frank Krauss, Andrea Lucaroni, Michelangelo Mangano, Tim Martin, Pierre Van Mechelen, Arthur Moraes, Steve Mrenna, Luca Mucibello, Emily Nurse, Doug Rank, Albert De Roeck, Gavin Salam, Torbjörn Sjöstrand, Peter Skands, Benoit Roland, Holger Schulz, Andrzej Sjodmok, Joe Virzi, Matous Vozak, Sebastian Wahrmund, Ben Wynne, Mohammad Zakaria, Oleg Zenin. SB is supported by Department of Science & Technology (DST) and University Grant Commission (UGC), New Delhi. Currently DK is supported by multiple grants from National Research Foundation (NRF), South Africa. RDF was supported in part by the U.S. Department of Energy (DoE).

Chapter 12

Phenomenology of Soft QCD: The Role of Minimum-Bias Measurements

Jan Fiete Grosse-Oetringhaus

CERN, 1211 Geneva 23, Switzerland

This chapter summarizes minimum-bias measurements at the Large Hadron Collider. In particular, the pseudorapidity density, the transverse-momentum spectra, the multiplicity distribution, the correlation of average transverse momentum and the multiplicity, and a measurement of minijets are presented. In addition to an overview of the results obtained to date at the LHC, the experimental challenges of defining particle and event sample and correcting to this sample are discussed.

1. Introduction

The study of minimum-bias (MB) physics comprises in its most general terms all signals which can be well extracted experimentally without the use of triggers which enhance certain, more rare, events in the data stream over the typical *average* event. The experimental apparatus should bias the least (hence minimum-bias or zero-bias) the ensemble of events. Contrary to this conceptually simple event selection, the phenomenology of minimum-bias events is very rich: the underlying processes are dominated by QCD in the non-perturbative regime (at small Q^2). Fragmentation and hadronization of the partons produced in the collision as well as multiple-parton interactions play an important role. Theoretically it is difficult to describe these mechanisms from first principles (see also Chapter 10) which makes the experimental study of these collisions crucial.

In order to study this regime, ideally the full phase-space distribution of produced particles and their correlations should be measured, which is given by the probability to find a number of particles N of type i and momentum vector p:

$$P_N(p_1^i, p_2^i, \ldots). \tag{1}$$

It is difficult to measure such a complete observable experimentally. Therefore, one starts more modest, by measuring the event-averaged number of produced particles neglecting their type as a function of a single kinematic property (e.g., as a function of the pseudorapidity $dN/d\eta$ or the transverse momentum dN/dp_{T}). These measures neglect correlations between the produced particles and characterize the average collision. Measuring the probability distribution of the number of produced particles $P(N)$ contains some degree of correlation between the particles. Similarly, the event dynamics is addressed by measuring the mean transverse momentum as a function of multiplicity. Both these observables show a particular sensitivity to MPIs, discussed below, and are of interest to be studied also at higher multiplicity (see Chapter 15). A direct access to the number of parton interactions can be obtained by studying so-called minijets defined as the particles originating from the same $2 \to 2$ scattering at low Q^2 of a few GeV/c. At these scales, the number of particles per parton is of the order of 1, and hence traditional jet finding methods are not applicable, instead statistical approaches can be used.

This chapter will first discuss the experimental challenges of minimum-bias measurements. Subsequently, results from the Large Hadron Collider will be presented. Finally, an outlook for the future of these types of measurements is given.

2. Experimental Challenges

The LHC detectors are complex multi-million channel devices designed to record traces of particles from a momentum of 100 MeV/c to several TeV/c at MHz collision rates. It is not surprising that this complexity actually results in a larger number of experimental challenges than, e.g., in historic bubble chamber experiments which had 4π coverage and very little material.

The detector effects that need to be corrected for stem from the fact that due to the detector material, dead areas in the detector and the efficiency of electronics and algorithms not every particle is reconstructed. Furthermore, interactions with material and decays of instable particles can create additional particles. Due to these inefficiencies, it may also happen that a certain event is not seen by the detector at all, resulting in a miscount of the total number of occurred collisions which also has to be corrected for.

The first non-trivial step is to define for which particles and for which collisions an observable is to be measured. The choices adopted by the LHC experiments are introduced in the following.

2.1. *Primary-particle definition*

The particles which shall be part of the result are called the *primary particles.* Due to the design of the LHC detectors, the measurements discussed in this chapter have been performed for charged particles. Most of the produced particles are instable.

Particles like the ρ which decay strongly, decay almost instantaneously and the decay products are therefore included among the primary particles. Weakly decaying particles need a special treatment, e.g., the K^0 and Λ. Those are neutral, thus not part of the charged primary particles, but due to their decay into charged particles they become part of the measured sample. These are referred to as *secondary particles* which need to be corrected for. Further sources of secondary particles are particles produced by interaction of primary particles with the detector material (e.g., $\gamma \to e^+e^-$).

The experiments at the LHC have adopted similar conventions for primary particles with minor differences. ATLAS[446] and LHCb[513] count particles produced in the collision or produced by decays of particles with a proper lifetime $\tau < 30$ ps and 10 ps, respectively. As there are no known particles with lifetimes in the range 10–30 ps,[514] these definitions are identical. ALICE[515] and CMS[516] use a definition which includes products of strong and electromagnetic decays, but excludes products of weak decays and hadrons originating from secondary interactions. Later CMS clarified the definition to include decay products of particles with proper lifetimes less than 1 cm.[447] In practice, these are identical to the definition using the proper lifetime given before. However, CMS measures only hadrons, therefore excluding leptons from their measurement. Lately, ATLAS revisited their definition [517] and excludes particles with a proper lifetime between 30 ps and 300 ps from their primary-particle definition. This change removes strange baryons from the sample for which the reconstruction efficiency was found to be very low. These differences in the definitions result in only few percent effects on the integrated yields but can be relevant in certain momentum regions. In direct comparisons, these have to be carefully considered.

2.2. *Event-sample definition*

Inelastic hadronic collisions are divided based on the occurring processes into diffractive and non-diffractive collisions. In diffractive collisions one (or both) of the incoming particles retain their quantum numbers except possibly the spin which are then called single-diffractive (and double diffractive), respectively. Historically, measurements were presented for non-single-diffractive (NSD) collisions excluding the single-diffractive (SD) contribution which was difficult to trigger on. Furthermore, measurements were performed for all inelastic collisions (diffractive + non-diffractive) requiring in particular for the SD component significant model-dependent corrections.

Results at the LHC were published for both, NSD and inelastic event classes. In addition, it was realized that the efficiency to measure events with low multiplicity, a region particularly dominated by SD collisions, is rather low, while almost all events with higher multiplicity are measured. Figure 1 shows exemplarily the selection efficiency of the ALICE experiment as a function of the number

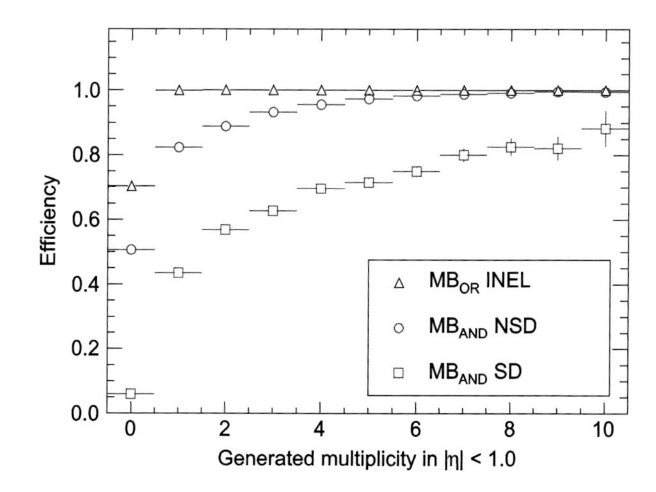

Fig. 1 Event selection efficiency of the ALICE experiment at $\sqrt{s} = 0.9\,\mathrm{TeV}$ for a single-arm trigger and inelastic collisions (triangles) and a double-arm trigger for non-single diffractive (circles) and single-diffractive collisions (squares). Figure from Ref. 515.

of charged particles N_{ch}. It can be seen that as soon as one particle enters the acceptance of the detector, the efficiency to record the event is 100% when a single-arm trigger is used. Such a trigger is typically used for the measurement of the inelastic event class. For the NSD event class, double-arm coincidence triggers (either during data-taking or in the analysis itself) are employed. These suppress a significant part of the SD contribution as illustrated in Fig. 1.

As the corrections for events which are not triggered upon are model-dependent, so-called *particle level* event classes have been added. These refer to events which have at least a certain number of particles within the detector acceptance (e.g., in $|\eta| < 2.5$ and $p_{\mathrm{T}} > 0.5\,\mathrm{GeV}/c$). These classes reduce the model dependence of the measurement significantly and thus the related uncertainties. However, they can only be compared to models which have a Monte Carlo implementation because the event selection has to be reproduced in the model.

2.3. *Corrections*

The measured distributions are corrected for tracking efficiencies and acceptance as well as contamination by secondary particles. In addition, the finite trigger and selection efficiencies are corrected for. These effects lead to a smearing of the measured quantities. Therefore, for certain observables like the $P(N_{\mathrm{ch}})$ where the spectrum is steeply falling, the measured distribution needs to be unfolded. While mathematically unfolding is an ill-posed problems, regularized unfolding which makes assumptions on the unfolded spectrum allows to recover the underlying distribution. A discussion of these methods can be found in Refs. 538–540.

3. Results

Despite the fact that the LHC is built as a discovery machine reaching very large integrated luminosities, it is also optimally suited for minimum-bias physics. The high-precision detectors as well as the versatility of the LHC where special runs with low instantaneous luminosity and special beam configurations are possible gave rise to an unprecedented data sample. While the first measurement was done with just 284 events,[518] later measurements used hundreds of millions of events.[521] Table 1 lists all relevant papers of the LHC experiment addressing the minimum-bias distributions discussed in this chapter. The precision and the reach in p_T and

Table 1. List of LHC minimum-bias results. In the results column, the letters denote $dN_{ch}/d\eta$ (A), dN_{ch}/dp_T (B), $P(N_{ch})$ (C), and $\langle p_T \rangle$ vs. N_{ch} (D). Particle-level event classes are denoted L^i where i is an index; their definitions are given below the table.

Experiment	$\sqrt{s}$ (TeV)	Events ($\times 10^6$)	Event classes	Results
ALICE[518]	0.9	0.0003	INEL, NSD	A
ALICE[515]	0.9, 2.36	0.15, 0.04	INEL, NSD	AC
ALICE[519]	0.9, 2.36, 7	0.05, 0.04, 0.24	L^a	AC
ALICE[520]	0.9	0.34	INEL, NSD	B
ALICE[521]	0.9, 2.76, 7	6.8, 65, 150	INEL	B
ALICE[522]	0.9, 2.76, 7	6.8, 65, 150	INEL	D
ALICE[523]	0.9, 2.76, 7	7, 27, 204	–	Minijets
ALICE[524]	0.9, 7	2.9, 2.7	L^b, L^c, L^d	AC
ALICE[525]	0.9, 2.76, 7, 8	7.4, 34, 404, 31	INEL, NSD, L^a	AB
ALICE[448]	13	1.5M	INEL, L^a	AB
ATLAS[526]	0.9	0.46	L^e	ABCD
ATLAS[446]	0.9, 2.36, 7	0.36, 0.006, 10	L^e, L^f, L^g, L^h, L^i	ABCD
ATLAS[527]	0.9, 7	0.36, 10	L^c, L^d	ABCD
ATLAS[528]	2.76	87†	INEL	B
ATLAS[517]	13	9	L^c, L^e	ABCD‡
ATLAS[529]	8	9	L^e, L^f, L^g, L^j, L^k	ABCD‡
ATLAS[530]	13	9	L^f	ABCD‡
CMS[516]	0.9, 2.36	0.07, 0.02	NSD	AB^{II}
CMS[531]	7	0.07	NSD	AB^{II}
CMS[532]	0.9, 2.36, 7	0.25, 0.02, 0.6	NSD	C^{II}
CMS[533]	0.9, 7	6.1, 0.8	L^c, L^d, L^l, L^m	A
CMS[447]	0.9, 7	6.8, 25†	NSD	B^{II}
CMS/TOTEM[534]	8	3.4	L^n, L^o	A^{II}
CMS[535]	13	0.17	INEL	A^{II}
CMS[536]	13	3.9	INEL, L^l, L^p	A
LHCb[513]	7	3	L^q	AC
LHCb[537]	7	3	L^r	ABC

(*Continued*)

Table 1. (*Continued*)

L^{a} At least 1 charged particle within $|\eta| < 1$
L^{b} At least 1 charged particle within $|\eta| < 0.8$ and $p_T > 0.15\,\text{GeV}/c$
L^{c} At least 1 charged particle within $|\eta| < 0.8$ and $p_T > 0.5\,\text{GeV}/c$
L^{d} At least 1 charged particle within $|\eta| < 0.8$ and $p_T > 1\,\text{GeV}/c$
L^{e} At least 1 charged particle within $|\eta| < 2.5$ and $p_T > 0.5\,\text{GeV}/c$
L^{f} At least 2 charged particle within $|\eta| < 2.5$ and $p_T > 0.1\,\text{GeV}/c$
L^{g} At least 6 charged particle within $|\eta| < 2.5$ and $p_T > 0.5\,\text{GeV}/c$
L^{h} At least 20 charged particle within $|\eta| < 2.5$ and $p_T > 0.1\,\text{GeV}/c$
L^{i} At least 1 charged particle within $|\eta| < 2.5$ and $p_T > 2.5\,\text{GeV}/c$
L^{j} At least 20 charged particle within $|\eta| < 2.5$ and $p_T > 0.5\,\text{GeV}/c$
L^{k} At least 50 charged particle within $|\eta| < 2.5$ and $p_T > 0.5\,\text{GeV}/c$
L^{l} At least 1 charged particle within $|\eta| < 2.4$ and $p_T > 0.5\,\text{GeV}/c$
L^{m} At least 1 charged particle within $|\eta| < 2.4$ and $p_T > 1.0\,\text{GeV}/c$
L^{n} At least 1 charged particle within $5.3 < |\eta| < 6.5$
L^{o} At least 1 charged particle within each $5.3 < \eta < 6.5$ and $-6.5 < \eta < -5.3$
L^{p} Additional event selections to enhance inelastic and diffractive contributions
L^{q} At least 1 charged particle within $2.0 < \eta < 4.5$
L^{r} At least 1 charged particle within $2.0 < \eta < 4.8$, $p_T > 0.2\,\text{GeV}/c$, $p > 2\,\text{GeV}/c$
† Includes events enhanced by a high p_T trigger
‡ Primary-particle definition excludes strange baryons
^{II} Primary-particle definition excludes leptons

multiplicity are unprecedented and the wealth of results is a legacy enabling the theory community to develop models with an accurate description of the non-perturbative QCD components which ultimately constitutes the bulk of the particles produced in LHC collisions. The following sections are only able to present a subset of these results with the aim of illustrating the different observables, what was learned and their potential.

3.1. *Pseudorapidity density $dN_{\text{ch}}/d\eta$ and transverse-momentum spectra $dN_{\text{ch}}/dp_{\text{T}}$*

The pseudorapidity density $dN_{\text{ch}}/d\eta$ and the transverse-momentum spectra $dN_{\text{ch}}/dp_{\text{T}}$ measure the average number of particles for a given event class. It is the most basic reduction of the complexity of a particle collision. Figure 2 presents a compilation of $dN_{\text{ch}}/d\eta$ from pre-LHC energies up to 8 TeV at the LHC. With increasing $\sqrt{s}$, both, the height of the central plateau and the variance of the distribution grow. The dip around $\eta \approx 0$ is an artifact of the transformation from rapidity to pseudorapidity. In addition to the measurements around mid-rapidity by ALICE, ATLAS and CMS, LHCb and TOTEM have studied the forward region.

The growth of $dN_{\text{ch}}/d\eta$ at $\eta = 0$ as a function of $\sqrt{s}$ is presented in Fig. 3. The dependence is described by a power-law as a function of $\sqrt{s}$ whose motivation

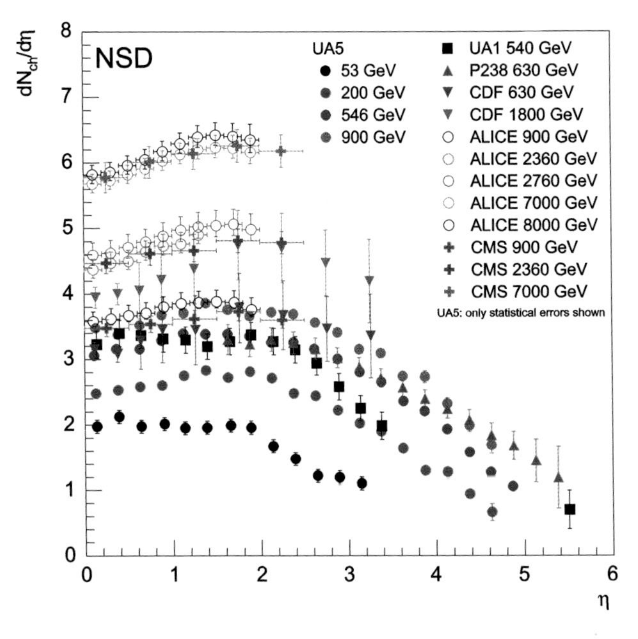

Fig. 2 The pseudorapidity density $dN_{\mathrm{ch}}/d\eta$ for NSD collisions over more than two orders of magnitude in $\sqrt{s}$. Data from Refs. 330, 515, 516, 525, 531, 541–544. As this figure presents only results for NSD collisions, no data for 13 TeV and from ATLAS and LHCb are shown.

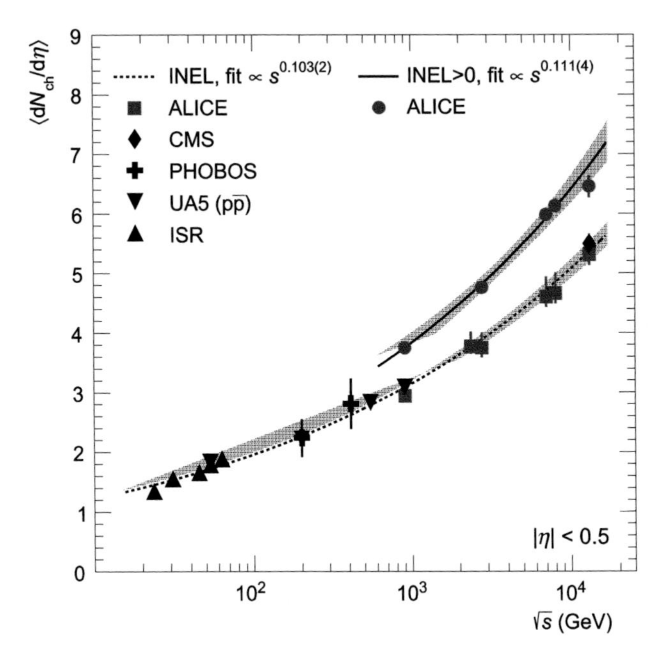

Fig. 3 The pseudorapidity density $dN_{\mathrm{ch}}/d\eta$ at $\eta = 0$ as a function of $\sqrt{s}$ for inelastic collisions and collisions with at least one particle within $|\eta| < 1$. Figure from Ref. 448.

is phenomenological. The increase at LHC energies was unexpectedly large[516,519] rendering many model predictions and tunes incorrect. This is interesting as the description of the increase of the average multiplicity in pQCD-inspired Monte Carlo models is sensitive to the $\sqrt{s}$ dependence of the lower momentum cut-off in the calculation of the $2 \to 2$ cross-section and the proton matter distribution which both affect strongly the number of parton interactions occurring in the same collision (see Chapter 10). The evolution of the MC tuning effort is illustrated in Fig. 4 which presents pre-LHC tunes and state-of-the-art Pythia tunes compared to early LHC results at 7 TeV. While the former deviate from the data by 25–30%, the latter accurately describes the data.

The transverse-momentum spectra $dN_{\mathrm{ch}}/dp_{\mathrm{T}}$ combines the measurement of the soft regime at low p_{T} with the hard regime at high p_{T} which can be calculated in pQCD. While early measurements could focus only on the regime up to a few GeV/c, distributions have later been measured up to 200 GeV/c.[447,528] An example is shown in Fig. 5.

3.2. *Multiplicity distribution* $P(N_{\mathrm{ch}})$

The multiplicity distribution $P(N_{\mathrm{ch}})$ gives the probability that a collision has a certain number of charged particles. Due to the limited acceptance of the LHC

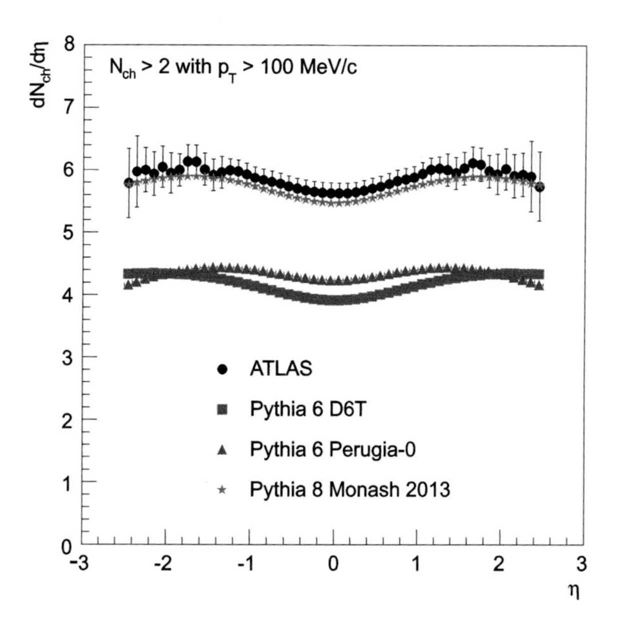

Fig. 4 Comparison of $dN_{\mathrm{ch}}/d\eta$ at $\sqrt{s} = 7$ TeV (circles, from Ref. 446) with pre-LHC tunes (Pythia 6[158] D6T (squares) and Perugia-0[434] (triangles)) and Pythia 8[192] Monash 2013[436] (stars) which has been tuned to LHC data. The unexpected increase in multiplicity is clearly visible by the large discrepancy between predictions and the data. The Monte Carlo simulation data are replotted from MCPLOTS.[441]

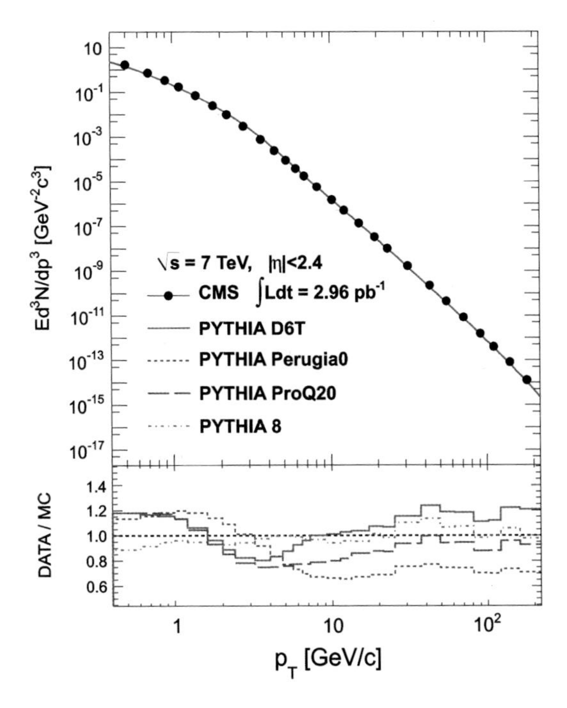

Fig. 5 Transverse-momentum spectra up to 200 GeV/c compared to several MC tunes which despite the impressive y axis range over 14 orders of magnitude describe the data within 20%. Figure from Ref. 448.

experiments, the measurement is typically performed in a limited pseudorapidity range (contrary to pre-LHC experiments which often measured this distribution in full phase space). Figure 6 presents a compilation of results from ALICE and CMS which measured $P(N_{\text{ch}})$ for NSD collisions in $|\eta| < 1.5$. In addition, to the good agreement between the two experiments, it can be seen that the width of the distribution grows significantly with $\sqrt{s}$ giving access to a very interesting high-multiplicity regime (see Chapter 15). Recent measurements reach out to 250 particles in $|\eta| < 2.5$.[530]

Historically, multiplicity distributions gave rise to a rich phenomenology.[540] KNO scaling[331] asserted that distributions at all $\sqrt{s}$ fall onto a universal curve under a transformation dividing by the average number of particles. Negative binomial distributions (NBDs) were successful to describe $P(N_{\text{ch}})$ at SPS energies[542] but failed at higher energies. Two-component approaches using two[545,546] (or even three[547]) NBDs could not survive up to LHC energies (see, e.g., Ref. 525). Nowadays, multiplicity distribution are a very sensitive probe of multiple parton interactions as collisions with large multiplicities are mostly composed of several parton interactions (see Section 3.4). Event generators fail to describe the tail of the multiplicity distribution without considering multiple parton interactions and the careful tuning of the related parameters.

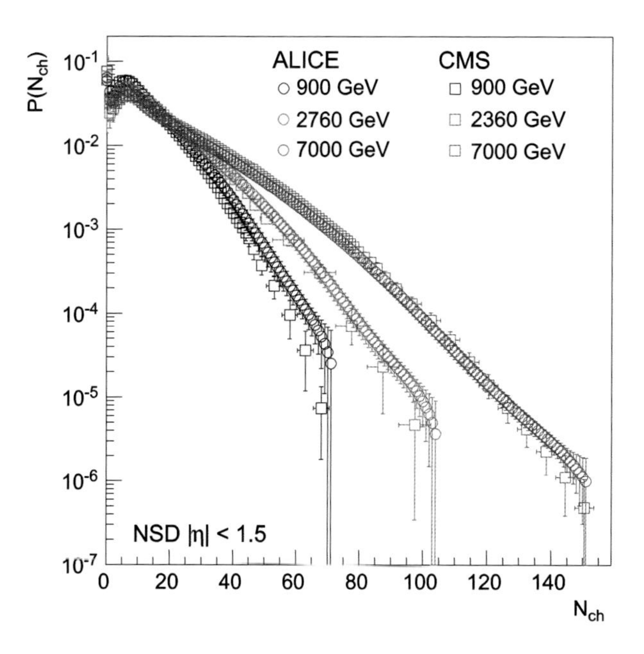

Fig. 6 Multiplicity distribution in $|\eta| < 1.5$ for NSD collisions from 900 GeV to 7 TeV illustrating the wide tail of the multiplicity distribution at LHC energies. Data from Refs. 525 and 532.

3.3. *Mean transverse-momentum evolution*

The evolution of the mean transverse momentum $\langle p_T \rangle$ as a function of N_{ch} measures the correlation of the momenta of the bulk of the produced particles with the multiplicity. It can differentiate if high-multiplicity events are *simple* superpositions of low-multiplicity collisions or if coherent effects between different parton interactions have a significant influence. Figure 7 presents a recent result which demonstrates the increase of $\langle p_T \rangle$ with growing N_{ch}. State-of-the art Pythia tunes describe this observable within 10% while the best description of this observable is provided by the EPOS LHC generator (see Chapter 19). In the Pythia model, the correct description of this observable requires the color-reconnection mechanism (see Chapter 10). Further insight of the growth of the $\langle p_T \rangle$ is obtained using so-called underlying event observables where the activity is studied transversely to the hardest object in the event (see Chapter 11).

3.4. *Minijets*

The observables presented above give an indirect access to the number of parton interactions through MC tuning. A direct way is the measurement of so-called minijets[523] which are constituted of particles stemming from the same $2 \to 2$ parton scattering, i.e., jets, but at p_T of a few GeV/c where traditional jet reconstruction

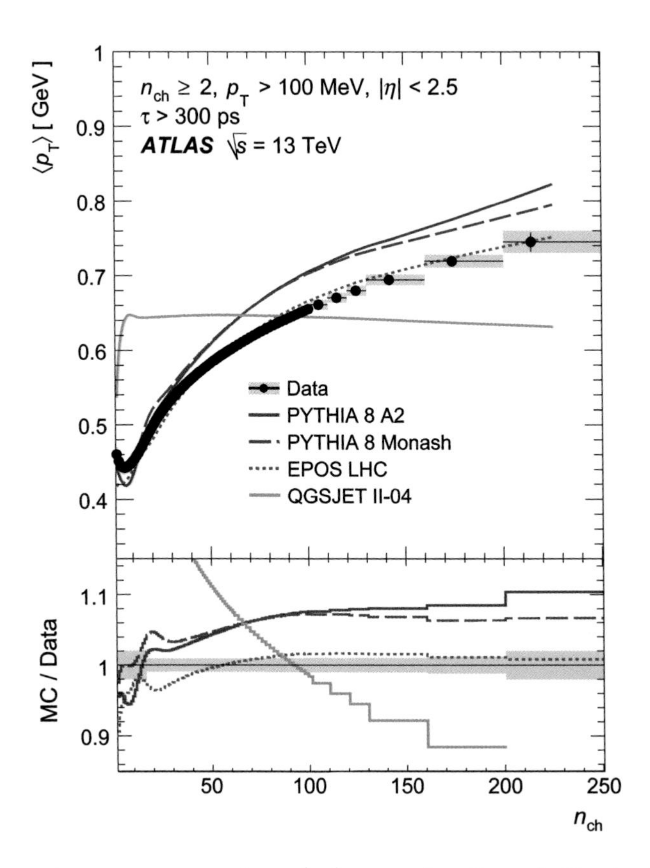

Fig. 7 Average transverse-momentum $\langle p_{\mathrm{T}} \rangle$ as a function of N_{ch} in $|\eta| < 2.5$ at $\sqrt{s} = 13\,\mathrm{TeV}$ compared to models. Figure from Ref. 530.

algorithms are not applicable. In this regime, each parton produces only 1–2 particles. Experimentally statistical methods like two-particle correlations allow nevertheless a measurement. In this method, the number of associated particles to a so-called trigger particle are extracted on the near- and away-side. From this the uncorrelated seeds can be calculated which are a measure of the number of independent clusters in a collision:[523]

$$\langle N_{\text{uncorrelated seeds}} \rangle = \frac{\langle N_{\text{trigger}} \rangle}{\langle 1 + N_{\text{assoc,near-side}} + N_{\text{assoc,away-side}}}. \tag{2}$$

Figure 8 illustrates that this quantity is proportional to the number of parton interactions in Pythia. Figure 9 shows the measured uncorrelated seeds for $\sqrt{s} = 0.9\,\mathrm{TeV}$ to $7\,\mathrm{TeV}$ as a function of N_{ch}. The results at different energies are very similar at fixed N_{ch} (left panel). In addition at large multiplicities a deviation from a linear trend is visible (shown in the right panel) hinting at a saturation in the number of parton interactions. It would be very interesting to extend this

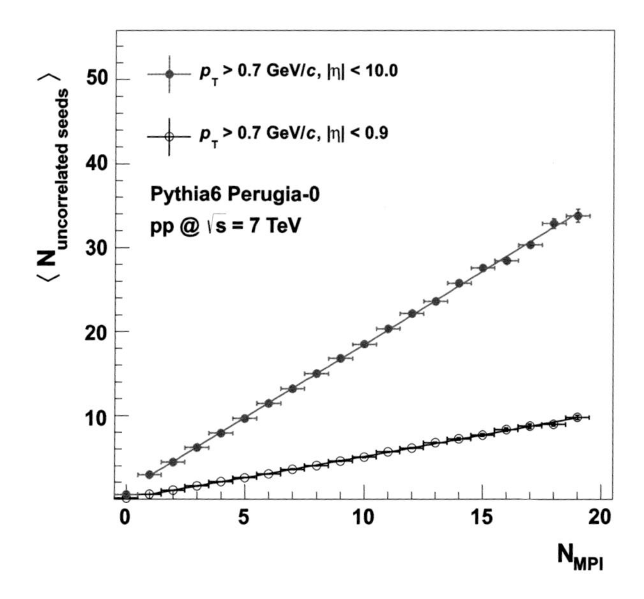

Fig. 8 Correlation of the uncorrected seeds and the number of parton interactions showing the clear proportionality of these two measures within Pythia. Figure from Ref. 523.

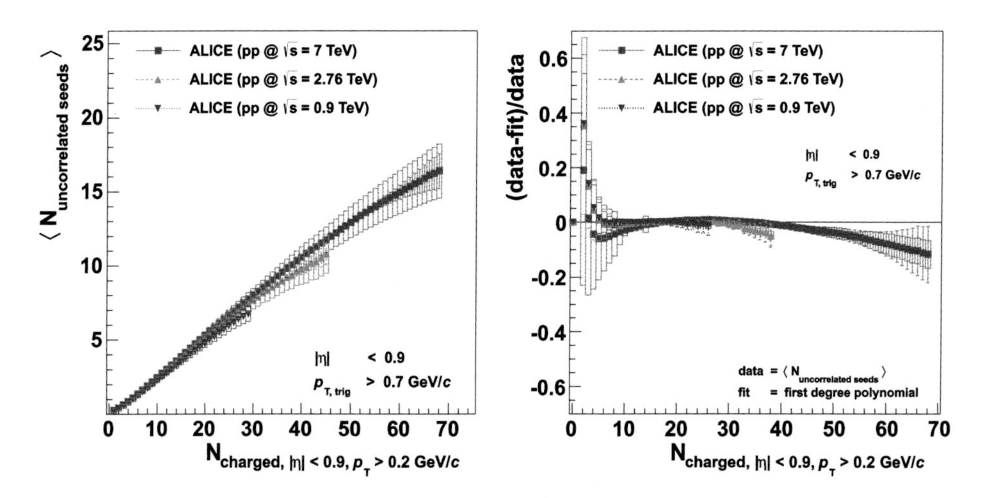

Fig. 9 Number of uncorrelated seeds as a function of N_{ch} for $\sqrt{s}$ from 900 GeV to 7 TeV (left panel). The uncorrelated seeds increase almost linearly with N_{ch} which is illustrated in the right panel where the difference to a linear fit is shown. At high N_{ch} a mild deviation from this linear trend is observed. Figures from Ref. 523.

measurement to higher N_{ch} and study if indeed above a certain N_{ch} no further increase of the number of parton interactions is observed.

4. Outlook

This chapter has presented selected results of minimum-bias results at the LHC, $dN_{\mathrm{ch}}/d\eta$, $dN_{\mathrm{ch}}/dp_{\mathrm{T}}$, $P(N_{\mathrm{ch}})$ and $\langle p_{\mathrm{T}} \rangle$ vs. N_{ch}. In addition, the minijet observable was introduced which gives direct access to the number of parton interactions. These measurements constitute a precise characterization of the average LHC collision. At the same time, they study rare collisions at high multiplicity. Future LHC data-taking will allow to extend this into the regime of very large multiplicity and very large number of parton interactions — a region full of interesting aspects of soft QCD.

Chapter 13

Tuning of MC Generator MPI Models

Andy Buckley* and Holger Schulz†

*School of Physics & Astronomy,
University of Glasgow, UK

†Institute for Particle Physics Phenomenology,
Durham University, UK

MC models of multiple partonic scattering inevitably introduce many free parameters, either fundamental to the models or from their integration with MC treatments of primary-scattering evolution. This non-perturbative and non-factorizable physics in particular cannot currently be constrained from theoretical principles, and hence parameter optimization against experimental data is required. This process is commonly referred to as MC tuning. In this chapter we summarize the principles and problems of MC tuning, and the still-evolving modern approach to both model optimization and estimation of modeling uncertainties.

1. Introduction

It is an unfortunate fact of life that the modeling approaches to multiple partonic scattering described in previous chapters and implemented in Monte Carlo event generator programs are not fully unambiguous. Not only are ansätze required for computation of the hadronic matter overlap and low-$p_\perp$ regularization of the partonic scattering cross-section, but also the secondary scattering must coherently connect to the other aspects of event modeling in general-purpose MC codes.

For example, MPI scattering must interface somehow with QCD evolution down to soft momentum-transfer scales, as described by parton showers and matrix element corrections, and to the color connections between the partonic scattering system and the beam remnants. Phenomenological hadronization models must also be modified to accommodate MPI as a source of partons, most notably via somewhat *ad hoc* "color reconnection" or "color disruption" mechanisms.

The result of this complexity is that MPI models not only contain degrees of freedom intrinsic to their own formulation, but also require extensions to the generator components concerned with the primary partonic scatter. As much of the physics involved is non-perturbative — and that which is not is only defined up to leading-order or leading-logarithmic accuracy — it is typical for more than ten model parameters to influence observables of interest, with little or no *a priori* prediction of their values. These parameters must somehow be "tuned" to describe MPI-sensitive observables in experimental data.

In this chapter, we describe the dominant modern approach to MC generator parameter optimization, from the technical machinery to the parameters and observables, and the methodology applied to both achieve convergence and avoid overfitting. As is always true for parameter estimation in physics, the resulting uncertainty is as important as the central value and hence we review the statistical methodology applied to estimate model uncertainties through tuning. Finally, we survey the road ahead for MC tuning and improved constraints on MPI modeling.

2. Tuning Methodology

From the outset we should be clear that tuning is not desirable. While currently a necessary part of the landscape of MPI modeling (and other hadron-collider event features), in an ideal world our models would have sufficiently few ambiguities that tuning will become unnecessary. We can dream! But for now, modeling flexibility is necessary to achieve the degree of data description required by experiment — at the significant cost of exchanging parameter fitting and uncertainty estimation for predictivity.

In this pragmatic compromise, it is preferable not to simply throw all possible parameters and data into a massive fit. Rather, well-motivated modeling components — typically those involving QCD at perturbative scales — are trusted to be predictive from first principles, while phenomenological models which are only unconstrained in the asymptotic limits of QCD are ripe for fitting. We use the word "tuning" to refer only to fits of the latter parameter type, and as far as possible avoid fitting true theory uncertainties such as scale choices in the perturbation expansion.

In the simplest case, tuning consists of finding the value of a single model parameter — say, the $p_\perp$ scale used to regularize the divergent secondary scattering cross-section — which gives the best agreement with a single data bin, e.g., a total cross-section for double-partonic scattering. For this, little technical machinery is required: a set of MC generator runs with different values of the parameter (either over the whole natural range of the model, or focused on a "known-good" region) are compared to the data, and the best-performing model point is chosen as the "tune". Perhaps a couple of iterations will be required. This is "manual tuning."

Extension of this scheme to include more data is simple but not entirely trivial. For example, a multi-bin observable such as an "underlying event" characterization of mean particle or energy flow away from the hard primary scattering products, simply requires that the goodness of fit be computed via an aggregation of the fit quality across the many bins. Computing the fit quality naturally introduces several questions:

(1) Which regions of the observable are most important to describe?
(2) Are there bins which the model fundamentally can/should not describe?
(3) Are these bins independent of each other, or correlated somehow?

Unfortunately, we do not have general answers to the first two of these points, which immediately makes a robust statistical foundation for tuning problematic. The third in principle can be solved by publication of bin-correlation data, but this has both been rare in MPI-sensitive measurements until now, and is arguably rendered moot by the first two issues. We shall return to this theme later.

Now a more technically troublesome extension: more *parameters*. In a two-dimensional (2D) parameter space, such as including a scaling parameter for the hadronic matter distribution in addition to the $p_\perp$ regularizer — a by-eye approach is still possible: as before, make MC runs for several combinations of the two parameters' values and compare to data. In addition, computing a goodness-of-fit measure for all points in a 2D grid can provide a useful visualization of the physics dependence, as illustrated in Fig. 1. But the increased computational cost is clear: if N trial points were required for a single parameter, $\mathcal{O}(N^2)$ will be required for two parameters, and this exponential scaling continues into the less visualizable spaces of three and more parameters. The severity of this problem is emphasized by the expense of a typical parameter-point evaluation: the required statistical precision means that $\mathcal{O}$(100 k–1 M) events are needed per colliding-beam configuration and per primary-scattering process type, hence the total time for a parameter point may be counted in *CPU-days*. A comprehensive grid-scan of the $\mathcal{O}$(10–20) parameters needed in a full generator is clearly unfeasible.

One possible approach is physical intuition, and this is undoubtedly helpful. Fortunately a 20-dimensional appreciation of model behavior is not necessary, since many modeling components approximately factorize and can be reasoned about in isolation. But even so, once more than a few interconnected parameters are involved, intuition is not enough to break fit-quality degeneracies, and it is impossible to be sure that a final intuitive tune is truly optimal. However, despite the availability of more technical machinery, the wholly intuitive approach is far from irrelevant.[436]

The alternative is to somehow make do with an incomplete sampling of the parameter space, and to use computational machinery to guide the fit. Done naïvely, this runs into problems of its own: a random sampling of a large-dimensional space gives little confidence that the best-seen parameter point is anywhere near to the global optimum. Attempting to systematically improve on the points visited so far would provide a better sampling of the space, but at the unsustainable cost of abandoning the implicitly parallelizable approach of independent MC

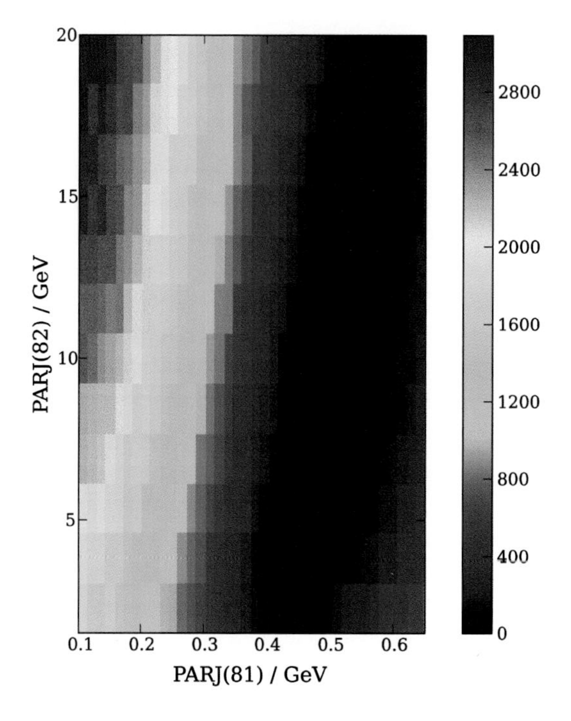

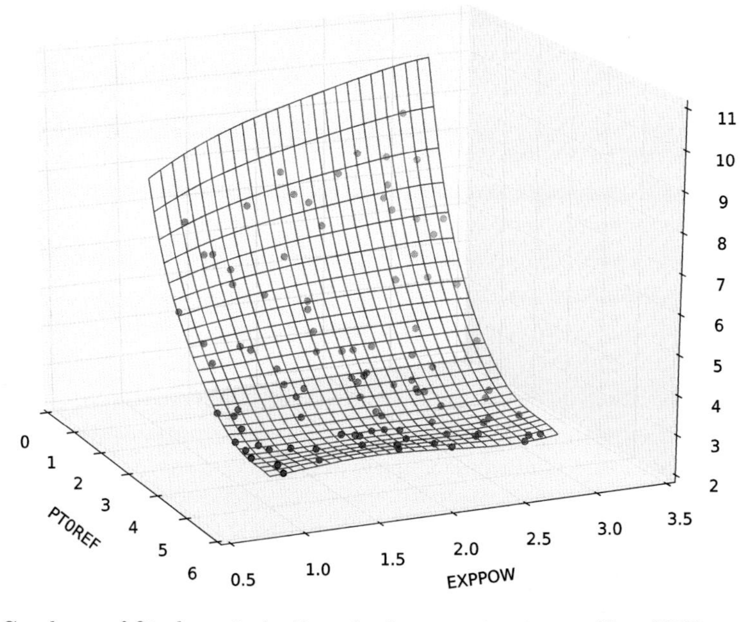

Fig. 1 Goodness-of-fit characterizations in 2-parameter tunes. Top: LEP event thrust in Pythia 6 parton shower; Bottom: LHC minimum-bias in Pythia 8 MPI model (with PROFESSOR χ^2 function fit to the sampled points).

runs. Even very sophisticated attempts to serially sample MPI model parameter spaces have proven intractable due to the high computational cost of the MC "function".[548] Instead, the method which has become most widespread, via the PROFESSOR toolkit,[439] is a hybrid of parallel sampling and serial optimization, via *parameterization* of the MC generator response to parameter variations. It is this approach to which we will dedicate most space in this summary. First, however, it is important that we acquaint ourselves with the typical parameters and observables that will be encountered while constructing an MC generator tune.

2.1. *Experimental observables and model parameters*

While our interest here is naturally biased toward the MPI-specific aspects of generator modeling, it is worthwhile to also discuss how the rest of the generator system is involved in tuning. In rough reverse order from final to initial state, the main components of an MC generator are hadronization, fragmentation, final-state QCD radiation, initial-state QCD radiation, and multiple partonic scattering. Taken together, these components readily comprise a parameter space with in excess of 30 dimensions. Even with computational tricks in the generator runs, sampling such a space requires extraordinary CPU resources. But there are usually significant factorizations between these major modeling steps, which can be exploited to make the tuning more tractable.

Typically we start by identifying these factorized blocks working backward from the final state, assuming that decays, hadronization (not including color reconnection[a]), and the final-state parton shower should be sufficiently independent of initial-state QCD effects that they can be tuned to e^+e^- data from LEP and SLD only. Using hadron collider data (at least in the first iteration) could bias this tuning via poor initial-state modeling, so it is typically left out, saving CPU time and complexity. Since hadronization models can account for the majority of MC generator parameters, it is not unusual to split this final-state tuning stage into several rounds, e.g., first concentrating on using identified hadron rates[514] to fix hadronization parameters such as strangeness suppression, then using identified particle energy spectra, event shapes, and jet rates[549–551] to constrain the final-state parton shower and fragmentation functions. As the parton shower is built on perturbative physics, it has few degrees of freedom: typically just the emission cutoff scale and perhaps the definition of α_S evolution. Treated in this combined way, the first PROFESSOR tunes of Pythia 6 achieved degrees of data description

[a]Color reconnection (CR), i.e., dynamic reconfiguration of final-parton color string/cluster topologies, is more a hadronization effect than an MPI one although it is often discussed as an aspect of secondary scattering. In principle it can therefore affect final-state observables, although there is as yet no strong evidence for this. More concerning is that soft-QCD tunes of CR may have inappropriate effects in high-$p_\perp$ hard-scattering processes, e.g., $t\bar{t}$ production.

for LEP event shapes, jet rates, and b-fragmentation that had previously eluded manual tunes.

Having addressed the tuning of final-state modeling, we now confront the combination of the initial-state parton shower, the MPI mechanism, and the color-reconnection mechanism: all key to description of particle production and energy flow observables at hadron colliders. The most important observables depend on the bias of one's physics interest, but in the general-purpose tunes which have received most attention, the emphasis in soft QCD has been on kinematics rather than flavor content. This due to experimental priorities: for most purposes at the LHC, secondary-scattering is a background to measurement of hard signal-process scattering, and hence the key requirement of a general-purpose MPI tune is to accurately model its contribution to charged track multiplicities and energy flows. Since these background contributions are divided into inclusive soft-QCD scattering in additional pp collisions (known as “pile-up”), and additional partonic scattering in signal events (the “underlying event”), the dominant observables in such tunes are those measured in inclusive “minimum bias” event selection, and those specific to the underlying event.

Minimum-bias physics measurements at hadron colliders have been dominated by charged-track observables, partially because the low-luminosity early phases of each data-taking run are crucial for calibration of detector tracking systems. The main data from these measurements, often grouped as “min-bias observables”, are charged-particle multiplicities and $p_\perp$ spectra within fiducial acceptance cuts (most obviously the tracker coverage in pseudorapidity, η), and the correlation of the average charged-particle $p_\perp$ with the event's fiducial track multiplicity. Broadly speaking, the distribution of charged-particle multiplicity N_{ch} with η is the canonical distribution used to indicate the inclusive amount of minimum bias particle production, particularly in the flat central region $|\eta| < 1$. This is governed by a correlated combination of MPI $p_\perp^0$ regularization parameter, the amount of hadronic matter overlap for large impact parameter (since $b \sim 1/Q$ and the typical scale Q of inclusive minimum-bias interaction is low), and perhaps scaling of the MPI partonic process via freedom in parameterization of the MPI α_{S}. The correlation of $\langle p_\perp \rangle$ with N_{ch} became an important tuning observable when it was noted that models without color-reconnection were unable to describe it, predicting too soft a particle production spectrum in higher-multiplicity scattering events. Color disruption mechanisms were added to the hadronization models of Pythia and Herwig to address this (see Chapter 10), naturally introducing extra degrees of freedom for tuning.

In the LHC era, additional fiducial cuts were introduced to minimum-bias data analyses as variations of “analysis phase space” to modify the sensitivity to different aspects of inclusive scattering physics. These include requirements on charged-particle fiducial multiplicity (e.g., from 1 to $\sim$20) and on minimum charged-particle $p_\perp$ (e.g., from 100 MeV to $\sim$10 GeV). The result is that a very large number of many-bin observables, generally with small uncertainties, is available from the Tevatron to the LHC, and from 300 GeV to 13 TeV beam energies.[379,446,517,526,529,530,532,543,552,553] In addition, ATLAS and CMS have published calorimetric measurements of energy flow as a function of η, including a minimum-bias trigger selection. These provide a counterpart to the track-specific measurements, including both a central overlap with tracking detector acceptance and extension to high-$|\eta|$, crucial for forward/diffractive and beam-connection physics. This availability of multiple independent measurements of each observable is important to avoid overfitting of a single measurement, and the different phase spaces enable, for example, degeneracy breaking between non-diffractive MPI and diffractive physics contributions to particle production.

Underlying event (UE) analyses are a specialization of the minimum bias observables to events where a more exclusive trigger is required, i.e., a genuinely high-scale scattering process such as hard jet or Z boson production. Typically the motivation of UE measurements is to specifically study the connection between this hard process and the associated secondary scattering, as a test of the eikonal MPI model and of the interaction between it and the perturbative QCD dressing of the primary scatter.

Since a hard primary partonic process will dominate the particle- and energy-flow characteristics of each event, UE analyses specifically analyze event regions expected to contain minimal hard process contamination — for example, regions azimuthally transverse to the axis of a balanced dijet event,[129,500,502] transverse to or in the direction of a hard leptonic Z,[492,503] or with identified jet activity "cut out" from anywhere in the event η–ϕ phase space.[381] The transverse regions are often further specialized to discriminate between the more and less active sides on a per-event basis, to provide additional resolution between MPI and parton shower activity.

The canonical UE observable is the evolution of the mean value of a minimum-bias event property like charged-particle multiplicity or $p_\perp$ sum within a sensitive phase-space region, as a function of the scale (usually a $p_\perp$) of the hard scattering process. This produces an extremely informative curve showing the smooth evolution of mean event properties from minimum-bias at low event scales (interpreted as peripheral hadronic collisions), up to very hard primary scatterings as $b \to 0$. Underlying event physics hence probes the same MPI mechanisms as minimum-bias (particularly high-activity MB phases spaces), but with a clear connection to the pedestal effect, which maps the matter overlap profile in detail, and an increased emphasis on the role of initial-state QCD radiation from the high-scale primary process.

Parton density functions play an important role in hadronic initial-state tuning, particularly MPI[b] since the partonic cross-section for multiple scattering at low-Q is

[b]Initial-state parton showers are largely insensitive to PDF detail, at least for a given α_S and perturbative order in QCD.[554]

strongly driven by the diverging low-x gluon PDF, which varies a great deal between different PDF fits. Additionally, PDF differences produce variations in the rapidity distribution of MPI partonic scattering. These effects, influencing the multiplicity of MPI scattering and $d\eta/dN_{\rm ch}$ distribution shapes respectively, are responded to in tuning by correlated shifts in the $p_\perp^0$ screening factor, matter distribution/overlap parameterization, and MPI $\alpha_{\rm S}$. The Pythia MPI model in particular is rather over-parameterized: the MPI rate can be more-or-less directly modified via $p_\perp^0$, $\alpha_{\rm S}$, *and* a partonic scattering scale-factor. Naïvely throwing all these parameters into a fit will likely produce degenerate or overfitted tunes: it is best to instead use a subset of at most three parameters, e.g., $\alpha_{\rm S}$ or scale-factor (or neither) but certainly not both in a single tune. The relationship between gluon luminosity (the integral of a PDF over MPI partonic scattering scales) and tuned $p_\perp^0$ may be seen in Fig. 2(a): a more divergent low-x PDF with a higher gluon luminosity is strongly correlated with a higher $p_\perp^0$ tune value and hence more screening of the divergence. Tunes, or at least their MPI component, are hence specific to a particular PDF — even, arguably, to the variation fits within a given PDF set.

A key aspect of MPI tuning in modern MC generators is to fit the $\sqrt{s}$ dependence of the model. The Jimmy MPI model, and early versions of Herwig++, attempted to describe all MPI activity with a fixed $p_\perp$ regularizer value at all energies, but this ultimately proved unworkable and a Pythia-like $\sqrt{s}$ dependence with a slow power-law dependence similar to total pp cross-section fits[555] was instead introduced. The analogy is not predictive, however, so this exponent is also a free tuning parameter for any tune interested in describing more than one collider energy. Experimental data from more than one energy is obviously needed to constrain this parameter, but the power-law ansatz has been found to work well and even been supported empirically by independent tunes of $p_\perp^0$ at different $\sqrt{s}$[435] as shown in Fig. 2(b).

Finally, a marginal aspect of tuning: the "primordial $k_\perp$". This quantity, implemented in several parton shower generators, is the $\mathcal{O}(1)$ GeV width of a function used to add a randomly sampled transverse momentum boost to the whole modeled scattering event. This is motivated almost entirely by the pragmatic desire for a good description of the Z $p_\perp$ differential cross-section, a precisely measured observable generated by initial-state recoils, which rises steeply from zero at $p_\perp(Z) = 0$ to a peak at a few GeV. The exact position of this peak is determined by resummation of large QCD logarithms, i.e., the process approximated by the parton shower. Comparisons to data with a "vanilla" parton shower almost invariably produce a peak at too small a $p_\perp$ value, and hence primordial $k_\perp$ smearing was introduced as the simplest possible mechanism to "correct" this flaw in data description.[c] While this nebulous shortcoming of parton shower models is somewhat

[c]While often justified via an uncertainty-principle "particle in a box" argument, the typical magnitude of the smearing width is an order of magnitude larger than expected from such an argument.

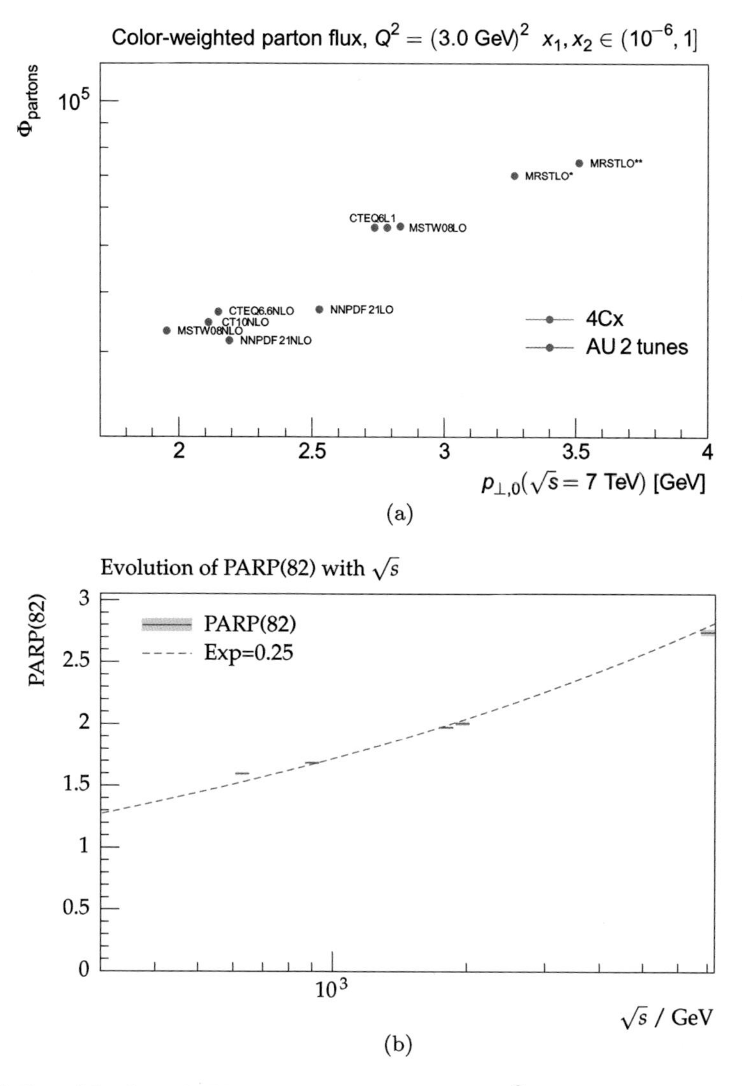

Fig. 2 (a) Tuned Pythia 8 MPI cross-section regularizer, $p_\perp^0$, as a function of low-x gluon luminosity from various PDFs. The correlation between higher gluon flux and higher $p_\perp^0$ (meaning more screening of that flux) is clear. (b) Empirical $p_\perp^0$ scaling via independent tunes to hadron collider data at different $\sqrt{s}$.

distressing, at present only the Z $p_\perp$ distribution is precisely enough measured at hadron colliders to be sensitive to this effect (and other very soft QCD modeling effects, such as the ISR shower cutoff scale and/or POWHEG real emission cutoff[556]). Primordial $k_\perp$ can hence be tuned virtually independently of other initial-state quantities, although it can equally be included in larger tunes, where it becomes a flat direction in the parameter space for all bins except those in the low-$p_\perp$ region of Z $p_\perp$ and related observables.

2.2. *Parameterization-based tunes*

Tuning via parameterization of MC generator behavior has a lengthy history.[549,550,557–560] The fundamental idea is to replace the expensive, probably multi-day, explicit evaluation of a proposed MC parameter point with a very fast, analytic approximation.

It is tempting to try to parameterize the shape of an observable as a whole as function of some input parameters by using splines or similar structures. This, however, is a non-trivial task as the functional form of an observable will in most cases not be parameterizable by simple functions. Similarly, parameterizing the entirety of a multi-bin goodness-of-fit (GoF) function proves fraught. Instead, a P-dimensional polynomial is independently fitted to the generator response, $\mathrm{MC}_b(\mathbf{p} = (p_1, \ldots, p_P))$, of each observable bin b. By doing so the potentially complicated behavior of observables is captured by a collection of simple analytical functions.

Having determined, via means yet undetailed, a good parametrization of the generator response to the steering parameters for each observable bin, it remains to construct a GoF function and minimize it. The result is a predicted parameter vector, $\mathbf{p}_{\mathrm{tune}}$, which should (modulo checks of the technique's robustness) closely resemble the best description of the tune data that the generator can provide.

In parametrization-based tuning, the run-time is dominated by the time taken to run the generator to produce inputs to the parameterization. This step is trivially parallelizable and large tunes can be tractable even with modest computing resources. The calculation of the parameterization rarely exceeds a few minutes, as does the subsequent numerical minimization step: this technique hence enables rapid tuning in response to new measurements, as well as systematic exploration of freedoms in the tuning procedure itself. The preparation of input data depends on the complexity of the task at hand, namely the sophistication of the generator and the dimension of the parameter space.

It is important to check the fidelity of the parameterization, and to ensure that the parameterization scan includes predictions surrounding the target data values. The construction of "envelope plots" is a good practice to check *a priori* that the chosen sampling range actually covers the data considered for tuning. For each bin of each observable the minimal and maximal value from the corresponding inputs is obtained and thus allows to spot mistakes and model limitations early on, as shown in Fig. 3. Using disjoint sets of input MC data for parameterization-building and testing allows the accuracy of the parameterization to be estimated.

Fitting model. To illustrate the method, we discuss the parameterization of the bin content MC_b using a general polynomial of second order:

$$\mathrm{MC}_b(\mathbf{p}) \approx \alpha_0^{(b)} + \sum_i \beta_i^{(b)} \, p'_i + \sum_{i \leq j} \gamma_{ij}^{(b)} \, p'_i \, p'_j. \tag{1}$$

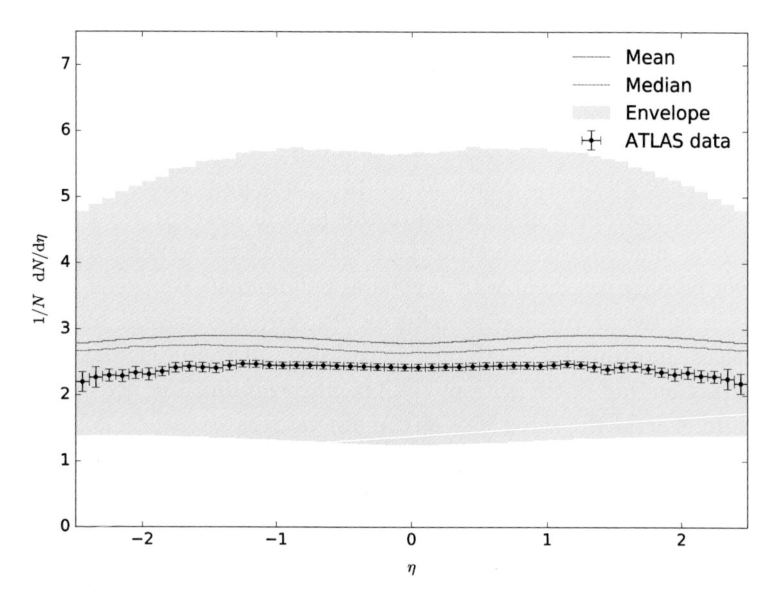

Fig. 3 Example of a tuning "envelope plot". Here the experimental data is clearly contained within the yellow band of the sampled model space, but that is not always the case.

The task at hand is to determine the coefficients $\alpha, \beta_i, \gamma_i$. Algorithmically, this is done by generating at least as many inputs $\mathrm{MC}_b(\mathbf{p})$ for different $\mathbf{p}$ as there are coefficients in the to-be-fit polynomial such that we are able to solve a system of linear equations. Since it is much more practical we cast the right-hand side into a scalar product:

$$\mathrm{MC}_b(\mathbf{p}) \approx f^{(b)}(\mathbf{p}) = \sum_{i=1}^{N_{\min}(P)} c_i^{(b)} \tilde{p}_i. \tag{2}$$

For a second-order polynomial of a 2D parameter space (x, y), the coefficient vector would have the form

$$\mathbf{c}^{(\mathbf{b})} = (\alpha, \beta_x, \beta_y, \gamma_{xx}, \gamma_{xy}, \gamma_{yy}), \tag{3}$$

and for each set of input points (x_i, y_i) we can write

$$\tilde{p}_i = (1, x_i, y_i, x_i^2, x_i y_i, y_i^2). \tag{4}$$

By doing so and denoting the set of $\tilde{p}_i$ as $\tilde{P}$ and the set of corresponding bin values as $\mathrm{MC}_\mathbf{b}$ we construct the matrix equation

$$\mathbf{MC}_b = \tilde{\mathbf{P}} \cdot \mathbf{c}^{(\mathbf{b})}, \tag{5}$$

which allows us to determine the set of coefficients $\mathbf{c}^{(\mathbf{b})}$ by inverting $\tilde{\mathbf{P}}$, i.e.,

$$\mathbf{c}^{(\mathbf{b})} = \mathscr{I}[\tilde{P}]\mathrm{MC}_\mathbf{b}. \tag{6}$$

In PROFESSOR, a singular value decomposition (SVD) algorithm (implemented in the Eigen3 library) is used to perform the matrix (pseudo)inversion $\mathscr{I}[\tilde{P}]$. The SVD method is equivalent to a desirable least-squares fit of the target polynomial to the input data. In the case of having as many input points as there are coefficients to be determined, the solution is exact. When providing more than $N_n(P)$ inputs, the system is over-constrained and therefore the fit will average out to some degree both statistical fluctuations and the fact that the true generator response will almost never be fully describable by a general polynomial. We prefer to oversample by at least a factor of two for robustness. As only the central bin values enter the SVD algorithm, but the statistical uncertainty of the input data is crucial to GoF construction and optimization, the bin *uncertainties* are fitted as a separate polynomial in exactly the same way as the bin values.

The dimension of the parameter space and the order of the polynomial to be fitted determine the dimension of the to be inverted matrix and thus the minimal number of required input datasets. Generating more inputs than minimally necessary is in this context equivalent to over-constraining a system of linear equations. Doing so has the benefit of being able to test the stability of the obtained best parameter point against two aspects. One being the order of polynomials chosen as higher order correlations can become important. Secondly, although parameterizations obtained from all available inputs should give the best prediction of the generator response in the whole of the parameter space, smaller subsets can yield different best parameter points which is indicative of the polynomial approximation breaking down (typically in a too large parameter space). The number of coefficients of a P-dimensional general polynomial of order n is

$$N_n(P) = 1 + \sum_{i=1}^{n} \frac{1}{i!} \prod_{j=0}^{i-1} (P + j). \tag{7}$$

How the number of parameters scales with P for second- and third-order polynomials is tabulated in Table 1 and shown in Fig. 4. The latter also shows how computational cost scales with P and n.

Table 1. Scaling of number of polynomial coefficients $N_n(P)$ with dimensionality (number of parameters) P, for polynomials of second ($n = 2$) and third-order ($n = 3$).

Number of parameters, P	$N_2(P)$ (second order)	$N_3(P)$ (third order)
1	3	4
2	6	10
4	15	35
8	45	165

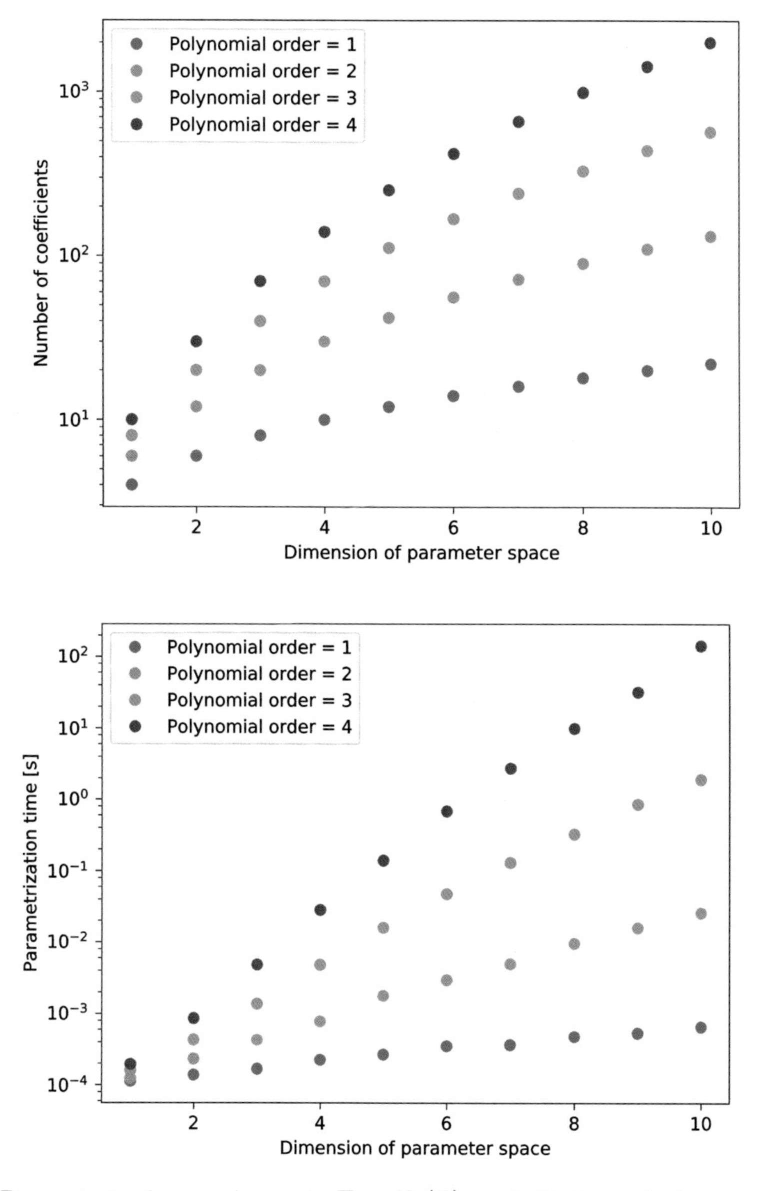

Fig. 4 Parameterization requirements. Top: $N_n(P)$ vs. P; Bottom: single core computation time vs. P.

PROFESSOR allows to calculate polynomials of in principle arbitrary order, including zeroth order, i.e., the constant mean value. To account for lowest-order parameter correlations a polynomial of at least second-order should be used as the basis for bin parameterization. In practice, a third-order polynomial suffices for almost every MC generator distribution studied to date, i.e., there is no correlated failure of the fitted description across a majority of bins in the vicinity of best generator behavior. An upper limit on the usable polynomial order is implicitly given by the double precision of the machinery effectively limiting the maximum usable order to about 11. It should be noted that other classes of single polynomials such as Chebychev or Legendre have no additional benefit in the PROFESSOR method. The quality of the resulting parameterization is exactly the same, with merely different groupings of the coefficients.

The set of input points for each bin are determined by randomly sampling the generator from N parameter space points in a P-dimensional parameter hypercube $[\mathbf{p}_{\min}, \mathbf{p}_{\max}]$ defined by the user. This definition requires physics input — each parameter p_i should have its upper and lower sampling limits $p_{\min,\max}$ chosen so as to encompass all reasonable values while avoiding discontinuities. In cases where bin values vary over many orders of magnitude general polynomials are obviously a poor choice for an approximate function. It is nonetheless possible to use PROFESSOR by simply taming the input data by parameterizing, e.g., the logarithm of bin values.

Goodness of fit function and optimization. With the parameterization of the generator at hand the optimization can be turned into a numerical task. By default an *ad hoc* GoF measure is minimized using Minuit's migrad algorithm with the typical features of being able to fix or impose limits on individual parameters during the fit. For the purpose of generator tuning with typically imperfect models it is often necessary to bias the GoF in order to force the description of certain key observables, e.g., the plateau of the underlying event or to switch off parts of histograms entirely when it is known that the underlying model is unable to describe the data at all. The relative importance of individual bins and observables can be set in PROFESSOR using weights, w_b, in the GoF definition:

$$\chi^2(\mathbf{p}) = \sum_{\mathcal{O}} \sum_{b \in \mathcal{O}} w_b \cdot \frac{(f^{(b)}(\mathbf{p}) - \mathcal{R}_b)^2}{\Delta_b^2(\mathbf{p})}, \tag{8}$$

where $\mathcal{R}_b$ is the reference data value for bin b. The error Δ_b is composed of the total uncertainty of the reference which is a constant for each point $\mathbf{p}$ and the parameterized statistical uncertainty of the Monte Carlo input for bin b. In practice we attempt to generate sufficient events at each sampled parameter point that the statistical MC error is much smaller than the reference error for all bins.

Further methods. Optimization in tuning is by no means limited to this setup. The Python language bindings to the core objects in PROFESSOR allow to easily construct arbitrary GoF measures and to use other minimizers that have Python bindings, such as MultiNest.[561]

Although being highly successful, the polynomial parameterizations have obvious limitations. The framework hence allows to use different parameterization methods for instance the ones available in `scikit-learn` and `Gaussian Processes` where no prior assumption on the functional form has to be made.

Professor also provides tools for the interactive exploration of the parameter space as GTK application as well as Jupyter notebooks. Both allow to load a previously calculated set of parameterizations and displays the corresponding histograms. GUI sliders (one per parameter) allow to conveniently set the parameter point to a new value resulting in the histogram being redrawn immediately. These tools help to build intuition into what effect parameters have on distributions.

3. Influential MC Generator Tuning Results

Parton shower MC generators have been tuned for as long as they have existed: the intrinsic limitations of their formal accuracy and the dependence of truncated perturbative calculations on unphysical scales means that some authorial intuition has long been needed to achieve a good description of key data observables "out of the box". At LEP, the experiments also sank effort into tuning of final-state parton showers and fragmentation. But the existence of a wider program of tuning, particularly for MPI modeling, started with the work of Rick Field and the CDF tunes of Pythia 6.[562]

The first such construction was Tune A, its name acknowledging immediately that it was bound not to be the last word on this limitless subject. Tune A was constructed specifically to provide a first reasonable modeling of the underlying event, based on CDF's first UE measurement.[128] However, it was soon noticed that its parameter choices, particularly for the PARP(64) parameter governing the ISR evolution scale and the PARP(91) primordial $k_\perp$ width, resulted in a Z $p_\perp$ spectrum whose peak was at too small a $p_\perp$ value. A new tune was needed,[562,563] and soon came in the form of Tune AW with a 2.1 GeVprimordial $k_\perp$ width and a reduced ISR scale — corresponding to a larger α_S and hence more initial-state radiation against which to recoil. Inevitably, Tune AW also hit an obstacle: the dijet azimuthal decorrelation distribution,[564] characterizing the extent to which, in generic Tevatron jet events, initial-state recoils between high-$p_\perp$ jets push the leading dijet system away from the 2-body back-to-back configuration. Having specifically boosted the amount of ISR to create Tune AW, it now needed to be reduced again (this time via Pythia's PARP(67) ISR starting scale parameter) to avoid creating too many hard multi-jet events: the resulting tune was christened Tune DW. Several other variations joined the swelling ranks of CDF Pythia 6 tunes, including variations in PDF choice and the energy dependence of MPI regularization. (This last issue, almost in isolation, was also applied to tuning of the Herwig MPI model, Jimmy, in anticipation of the energy leap from Tevatron to LHC.)

The pattern at this point had become clear: the initial-state system of MPI, ISR, and intrinsic $k_\perp$ — as well as developments in matrix element matching and merging — was too complex to be entirely optimized by hand. Every single-parameter change would both address the modeling problem at which it was aimed, and break several other distributions. The Professor tuning effort arose at this point, to apply the computational methods described above to this optimization problem. The first and only tunes to bear the PROFESSOR label were the Prof0-Q2 and Prof0-pT tunes[439] of Pythia 6, for its virtuality ordered parton shower and newer $p_\perp$-ordered parton shower respectively. Both tunings were "global", in the senses that they covered all aspects of the generator from final-state showering and hadronization, to the initial-state effects covered by the CDF tunes[d] as well as the widest available dataset from LEP hadron spectra to event shapes, and to Tevatron minimum-bias and underlying event data. More influentially, the PROFESSOR machinery was immediately used within ATLAS to produce its own tune series first based on CDF data and then including the early ATLAS data in the AMBT and AUET tune series between 2009 and 2012. Also influential were the hand-tuned "Perugia" family by Peter Skands, particularly since they included systematic variations of the parton showers useful for uncertainty estimation.

As suggested by the story so far, the initial LHC tuning community focus was concentrated on Pythia 6. Additional work was performed at this time, largely via the PROFESSOR technique, to tune the SHERPA fragmentation model and the Jimmy MPI mechanism for the Herwig 6 generator. The next major developments in MPI tuning were the shift during LHC Run 1 to the newer C++ family of generators. The Pythia 8, Herwig++, and SHERPA generators were all tuned using the PROFESSOR tools within their development collaborations, with the most notable outputs being the Pythia 8 Tune 4C, and Herwig++'s UEEE tune series which introduced tuned modeling of MPI energy evolution and color-reconnection in response to the evidence that MPI observables could not be successfully tuned within the existing model space without such mechanisms.

ATLAS and CMS tuning of Pythia continued, with ATLAS's A1, A2 and AU2 tunes being heavily used in the Run 1 simulation leading up to the Higgs boson discovery, while CMS' "Z"-tune variants on the AMBT series were used for the same purpose on that experiment. Each experiment focused on its own growing collection of soft-QCD data analyses as the LHC energy increased. At the end of LHC Run 1, Skands and collaborators provided the "Monash" global tunes of Pythia 8,[436] which ATLAS modified into the "A14" tunes for use in Run 2 modeling, incorporating high-$p_\perp$ Z and $t\bar{t}$ observables into the fit to serve the needs of BSM searches.[565] CMS, meanwhile, constructed its own Run 2 series, the CUET and CDPST tunes,[152] the latter of which is unique in being tuned to hard double-partonic scattering data. At the time of writing, these experiments' and authors' tunes of Herwig++ and SHERPA are the most widespread general purpose tunes at the LHC.

Recently most LHC tuning effort has been focused on configurations most suitable for use with matching and merging event generators in which the parton

---

[d]The Prof0-pT tune in fact provided the first final-state tune of the Pythia 6 $p_\perp$-ordered shower, which had been previously used with the virtuality-ordered settings.

shower is interleaved with collections of higher-order matrix elements. The results have been specialist tunes such as ATLAS' AZ and AZNLO[566] (specifically for description of the Z $p_\perp$, to be used in W-mass measurement) and ATTBAR.[567] The LHC split between tunes for minimum-bias data description and underlying-event description has also still to be resolved, perhaps by inclusion of more advanced diffractive physics models although efforts along those lines have yet to prove fully satisfying Fig. 5.

4. Tuning Uncertainties

Producing optimal fits of MC models to data is valuable, but not the whole story. In particular for experimentalists' usage of these simulations, it is crucial that the *uncertainties* in a tuned model also be quantified. This permits, in varying degrees of sophistication, numerical treatment of modeling uncertainties as nuisance parameters in fits of physics both from the Standard Model and from beyond it. Estimates of tuning uncertainty also proved valuable in the run-up to LHC operations, when tune fits to Tevatron and other low-energy data permitted a quantitative estimate of the range of underlying event activity to be expected at the new 13 TeV collider — this extrapolation of uncertainty is visible in Fig. 6(a).

Simple uncertainty estimates can be produced in various *ad hoc* ways. First, an approach analogous to scale choices in perturbative calculation: simply pick some parameter variations that seem "reasonable" — typically factors of two — and release them as variation tunes. A step up in sophistication is to make such changes to key parameters, e.g., enforcing more or less initial-state radiation, and then re-tuning the remaining $(P - 1)$-parameter system to infer how much other parameters can compensate for the forced move away from the global optimum. As with scale-setting, there is a degree of artistry to this: parameter changes which result in very large or very small changes to observables may be judged as "unreasonable" and be modified accordingly.

The logical conclusion of this is to decide that the ultimate arbiter of how large a parameter change should be is that it produces a model variation comparable to the measured experimental uncertainty on the observables, such that the union of all parameter variations envelopes all or at least most data uncertainties. This approach is the philosophy adopted by several recent PROFESSOR-based tunes, such as the A14 series,[565] which provide "eigentune" variations to complement the global fit — an example is shown in Fig. 6(b).

The additional detail in eigentune construction is that there are an infinity of ways in which to make model variations "cover" all data, so we prefer a set of variations which are maximally decoupled from one another. This can be achieved by use of a second-order approximation to the χ^2 valley around the global optimum: in general this will be an ellipsoid in the parameter space, and the principal basis of this ellipsoid defines $2P$ principal vectors along which to make parameter variations.

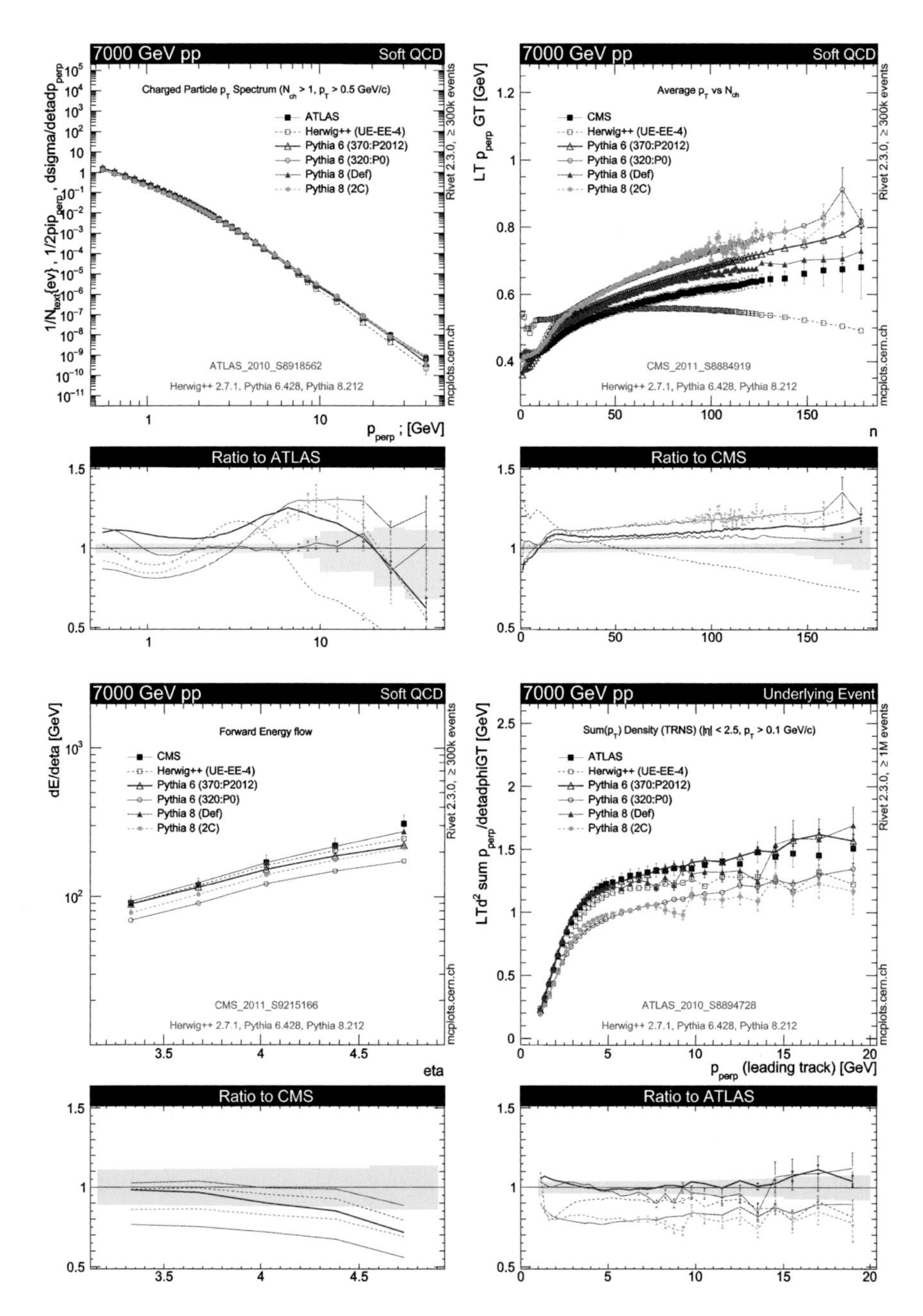

Fig. 5 Performance of MC tunes on MPI-sensitive observables, showing Pythia 6 (two "Perugia" tunes), Pythia 8 (the author Tune 2C and default Monash tune), and Herwig++ (UEEE-4 tune). The Perugia 0 and 2C tunes are based on pre-LHC Tevatron data, and the others include early LHC data. The top row shows LHC minimum-bias $p_\perp$ spectrum and $\langle p_\perp \rangle$ vs. $N_{\rm ch}$ observables, and the bottom row shows tune performance on minimum-bias energy flow (left), and underlying event $\sum p_\perp$ observables. The inclusion of LHC data in all cases improves the data description.

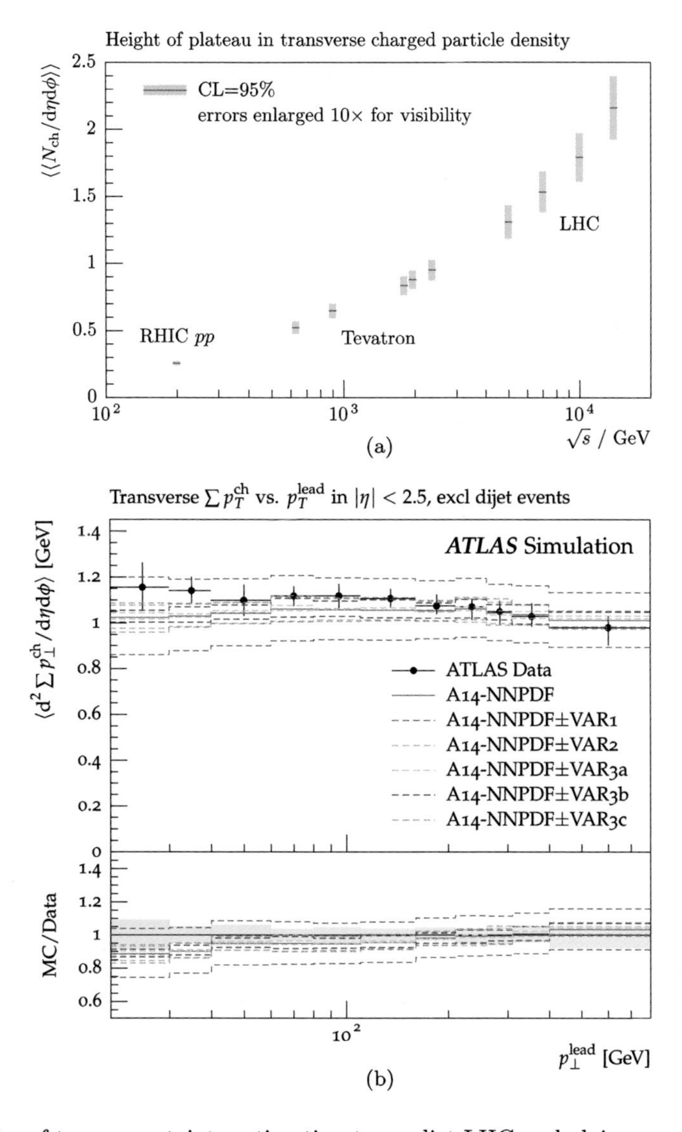

Fig. 6 Use of tune uncertainty estimation to predict LHC underlying event activity.

These vectors are obtained as the eigenbasis of the covariance matrix computed at the global tune point.

If the test statistic were truly distributed according to the χ^2 distribution, the required deviation could be calculated analytically from the $\chi^2(k, x)$ distribution for degree of freedom $k = N_{\text{bins}} - P$, by making pairs of deviations along the P eigenvectors until the χ^2 corresponding to a p-value of, e.g., 1σ is found: for a perfect χ^2 statistic and a best-fit value of $\chi^2_{\text{best}} \sim k$, this should require a $\Delta\chi^2$ shift of $\sim 2k/3$. But in practice we find that such a construction fails the "reasonableness test": the resulting variations are far too small. Empirically, the typical distribution in global LHC fits has a much larger mean than expected for the number of degrees of freedom, and a narrow spread incompatible with the χ^2 distribution's relation between mean and variance. Instead the strategy of producing empirical model variations which cover the experimental uncertainties has been adopted in tune sets such as A14, requiring a $\Delta\chi^2 \sim N/2$ to produce experimentally useful systematic variations. Other, less empirical, approaches are still being explored.

It is thought that this deviation from statistical expectation largely stems from the fact that we do not yet have models capable of reproducing all data observables: "the truth" is not contained in the model space. In addition, the data available so far for MPI tuning have not included detailed bin-to-bin correlations from systematic uncertainties, which in principle could be eliminated by a nuisance parameter fit — this further distorts the goodness-of-fit measure away from the χ^2 distribution. Finally, there is the ever-present risk of underestimated experimental uncertainties. There is hence still potential to improve tuning methodology, by construction of more robustly motivated systematic variations and reducing tuning uncertainties by fully correlated likelihood construction across multiple measurements.

In addition to the eigentune method described here, a full assessment of tuning uncertainties should coherently encompass statistical uncertainties in the PROFESSOR parameterization construction (estimated by making many semi-independent parameterizations, from subsets of the available MC runs), and correlations in χ^2 construction. Accurate treatment of these effects may eventually permit tune uncertainties to be treated on the same statistical level as those of parton density function fits.

5. Outlook

Tuning has proven to be an important activity in the development of MC models of initial-state QCD at the LHC, both by providing the experiments with unprecedentedly well-honed simulations of collision events (including pile-up), but also by providing a mechanism by which to unambiguously identify when a model's limitations are fundamental. At the same time, the development of tuning machinery for the LHC has provided ways to quantitatively estimate model and tune uncertainties, and in principle to reduce them — although the statistical foundation still requires development.

Technically, the PROFESSOR framework has seen recent advances, originally developed for BSM physics studies but readily applied to QCD MC tuning: the most obvious of these are the inclusion of non-polynomial functional forms such as neural nets, support vector machines, and methods based on decision trees. Work on using Gaussian processes, along with more refined statistical testing of parameterization fidelity, offer the possibility of yet more accurate MC parameterization for use in fitting. In parallel, a serial Markov chain approach to tuning, based on Bayesian parameter optimization has been developed and appears interesting, if yet unproven on the large-scale problem of initial-state QCD tuning where the MC runs are very computationally expensive.[568]

The most painful price paid for the increasing LHC demands of simulation accuracy has been the proliferation of tunes specific to process types: underlying event vs. inclusive minimum-bias, or QCD-singlet vs. colored hard-processes. Such fragmentation is undesirable because it implies a lack of predictivity in the models: if we cannot trust an MPI+shower model to simultaneously describe minimum-bias and underlying event, how confident can we be about its extrapolation to more rarified regions of phase-space? The resolution of this problem, and the coherent integration of diffractive processes and hence the connection between fiducial and total inelastic scattering cross-sections, must be the main challenges for development and tuning of MC models in the coming years. While technology has helped the development of tunes through the early phase of the LHC, in the end it must be coupled to physics insights to achieve the goal of truly comprehensive description of hadronic initial-state interactions.

Chapter 14

Multiparton Interactions, Small-x Processes and Diffraction

F. Hautmann* and H. Jung†

*Rutherford Appleton Laboratory and University of Oxford,
Elementaire Deeltjes Fysica, Universiteit Antwerpen, Belgium

†Deutsches Elektronen-Synchroton DESY,
Notkestraße 85, 22607 Hamburg, Germany

The connection between multiparton interaction, diffractive processes and saturation effects is discussed. The relation of the rise of the gluon density at small longitudinal momentum fractions x with the occurrence of saturation, diffraction and multi-parton interaction is being studied both experimentally and theoretically. We illustrate key ideas underlying recent progress and stress the role of different theoretical approaches to small-x QCD evolution in investigations of multiparton interactions.

1. Introduction

The contribution of multi-parton interactions (MPIs) to high-energy hadronic collisions has been considered since the early days of the QCD parton model.[6,30,32,163,339,341,569] In the absence of a first-principle systematic approach to go beyond single parton interaction in the framework of QCD factorization formulas, progress on MPI has since been driven mainly by Monte Carlo modeling[35,36,365,487] — see the recent comprehensive overview[280] of MPI developments from the standpoint of the Pythia Monte Carlo event generator. Within this context, experimental signals of MPI have emerged from comparison of Monte Carlo simulations with collider measurements at the Sp$\bar{p}$S, HERA, Tevatron, LHC for production of multi-jets, multi-leptons, photons, heavy flavors.[71,99,100,102,147,148,174,570]

The relevance of MPI for LHC phenomenology[88,89,198,268] has spurred efforts in the last few years to investigate the theoretical basis of multiple parton scattering from the point of view of perturbative QCD and factorization.[3,5,9,10,14,18,21–23,67,82,84,85] Besides, methods have been suggested for

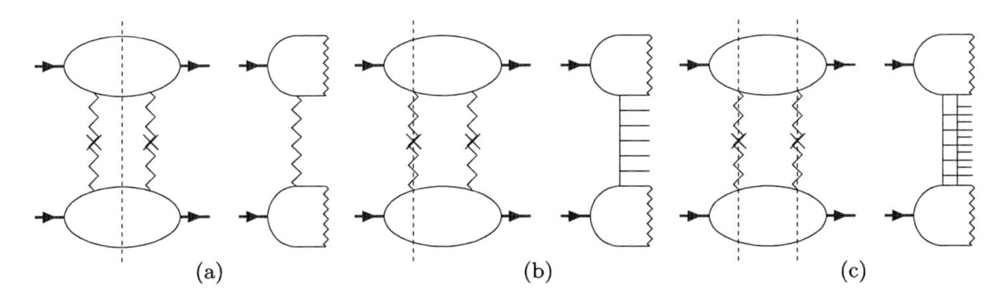

Fig. 1 Sketch of MPIs, indicating the different ways to cut the two-pomeron diagrams: (a) diffraction (b) single multiplicity (saturation) and (c) double multiplicity (multi-parton interaction). Figure reprinted with permission from Ref. 576. Copyright 2005 by Springer Nature.

estimating the ratio of MPI to single parton scattering contributions from data.[43,44,75,87,90,91,95,96,110,112,175,177,205,248,249,267,274,275,571–574]

The region of small longitudinal momentum fractions x plays a particularly important role in the context of MPI[575] because with decreasing x parton densities grow, and with high parton densities the probability for significant contributions beyond single parton interaction increases. MPI may be expected to affect the detailed structure of the exclusive components of final states even when inclusive cross-sections are not influenced. Indeed, MPI signals are sought for experimentally in multi-differential cross-sections and final-state correlations. Therefore, the exclusive structure of multi-particle final states associated with small-x processes is particularly crucial to the discussion of MPI.

MPI are also naturally connected with aspects of small-x physics such as diffraction and saturation. It is instructive to think of this from the viewpoint of the AGK cutting rules[38,576–578] in the Regge picture of hadronic interactions. Figure 1 provides a schematic illustration of the relationship between diffraction, saturation and multiparton interaction in terms of Regge theory cut diagrams. The graph in Fig. 1(a) depicts a diffractive cut, while the graphs in Figs. 1(b) and 1(c), corresponding to different cuts, depict respectively single-multiplicity interactions with saturation corrections and double-multiplicity interactions. The AGK rules[38] connect the different processes in Fig. 1.[576] This connection also has an analogue in the partonic Monte Carlo models for MPI.[35,487]

In this chapter we give a concise account of the role of small-x processes in the physics of multiparton interactions. We concentrate on general concepts rather than describing specific results, with a view to pointing the reader to broad areas of interplay of small-x dynamics and MPI. Section 2 discusses theoretical approaches to the evolution of small-x final states and their potential impact on MPI. Sections 3–5 address different aspects of small-x physics relevant to MPI: diffraction, saturation and multi-jet production. We summarize in Section 6.

2. MPI and the Evolution of Small-x Final States

In general, experimental measurements do not allow one to observe multiple interactions explicitly. What can be observed in the experiment are particles and jets, distributed in phase space, and what can be measured are correlations among them as well as multiplicities. Such correlations and multiplicities of particles and jets can then be interpret within different theoretical frameworks.

MPI become increasingly important with energy as parton densities grow with decreasing x. Even though they may not influence the inclusive rates for hard processes with a large p_T momentum transfer, MPI can contribute significantly to highly differential cross-sections, sensitive to the detailed distribution of multi-particle final states produced by parton evolution.

The evolution of parton cascades based on collinear factorization and DGLAP evolution,[392–394,579] implemented in Monte Carlo event generators such as Pythia,[113] Herwig,[580] Sherpa[157] (called *DGLAP shower*[a] in the following), is known to describe measurements well over a large range of observables. However, when longitudinal momentum fractions x become small and parton densities increase new effects are expected.

In fact, new QCD dynamics is known to arise when trying to push the parton evolution picture to higher and higher energies $\sqrt{s}$. On one hand, soft-gluon emission currents[582,583] are modified by terms that depend on the total transverse momentum transmitted down the initial-state parton decay chain.[390] Correspondingly, high-energy factorization formulas apply which are valid at fixed transverse momentum.[584–587] On the other hand, the structure of virtual corrections at high energy implies, besides Sudakov form factors, transverse-momentum dependent (TMD) — but universal — splitting functions and new "non-Sudakov" form factors,[390,391,588] which are necessary to take into account soft-gluon coherence not only for collinear-ordered emissions but also in the non-ordered region that opens up at high $\sqrt{s}/p_T$. These finite TMD corrections to parton branching are implemented in the CCFM evolution equations,[390,391,589,590] and, in the high-energy factorization framework, are found to have important implications for multiplicity distributions and the structure of angular correlations in final states with high multiplicity.[591–594] The CCFM evolution equations may be thought of as forming a bridge between the small-angle DGLAP[392–394,579] and high-energy BFKL[171,172,595] regimes.

In phenomenological analyses which perform comparisons of experimental measurements for multi-particle final states and correlations with Monte Carlo calculations based on *DGLAP shower* event generators such as Pythia, Herwig, Sherpa, it is found that MPI are needed to describe measurements such as soft

---

[a]This terminology is used for brevity. It is a misnomer though, as DGLAP is an inclusive equation. See also note at the end of this section, and Ref. 581 for a related discussion.

particle spectra in minimum bias events, the underlying events in jet production as well as (de-)correlations in multi-jet events, including 4-jet, $b\bar{b}+jj$, $W+j$ events.

However, in calculations based on high-energy factorization[584,585] and CCFM[390,391,589,590] (or BFKL[171,172,595]) transverse momentum-dependent parton densities (unintegrated parton density functions uPDFs), the correlation between partons and particles in the final state is different from those predicted by the *DGLAP shower*, since finite transverse momenta in the initial state are included from the beginning. The small-x behavior of the (unintegrated) parton densities obtained from CCFM evolution (e.g., Ref. 596) is different compared to the one from DGLAP distributions, and therefore the amount of multiple partonic interactions might also be different. Multi-parton radiation in CCFM evolution is allowed in an angular-ordered region of phase space, which is determined by small-x gluon coherence.[592] This makes a new scenario possible, in which the higher transverse momenta in the parton cascade, compared to a Monte Carlo simulation based on *DGLAP shower*, can produce final states similar to what is obtained from MPI.

The Monte Carlo event generator CASCADE,[597,598] based on CCFM uPDFs and high-energy factorization, includes initial state parton showering according to the CCFM evolution equation (called *CCFM shower*[b] in the following) as well as final state parton shower and hadronization. Studies[599–601] of jet production at high rapidity based on simulations with CASCADE (without MPI) show that the energy flow outside the jets is significantly larger than what is obtained from Pythia without MPI. In Ref. 602, the relation between Monte Carlo simulations using MPI and high-energy factorization has been studied using mini-jets. It was found that jet distributions are similar in both approaches, suggesting that at least part of the effects that are attributed to MPI when comparing Monte Carlo calculations based on *DGLAP shower* with data are already contained in the single scattering when using uPDFs and high-energy factorization.

3. Diffractive Dissociation and MPI

In the scattering matrix formalism,[603] hadronic diffraction is thought of as arising from fluctuations in the scattering amplitude. Regge theory incorporates high-mass diffraction through pomeron exchange with triple (and multiple) pomeron couplings.[604–606] Both these ideas were given partonic interpretations, respectively in Refs. 607 and 418. The partonic interpretation of hard diffraction leads to the notion of diffractive parton densities.

Diffractive processes in deep inelastic scattering (DIS) have been measured in great detail at HERA. Based on the factorization theorem for diffractive

[b]Unlike DGLAP, CCFM is an exclusive equation. The terminology of *DGLAP shower* and *CCFM shower* is somewhat misleading. Strictly speaking, only the latter is defined.

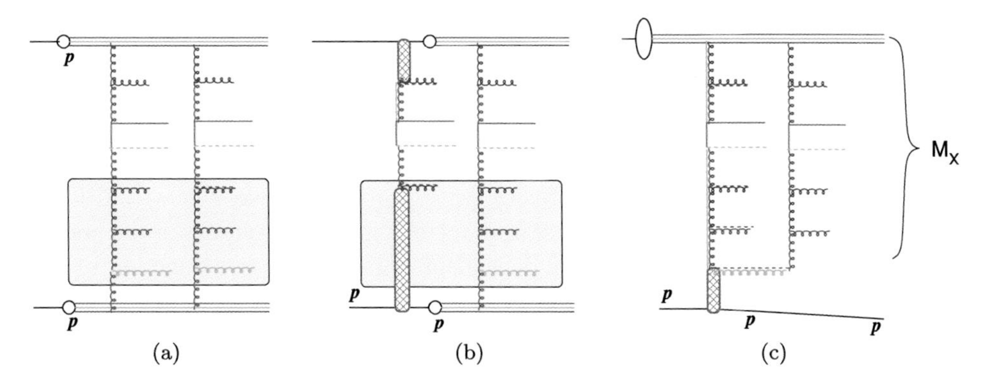

Fig. 2 Left: Non-diffractive multiparton interactions. Middle: Sketch of a diffractive process together with a non-diffractive process, destroying the rapidity gap. Right: multiparton interactions inside the dissociative system.

DIS,[125,608–611] diffractive parton densities[612] have been obtained. However, using these diffractive parton densities diffractive jet production in hadron–hadron collisions is predicted to be of the order of 10 times larger than the measurements.[424,425,613,614] Therefore, whilst deep inelastic diffraction can be understood in terms of color transparency (see Refs. 615–619 for structure functions and Refs. 620 and 621 for jet and charm production), new (absorptive) effects occur in hadron–hadron diffraction. In most of the current phenomenological models, these are embodied in gap suppression factors, or gap survival probabilities.

Multiparton interactions in addition to the diffractive process can destroy the experimental signature of diffractive interactions, i.e., a region in rapidity which is devoid of energy deposition (rapidity gap, see Fig. 2(b)). Estimates on the probability of multiple interactions lead to a suppression factor which is in the correct order of magnitude. Recently a relation between multiple parton scattering and hard diffractive processes has been implemented in Pythia.[426] This model, based on diffractive parton densities (as obtained from HERA) together with MPI, allows one to predict hard diffractive processes in hadron–hadron collisions without introducing artificially gap suppression factors. It is very encouraging that predictions coming from this new model[426] follow the trend of the measurements of hard diffractive dijet production at the LHC.

In a process as shown in Fig. 2(a), the density of partons in the forward or backward regions (indicated by the yellow box) is different from those of the processes shown in Fig. 2(b), because one interaction chain allows no particle radiation, while the other does. Experimentally it is difficult to identify such processes, but they play a role when adjusting parameters to describe the underlying event and multiparton interactions. It will be important that in forthcoming Monte Carlo tuning diffractive processes are included, since this might lead to different parameters describing particle multiplicity.

While multiparton interactions can destroy a rapidity gap, multiparton interactions can also occur within the diffractive dissociation system (Fig. 2(c)), if the diffractive system has a large enough mass M_X. At MPI@LHC 2017 the effect of MPI inside the diffractive system M_X was shown for the first time[622] in a measurement of $dn/d\eta$ for event samples enhanced with single diffractive dissociation. Investigating such processes (with hard diffractive events), MPI can be studied in a very new environment and one can investigate in detail the energy dependence of MPI (and its energy-dependent parameters) as a function of the centre-of-mass energy of the pomeron–proton system M_X (indicated in Fig. 2(c)).

A further mechanism relating inelastic diffraction and MPI is discussed in Refs. 95, 96 and Chap. 5, based on two-pomeron contributions to the generalized double parton distribution function.[85]

4. Saturation Effects and MPI

In the context of total hadron–hadron cross-sections[623] and their rise with energy,[197] unitarity constraints are given, at fixed impact parameter, by the "black disc" limit of the scattering matrix.[624,625] In this limit, the elastic cross-section equals half the total cross-section. Such saturation effects are important, for instance, for the observed behavior of the diffractive cross-section.[626] In the Regge theory picture, saturation implies the breakdown of the simple Regge pole behavior and the onset of multiple pomeron exchange.[627]

Unitarity arguments are also used to treat saturation of parton densities in the case of hard processes in the limit of high energies (or large nuclei),[235,366,628–630] with corresponding nonlinear evolution equations.[631–635] This saturation formalism is given in terms of color-dipole degrees of freedom rather than parton degrees of freedom. (See also Refs. 115 and 636 for an alternative, dipole-based perspective on saturation.) It has limited applicability in the region of high transferred momenta well above the saturation scale. See, e.g., discussion in Refs. 637 and 638. Therefore, even though the saturation formalism is constructed in the case of hard processes, it is generally not relevant to jet physics at high p_T.

On the other hand, for energies high enough the scale at which saturation effects become important can rise into the region of transverse momenta p_T large compared to $\Lambda_{\rm QCD}$, say, $p_T \sim \mathcal{O}(5 \text{ GeV})$ — potentially, a region of weak-coupling but non-perturbative dynamics, sitting just above the saturation scale. This was the scenario proposed in Ref. 81. It was emphasized in Ref. 81 that in this scenario a connection arises between saturation and MPI. The connection could be seen in terms of two aspects of the Monte Carlo model:[35,365,487] the contribution of multiple interactions, and the screening of the jet production cross-section.

One of the arguments leading to the introduction of MPI in Monte Carlo event generators[35,36,487] was the observation that the $2 \to 2$ partonic cross-section integrated above $p_{T\,\min}$ over the transverse momentum of the final partons can become larger than the inelastic pp cross-section for values of $p_{T\,\min} \gg \Lambda_{\rm QCD}$. Since the $2 \to 2$ cross-section is a jet — and not an event — cross-section, the ratio $\sigma(p_T > p_{T\,\min})/\sigma_{\rm inel}$ was interpreted as the average number of interactions per event.

The $2 \to 2$ partonic cross-section for small transverse momenta is $d\sigma/dp_T \sim \alpha_s^2/p_T^4$, and diverges for $p_T \to 0$. This behavior is related to the non-perturbative nature of the process for small p_T and is regulated in models based on collinear factorization by introducing an additional cutoff parameter, such that $d\sigma/dp_T \sim \alpha_s^2/p_T^4 \to \alpha_s^2(p_T^2 + p_{T\,0}^2)/(p_T^2 + p_{T\,0}^2)^2$, with an arbitrary but finite parameter $p_{T\,0}$. This parameter $p_{T\,0}$ separates the perturbative from the non-perturbative regions and has to be determined from fits to experimental data. At LHC energies, this parameter was determined to be of the order of 2–3 GeV, far above $\Lambda_{\rm QCD}$. In general, $p_{T\,0}$ depends on the centre-of-mass energy $\sqrt{s}$.

In Ref. 81, it was proposed to measure the leading mini-jet or leading track cross-section for a direct investigation of the behavior of the cross-section from the perturbative to the non-perturbative region at small p_T. In Ref. 639, a measurement of leading track and leading mini-jet cross-sections was reported which shows directly how the cross-section is saturated at small p_T. The observation of the turnover of the cross-section from a $\sim 1/p_T^4$ behavior to a constant value, independent of Monte Carlo modeling, is an important result.

For interpretation of this result, see discussions in Refs. 134 and 135, which examine impact-parameter unitarity constraints for minijets, and in Refs. 640–645, which use nonlinear evolution equations for parton distributions at fixed transverse momentum.

5. MPI in Multi-jet Production

Mueller–Navelet jets,[646] having comparable transverse momenta and large rapidity separation (Fig. 3(a)), have long been investigated as a probe of BFKL[171,172,595] dynamics. While experimental measurements of forward–backward jets at the LHC[647–650] do not point to any striking BFKL signature, and Pythia Monte Carlo simulations compare well with data, interest has grown in examining MPI contributions (included in the Pythia simulations) to such processes (Fig. 3(b)). Studies of the simplest term from double parton scattering in a collinear framework[651] and a transverse momentum-dependent framework[652] suggest that the latter leads to smaller double-scattering contribution.

A simulation, including a full treatment of the final state, based on the *CCFM shower* will be of great importance. The CASCADE Monte Carlo event generator (based on CCFM parton showers) is not designed at present to treat this process, since this process involves flavor channels not yet included in the Monte Carlo. Very recently a first step towards a determination of k_T-dependent parton densities for

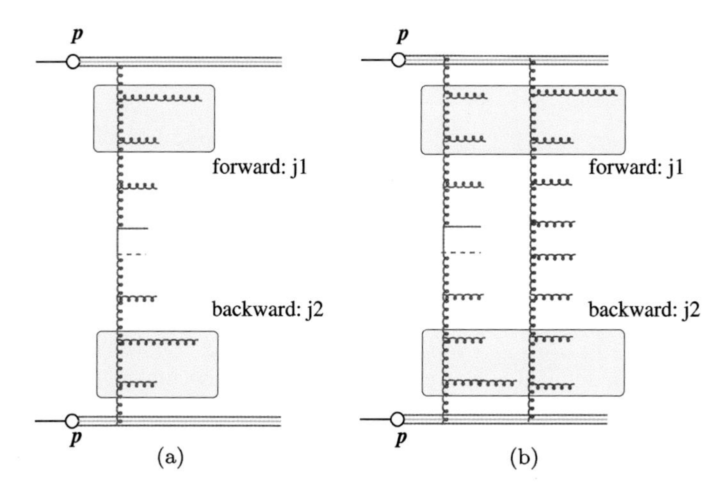

Fig. 3 (a) Forward–backward jet production. (b) Forward–backward jet production from MPI.

all flavors has been reported in Refs. 581 and 653, which could be used to obtain a more complete simulation of the final state in CASCADE.

At present, experimental signals for hard double parton scattering come from studies of correlations in 4-jet, $b\bar{b}$+2-jet and W/Z+jet events. In the case of double parton scattering, the jets or $b\bar{b}$ pair or W/Z bosons which come from one interaction will be decorrelated from those coming from the second interaction. However, such a decorrelation could also originate from significant transverse momenta of the initial-state partons within a single interaction. This can be studied by using multi-leg matrix element calculations (for example at least $2 \to 6$ for the 4-jet case in order to allow initial transverse momenta of the $2 \to 4$ subprocess) or by using, in the high-energy factorization framework,[584,585] off-shell matrix elements for $2 \to 4$ together with uPDFs. One such study has been performed in Ref. 151 and an interesting observation was made: the prediction from a single chain interaction for the azimuthal angular correlation ΔS between two jet pairs is already very close to the experimental measurements of Ref. 100.

An important point in interpreting this result, however, concerns the role of parton showers. In Fig. 4 a schematic picture is given for a generic hard process described by high-energy factorization. Since the transverse momenta of the incoming partons can have any kinematically allowed value, and especially the k_T can be larger than the p_T of the partons of the matrix element process (indicated by ME in Fig. 4), the simulation of the recoils (in form of an explicit parton shower) cannot be neglected as they might significantly contribute. This is different from the case of simulations based on a *DGLAP shower*, where the ME-partons constitute the partons with the highest p_T. Predictions for jets based on approaches including the initial-state transverse momentum should take into account the recoils besides the uPDFs. Thus, before a conclusion on the size of MPI

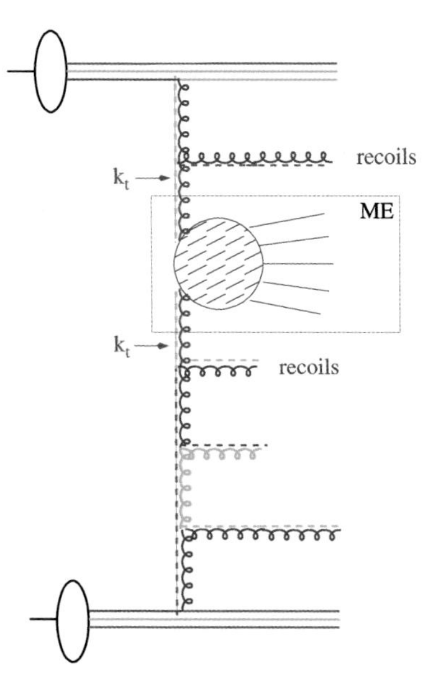

Fig. 4 Sketch for recoil treatment in high-energy factorization.

in the high-energy factorized calculation can be drawn, a full simulation including parton showers is needed.

6. Summary

At high energies, when the density of partons is large, multiple partonic interactions can occur with non-negligible rates. The relation of the rise of the parton density at small longitudinal momentum fractions x and the occurrence of saturation, diffraction and multi-parton interaction is being studied both experimentally and theoretically.

MPI affect primarily highly differential cross-sections and the detailed distribution of multi-particle final states produced by parton evolution. Theoretical predictions for MPI observables are thus sensitive to the theory of final states associated with small x. We have discussed the role of different approaches to small-x QCD evolution in the interpretation of measurements in terms of multi-parton interaction.

Small-x aspects of MPI influence both soft-interaction and hard-interaction processes. We have illustrated this by discussing their effects in diffraction, saturation, multi-jet production.

Acknowledgments

We are grateful to the editors for careful reading of the manuscript and advice. We thank M. Strikman and J. Gaunt for useful discussions.

Chapter 15

High Multiplicity Collisions

Michele Floris*,† and Wei Li‡

*CERN, CH-1211 Geneva 23, Switzerland

†University of Derby, Kedleston Rd, Derby DE22 1GB, UK

‡Rice University, 6100 Main St, MS-315, Houston, TX 77005, USA

This chapter summarizes measurements of (identified) hadrons spectra and correlations in high-multiplicity pp events at the Large Hadron Collider (LHC). A particular emphasis is put on the similarities with corresponding results from heavy-ion collisions and on the implications for the understanding of hadronic interactions at high energy.

1. Introduction

The study of high-multiplicity events in pp and pA collisions has attracted wide interests from the high-energy and nuclear physics communities over the past years. In pp collisions, the selection of events with large hadronic final-state multiplicities biases the sample towards a larger average number of Multiple Parton Interactions (MPIs), as confirmed with the measurement of the number of independent scattering centers ("uncorrelated seeds",[523,654] see also Chapter 12). Moreover, recent measurements at the LHC revealed striking similarities between high-multiplicity events in small (pp, pA) and large (AA) colliding systems, for what concerns phenomena that were considered to be hallmarks of a strongly-interacting, fluid-like QCD medium ("quark-gluon plasma") formed in heavy-ion collisions. A central aspect for the description of heavy-ion data was the observation of "collectivity", that is long-range correlations among particles produced over the full phase space (in azimuthal angle and rapidity) in the event. This supported the interpretation in terms of the creation of a locally thermalized, strongly interacting system, which can be described with relativistic viscous hydrodynamics. Establishing if the same kind of collectivity is present in small systems was one of the main goals of the recent experimental efforts at the LHC.

These observations in pp and pA collisions clearly suggest that a high-multiplicity event is not a trivial overlap of independent parton–parton collisions. However, the precise physical origin of these observations has not been completely understood yet. Extensive experimental and theoretical research programs are ongoing, aimed at addressing some crucial questions, e.g.: are the observed collective effects a consequence of strong final-state rescatterings in the limit of high density MPIs? Is there a unified description of the underlying physics in high-multiplicity pp, pA and AA collisions? It is becoming clear that traditional parton shower Monte Carlo models are incomplete, as they do not contain any mechanism that can reproduce these recent observations. At the same time, models of heavy-ion collisions typically employ a macroscopic (thermodynamic and hydrodynamic) approach, and thus lack a description of the microscopic degrees of freedom of the system. Studying the evolution from a small and dilute (pp) to a large and dense (Pb–Pb) systems is the prime way to develop a unified understanding of all hadronic collisions.

In this chapter, recent experimental results in high-multiplicity pp and p–Pb collisions at the LHC are reviewed. Measurements of multi-particle correlations and transverse momentum distributions of identified particles are discussed, which have been proven to be powerful tools in unraveling the detailed underlying dynamics of particle production in hadronic collisions. The main focus of the discussion is on pp collisions, but p–Pb results are presented whenever relevant.

2. Event Multiplicity Classification

Before discussing the experimental results, a few comments on the choice of the multiplicity estimators in classifying the events are made in this section. In pp and pA collisions, the number of elementary (parton–parton or nucleon–nucleon) interactions is small, so that the overall event activity is easily affected by fluctuations in the collision and particle production processes. This is in contrast with heavy-ion collisions, where the multiplicity distribution has a strong connection to the collision geometry: events producing larger multiplicities correspond to those with smaller impact parameters (more central).[655] As a consequence, multiplicity distributions in regions separated by several units of pseudorapidity are well correlated. On the other hand, a high-multiplicity pp collision typically shows an increase in the event activity localized in pseudorapidity (η), and the correlation of multiplicities in regions separated by more than ∼0.5 units in η is loose. Since different physics mechanisms can produce multiplicity correlations over long range or short range in η, the choice of the multiplicity estimator may affect the final observables. Naively, the multiplicity of a pp collision can change because of a larger number of MPIs, or as a consequence of fluctuations, both in the hardness of the parton scattering process and in the parton fragmentation. It is thus clear that different multiplicity estimators may impose different biases on these effects, with important consequences on the qualitative trend of the experimental results.

Two main strategies have been employed in the literature:

(1) Separate η regions for the multiplicity estimator and the observable of interest (typically a forward multiplicity estimator for an observable studied around mid-rapidity).
(2) The same η region for the multiplicity estimator and the observable of interest.

The first strategy has the advantage that fluctuations in the multiplicity estimator are averaged out, since they are not correlated with the observable of interest. However, the multiplicity reach is limited because of the weak correlation between different pseudorapidity regions. The second strategy, conversely, allows increasing the reach in multiplicity, but it is more sensitive to biases stemming from fluctuations and auto-correlations.

So far, the first approach has been used in the ALICE experiment, because the relatively narrow acceptance at mid-rapidity ($|\eta| < 0.8$) makes the biases stronger for the second method. The ATLAS and CMS collaborations have used the second approach, exploiting the wide acceptance of their detectors ($|\eta| \lesssim 2.4$). It is however clear that repeating the same measurement with different estimators allows controlling for the biases induced by fluctuations and contributes to pinpoint the origin of the phenomena that are discussed in the next sections.

3. Identified Particles and Strangeness

The production of the bulk of hadrons happens via non-perturbative processes, which have been modeled in a variety of approaches, often implemented in Monte Carlo event generators (see for instance the discussion in Chapters 10 and 17–19). Measurements of identified particles are important for the validation and tuning of these models.

In the general purpose parton-shower Monte Carlo generators, hadronization is controlled by a number of free parameters (about 20), which need to be adjusted (tuned) using experimental data (Chapters 10, 17 and 19). In these models, the production of different hadron species is driven by their masses and constituent quarks. Moreover, the momentum distributions and abundances of identified particles are sensitive to the mechanisms used to describe the color flow and possible interactions between color charges produced in different partonic collisions (e.g., color reconnection, see Chapter 10). Combining measurements of low momentum (soft) light flavor identified particles with hard observables (high-momentum particles or heavy flavors) allows disentangling (semi)hard and non-perturbative contributions to particle production.

In heavy-ion collisions a macroscopic (thermodynamic) theoretical approach is frequently used to describe hadron production. Identified particle spectra have always held a central role in these studies, as a tool to constrain the thermal properties and expansion dynamics of the system.[656,657] Strange particles, in particular, were predicted to be enhanced in AA collisions (relative to pp collisions), as a consequence of the creation of a quark–gluon plasma.[658,659] This enhancement was actually observed,[660–662] even if it is no longer considered as an unambiguous signature of deconfinement. The main tool used to describe the data in this context is the "Statistical Hadronization Model" (SHM, also known as "thermal model", see Refs. 663 and 664 and references therein). Following a thermal equilibrium ansatz, the model allows computing the abundances of different hadron species with a small number of free parameters (typically 3 to 5 depending on the implementation). The most basic version assumes a grand-canonical ensemble. The relative abundance of light flavor identified hadrons have been found to be in food agreement with the grand-canonical SHM: the enhancement of strangeness production to the extent that it is compatible with full (grand-canonical) thermal equilibrium was considered to be a defining feature of heavy-ion collisions. The term "hadrochemistry" is used to refer to the (relative) abundances of different hadron species.[a] Somewhat surprisingly, the SHM has been shown to be rather successful in pp (and even e^+e^-) collisions. The only exception is the production of strange particles, which was seen to be suppressed in these small systems with respect to the thermal equilibrium expectation.[665]

The LHC opened a new era in the study of identified particles at the energy frontier, with systematic measurements of unprecedented precision, as a function of energy (from 0.9 to 13 TeV), multiplicity, and colliding system (pp, p–Pb, Pb–Pb). The excellent capabilities of the LHC detectors in terms of tracking and particle identification (PID) were instrumental to this achievement.

The measurement of the total abundance of particles, and in general the study of soft QCD processes, require the measurement of tracks at very low transverse momentum ($p_{\rm T} \sim 100$ MeV).

PID detectors are crucial for the identification of stable particles (π, K, p, and light nuclei). They also help improving the signal-over-background ratio of resonances and weakly-decaying hadrons measured through their decay products. Short-lived resonances are typically identified simply through the reconstruction of the invariant mass of their decay daughters. Weakly-decaying particles can be identified exploiting their characteristic decay topology. The latter technique requires a good resolution on the distance of closest approach of the daughter tracks to the primary collision vertex.

The ALICE experiment[666,667] was designed to have excellent PID and low-$p_{\rm T}$ tracking capabilities. It is thus very well suited for identified particles studies. It performed a wide range of measurements on light flavor hadron production,

[a] In this paper, with "particle abundance" we indicate the $p_{\rm T}$-integrated rapidity density dN/dy at $y \simeq 0$.

which include stable and electromagnetically-decaying hadrons ($\pi^{\pm}$, $K^{\pm}$, π^0, η, p, d), strange particles (Λ, K^0_S, Ξ, Ω), meson (ϕ, K^*) and baryon ($\Lambda(1520)$, Ξ^*) resonances, light nuclei (deuteron, triton and ^{3}He).[444,449,522,668–682] Results on the production of stable hadrons (π, K, p, identified via the specific energy loss in silicon) and weakly decaying strange particles (topological identification) have also been published by the CMS collaboration.[450,683–685] Contributions from ATLAS and LHCb to these studies have been limited so far to minimum-bias measurements.[686–691] We emphasize that it would be important to complement existing results with multiplicity-dependent measurements at forward rapidity, an area where the LHCb experiment has unique capabilities.

In the rest of this section, we focus on (mostly multiplicity-dependent) measurements in pp collisions, but also discuss p–Pb collisions whenever relevant. In general, results from pp and p–Pb as a function of multiplicity are strikingly similar, and smoothly approach corresponding measurements in Pb–Pb collisions. We start with a brief summary of existing minimum-bias measurements, which represent the baseline for the understanding of high multiplicity results. The bulk of this section then discusses two complementary aspects of multiplicity-dependent measurements: the evolution of total particle abundances and of p_T distributions. We conclude with a discussion on the interplay of soft and hard processes, presenting results on heavy-flavor production.

3.1. *Early minimum-bias measurements*

Measurements of spectra and multiplicity distributions of charged and identified hadrons were among the first results to be released after the start-up of the LHC. These early studies allowed testing predictions of various soft QCD models, in particular calculations from Monte Carlo event generators tuned to lower energy data. They provided important data for the subsequent model tuning.[450,668,692] The measurements showed that a reasonable description could be achieved for charged hadrons and pions, and, to some extent, for kaons and protons. Large discrepancies were reported in the data-model comparison for strange particles and heavier baryons. As an example, a comparison of Ξ and Ω baryons to the Pythia Perugia 2011 tune[434,668] is shown in Fig. 1. This tune used early LHC measurements (including the production of Λ baryons), but it under-predicts the data in Fig. 1 by a factor between 2 and 4, for the Ξ and Ω baryons respectively. The initial expectation was that further model tuning would mitigate this discrepancy, but the description of strange particles remains challenging for most of the general purpose Monte Carlo models. An exception is represented by the EPOS model (Chapter 19), which describes satisfactorily most of the existing measurements under the assumption that the system created in pp collisions undergoes a "collective hydrodynamic expansion" (see below).[693] This is a concept inherited from heavy-ion collisions. It assumes the creation of a nearly-thermalized macroscopic system, which evolves

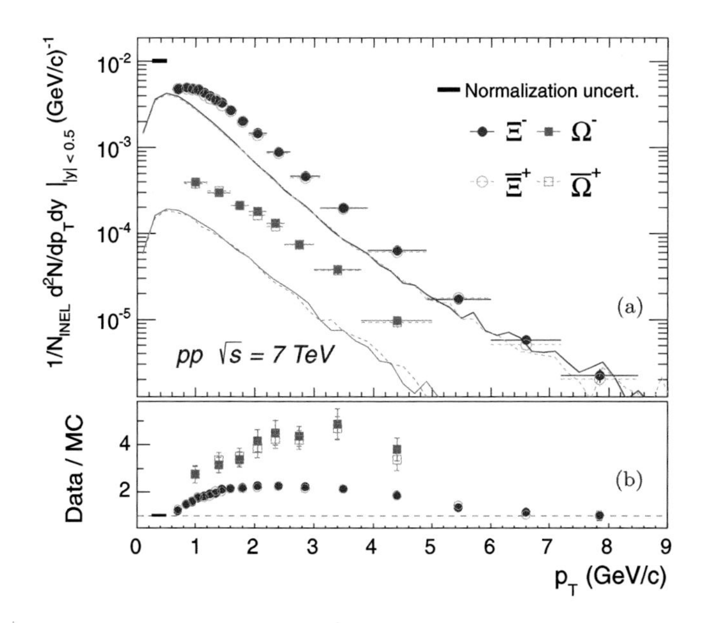

Fig. 1 p_{T}-differential yields of Ξ and Ω baryons in pp collisions at $\sqrt{s} = 7$ TeV, compared to the Pythia Perugia 2011 tune.[669]

under the effect of thermal pressure gradients, a still controversial assumption in pp collisions (see Ref. 694 for a review and Refs. 695 and 696 for some recent developments). These comparisons suggest that important physics mechanisms are missing from the standard parton-shower Monte Carlo models,[442] a conclusion also supported by recent multiplicity-dependent measurements.

3.2. *Particle abundances*

Measurements of identified particle abundances as a function of multiplicity revealed non-trivial trends, both in pp and in pA collisions. The most striking result in this respect is the measurement of an enhancement in the production of strange particles, relative to non-strange light-flavor particles.

Figure 2 shows the p_{T}-integrated yield (dN/dy) ratios of Λ, K^0_{S}, Ξ, and Ω to pions $(\pi^+ + \pi^-)$, measured in $|y| < 0.5$ as a function of the charged particle multiplicity $\langle dN_{\mathrm{ch}}/d\eta\rangle$.[443] The relative strangeness abundance increases smoothly as a function of $\langle dN_{\mathrm{ch}}/d\eta\rangle$ in pp and p–Pb collisions, and seems to tend towards the values measured in Pb–Pb collisions. The increase is more pronounced for particles with a larger strangeness content, i.e., the relative abundance of the Ω baryon increases faster than the one of the K^0_{S} or Λ particles. The same trend is seen in pp and p–Pb, despite the different collision energies, suggesting a common production mechanism only driven by the local (short-range in η) multiplicity density. It has also been suggested that strangeness production follows a universal scaling as a

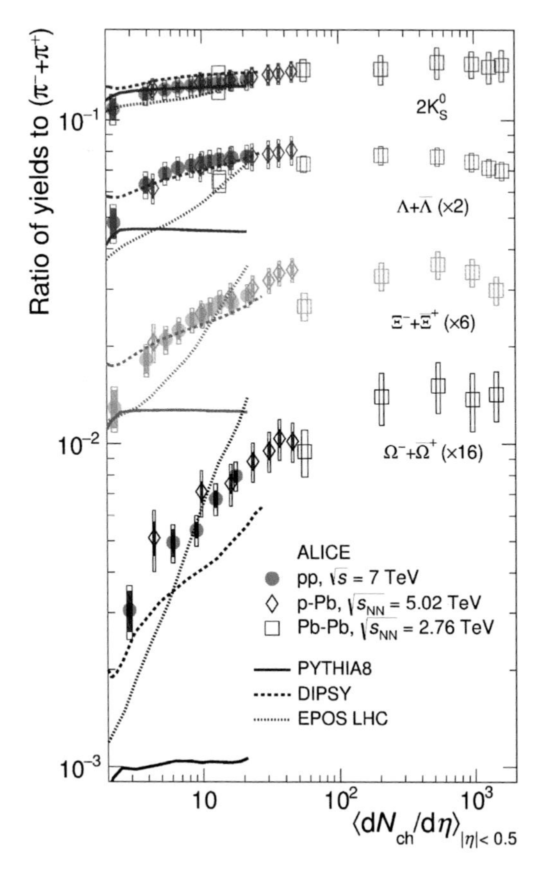

Fig. 2 p_{T}-integrated yield ratios of strange particles to pions $(\pi^+ + \pi^-)$, measured in $|y| < 0.5$ as a function of $\langle dN_{\mathrm{ch}}/d\eta\rangle$ from Ref. 443.

function of the transverse energy density for all systems.[697,698] Measurements with smaller uncertainties at higher multiplicity and at different energies within the same colliding system are needed to confirm these observations. These results are surprising for two reasons. First of all, strangeness enhancement was thought to be a defining feature of heavy ion collisions. Secondly, the most commonly used pp Monte Carlo event generators (such as Pythia) do not contain any mechanism that can lead to an enhancement of strange particles as a function of multiplicity (Chapter 10).

In the ALICE measurements, multiplicity was selected using an array of scintillator detectors at forward rapidity ($2.8 < \eta < 5.1$ and $-3.7 < \eta < -1.7$). This choice avoids trivial biases on the hadrochemistry: using the charged particle multiplicity at mid-rapidity to define the event classes would artificially enhance primary charged particles over neutral and weakly decaying particles. At the same time, each pp and p–Pb point in Fig. 2 is obtained from a set of events with a relatively wide $dN_{\mathrm{ch}}/d\eta$ distribution at mid-rapidity.

Figure 2 also shows calculations from several commonly-used Monte Carlo models. Pythia does not predict any strangeness enhancement with multiplicity. This model is based on the assumption that the fragmentation parameters of the Pythia strings do not depend on the colliding system nor on the environment of the collision (jet universality).[442] The disagreement with this data challenges that assumption and indicates that the model is not complete[442] (see also Chapter 10). The other two models depicted in the figure (EPOS-LHC and DIPSY) show a relative increase of strange particles with multiplicity, but they do not quantitatively reproduce the data, and also reveal some important qualitative differences. In the EPOS-LHC model,[409] MPIs give rise to a dense "core" and a dilute "corona". The "core" is described macroscopically as an expanding fluid (possibly a quark-gluon plasma), using relativistic hydrodynamics. At the end of the evolutions particles are produced using a statistical hadronization model.[409] Particles from the "corona" are produced via string fragmentation. In this scenario, the dynamics as a function of multiplicity comes from the interplay between the fully-equilibrated "core" and the diluted "corona". The relative abundance of Λ baryons seems to increase faster than the one of K^0_S, a feature not seen in the data. EPOS-LHC is a simplified version of the full EPOS model and employs, in particular, a parameterized description of the hydrodynamic evolution. Recent calculations obtained with the full version of the model (EPOS 3, see Chapter 19) give a satisfactory description of the existing data. Besides the improved description of the hydrodynamic evolution, an important element in these calculations are interactions between the hadrons produced in the collisions before the final decoupling (hadronic phase), which lead to sizable effects, such as a reduction in the number of baryons because of baryon–antibaryon annihilations. The effect of the hadronic phase was recently advocated to explain differences between central Pb–Pb data and the statistical model at the LHC.[699] However, the actual impact of this phase is still controversial. The study of particle ratios as a function of system size can help to constrain it, leading to a more realistic description of the late stages of hadronic collisions. In the DISPY model,[700] densely packed strings in the transverse plane are allowed to recombine, forming color ropes, and exert a transverse pressure on the neighboring strings. This idea could provide a possible microscopic interpretation of the evolution of the "core" used in EPOS. This model gives a fair description of the trend of strange particles. However, DIPSY also predicts a mild increase in the p/π ratio, in contrast with the data where a flat trend is seen as a function of multiplicity.[443] It remains to be seen if a better agreement with the data can be reached by tuning the current event generator, or if additional processes will have to be invoked (for instance, the effects in the hadronic phase). For a more complete discussion of event generators, see Chapters 17–19.

Results on π, K, p production in pp and p–Pb collisions were also published by the CMS experiment.[683,685,701] These measurements suggest a flatter trend as a function multiplicity for the ratio K/π than the one measured by the

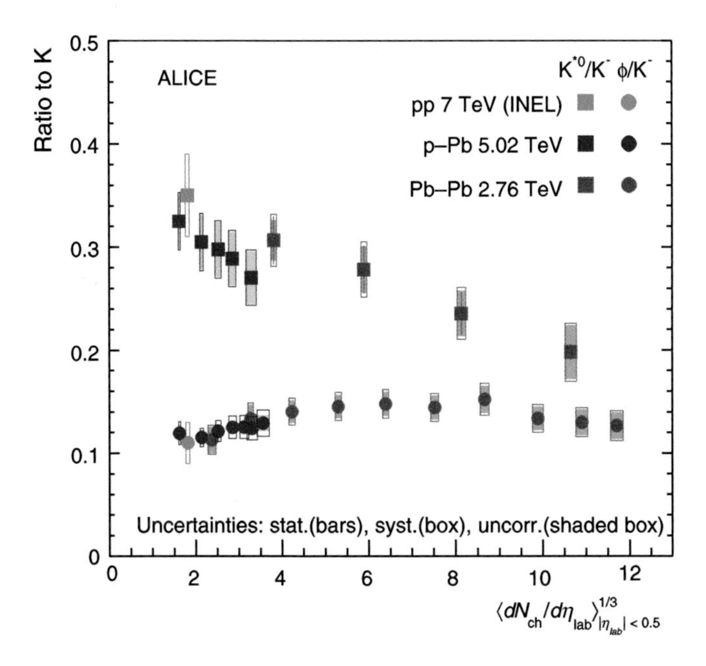

Fig. 3 Ratio of K^* and ϕ to charged K measured in the three colliding systems, as a function of the cubic root of the average charged particle density ($\langle dN_{\rm ch}/d\eta\rangle^{1/3}$).[678]

ALICE collaboration. The difference could be explained by the different choice of multiplicity estimator, which in the case of the CMS measurements is the number of tracks in $|\eta| < 2.4$.

Another intriguing observation is the reduction in the relative abundance of the K^* resonance in high multiplicity events, which was initially observed in nucleus–nucleus collisions,[670,672,678,702] and has now also been reported in p–Pb[678] as shown in Fig. 3, where the cubic root of the average charged particle density ($\langle dN_{\rm ch}/d\eta\rangle^{1/3}$) is used as a proxy for the path-length of the K^* in the medium. Preliminary ALICE results (not shown in Fig. 3) suggest that the same effect may be present in pp data.[703] This observation is interpreted in nuclear collisions as the consequence of the re-scattering of the K^* decay products in the hadronic phase during the late stages of the system evolution.[672,704,705]

The picture emerging from these recent measurements is that of a much more complex underlying dynamics of pp and pA collisions than previously believed, with sizeable initial and final state interactions, arising as a consequence of MPIs.

3.3. *Transverse momentum distributions*

The $p_{\rm T}$ distributions of light flavor hadrons are known to become harder with increasing mass and multiplicity in pp and p–Pb collisions. Indeed, the increase of $\langle p_{\rm T}\rangle$ with multiplicity was one of the key observation leading to the development of MPI models in the Monte Carlo event generators (see the discussion in Chapter 10 and references therein). In Pythia, for instance, it is understood as the consequence of MPIs and of the color reconnection mechanism, which joins different MPIs, hence increasing the $\langle p_T \rangle$ and reducing the multiplicity per partonic collision.[442,706] The $\langle p_T \rangle$ is also expected to increase with mass in the Monte Carlo event generators, because lighter particles are produced more often as decay products, and have hence a $\langle p_T \rangle$ smaller than particles produced directly in the string fragmentation process.[442]

In heavy-ion collisions the increase of $\langle p_T \rangle$ with multiplicity, more pronounced for heavier particles, has been interpreted as a consequence of the creation of a strongly-interacting system following hydrodynamic behavior, possibly as the consequence of local thermal equilibrium. A system close to thermal equilibrium, in fact, expands under the effect of thermal pressure gradients, with hadrons moving in a common velocity field. This boosts (blue-shifts) the transverse momentum distributions, in a more pronounced way for heavier particles. More central collisions reach higher initial energy densities, and expand more explosively. This effect, known as "radial flow", leads to an increase of $\langle p_T \rangle$ with mass and multiplicity. Another very non-trivial consequence of the collective expansion interpretation is the "anisotropic flow", discussed in details in Section 4. It was recently noted[706] that the color reconnection mechanism implemented in Pythia introduces a boost at the partonic level, leading to very similar trends on the p_T distributions as those expected for radial flow in the collective expansion interpretation ("flow-like effects").

Figure 4 shows a recent collection of $\langle p_T \rangle$ as a function of mass in minimum-bias pp collisions (from Ref. 444), compared to a recent Pythia 6 tune (Perugia 2011[434]) and to Sherpa.[157] The models show a weak increase of the $\langle p_T \rangle$ with mass, and under-predict the $\langle p_T \rangle$ of heavier particles by about 20%. A detailed comparison reveals large differences of the spectra with the models, as also mentioned above (Section 3.1). A similar conclusion holds when comparing the data to state-of-the-art versions of these models.[442] Assuming the formation of a small thermalized and collectively-expanding system also in minimum-bias pp collisions improves the description of the $\langle p_T \rangle$ and of particle spectra in general[693] (see also the discussion in Section 3.1).

A comparison to lower energy data is also shown in Fig. 4. A similar trend with mass, but lower overall $\langle p_T \rangle$ values, are measured at the lower energies. The $\langle p_T \rangle$ increase with collision energy could be due to an increase in mini-jet production or could be another indication of flow-like effects (see Ref. 444 and references therein).

The multiplicity dependence of the charged hadrons $\langle p_T \rangle$ is shown in Fig. 5 for all colliding systems studied at the LHC.[522] The charged hadrons used for the selection of the multiplicity class and for the estimate of $\langle p_T \rangle$ are measured in the region $|\eta| < 0.3$. The trend in pp, p–Pb, and Pb–Pb collisions is significantly

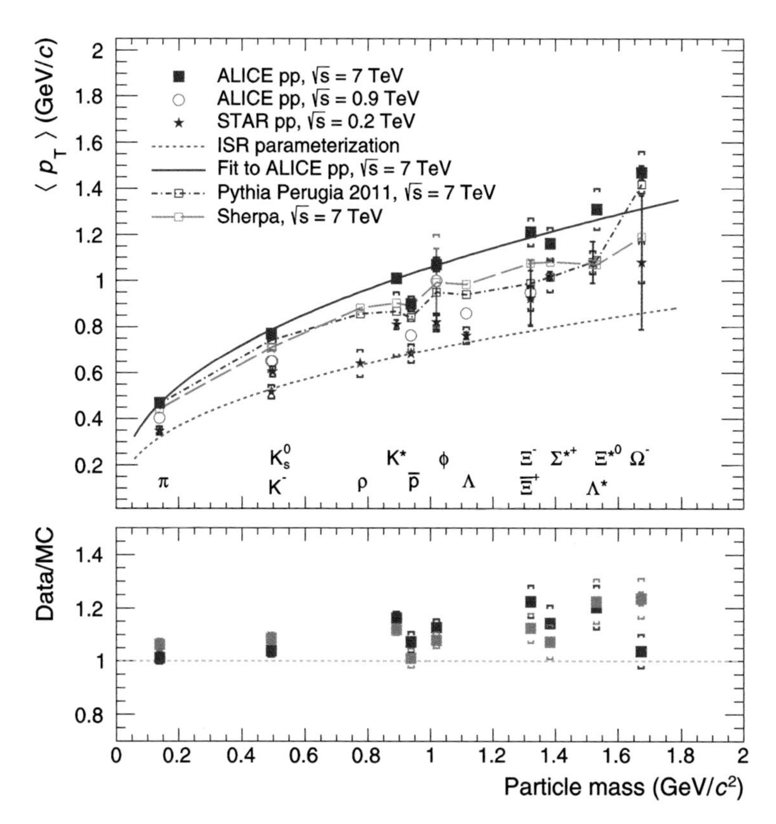

Fig. 4 Average p_{T} of various particle species as a function of mass, in pp collisions at $\sqrt{s} = 0.2$, 0.9 and 7 TeV mean pt.[444]

different. The $\langle p_{\mathrm{T}} \rangle$ increases with multiplicity faster in the small systems than in Pb–Pb collisions, where it is almost flat within the multiplicity range shown in the figure. This suggests that the main mechanism leading to a multiplicity increase in Pb–Pb collisions is the increase in the number of scattering centers (e.g., participant nucleons), without any significant bias on the hardness of individual nucleon-nucleon collision. Conversely, in pp and p–Pb collisions, the increase in MPIs seems to be accompanied by an increase in the hardness of the collisions, possibly as a consequence of the color reconnection mechanism mentioned above. The difference between colliding systems could also be partly due to biases imposed by the multiplicity selection in the small systems: selecting the multiplicity in the same (narrow) η region as the particles used to compute $\langle p_{\mathrm{T}} \rangle$ may introduce a bias towards harder partonic collisions, which fragment in a larger number of hadrons on average (see also the discussion in Section 2 and Refs. 707 and 708).

Additional information can be obtained by studying the multiplicity dependence of identified particle spectra. Figure 6 shows the $\langle p_{\mathrm{T}} \rangle$ of π, K and p as a function of multiplicity, measured by the CMS collaboration[683] in pp collisions at $\sqrt{s} = 7$ TeV.

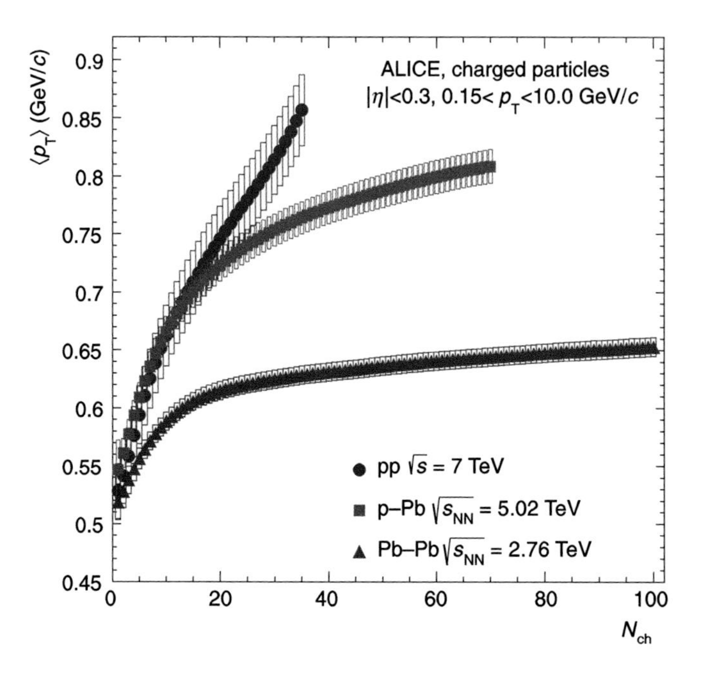

Fig. 5 Average p_T of charged particles as a function of charged particle multiplicity in three colliding systems (pp at $\sqrt{s} = 7$ TeV, p–Pb at $\sqrt{s} = 5.02$ TeV and Pb–Pb at $\sqrt{s} = 2.76$ TeV).

As expected, the $\langle p_T \rangle$ increases with multiplicity. The increase is more pronounced for the heavier protons than it is for the lighter pions. A similar observation was made in p–Pb collisions at $\sqrt{s_{NN}} = 5.02$ TeV, as shown in Fig. 7, which depicts a compilation of results from the ALICE collaboration.[678] As mentioned, this behavior is expected in heavy-ion collisions in case of a collective hydrodynamic expansion.

The full p_T distributions of identified particles contain more information than the $\langle p_T \rangle$, and can be used to further test this "collectivity" hypothesis. The momentum boost imposed by the collective expansion leads to characteristic mass-dependent changes in the p_T distributions, which can be described with the "blast wave" parameterization.[656,709] This parameterization, inspired by hydrodynamics, assumes that all particle species are emitted from an expanding system within a common velocity field. A simultaneous fit of several identified particles allows estimating the freeze-out temperature, T_{kin}, and average transverse expansion velocity, $\langle \beta_T \rangle$, of the system. The ALICE collaboration performed a combined fit of π, K, p, and Λ in p–Pb collisions[710] with the blast wave parameterization. The fit was found to describe well the low p_T part of the spectrum ($p_T < 2 - 3$ GeV/c depending on the species). Remarkably, this fit also correctly predicted the p_T distribution of multistrange baryons,[679] as shown in Fig. 8. The $\langle \beta_T \rangle$ increases

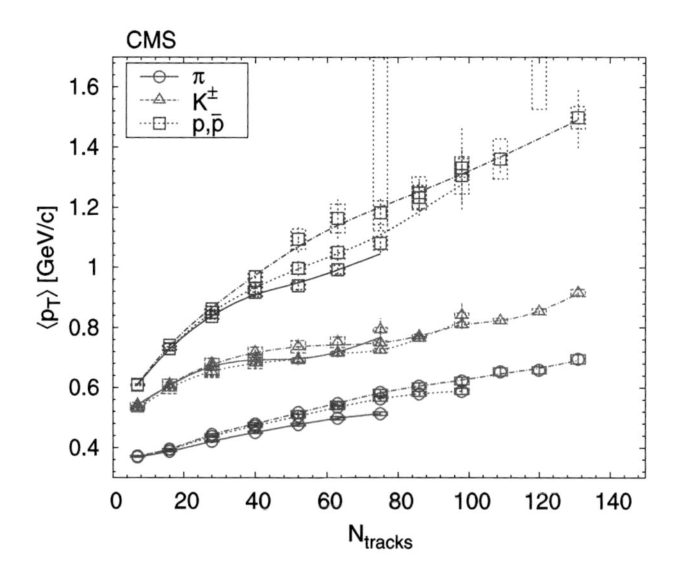

Fig. 6 Average p_T of pions, kaons and protons in the range $|y| < 1$ as a function the number of charged particles produced in $|\eta| < 2.4$, at energies $\sqrt{s} = 0.9$, 2.76 and 13 TeV, measured by the CMS experiment.[683] Lines are drawn to guide the eye and to indicate the centre-of-mass energy (solid 0.9 TeV, dotted 2.76 TeV, dash-dotted 7 TeV).

with multiplicity in all colliding systems, indicating the creation of a denser, more explosive system (within the hydrodynamic interpretation). The $\langle\beta_T\rangle$ values were also found to be larger in p–Pb than in Pb–Pb collisions for similar T_{kin} and multiplicity. A similar conclusion was reached by the CMS collaboration as shown in Fig. 9, which depicts the results of a fit to the K^0_S and Λ transverse momentum distribution.[684] The CMS analysis shows a clear ordering in the $\langle\beta_T\rangle$ expansion velocity between pp, p–Pb, and Pb–Pb collisions, with a more violent expansion in the smaller colliding systems (pp or p–Pb), when comparing events with similar final-state multiplicities. This can be understood in the context of a collectively expanding systems: for the same total initial-state entropy, larger transverse pressure gradients can be expected in pp/p–Pb, as a consequence of the smaller initial transverse size as compared to Pb–Pb collisions.[711] Preliminary ALICE data suggest similar T_{kin} and $\langle\beta_T\rangle$ values in pp and p–Pb collisions, with an apparent tension with the CMS data.[712] It should however be noted that estimates extracted with the blast wave parameterization are to be regarded as qualitative at best. The values, in fact, are known to be sensitive to the considered particle species and to the fit range.[713] Differences between the ALICE and CMS results could be explained by the different multiplicity estimators used, by the different set of particles, or by a different p_T range of the fit.

The boost-induced changes in the p_T distributions are also reflected in the p_T-differential ratios of baryons over mesons. Figure 10 shows the ratio Λ/K^0_S for the

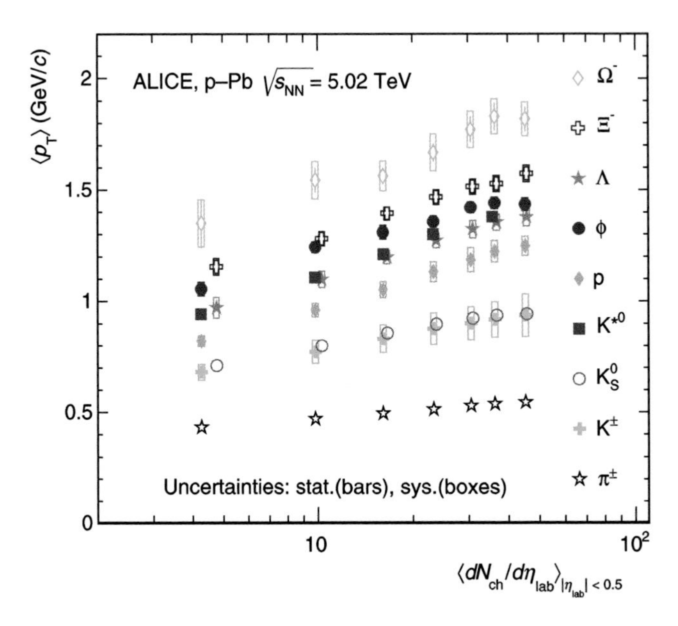

Fig. 7 Average p_T of π, K, p, ϕ, Λ, Ξ, Ω as a function of multiplicity, measured by the ALICE experiment in p–Pb collisions at $\sqrt{s} = 5.02$ TeV.[678]

three colliding systems. An increase at intermediate transverse momentum ($p_T \sim$ 3 GeV/c), with a corresponding depletion at low p_T is seen. The qualitative trend is the same in all systems. A collective hydrodynamic expansion would naturally explain the rise of this ratio up to intermediate p_T and its multiplicity dependence: the more explosive expansion expected at high multiplicity leads to stronger mass-dependent shifts in the p_T distributions. The decrease at higher p_T most likely comes from the interplay of soft and hard processes, and has sometimes been interpreted in heavy-ion collisions as a consequence of hadronization via coalescence of partons from the quark–gluon plasma with semi-hard mini-jets.[714]

While these results provide some support to the hypothesis of collective expansion, a solid conclusion requires full-fledged hydrodynamic calculations and the descriptions of many different observables within the same theoretical framework (including, in particular, the azimuthal anisotropy observables discussed in Section 4). State-of-the-art hydrodynamic calculations give a good description of the existing data[696,715,716] (see also Chapter 19). This led to a critical re-evaluation of the current interpretation of heavy-ion data, in particular for what concerns the assumption that the system quickly approaches thermal equilibrium: it was in fact found that for pp and pA collisions the system is far from equilibrium for most of its evolution, and yet hydrodynamics gives a good description of collective expansion phenomena.[695,717] The fact that viscous hydrodynamic calculations successfully describe the data, therefore, does not seem to imply the creation of a nearly-thermalized system.[718] It should also be noted that any theoretical

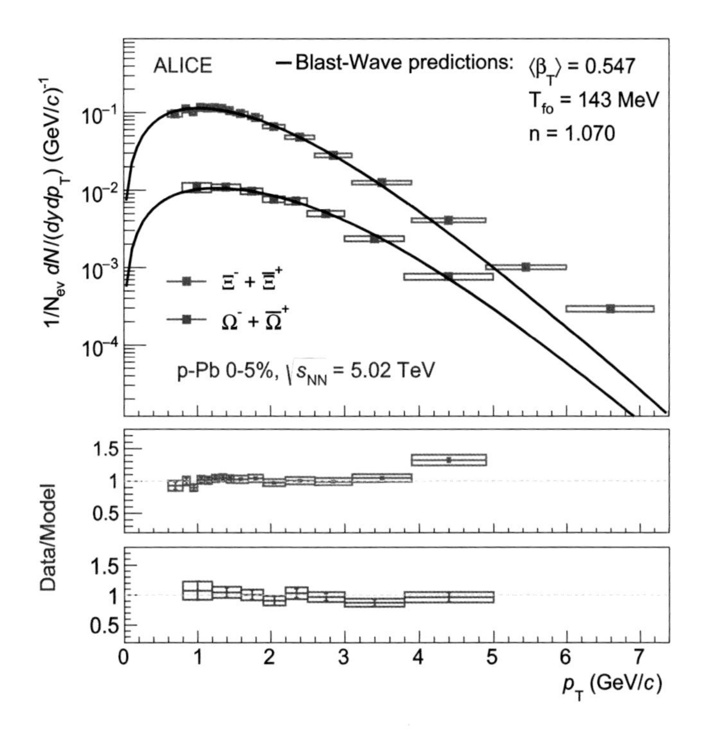

Fig. 8 Comparison of the $p_{\rm T}$ distributions of Ω and Ξ baryons with the Blast-Wave parameterization in high-multiplicity p–Pb collisions at $\sqrt{s} = 5.02$ TeV.[679] The Blast-Wave parameters have been fixed with a fit to π, K, p and Λ hadrons.[710]

framework which leads to a boost in the particle producing sources can lead to a mass ordering. The color-reconnection mechanism implemented in Pythia, as discussed above, is an example of such an alternative mechanism (see also the discussion in Chapter 18).

3.4. *Heavy flavor and hard processes*

Measurements of hard processes as a function of multiplicity allow studying the impact of (semi)hard partonic interactions and its interplay with soft particle production.

The ALICE collaboration has measured the production of J/ψ[719] and D[720] mesons as a function of multiplicity in pp collisions at $\sqrt{s} = 7$ TeV. The data are shown in Fig. 11 in the form of "self normalized yields", namely yields normalized to the corresponding minimum-bias result in each multiplicity bin. The yields of heavy flavors are seen to increase at a faster rate than the charged particle multiplicity. The increase with multiplicity is faster than linear for D mesons, which however are measured in a limited $p_{\rm T}$ window: part of the trend as a function of multiplicity could be introduced by the change of momentum distributions rather than by a change in the total production cross-section of heavy flavors. In the case of the J/ψ,

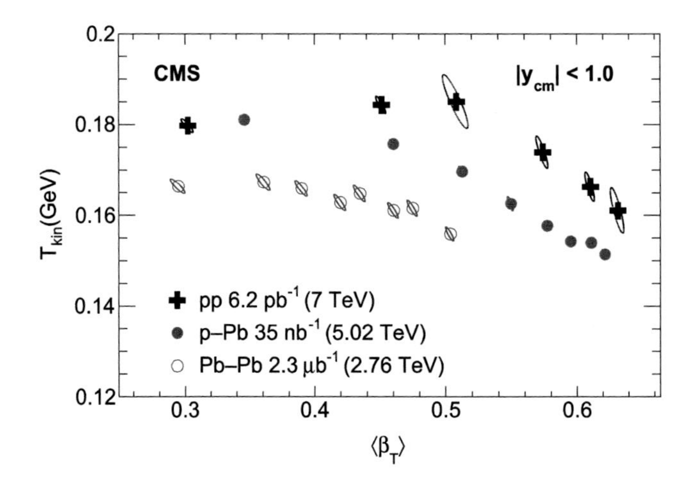

Fig. 9 $\langle\beta_{\rm T}\rangle$ and $T_{\rm kin}$ extracted from Blast-Wave fits to the CMS data in pp, p–Pb and Pb–Pb collisions (see text for details).[684]

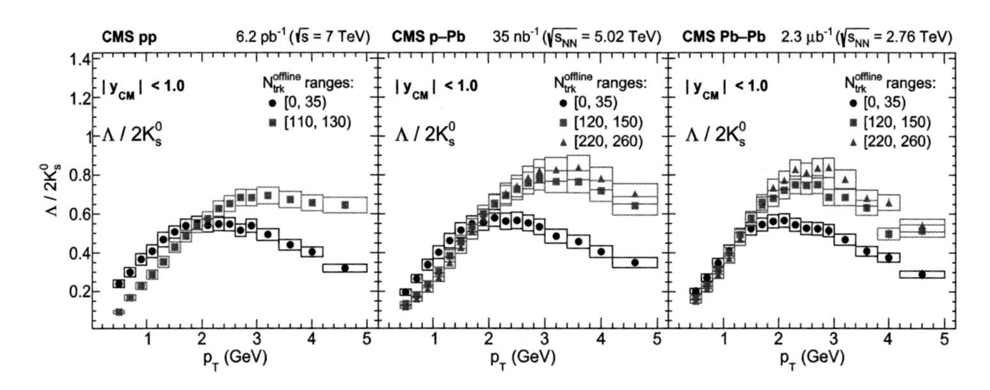

Fig. 10 Particle yield ratio Λ/K^0_S as a function of $p_{\rm T}$, measured in various multiplicity classes in pp, p–Pb, and Pb–Pb collisions (see text for details).[684]

it is not clear if the trend is faster than linear within current uncertainties. The same dependency is seen for non-prompt J/ψ production, that is J/ψ coming from decay of B mesons, as discussed in Ref. 719.

A variety of theoretical approaches have been suggested to describe the heavy-flavor results, as discussed in details in Ref. 720. Hard processes can contribute significantly to the multiplicity of the event, leading to auto-correlation effects. Moreover, it is known that hard processes in pp collisions are associated with a higher-than-average multiplicity (pedestal effect). It seems clear that, in order to describe this data in Monte Carlo event generators, a complex interplay of MPIs, fluctuations in the hardness of the collisions and a mechanism to control the evolution of charged particle multiplicity are needed. Based on recent theoretical and

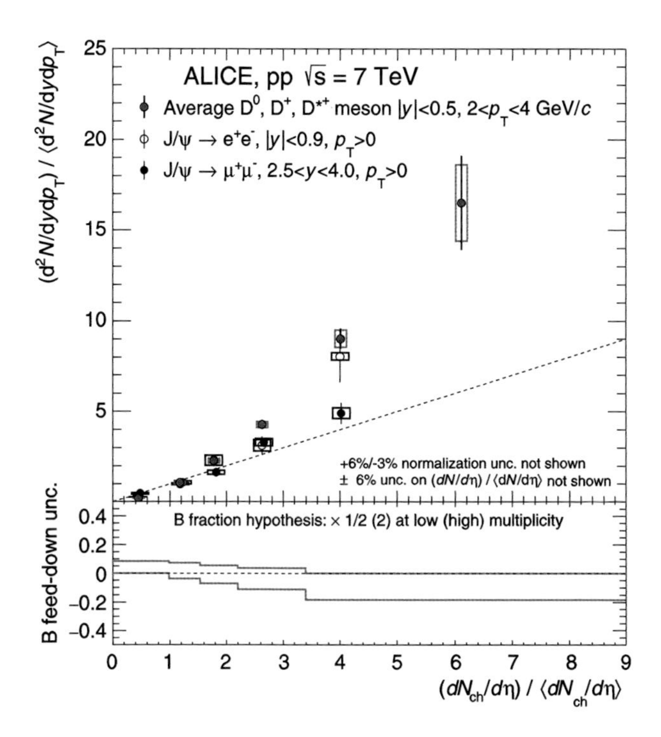

Fig. 11 Self-normalized yields of D and J/Ψ mesons, as a function of the self normalized min-rapidity multiplicity density, in pp collisions at $\sqrt{s} = 7$ TeV.[720]

experimental developments, it is becoming widely accepted that coherence effects within the densely-packed "strings" created in high multiplicity pp collisions are at the origin of most of the phenomena reported in this chapter. The data can likely be described with a superposition of two components: processes affected by this dense "core" and processes which escape without any interaction with it. Hard QCD processes are less affected by the dense core than the bulk of produced particles. The study of the multiplicity dependence of heavy flavors and, in general, hard processes can allow understanding the interplay of these two components.

4. Collective Long-Range Correlations in pp Collisions

Multi-particle angular correlations in minimum bias pp collisions have been measured over a wide range of collision energies before the LHC.[332,721,722] Short-range correlations (with a typical width of about 1 unit in pseudorapidity, η) have been revealed, which are characteristics of cluster-like correlations from decays of resonances and minijet fragmentation. In relativistic nucleus–nucleus collisions, of particular interest are the studies of long-range azimuthal particle correlations in η, which suggested collective behavior of particles emitted from a strongly interacting fluid-like quark–gluon medium, described by nearly ideal hydrodynamic models.[723–726] This section gives an overview of recent experimental studies of long-range particle correlations in pp collisions as a function of multiplicity at the LHC, and on the exploration of possible collective phenomena in small hadronic collisions.

4.1. *Two-particle $\Delta\eta$–$\Delta\phi$ correlations*

The measurement of two-particle $\Delta\eta$–$\Delta\phi$ correlations is an important technique for studying the properties and dynamics of particle production, especially the mechanism of hadronization and possible collective phenomena, in high-energy collisions of protons and nuclei. Here, $\Delta\eta$ and $\Delta\phi$ are the difference in pseudorapidity and azimuthal angle (in radians). The two-particle correlation functions are typically constructed by selecting "trigger" and associated particles within given p_T and η ranges (which can be either the same or different for trigger and associated particles) and obtaining the per-trigger-particle associated yield defined as

$$\frac{1}{N_{\rm trig}}\frac{d^2N^{\rm pair}}{d\Delta\eta d\Delta\phi} = B(0,0) \times \frac{S(\Delta\eta,\Delta\phi)}{B(\Delta\eta,\Delta\phi)}. \tag{1}$$

The signal pair distribution, $S(\Delta\eta,\Delta\phi)$, represents the yield of particle pairs normalized by the total number of trigger particles ($N_{\rm trig}$) from the same event[b] ($N^{\rm same}_{\rm pair}$),

$$S(\Delta\eta,\Delta\phi) = \frac{1}{N_{\rm trig}}\frac{d^2N^{\rm same}_{\rm pair}}{d\Delta\eta d\Delta\phi}. \tag{2}$$

The mixed-event pair distribution,

$$B(\Delta\eta,\Delta\phi) = \frac{1}{N_{\rm trig}}\frac{d^2N^{\rm mix}_{\rm pair}}{d\Delta\eta d\Delta\phi}, \tag{3}$$

is constructed by pairing the trigger particles in each event with the associated particles from several randomly selected events that have similar characteristics such as collision vertex and event multiplicity. Here, $N^{\rm mix}_{\rm pair}$ denotes the number of pairs taken from the mixed events. The ratio $B(0,0)/B(\Delta\eta,\Delta\phi)$ accounts for the random combinatorial background as well as for pair-acceptance effects, with $B(0,0)$ representing the mixed-event associated yield for both particles of the pair going in approximately the same direction and thus having full pair acceptance.

The two-particle angular correlation function in minimum bias (e.g., non-single diffractive) pp events for particles with low or intermediate p_T is characterized by short-range, cluster-like correlations, as shown in Fig. 12 (left) for pp collisions at $\sqrt{s} = 7$ TeV. The narrow peak around $(\Delta\eta,\Delta\phi) \sim (0,0)$ comes from the showering

[b]The CMS collaboration performs this normalization event-by-event and then averages over all events, while the ALICE collaboration does it inclusively after obtaining the pair distribution of all events.

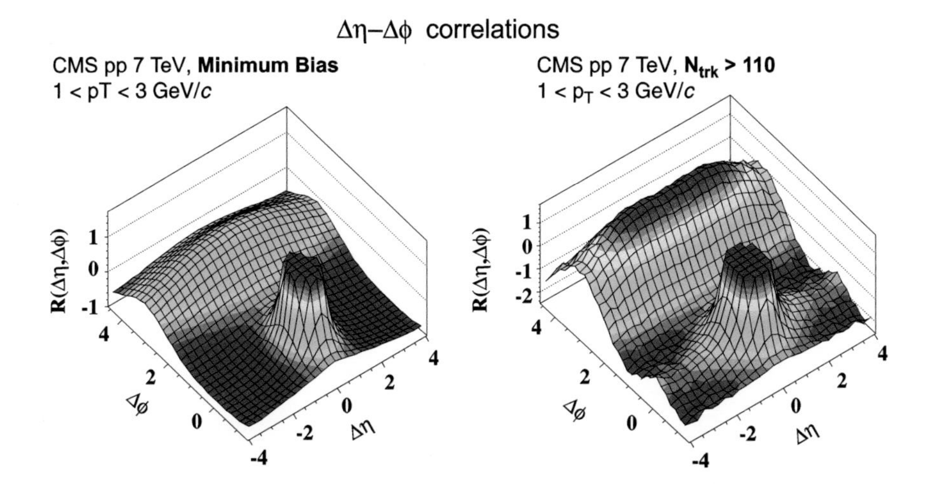

Fig. 12 Two-particle $\Delta\eta$–$\Delta\phi$ correlation function for particles having $1 < p_{\rm T} < 3$ GeV/c in minimum bias (left) and high-multiplicity (right) pp collisions at $\sqrt{s} = 7$ TeV.[282]

and hadronization of a leading parton or mini-jet, while the long-range away-side ($\Delta\phi \sim \pi$) structure in $\Delta\eta$ represents correlations of particles from the recoiling partons. As the two partons are produced back-to-back in ϕ but do not necessarily have opposite rapidities, the recoiling patrons are distributed over an elongated structure covering the full acceptance in $\Delta\eta$. Qualitative features of short-range two-particle angular correlations in minimum bias pp events are described by Monte Carlo event generators like Pythia[158] and models of hadronization via intermediate cluster decays.[722]

Surprisingly in 2010, a new feature of two-particle angular correlations in pp collisions was discovered at the LHC when selecting on rare events with "high multiplicities" (large number of final-state particles).[282] The correlation function for high-multiplicity pp events having charged multiplicity $N_{\rm trk} > 110$ (with $p_{\rm T} > 0.4$ GeV and $|\eta| < 2.4$) is shown in Fig. 12 (right), measured by the CMS collaboration. Besides the short-range correlation structure seen in minimum bias events, a long-range structure in $\Delta\eta$ emerges on the near side ($\Delta\phi \sim 0$) of the correlation function, which extends over at least 5 units in $\Delta\eta$. This ridge-like structure, not present in minimum bias pp collisions or any previous pp MC event generators, is reminiscent of the long-range two-particle correlation structure in nucleus–nucleus (AA) collisions, which is attributed to collective flow effects from hydrodynamic expansion of a strongly interacting quark-gluon plasma. Therefore, the observation of this "near-side ridge" may represent a first indication of collective flow in pp collisions. However, the formation of a hydrodynamic medium in small system like pp was not expected as the system size and lifetime may not be sufficient for developing any collective behavior via final-state interactions. The physical origin of the pp ridge has been under intense debate over the past years. A variety of theoretical interpretations have been proposed to explain the observed ridge structure in high-multiplicity pp events (see Ref. 727 for an early review).

In 2012, a long-range near-side ridge structure was also discovered in proton–lead (p–Pb) collisions for high-multiplicity events.[728–730] A review on experimental and theoretical progress in understanding the ridge in p–Pb collisions can be found in Ref. 694.

4.2. *Azimuthal anisotropies (v_n) from long-range correlations*

In relativistic heavy-ion collisions, the presence of an azimuthal anisotropic momentum distribution in the emission of final-state particles is commonly attributed to the strong collective flow behavior of a strongly coupled, hot and dense medium, which can be described by nearly ideal hydrodynamics. For a non-central heavy-ion collision, the overlap region of the two colliding nuclei has a lenticular shape. Strong rescattering of the partons in the initial state may lead to the build-up of anisotropic pressure gradients, which drive a collective anisotropic expansion. The expansion is fastest along the largest pressure gradient, i.e., along the short axis of the lenticular region. Therefore, the eccentricity of initial-state collision geometry results in an anisotropic azimuthal distribution of the final-state hadrons. In general, the anisotropy can be characterized by the Fourier harmonic coefficients, which can be measured via two- and multi-particle correlations. For instance, two-particle azimuthal correlations in heavy-ion collisions are typically characterized by its Fourier components, $\sim 1 + 2\sum_n v_n^2 \cos(n\Delta\phi)$, where v_n corresponds to anisotropy harmonics of the final-state single-particle azimuthal distribution.[731] In the context of hydrodynamics, the second (elliptic flow, v_2) and third (triangular flow, v_3) Fourier harmonics most directly reflect the medium response to the initial collision geometry and to its fluctuations, respectively. Detailed studies of elliptic and triangular flow provide insight into fundamental transport properties of the medium.[732–734] The long-range correlations and anisotropy Fourier harmonics have been extensively studied in Pb–Pb collisions at the Large Hadron Collider (LHC)[735–742] and found to be in good agreement with hydrodynamic models.

The extraction of collective v_n harmonics in pp collisions (possibly of the same origin as in heavy-ion collisions), on the other hand, is highly non-trivial due to the presence of strong back-to-back jet correlations, which dominates the long-range correlation structure on the away side (shown in Fig. 12). To properly extract the v_n harmonics possibly with a collective origin, contribution from short-range non-collective correlations must be eliminated. Two different approaches have been proposed. As first applied by the ALICE collaboration in the study of p–Pb correlation data,[730] the lowest multiplicity events can be used as an

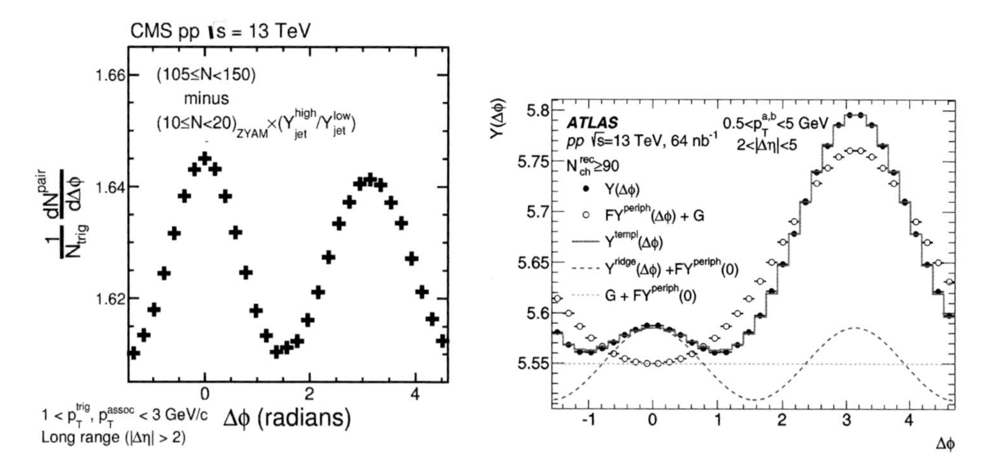

Fig. 13 Left: the 1D $\Delta\phi$ correlation function for the long-range regions in high-multiplicity pp events at $\sqrt{s} = 13$ TeV, after subtracting scaled results from low-multiplicity events.[451] Right: template fit to the 1D $\Delta\phi$ correlation function in high-multiplicity pp events at $\sqrt{s} = 13$ TeV.[743]

estimate of contribution of short-range correlations and be subtracted from the overall correlations, under the assumption that they contain no collective v_n signals. Alternatively, a template fit method was proposed by the ATLAS collaboration.[283] Instead of assuming a zero collective signal in the lowest multiplicity events, the correlation function at low multiplicities together with a Fourier flow component is used to fit the correlation function at high multiplicities. Unfortunately, both methods have their limitations, which led to large model dependence of v_n measurements for low multiplicity pp events. However, for the very-high-multiplicity region, the two methods tend to converge and extracted v_n values may differ by only 10–15%. This is a strong indication of the presence of collective long-range correlations in the events. Example correlation functions projected onto 1D as a function of $\Delta\phi$ in pp collisions at $\sqrt{s} = 13$ TeV are shown in Fig. 13, obtained with both subtraction procedures. Long-range azimuthal correlations from back-to-back jets are dominated by a large v_1 component, as shown in Fig. 13 (right). In the low-multiplicity subtraction method shown in Fig. 13 (left), the correlation function is dominated by a v_2-like structure after jet correlations are largely removed.

In Fig. 14, the results of v_2 from the template fitting procedure in pp collisions are shown as a function of event multiplicity and particle transverse momentum. The p–Pb data are also shown for comparison. As a function of p_{T}, the v_2 value in both pp and p–Pb exhibits a trend of first rising and then falling at $p_{\mathrm{T}} \approx 3$ GeV/c. This trend of p_{T} dependence is similar to that in AA collisions, which is characteristic of hydrodynamic flow effect. As a function of event multiplicity, a v_2 value of about 6% is observed for high-multiplicity pp events, which is smaller than the typical v_2 values found in p–Pb collisions. As event multiplicity decreases, the v_2 value for

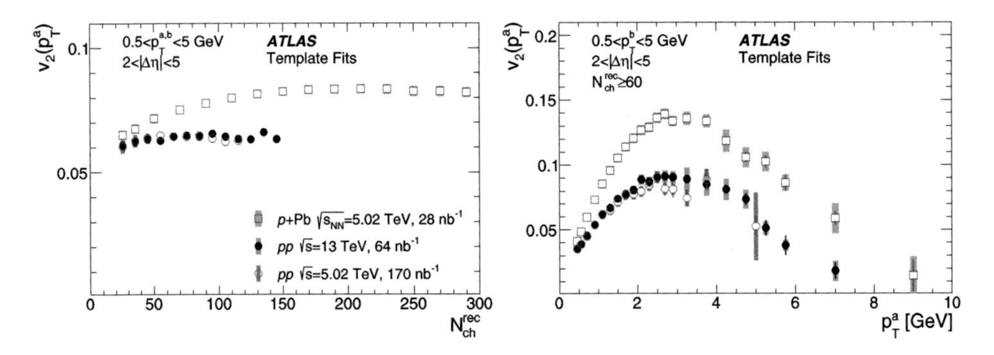

Fig. 14 The v_2 result from the template fitting procedure as a function of event multiplicity (left) and p_T (right) in pp collisions $\sqrt{s} = 5.02$ and 13 TeV, and p–Pb collisions at $\sqrt{s_{NN}} = 5.02$ TeV.[743]

p–Pb data shows a decreasing trend while it remains largely constant for pp data. However, as mentioned earlier, due to limitations in the procedure of subtracting short-range correlations, the trend of v_2 in pp collisions toward very low multiplicity region still has large uncertainties.

4.3. *Identified particle azimuthal correlations*

The dependence of the elliptic and higher-order flow harmonics on particle species can shed further light on the nature of the observed long-range correlations. The v_2 data as a function of p_T for identified K^0_S and $\Lambda/\overline{\Lambda}$ particles are extracted for pp collisions at $\sqrt{s} = 13$ TeV by the CMS collaboration.[451] Figure 15 shows the results for a low ($10 \leq N_{trk} < 20$) and a high ($105 \leq N_{trk} < 150$) multiplicity range before applying the jet correction procedure. The v_2 results for inclusive charged particles, shown for comparison, are dominated by contributions of pions, especially at low p_T (e.g., < 1 GeV/c).

At low multiplicity (Fig. 15), the v_2 values are found to be similar among charged particles, K^0_S and $\Lambda/\overline{\Lambda}$ hadrons across most of the p_T range within statistical uncertainties, similar to the observation in p–Pb collisions at $\sqrt{s_{NN}} = 5$ TeV.[744] This is consistent with the expectation that back-to-back jets are the dominant source of long-range correlations on the away side in low-multiplicity pp events, an assumption applied in the low-multiplicity subtraction method.

Moving to high-multiplicity pp events ($105 \leq N_{trk} < 150$, Fig. 15), a clear deviation of v_2 among various particle species is observed. In the lower p_T region of $\lesssim 2.5$ GeV/c, the v_2 value of K^0_S is greater than that of $\Lambda/\overline{\Lambda}$ at a given p_T value. Both are consistently below the inclusive charged particle v_2 values. Since most charged particles are pions in this p_T range, the result suggests that particle species with a lighter mass exhibit a stronger azimuthal anisotropy signal, when

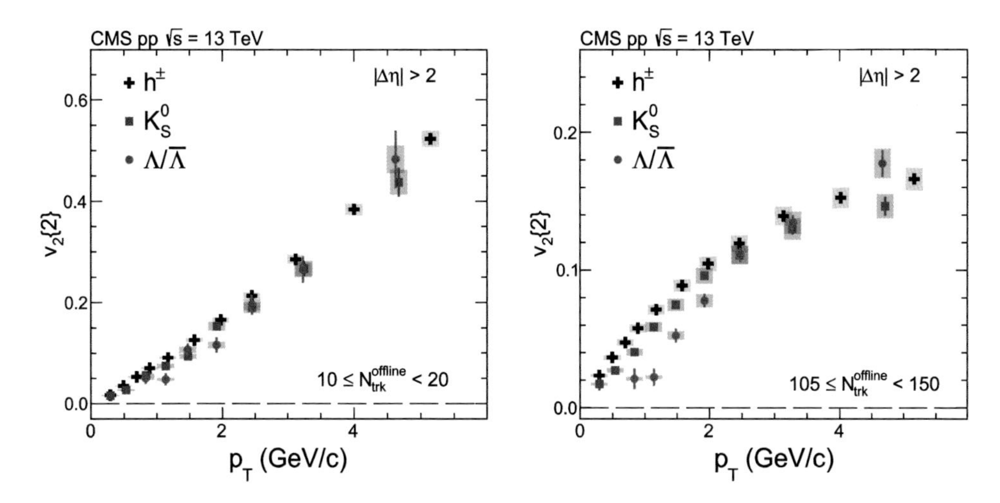

Fig. 15 The v_2 results for inclusive charged particles, K^0_S and $\Lambda/\overline{\Lambda}$ particles as a function of p_T in pp collisions at $\sqrt{s} = 13$ TeV, for $10 \leq N_{trk} < 20$ (left) and $105 \leq N_{trk} < 150$ (right).[451]

compared at a given p_T value. A similar trend was first observed in AA collisions at RHIC and the LHC,[745,746] and later also seen in p–Pb collisions at the LHC.[744,747] This behavior is consistent with the interpretation that particles are emitted from a collectively expanding source with a common velocity. Therefore, heavier particles receive a larger boost in their p_T. As discussed in Section 3, similar mass ordering is also seen in identified particle spectra as a function of p_T. The mass ordering of p_T spectra and v_2 have been considered to be a key signature of collectivity and a hydrodynamic expanding medium.[748,749] However, it was shown that an alternative model of collectivity, motivated by the gluon saturation model of initial gluon interactions, can also describe the data[750] (see Chapter 18 for detailed discussions). At $p_T > 2.5$ GeV/c, the v_2 values of $\Lambda/\overline{\Lambda}$ particles are larger than those of K^0_S particles. This reversed ordering of K^0_S and $\Lambda/\overline{\Lambda}$ at high p_T is similar to what was previously observed in p–Pb and Pb–Pb collisions.[744]

After applying the correction for jet correlations with the subtraction of low-multiplicity data, the v_2^{sub} results as a function of p_T for $105 \leq N_{trk} < 150$ are shown in Fig. 16 (left) for the identified particles and charged hadrons. The v_2^{sub} values for all three types of particles are found to increase with p_T, reaching 0.08–0.10 at $2 < p_T < 3$ GeV/c, and then show a trend of decrease for higher p_T values. The particle mass ordering of v_2 values in the lower p_T region is more pronounced after applying jet correction procedure. A remarkable scaling behavior first discovered in AA collisions is that, if divided by the number of constituent quarks, n_q, and plotted as a function of transverse kinetic energy per quark, KE_T/n_q, the v_2 values for baryons ($n_q = 3$) and mesons ($n_q = 3$) will fall on a universal curve.[745,751,752] This may be an evidence of collectivity at the

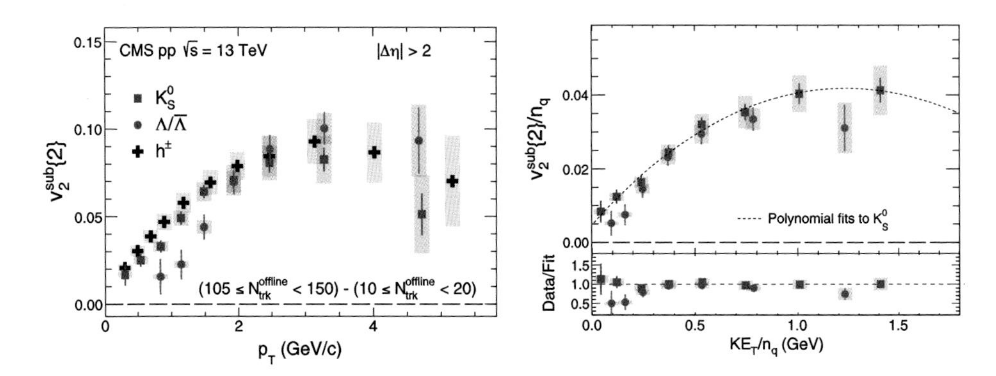

Fig. 16 Left: the $v_2^{\rm sub}$ results of inclusive charged particles, K_S^0 and $\Lambda/\overline{\Lambda}$ particles as a function of p_T for $105 \leq N_{\rm trk} < 150$, after correcting for back-to-back jet correlations estimated from low-multiplicity data. Right: the n_q-scaled $v_2^{\rm sub}$ results for K_S^0 and $\Lambda/\overline{\Lambda}$ particles as a function of KE_T/n_q. Ratios of $v_2^{\rm sub}/n_q$ for K_S^0 and $\Lambda/\overline{\Lambda}$ particles to a smooth fit function of data for K_S^0 particles are also shown.[451]

partonic level.[753–755] A similar scaling by the number of constituent quarks (NCQ) is also observed for identified particle v_2 data in p–Pb collisions.[744] The study of NCQ scaling of v_2 data for high-multiplicity pp events is shown in Fig. 16 (right). The dashed curve corresponds to a polynomial fit to the K_S^0 data. The ratio of n_q-scaled $v_2^{\rm sub}$ results for K_S^0 and $\Lambda/\overline{\Lambda}$ particles divided by this polynomial function fit is also shown in Fig. 16 (bottom). An approximate scaling is seen for $KE_T/n_q \gtrsim 0.2$ GeV within about ±10%, similar to that observed in AA collisions.[745,751,752]

4.4. *Multi-particle azimuthal correlations*

To directly explore the possible collective nature of long-range ridge correlations, the multi-particle cumulant technique was proposed to measure the strength of the azimuthal anisotropy by correlating four, six or event more particles, simultaneously.[756–758] Non-collective correlations involving only a few particles (e.g., jet-like correlations) will give reduced correlation strength for higher-order cumulants, while long-range collective signals would be invariant for different order of cumulants. This approach has been widely applied in studying AA collisions.[737,738,759–763]

These cumulants, $c_n\{m\}$, are calculated as follows:

$$\begin{aligned}
c_n\{2\} &= \langle\langle 2\rangle\rangle, \\
c_n\{4\} &= \langle\langle 4\rangle\rangle - 2\cdot\langle\langle 2\rangle\rangle^2, \\
c_n\{6\} &= \langle\langle 6\rangle\rangle - 9\cdot\langle\langle 4\rangle\rangle\langle\langle 2\rangle\rangle + 12\cdot\langle\langle 2\rangle\rangle^3, \\
c_n\{8\} &= \langle\langle 8\rangle\rangle - 16\cdot\langle\langle 6\rangle\rangle\langle\langle 2\rangle\rangle - 18\cdot\langle\langle 4\rangle\rangle^2 + 144\cdot\langle\langle 4\rangle\rangle\langle\langle 2\rangle\rangle^2 - 144\langle\langle 2\rangle\rangle^4,
\end{aligned} \tag{4}$$

where two- and multi-particle azimuthal correlations are evaluated as follows:

$$\begin{aligned}
\langle\langle 2\rangle\rangle &\equiv \langle\langle e^{in(\phi_1-\phi_2)}\rangle\rangle,\\
\langle\langle 4\rangle\rangle &\equiv \langle\langle e^{in(\phi_1+\phi_2-\phi_3-\phi_4)}\rangle\rangle,\\
\langle\langle 6\rangle\rangle &\equiv \langle\langle e^{in(\phi_1+\phi_2+\phi_3-\phi_4-\phi_5-\phi_6)}\rangle\rangle,\\
\langle\langle 8\rangle\rangle &\equiv \langle\langle e^{in(\phi_1+\phi_2+\phi_3+\phi_4-\phi_5-\phi_6-\phi_7-\phi_8)}\rangle\rangle.
\end{aligned} \tag{5}$$

Here $\langle\langle\cdots\rangle\rangle$ represents the average over all combinations of particles from all events. The collective elliptic flow harmonics are related to the multi-particle cumulants by

$$v_n\{4\} = \sqrt[4]{-c_n\{4\}}, \tag{6}$$

$$v_n\{6\} = \sqrt[6]{\frac{1}{4}c_n\{6\}}, \tag{7}$$

$$v_n\{8\} = \sqrt[8]{-\frac{1}{33}c_n\{8\}}. \tag{8}$$

As can be seen in Eq. (6), an alternating sign of the $c_n\{m\}$ cumulants is expected with increasing order m in presence of correlations of collective origin.

Measurements of multi-particle cumulants recently performed in p–Pb collisions[764–767] showed that the elliptic anisotropy, v_2, signals extracted from four-, six-, eight-, and all-particle correlations are nearly identically within uncertainties.[767] This provided strong evidence for the collective nature of the ridge in high-multiplicity p–Pb collisions.

The first measurement of four- and six-particle cumulants in pp collisions at various energies was reported by the CMS collaboration,[451] as shown in Fig. 17 for $c_2\{4\}$ and $c_2\{6\}$, as a function of event multiplicity for charged particles, averaged over $0.3 < p_{\mathrm{T}} < 3.0$ GeV/c and $|\eta| < 2.4$. The p–Pb data at $\sqrt{s_{NN}} = 5$ TeV are also shown for comparison. A striking feature observed for the $c_2\{4\}$ data in pp collisions at $\sqrt{s} = 13$ TeV is the sign change from positive to significant negative values as event multiplicity increases. According to Eq. (8), a collective $v_2\{4\}$ signal can be extracted only if the $c_2\{4\}$ value is negative. Therefore, the observed negative $c_2\{4\}$ value in Fig. 17 is an evidence of collective long-range correlations in pp collisions (although it is argued in Ref. 768 that a negative sign of $c_2\{4\}$ alone is neither a necessary nor a sufficient condition of collective correlations). The p–Pb data also show significant negative $c_2\{4\}$ values, with larger magnitudes than those in pp collisions. A positive sign is seen for $c_2\{6\}$, as expected for correlations of both collective and non-collective origin.

Furthermore, the v_2 signals are extracted from four- and six-particle cumulants in pp collisions at $\sqrt{s} = 13$ TeV, as a function of event multiplicity, shown in Fig. 18 (left). Within experimental uncertainties, the v_2 from four- and six-particle correlations are consistent with each other over the measured multiplicity range. Both of them are also agree with the v_2 value extracted from long-range two-particle

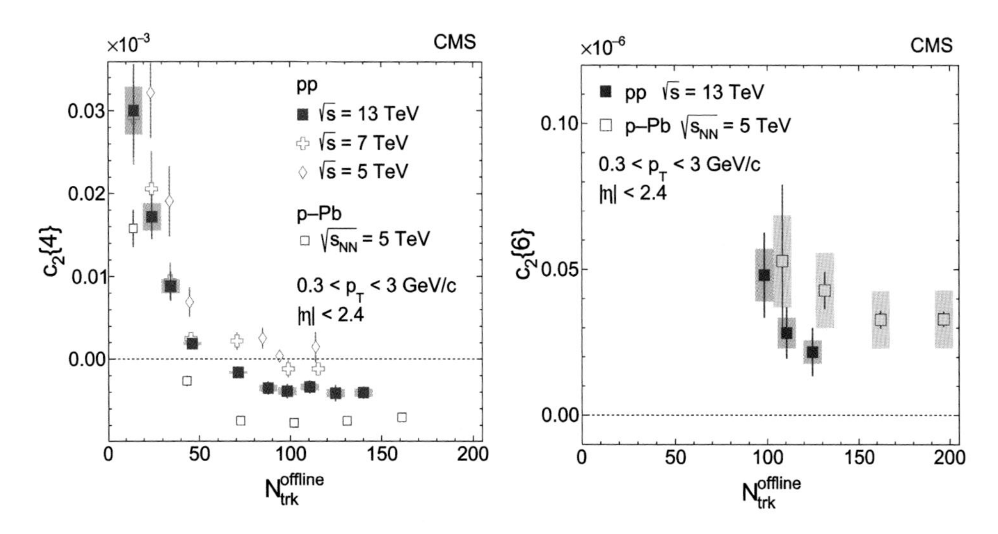

Fig. 17 The $c_2\{4\}$ (left) and $c_2\{6\}$ (right) values as a function of event multiplicity for charged particles, averaged over $0.3 < p_T < 3.0$ GeV/c and $|\eta| < 2.4$, in pp collisions at $\sqrt{s} = 5$, 7, and 13 TeV.[451] The p–Pb data at $\sqrt{s_{NN}} = 5$ TeV are also plotted for comparison.

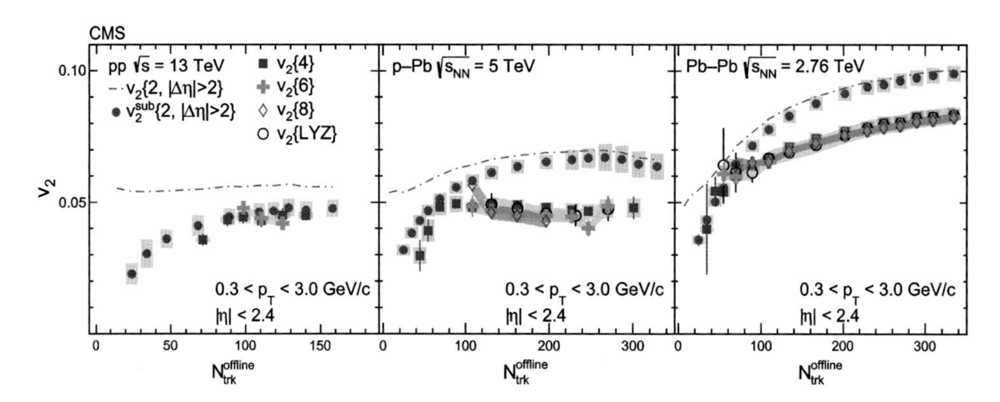

Fig. 18 The v_2 result from two- and multi-particle correlations as a function event multiplicity for charged particles, averaged over $0.3 < p_T < 3.0$ GeV/c and $|\eta| < 2.4$, in pp collisions at $\sqrt{s} = 13$ TeV (left), p–Pb collisions at $\sqrt{s_{NN}} = 5$ TeV (middle), and Pb–Pb collisions at $\sqrt{s_{NN}} = 2.76$ TeV (right).[451]

correlations. Similar to the observations in p–Pb and Pb–Pb (also shown in Fig. 18), this results provides a strong evidence for the creation of a collective system in pp collisions at high multiplicities. It should be stressed that although the observed scaling in Fig. 18 is a necessary outcome of a hydrodynamic framework, it is not sufficient proof that hydrodynamics is the correct underlying theory.

While ample evidence for collectivity in very-high-multiplicity pp events has been presented, the trend of collectivity toward low-multiplicity region remains an open question. The main limitation comes from contamination of short-range non-collective correlations, such as jets. As discussed earlier, the v_2 signal extracted from long-range two-particle correlations shows either a constant or declining trend as event multiplicity decreases, depending on the specific assumption made in the procedure of removing short-range correlations. Without further information, it is not possible to eliminate this model dependence in the two-particle correlation method. The multi-particle cumulant approach has the advantage of strongly suppressing the short-range correlations, but significant remaining effects may still be present for very-low-multiplicity events. Recently, progress has been made in developing a multi-particle cumulant method with particles selected from subevents separated by certain pseudorapidity gaps.[768,769] Specifically, two- and three-subevent methods have been studied with great details, which are evaluated as follows:

$$\begin{aligned} c_n\{2\}_{a|b} &= \langle\langle 2\rangle\rangle_{a|b}, \\ c_n\{4\}_{2a|2b} &= \langle\langle 4\rangle\rangle_{2a|2b} - 2\cdot\langle\langle 2\rangle\rangle_{a|b}^2, \\ c_n\{4\}_{2a|b,c} &= \langle\langle 4\rangle\rangle_{2a|b,c} - 2\cdot\langle\langle 2\rangle\rangle_{a|b}\langle\langle 2\rangle\rangle_{a|c}, \end{aligned} \tag{9}$$

where azimuthal correlations of particles taken from subevents separated by η gaps are evaluated as follows:

$$\begin{aligned} \langle\langle 2\rangle\rangle_{a|b} &\equiv \langle\langle e^{in(\phi_1^a-\phi_2^b)}\rangle\rangle, \\ \langle\langle 4\rangle\rangle_{2a|2b} &\equiv \langle\langle e^{in(\phi_1^a+\phi_2^a-\phi_3^b-\phi_4^b)}\rangle\rangle, \\ \langle\langle 4\rangle\rangle_{2a|b,c} &\equiv \langle\langle e^{in(\phi_1^a+\phi_2^a-\phi_3^b-\phi_4^c)}\rangle\rangle, \end{aligned} \tag{10}$$

Here, index a, b and c denote the subevents. For the case of two-subevent method, subevents a and b are divided at $\eta = 0$, while for the three-subevent method, the full η coverage is equally divided into subevents a, b and c.

First results of multi-particle cumulant v_2 with the subevent method were studied by the ALICE[770] and ATLAS[771] collaborations. Figure 19 (left) presents the ATLAS four-particle cumulant result, $c_2\{4\}$, using the three-subevent method, in pp collisions at $\sqrt{s}$ = 5.02 and 13 TeV, and p–Pb collisions at $\sqrt{s_{NN}}$ = 5.02 TeV, for $0.3 < p_T < 3$ GeV/c as a function of event multiplicity. Compared to the result from the standard cumulant method without subevents in Fig. 18, the $c_2\{4\}$ result in pp collisions with subevents shows significant negative values down to even lower multiplicity range of about 60. This is likely a consequence of short-range correlations being better suppressed by the implementation of subevents in the cumulant calculation. A similar behavior is also observed for p–Pb collisions. While the subevent cumulant method gives consistent results as the standard method in the high-multiplicity region, it extracts negative $c_2\{4\}$ values down to a much lower multiplicity region. The extracted $v_2\{4\}$ values from the four-particle subevent cumulant are shown in Fig. 19 (right) for pp collisions at $\sqrt{s}$ = 13 TeV, down to event multiplicity region of about 60. They are found to be comparable to v_2

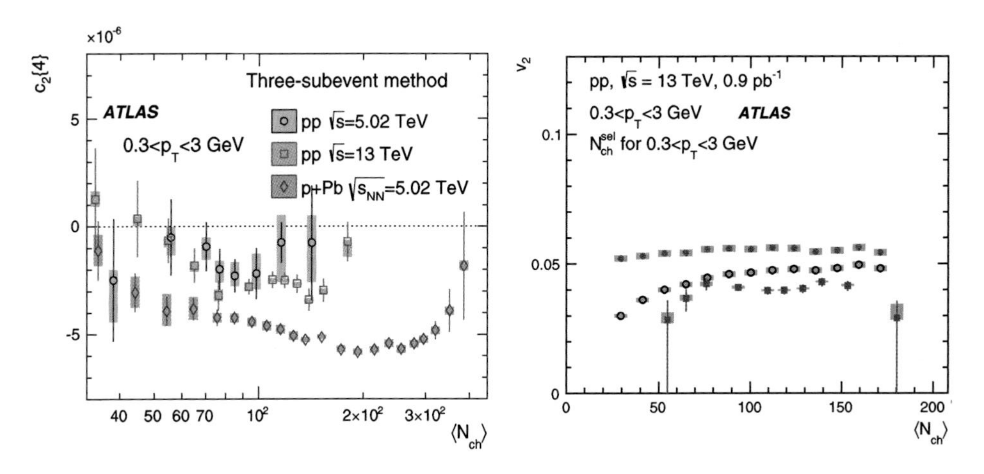

Fig. 19 The $c_2\{4\}$ (left) and $v_2\{4\}$ results for charged particles with $0.3 < p_{\mathrm{T}} < 3.0$ GeV/c from the three-subevent method as a function of multiplicity in pp collisions at $\sqrt{s} = 5.02$ and 13 TeV, and p–Pb collisions at $\sqrt{s_{NN}} = 5$ TeV.[771]

values obtained from long-range two-particle correlations after removing short-range correlations via the template fit and low-multiplicity subtraction methods.

Overall, multi-particle cumulant methods have a great potential of exploring the collective behavior of events with low multiplicities where short-range non-collective correlations are significant. It is promising to apply the method to extract even higher-order cumulant $v_2\{m\}$, $m \geq 6$ so that stronger evidence of collectivity in small collision systems may be obtained. Finally, the method can also be used to extract higher order Fourier harmonics such as v_3 with four-particle correlations, and event-by-event correlations between different orders of Fourier harmonics. A first attempt has been carried out by CMS in a measurement of v_2 and v_3 correlation using four-particle cumulant (also called symmetric cumulant[772]) without subevents.[773] In the context of hydrodynamic models, the correlation of v_2 and v_3 provides key insights to the initial-state geometry fluctuations. Detailed studies of multi-particle cumulants in the future with subevent methods should shed new light on the collective phenomena in pp collisions.

5. Summary

Studies of high-multiplicity pp and p–Pb events at the LHC provide strong evidence of novel QCD dynamics in small systems, originating from the large density of color charges. In this chapter, main experimental observations leading to this conclusion are summarized: (i) the relative abundances of identified particle species changes with multiplicity, in particular the relative strangeness content increases with increasing multiplicity and the abundance of heavy-flavor particles increases faster than that of light flavor ones; (ii) the p_{T}-distribution of identified particles become harder with increasing multiplicity, with heavier particles showing a more pronounced increase in the average p_T; (iii) long-range multiparticle correlations are observed, with an ordering in mass suggestive of a collective particle production. These observations are reminiscent of phenomena that are considered to be characteristics of relativistic heavy-ion collisions, which are interpreted as a consequence of the formation of a strongly interacting, locally thermalized quark–gluon plasma.

Intense activities in the theoretical community are ongoing, aimed at understanding the origin of these phenomena in small systems. On one hand, theoretical tools used for the description of heavy-ion collisions have been extended to small colliding systems. The data are found to be consistent with the assumption of the creation of collective medium, whose expansion can be described using viscous relativistic hydrodynamics and which hadronizes statistically (see a review in Ref. 694). However, key questions like whether the system has rapidly reached thermal equilibrium and whether hydrodynamic description is applicable with a sub-fermi system size are still debated. It was also suggested recently that hydrodynamic modeling of systems far from equilibrium can successfully describe the data and this may lead to a paradigm shift in the understanding of heavy-ion data.[695] On the other hand, alternative explanations of the data have been proposed, particularly in the framework of gluon saturation models coupled with hadronization via colour strings, which provide a viable mechanism to describe the data (Chapter 18). Future efforts from both experimental and theoretical communities are vital in understanding the novel dynamics of high-multiplicity events.

At the same time, the widely used parton-shower Monte Carlo models such as Pythia, did not predict any of these observations in high-multiplicity events. In particular, the assumption that the fragmentation parameters of the Pythia strings do not depend on the colliding system nor on the environment of the collision (jet universality) is challenged by these observations.[442] Work to extend these models to take into account the effect of densely packed strings in the transverse plane is in progress (Chapters 10 and 17). If successful, these efforts could provide a microscopic interpretation of the collective effects discussed in this chapter.

In summary, the study of high-multiplicity pp and pA collisions has opened up a new avenue for the exploration of novel many-body QCD dynamics. The experimental observations reported in this chapter are leading to a paradigm shift in the description of hadronic collisions and challenge conventional models of soft QCD processes (for what concerns the universality of fragmentation) and of the quark–gluon plasma (for what concerns the degree of thermal equilibrium reached by the system).

Chapter 16

Experimental Results on Event Shapes at Hadron Colliders

Antonio Ortiz

Instituto de Ciencias Nucleares, Universidad Nacional Autónoma de México, Apartado Postal 70-543, México D.F. 04510, México

In this chapter, a review on event shapes at hadron colliders, mainly focused on experimental results, is presented. Measurements performed at the Tevatron and at the LHC, for the soft and hard regimes of QCD, are reviewed. The potential applications of event shapes for unveiling the origin of collective-like phenomena in small collision systems as well as for testing pQCD predictions are discussed.

1. Introduction

In the hard regime of quantum chromodynamics (QCD), jets are produced from parton (quarks or gluons) scatterings at large transverse momentum (p_{T}). Since in that case the coupling constant (α_{s}) is small, precise perturbative QCD (pQCD) calculations are doable. In contrast, for processes involving low momentum transfer QCD-inspired phenomenological models are built up.

Event shapes can be used to test QCD mainly because, by construction, they are collinear and infrared safe observables, i.e., they do not change their value when an extra soft gluon is added or if a parton is split into two collinear partons. This is a necessary condition for the cancellation of real and virtual divergences associated with such emissions, and therefore for making finite pQCD predictions.[774]

Several studies using event shapes were performed in e^+e^- annihilations. For example, in e^+e^- at $\sqrt{s} = 22 - 44\,\mathrm{GeV}$,[775] pQCD calculations combining next-to-leading-order $\mathcal{O}(\alpha_{\mathrm{s}}^2)$ and next-to-leading logarithms (NLLA) precisions were used for extracting the energy dependence of α_{s} in the following way. Starting from the general $\mathcal{O}(\alpha_{\mathrm{s}}^2)$ expression for a given event shape observable, y:

$$R(y) = 1 + A(y)\left(\frac{\alpha_{\mathrm{s}}}{2\pi}\right) + B(y)\left(\frac{\alpha_{\mathrm{s}}}{2\pi}\right)^2, \tag{1}$$

where $R(y)$ is the cumulative cross-section of y normalized to the lowest order Born cross-section. The coefficients A and B (know from $\mathcal{O}(\alpha_s^2)$ matrix elements[776]) were used in the analogous expression from NLLA calculations:

$$R(y) = \left(1 + C_1\left(\frac{\alpha_s}{2\pi}\right) + C_2\left(\frac{\alpha_s}{2\pi}\right)^2\right)\exp\left(Lg_1\left(\frac{\alpha_s}{2\pi}L\right) + g_2\left(\frac{\alpha_s}{2\pi}L\right)\right) \tag{2}$$

where $L = \ln(1/y)$. From fitting the predictions to the data (corrected to the parton level) α_s was determined.[777] These features allowed event shapes to be among the most studied QCD observables,[778] both theoretically and experimentally, not only in e^+e^- but also in deep inelastic scattering.

For example, sphericity was used at SLAC (Stanford Linear Accelerator Center) to prove to the existence of jets in e^+e^- processes at centre-of-mass energies up to 7.4 GeV.[779] Moreover, event shape variables were valuable tools which allowed the gluon discovery in processes where three prolonged jets were produced at energies up to 31.5 GeV in the centre-of-mass.[780–782] Since event shape variables are defined with partons and they are measured in terms of visible particles, it is possible to use them for studying corrections due to hadronization effects.[783,784]

In $e^+e^- \to q\bar{q} \to$ hadrons events, the particle production is studied with respect to the (unknown) $q\bar{q}$ axis, which can be approximated by the event shape axis. Since there is no preferred direction for the axis, the three components of the particle momentum are required in the calculation. In contrast, at hadron colliders the event shape axis is searched in the plane perpendicular to the beam axis. Moreover, the total momentum perpendicular to the plane formed by the main hard scattering (event shape axis) and the beam axis is sensitive to soft physics allowing to test QCD-inspired phenomenological models.

Since at hadron colliders, pQCD calculations are available for a vast number of event shapes,[785,786] they have been used to study different aspects of QCD.[495,787–790] In addition, the shape of the events has also been proposed for discriminating events where black-holes (BH) are produced, or in supersymmetry (SUSY) searches.[791] For example, BH events are characterized by higher sphericity than that expected in SUSY processes.[792]

At hadron colliders, QCD has been extensively tested in the perturbative regime, where precise calculations can be done. This is not the case of non-perturbative QCD, therefore, the soft regime represents an opportunity for the discovery of new phenomena. In this regard, the study of atypical events, like the "hedgehogs"[a] observed in $p\bar{p}$ collisions at $\sqrt{s} = 1.8$ TeV by the Collider Detector at FERMILAB (CDF), was proposed to unveil the nature of soft particle production at the Large Hadron Collider (LHC) energies.[793] Similar events were also reported

[a]High transverse energy (321 GeV) isotropically distributed within a large acceptance ($\eta < 4$).

at lower energies by the UA1 collaboration.[794] Clearly, "hedgehog" enriched samples could be easily achieved by running an event shape selection imposing a low p_T threshold.

Moreover, motivated by the discovery of collective-like phenomena in small collision systems (pp and pA collisions) at the LHC,[282,729,730,795] the study of event shapes for soft physics has been encouraged. Since for the heavy-ion program it is important to establish whether or not the strongly interacting Quark–Gluon–Plasma (sQGP) is formed in small systems, where as will be discussed below, the event shape analysis is a promising tool for controlling the jet bias in high multiplicity pp events.[796,797]

In this paper, a review of event shape measurement at hadron colliders will be presented covering the soft and hard regimes. Data from different experiments will be used, going from $\sqrt{s} = 0.9\,\mathrm{TeV}$ up to the largest energies reached by the Large Hadron Collider.

The paper is organized as follows. In Section 2 the definitions of the different event shape variables are presented, the applications in the analyses of pp data are discussed in Sections 3 and 4 for the hard and soft regimes of QCD, respectively. An outlook is then displayed in Section 5.

2. Definitions

As discussed in the introduction, at hadron colliders event shapes are defined in the transverse plane, i.e., in the plane perpendicular to the beam axis.[b] The objects (particles or jets) which participate in the calculation are restricted to some kinematic regions imposed by the detectors. For example, in the case of the transverse sphericity (S_T)[798] defined by the ALICE collaboration, only charged particles at mid-pseudorapidity ($|\eta| < 0.8$) and with transverse momenta greater than $0.5\,\mathrm{GeV}/c$ are considered. For the transverse sphericity to be defined, the transverse momentum matrix, $\mathbf{S}$, should be first diagonalized:

$$\mathbf{S} = \frac{1}{\sum_i p_{\mathrm{T},i}} \sum_i \frac{1}{p_{\mathrm{T},i}} \begin{pmatrix} p_{\mathrm{x},i}^2 & p_{\mathrm{x},i}p_{\mathrm{y},i} \\ p_{\mathrm{y},i}p_{\mathrm{x},i} & p_{\mathrm{y},i}^2 \end{pmatrix}, \tag{3}$$

where, $p_{\mathrm{T},i}$ is the transverse momentum of the ith particle, being $p_{\mathrm{x},i}$ and $p_{\mathrm{y},i}$ the components along the x and y axes, respectively. The transverse sphericity is then defined in terms of the eigenvalues, $\lambda_1 > \lambda_2$, as follows:

$$S_\mathrm{T} \equiv \frac{2\lambda_2}{\lambda_1 + \lambda_2}. \tag{4}$$

[b]It is worth mentioning that the ATLAS collaboration has also measured sphericity, but using jets and the full momentum tensor of the event, i.e., considering the component along the beam axis.[787]

Using the ratio of the smaller and larger eigenvalues, another event shape variable named $\mathscr{F}$-parameter can be defined as

$$\mathscr{F} \equiv \frac{\lambda_2}{\lambda_1}. \tag{5}$$

On the other hand, transverse spherocity, originally proposed here[785] and recently studied in Ref. 797 is defined for a unit vector $\hat{\mathbf{n}}_{\mathbf{s}}$ which minimizes the ratio:

$$S_0 \equiv \frac{\pi^2}{4} \min_{\hat{\mathbf{n}}_{\mathbf{s}}} \left(\frac{\sum_i |\mathbf{p}_{\mathrm{T},i} \times \hat{\mathbf{n}}_{\mathbf{s}}|}{\sum_i p_{\mathrm{T},i}} \right)^2. \tag{6}$$

Concerning the thrust-related variables, the transverse thrust τ_{T}, defined as

$$\tau_{\mathrm{T}} \equiv 1 - \max_{\hat{\mathbf{n}}_{\boldsymbol{\tau}}} \frac{\sum_i |\mathbf{p}_{\mathrm{T},i} \cdot \hat{\mathbf{n}}_{\boldsymbol{\tau}}|}{\sum_i p_{\mathrm{T},i}} \tag{7}$$

considers the unit vector $\hat{\mathbf{n}}_{\boldsymbol{\tau}}$ (thrust axis) that maximizes the sum, and thereby minimizes τ_{T}. The transverse thrust is more sensitive to the modeling of two- and three-jet topologies, while it is less sensitive to QCD modeling of larger jet multiplicities.[789]

Using the direction given by $\hat{\mathbf{n}}_{\boldsymbol{\tau}}$, the transverse region can be separated into an upper side $\mathscr{C}_{\mathrm{U}}$ (lower side $\mathscr{C}_{\mathrm{L}}$) consisting of all particles which satisfy $\mathbf{p}_{\mathrm{T}} \cdot \hat{\mathbf{n}}_{\boldsymbol{\tau}} > 0$ ($\mathbf{p}_{\mathrm{T}} \cdot \hat{\mathbf{n}}_{\boldsymbol{\tau}} < 0$). Moreover, the thrust axis $\hat{\mathbf{n}}_{\boldsymbol{\tau}}$ and the beam direction $\hat{\boldsymbol{z}}$ together define the event plane in which the primary hard scattering occurs. Using this new axis, $\hat{\mathbf{n}}_{\mathbf{m}} = \hat{\mathbf{n}}_{\boldsymbol{\tau}} \times \hat{\boldsymbol{z}}$, the transverse thrust minor is defined as:

$$T_{\mathrm{min}} \equiv \frac{\sum_i |\mathbf{p}_{\mathrm{T},i} \cdot \hat{\mathbf{n}}_{\mathbf{m}}|}{\sum_i p_{\mathrm{T},i}}. \tag{8}$$

The observable T_{min} is a measure of the out-of-plane transverse momentum and varies from zero, for an event entirely in the event plane, to $2/\pi$ for a cylindrically symmetric event.[495]

There are some common features among the event shape variables defined before.

- The event shapes are linear in particle momenta, therefore is a collinear safe quantity in pQCD.
- The lowest limit of the variables corresponds to the "pencil-like" structure (jetty-like structure for high multiplicity events).
- The highest limit of the variables corresponds to the isotropic structure.

To illustrate the features of the different event shapes, Fig. 1 displays sketches of events as seen from the plane perpendicular to the beam axis. The left-hand side figure exhibits the case when only particles belonging to a back-to-back jet are considered in the calculations of different event shapes. The different axes

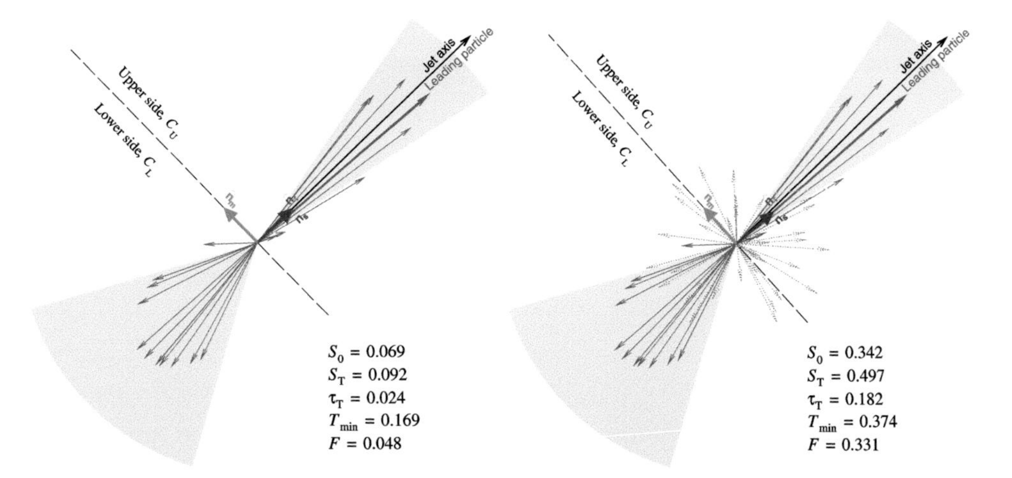

Fig. 1 Sketch of a back-to-back jet without (left) and with (right) the underlying event, the information is projected in the plane perpendicular to the beam axis. The solid arrows indicate the particles within the jet, while, the dotted ones indicate particles belong to the underlying event. All the relevant event shape axes are also illustrated: spherocity (blue arrow), thrust (red arrow) and thrust minor (green arrow). The values of different event shapes are also displayed.

which were already discussed are also displayed. As expected, the thrust ($\hat{\mathbf{n}}_{\boldsymbol{\tau}}$) and spherocity ($\hat{\mathbf{n}}_{\mathbf{s}}$) axes are almost parallel to the jet axis. In addition, transverse sphericity, spherocity, thrust and $\mathscr{F}$ take values which are very close to zero, whereas, the variable which gives the out-of-plane $p_{\rm T}$ amounts to 0.169. When the underlying event (UE), i.e. the soft component which accompanies the hard scattering, is taken into account in the calculation (right-hand side figure) of the event shapes: $S_{\rm T}$, $\tau_{\rm T}$ and $\mathscr{F}$ are strongly affected, in contrast to S_0 and $T_{\rm m}$. Moreover, the thrust and spherocity axes are now more aligned to the leading particle. It is worth recalling that for transverse spherocity is harder to reach the isotropic limit than for sphericity,[785] therefore, the discrimination power between isotropic soft events and symmetric multi-jet events might be the highest for spherocity.

Finally, the total jet broadenig ($B_{\rm tot}$) is defined as follows:

$$B_{\rm tot} \equiv B_{\rm U} + B_{\rm L}, \tag{9}$$

where U (respectively, L) refers to the upper (respectively, lower) side. The jet broadening variable in each region is defined as

$$B_{\rm X} \equiv \frac{\sum_{i\in\mathscr{C}_{\rm X}} p_{{\rm T},i}\sqrt{(\eta_i-\eta_{\rm X})^2+(\phi_i-\phi_{\rm X})^2}}{2\sum_{i\in\mathscr{C}_{\rm X}} p_{{\rm T},i}}, \tag{10}$$

where the ith particle is within the $\mathscr{C}_{\rm X}$ ($\mathscr{C}_{\rm U}$ or $\mathscr{C}_{\rm L}$) side. The pseudorapidity and azimuthal angles of the axes for the upper and lower sides are defined as follows:

$$\eta_{\rm X} \equiv \frac{\sum_{i\in\mathscr{C}_{\rm X}} p_{{\rm T},i}\eta_i}{\sum_{i\in\mathscr{C}_{\rm X}} p_{{\rm T},i}}, \tag{11}$$

$$\phi_{\rm X} \equiv \frac{\sum_{i\in\mathscr{C}_{\rm X}} p_{{\rm T},i}\phi_i}{\sum_{i\in\mathscr{C}_{\rm X}} p_{{\rm T},i}}. \tag{12}$$

3. Hard QCD Sector

This section aims to show that event shapes are useful tools to test the validity of QCD prediction, and at the same time, to provide confidence in the current MC models for the description of the Standard Model processes and their application for determining background in new physics searches at hadron colliders.

At the Large Hadron collider, the ATLAS collaboration has reported a study as a function of the scalar sum of the jet transverse momenta which is defined as:

$$\frac{1}{2}H_{{\rm T},2} \equiv \frac{1}{2}(p_{{\rm T},1} + p_{{\rm T},2}) \tag{13}$$

where the subscript $i = 1, 2$ refers to the leading and sub-leading jet in the event.[787] Overall, the modeling of data by Pythia 6[158] (tune Perugia 0[434]) and Alpgen[186] are more accurate than that by Herwig++.[484] For example, Fig. 2 shows the transverse thrust minor distribution which within uncertainties is well described

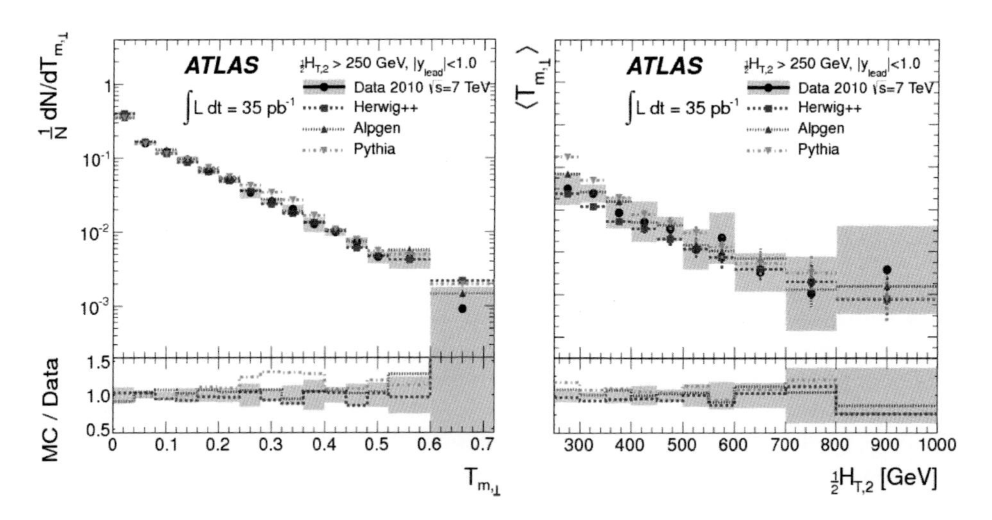

Fig. 2 Transverse thrust minor distribution for hard events ($\frac{1}{2}H_{{\rm T},2} > 250\,{\rm GeV}/c$) (left) and average transverse thrust minor as a function of $\frac{1}{2}H_{{\rm T},2}$ (right). Results for pp collisions at $\sqrt{s} = 7\,{\rm TeV}$ are compared with different Monte Carlo generators. Figures reproduced from Ref. 787.

by the different models, albeit, the average thrust minor predicted by Pythia is slightly higher at low $\frac{1}{2}H_{\mathrm{T},2}$ than in data. For the different event shapes, their mean values decrease with $\frac{1}{2}H_{\mathrm{T},2}$ and the trend is well modeled by the Monte Carlo (MC) generators. A similar study has been performed by the CMS collaboration.[799] Although, the MCs used to compare the data are not the same to those used by ATLAS,[787] the conclusions are indeed similar.

The study was further extended using a larger dataset, in that case event shapes have been measured using jets as inputs in multi-jet events.[790] It has been shown that event shapes that are more sensitive to the longitudinal energy flow show larger discrepancies between data and MC simulations. As an example, Fig. 3 shows the jet broadening distribution for various intervals of leading jet p_{T}. As discussed in Ref. 790, this observable is insensitive to the underlying event and hadronization details, however, a precise modeling of the matrix element calculations and parton showering is crucial in order to improve the description of the data. In this context, within uncertainties the MadGraph[800] matrix-element (ME) calculator combined with Pythia 6 consistently reproduces all the distributions.

At the Tevatron, the CDF experiment used τ_{T} and T_{min} to built a new quantity less dependent on the UE aimed to have a more meaningful comparison between NLO + NLL parton-level predictions and data.[495] To illustrate the idea, one sees that event shapes can be approximately written in terms of the hard and soft components as

$$\tau_{\mathrm{T}} \approx \frac{\sum_i p_{\mathrm{T},i}^{\mathrm{hard}} - \max\limits_{\hat{\mathbf{n}}_\tau} \sum_i p_{\mathrm{T},i}^{\mathrm{hard}} |\cos\phi_i^{\mathrm{hard}}|}{\sum_i p_{\mathrm{T},i}^{\mathrm{hard}} + \sum_j p_{\mathrm{T},j}^{\mathrm{soft}}} + \frac{\sum_j p_{\mathrm{T},j}^{\mathrm{soft}}(1 - |\cos\phi_j^{\mathrm{soft}}|)}{\sum_i p_{\mathrm{T},i}^{\mathrm{hard}} + \sum_j p_{\mathrm{T},j}^{\mathrm{soft}}}, \quad (14)$$

$$T_{\mathrm{min}} = \frac{\sum_i p_{\mathrm{T},i}^{\mathrm{hard}} |\sin\phi_i^{\mathrm{hard}}|}{\sum_i p_{\mathrm{T},i}^{\mathrm{hard}} + \sum_j p_{\mathrm{T},j}^{\mathrm{soft}}} + \frac{\sum_j p_{\mathrm{T},j}^{\mathrm{soft}} |\sin\phi_j^{\mathrm{soft}}|}{\sum_i p_{\mathrm{T},i}^{\mathrm{hard}} + \sum_j p_{\mathrm{T},j}^{\mathrm{soft}}}, \quad (15)$$

where ϕ^{hard} (respectively, ϕ^{soft}) represents the azimuthal angle between the thrust axis and the hard (respectively, soft) component. The so-called transverse thrust differential can be built using the weighted difference between the mean values of the thrust and thrust minor: $\alpha\langle T_{\mathrm{min}}\rangle - \beta\langle\tau_{\mathrm{T}}\rangle$, being $\alpha = 1 - 2/\pi$ and $\beta = 2/\pi$, as follows:

$$D(\langle T_{\mathrm{min}}\rangle, \langle\tau_{\mathrm{T}}\rangle) \equiv \gamma_{\mathrm{MC}}(\alpha\langle T_{\mathrm{min}}\rangle - \beta\langle\tau_{\mathrm{T}}\rangle), \quad (16)$$

where the additional correction factor γ_{MC} was obtained from Pythia 6. Since for the soft underlying event, the following approximations are valid:

$$1 - \tau_{\mathrm{T}}^{\mathrm{soft}} \approx T_{\mathrm{min}}^{\mathrm{soft}} \approx 2/\pi, \quad (17)$$

then, in $D(\langle T_{\mathrm{min}}\rangle, \langle\tau_{\mathrm{T}}\rangle)$ the UE component approximately cancels out.

Figure 4 shows the transverse thrust distribution measured in $\mathrm{p\bar{p}}$ collisions at $\sqrt{s} = 1.96\,\mathrm{TeV}$ compared with NLO + NLL and Pythia 6 predictions, clearly NLO + NLL calculations deviate significantly from data and MC because they do

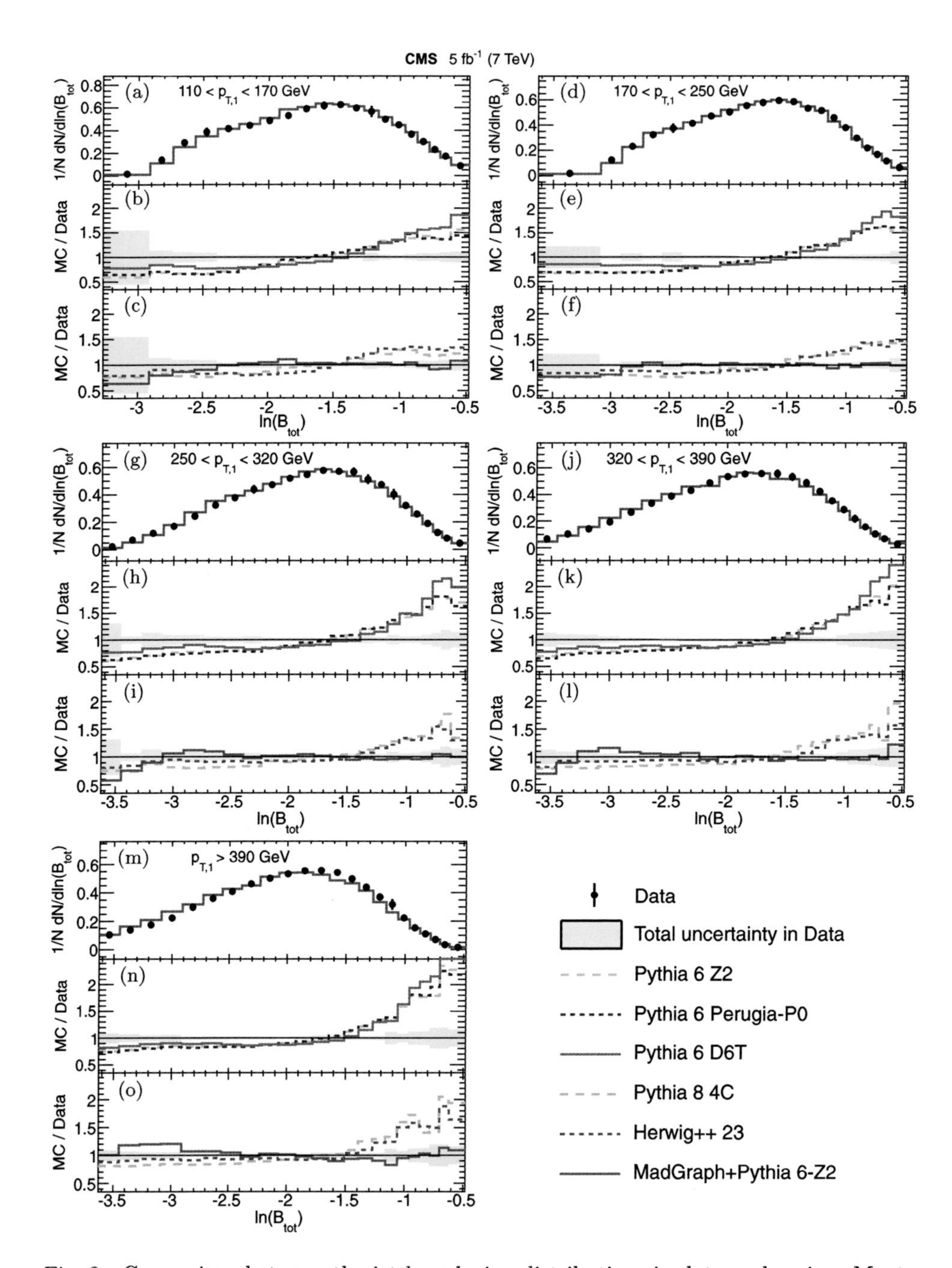

Fig. 3 Comparison between the jet broadening distributions in data and various Monte Carlo generators. Going from top to bottom, the leading jet p_{T} increases. Figures reproduced from Ref. 790.

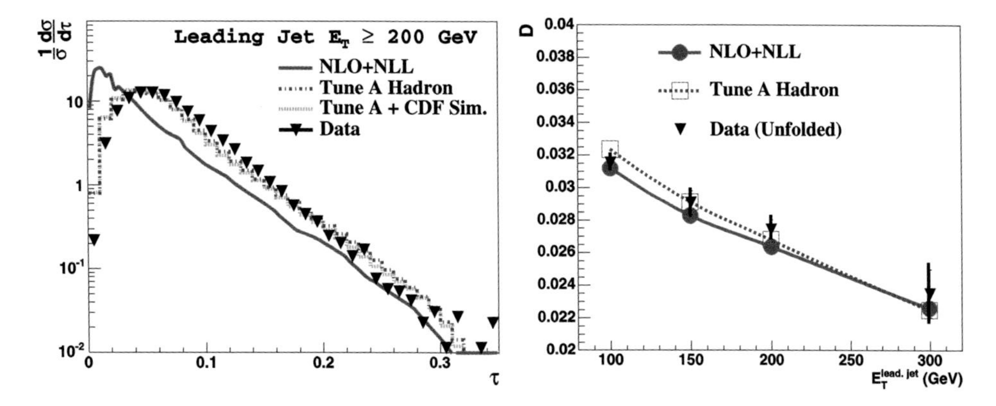

Fig. 4 The uncorrected distributions of transverse thrust for leading jet transverse energy greater than 200 GeV (left) and the thrust differential as a function of the transverse energy of the leading jet (right) measured in $p\bar{p}$ collisions at $\sqrt{s} = 1.96\,\text{TeV}$ are compared with a parton level NLO + NLL calculations and with Pythia 6 at the hadron level. Figures reprinted with permission from Ref. 495. Copyright 2011 by the American Physical Society.

not incorporate neither hadronization nor the underlying event. However, Pythia 6 and the NLO + NLL calculations succeed in describing the data on the thrust differential.

Studies using event shapes in Z+jets final states in pp collisions at the LHC have also been reported.[788,801] It has been shown that Pythia agrees better with data only for large transverse momentum of Z ($> 150\,\text{GeV}/c$). On the other hand, the MC models that combine multi-parton QCD LO ME interfaced to parton shower evolution tend to describe the data over a broader range of p_T-Z. This shows the importance of additional corrections from LO and NLO ME formulations.

4. Soft QCD: Prospects for Small Systems

The main difference with respect to the studies using hard events, is that for soft physics the events shape calculation considers charged particles (tracks) as input. For example, the ALICE collaboration reported for the first time a study of transverse sphericity as a function of multiplicity considering tracks with transverse momentum above $0.5\,\text{GeV}/c$ and within its Time Projection Chamber acceptance, $|\eta| < 0.8$.[798]

The study of soft physics in pp and p–Pb collisions is attractive because in high multiplicity events unexpected new collective-like phenomena were recently discovered at the LHC. In particular, for high-multiplicity proton–proton and proton–lead collisions, radial flow signals,[673,710] long-range angular correlations,[451,729,747] and the strangeness enhancement[443,679,684] have been reported. Those effects are well-known in heavy-ion collisions, where they are attributed to the existence of the sQGP.[802] Understanding the phenomena is crucial because, for heavy-ion physics, pp and p–Pb collisions have been used as the baseline ("vacuum") to extract the genuine sQGP effects. However, it is worth mentioning that no jet quenching effects have been found so far in p–Pb collisions,[803] suggesting that initial state effects could also play a role in producing collective-like behavior in small collision systems.[706,804–809]

The issue with the high multiplicity pp collisions is that a non-negligible part of the events may have a hard scattering origin, and, it has been demonstrated that jets complicate the interpretation of the new phenomena in terms of collectivity.[796,797,810] Fortunately, the event shape analysis is a promising tool for controlling the jet bias in high multiplicity pp events.[796,797]

A simulation study using inclusive pp collisions (without any selection on multiplicity) has already unveiled a difference in the particle composition when one studies the jetty-like and isotropic events, separately. For example, Fig. 5 shows the p_{T}-integrated particle ratios (K/π, p/π and ϕ/π) calculated for pp collisions at $\sqrt{s} = 7\,\mathrm{TeV}$ using Pythia 8.[192] For p_{T} below $2\,\mathrm{GeV}/c$, the ratios exhibit a depletion going from low to high S_0, whereas, for larger p_{T} the ratios increase with spherocity. For p_{T} above 5–6 GeV/c the proton-to-pion ratio in high S_0 events gets more similar to that for low S_0. The particle ratios which involve strange hadrons deviate each other, i.e., the values for jetty-like and isotropic events, for $p_{\mathrm{T}} > 2\,\mathrm{GeV}/c$. These results are compared to those where the inelastic pp interactions have a partonic $\hat{p}_{\mathrm{T}}$ of 6 or $30\,\mathrm{GeV}/c$, being this the most energetic pQCD process within the event. The sample with $6\,\mathrm{GeV}/c$ jets is in a qualitatively good agreement with results for spherical events suggesting that this tool can be used to isolate the underlying event. While considering events with $30\,\mathrm{GeV}/c$ jets, the particle ratios are significantly below those for soft events. It is worth noticing that a similar effect has been observed in the hadrochemistry measured in the so-called

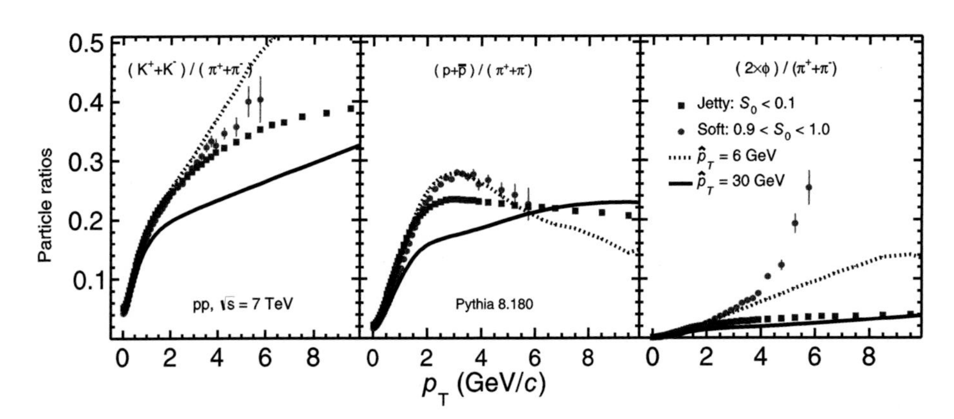

Fig. 5 Particle ratios as a function of p_{T} for jetty-like (squares) and soft (circles) events obtained in inclusive pp collisions simulated with PYTHIA 8. Results are compared to other simulations where jets with transverse momentum ($\hat{p}_{\mathrm{T}}$) of 6 (dotted line) and 30 (solid line) GeV/c are produced. Figure reproduced from Ref. 796.

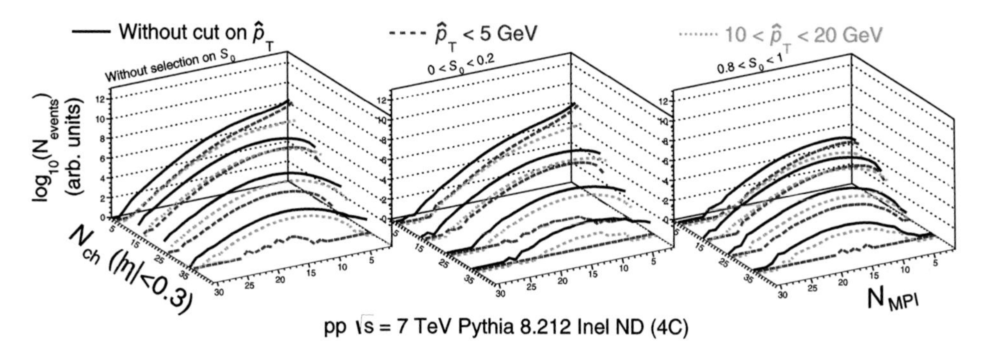

Fig. 6 Distributions of the number of multi-parton interactions as a function of the event multiplicity and leading partonic p_T ($\hat{p}_T$) for non-diffractive (ND) inelastic pp collisions at $\sqrt{s} = 7\,\text{TeV}$. From left to right, inclusive, jetty-like events (low spherocity) and isotropic events (high spherocity) are shown. Figure reprinted with permission from Ref. 813, Copyright 2016, with permission from Elsevier.

"bulk" (outside the jet peak) and the jet regions in p–Pb and Pb–Pb collisions at the LHC.[811,812]

Another potential application is related with the possibility of using the event shapes for selecting events with different number of multi-parton interactions (MPI).[797,813] Figure 6 shows the distribution of number of MPI (N_{MPI}) as a function of multiplicity for non-diffractive pp collisions and for two extreme spherocity bins, $S_0 < 0.2$ (jetty-like) and $S_0 > 0.8$ (isotropic). Also shown are the results for two intervals of the leading parton transverse momentum ($\hat{p}_T$) of the event: $\hat{p}_T < 5\,\text{GeV}$ and $10 < \hat{p}_T < 20\,\text{GeV}$. We observe that the average number of multi-parton interactions increases with increasing multiplicity, slightly more for isotropic events. At low multiplicity the dominant events are those of low N_{MPI} and with $\hat{p}_T < 5\,\text{GeV}/c$. On the other hand, in high multiplicity events the number of events with $10 < \hat{p}_T < 20\,\text{GeV}$ is larger than that for $\hat{p}_T < 5\,\text{GeV}$. In Pythia, high multiplicity events are therefore related with high N_{MPI} activity, where the occurrence of hard partonic scatterings is higher than in low N_{MPI} events.[798] It is worth noticing that the event is isotropic or jetty-like within the restricted η-range which is considered for the calculation of the spherocity. Therefore, by selecting isotropic (jetty-like) events we can study samples with enriched (reduced) underlying event activity within the acceptance under consideration.

The first measurement at the LHC on event shapes using MB events came from the ALICE collaboration, which reported the average transverse sphericity as a function of the event multiplicity. Both quantities were computed at mid-pseudorapidity ($|\eta| < 0.8$) using pp data at $\sqrt{s} = 0.9$, 2.76 and 7 TeV.[798] The mean sphericity as a function of multiplicity was studied for different event classes selected using a cut on the maximum p_T of the event (leading particle), being the "soft" ("hard") events those without (with) the leading particle having p_T above $2\,\text{GeV}/c$. In Fig. 7, the results are presented for the aforementioned event classes

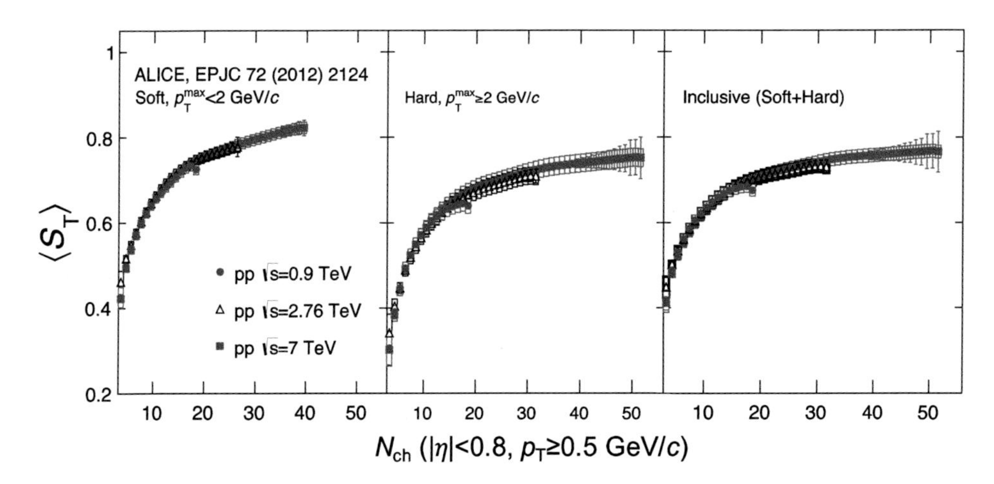

Fig. 7 Mean transverse sphericity versus multiplicity for inclusive (right), "hard" (middle) and "soft" (left) pp collisions at $\sqrt{s}$ = 0.9 TeV, 2.76 and 7 TeV. The statistical and systematic uncertainties are displayed as error bars and empty boxes, respectively. Data have been taken from Ref. 798.

along with the inclusive case ("all"). Like in other observables recently reported in the context of collectivity in small systems,[673,683] the average sphericity does not show dependence on $\sqrt{s}$. Instead, the physics seems to be encoded in the multiplicity. For "soft" events the rise is significantly steeper than in "hard" events, however, no indication of a reduction of the average transverse sphericity is drawn at high multiplicity for inclusive and "hard" pp collisions. The comparison of these results with MC predictions (Pythia 6, Pythia 8 and Phojet[404]) gives interesting insight, because as shown in Fig. 8 the MC average sphericity exhibits a reduction at high multiplicity, in inclusive and "hard" pp collisions, whereas the sphericity in data stays constant or shows a little increase with multiplicity. In QCD-inspired MC generators, the prime mechanism to produce high multiplicity pp collisions is related with the partonic interactions with large momentum transfer, therefore, the reduction may indicate an increase on the production of back-to-back jets. The deviation of the data from QCD-inspired predictions can be explained in the framework of the string percolation model[814] where the area covered by the strings increases with the size (multiplicity) and energy of the system leaving less room for hard scatterings.[815] Elliptic flow has been also proposed to explain the differences observed at high multiplicity.[816] A complementary study using jets was further performed by the CMS collaboration, supporting the findings reported by ALICE. Namely, the deviations from PYTHIA prediction at high multiplicity could be due to an apparent reduction and softening of the jet yields.[154]

The ATLAS collaboration reported a similar study but using transverse thrust, thrust minor and transverse sphericity, each defined for events having at least six charged particles ($|\eta| < 2.5$ and $p_T > 0.5\,\mathrm{GeV}/c$) for the calculation of the event

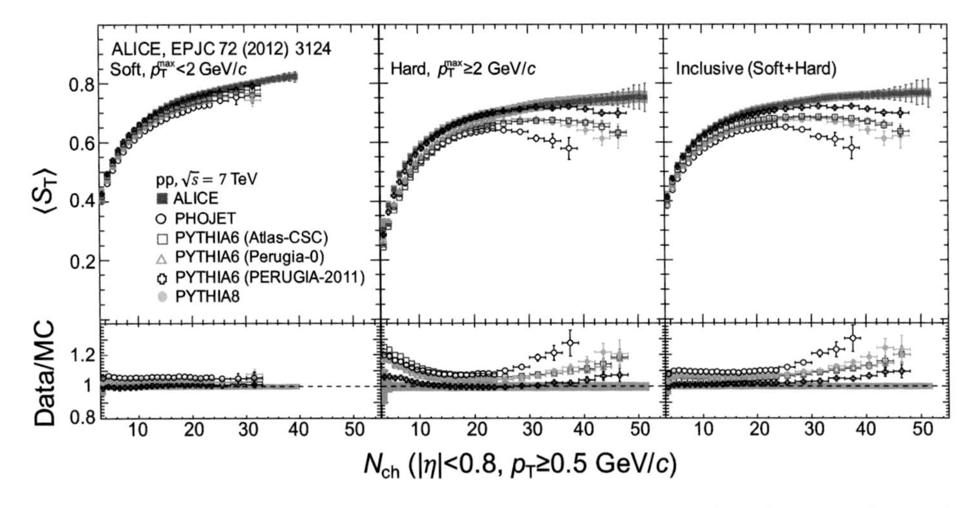

Fig. 8 Mean transverse sphericity versus multiplicity for inclusive (right), "hard" (middle) and "soft" (left) pp collisions at $\sqrt{s} = 7\,\text{TeV}$. The ALICE data are compared with different MC models: Phojet, Pythia 6 (tunes: ATLAS-CSC, Perugia-0 and Perugia-2011) and Pythia 8. The statistical errors are displayed as error bars and the systematic uncertainties as the shaded area. Figure reproduced from Ref. 798.

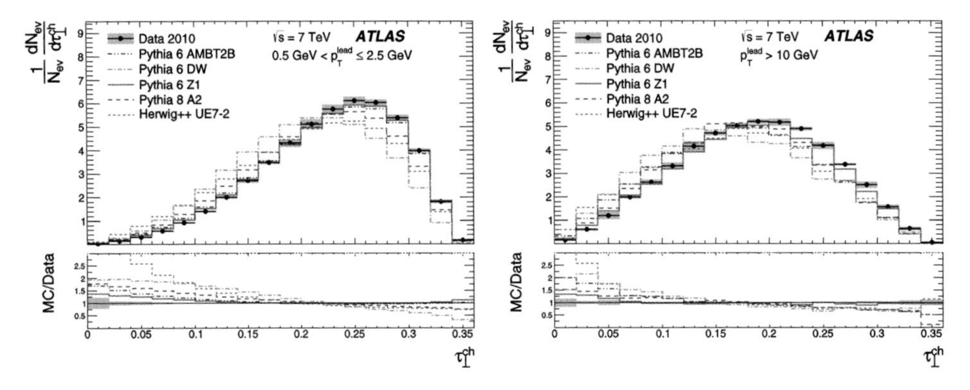

Fig. 9 Transverse thrust distributions for pp collisions at $\sqrt{s} = 7\,\text{TeV}$. The results are presented for "soft" (left) and "hard" (right) events selected using a cut on the p_T of the leading particle. Data are compared with different MC models: Pythia 6, Pythia 8 and Herwig++. Figures reproduced from Ref. 817.

shape.[817] The distributions shown in Fig. 9 indicate a prevalence of spherical events in the lower leading p_T intervals, and then a slight shift toward less spherical events and a broadening of the distributions is observed for events having larger leading particle p_T (>10 GeV/c). The different event generators (Pythia 6,[158] Pythia 8[192] and Herwig++[484]) which are compared with data share a similar feature, they overestimate (underestimate) the production of low (high) spherical events. Overall, the Pythia 6 tune Z1, tuned to the underlying event distributions at the LHC, agrees

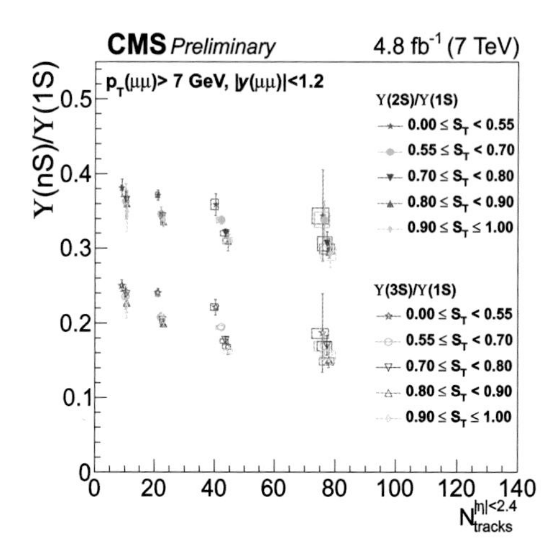

Fig. 10 Υ(2S)/Υ(1S) and Υ(3S)/Υ(1S) as a function of the event multiplicity ($|\eta| < 2.4$ and $p_{\rm T} > 0.4\,{\rm GeV}/c$) in different intervals of sphericity. The Υ states satisfy $p_{\rm T} > 7\,{\rm GeV}/c$ and $|y| < 1.2$. Figure reproduced from Ref. 818.

the best with most of the distributions. For the different event shapes studied by ATLAS, the evolution toward a more spherical event shape with increasing multiplicity is rapid initially and slows at higher multiplicities.

The event shape analysis at hadron colliders has been shown to be feasible for soft physics. Now the next step is to study the events cutting on the shape. In this way, at high multiplicities jets and the underlying event (or maybe sQGP enhanced samples) can be analyzed separately. The first attempts have been reported by the CMS collaboration who has measured observables traditionally used to probe the sQGP formation in heavy-ion collisions. Using pp data at $\sqrt{s} = 7\,\text{TeV}$, the ratios: Υ(2S)/Υ(1S) and Υ(3S)/Υ(1S) have been measured as a function of the charged particle multiplicity and for different sphericity classes.[818] The ratios exhibit a reduction with increasing multiplicity, being the Υ(2S)/Υ(1S) ratios significantly higher than those for Υ(3S)/Υ(1S). Moreover, in Fig. 10 we can see their sphericity dependences which show that the ratios are systematically lower (effect of $\approx 40\%$) in spherical events than in jetty-like events. The sphericity dependence of the ratios resemble features of the sequential suppression of the Υ(nS) states observed in heavy-ion collisions.[819]

So far, the results are encouraging and therefore, further studies are desirable.

5. Outlook

The event shape variables have been widely studied using hard events of p$\bar{\rm p}$ and pp collisions at the Tevatron and at the LHC, respectively. The vast amount of data has allowed to test pQCD predictions. Overall, a good agreement between data and theory is observed, the deviations are well understood as due to the lack of precise modeling of the matrix element calculations and parton showers. Moreover, the measured event shapes which were constructed to be less sensitive to the underlying event agree much better with the theoretical expectations.

For soft physics, the recent discovery of sQGP-like signatures in small systems calls for new analysis techniques in order to control the pQCD processes in high multiplicity events. As shown in this review, an analysis combining event shapes and multiplicity is a potential new direction in the study of the origin of the most intriguing phenomena seen at the LHC. Two examples were presented. The event shape distributions at high multiplicity which showed that models underestimate the amount of isotropic events and overestimate the production of jetty-like events. Suggesting that in nature soft physics may dominate at high multiplicities. The preliminary study using Υ(nS) states shows different Υ(2S)/Υ(1S) and Υ(3S)/Υ(1S) ratios for jetty-like and isotropic events. Further studies are needed to established whether or not this is connected with the sQGP formation. Finally, the Monte Carlo results presented here encourage the development of analogous analyses using data.

Acknowledgments

The author acknowledges Guy Paić and Peter Christiansen for the critical reading of the manuscript and the valuable discussion and suggestions. Support for this work has been received by CONACyT under the grant number 280362 and PAPIIT-UNAM under Project No. IN102118.

Chapter 17

Dipoles in Impact Parameter Space and Rapidity

Gösta Gustafson and Leif Lönnblad

Department of Astronomy and Theoretical Physics, Lund University, Sweden

We describe the DIPSY model for initial state evolution and the implementation of this in an event generator for producing fully exclusive hadronic final states. The model is formulated in impact parameter space and rapidity. Including non-leading and nonlinear effects, it gives a unique picture of fluctuations and correlations in the initial state parton evolutions. It also gives an impact-parameter-dependent description of the final state, allowing for new insights into the hadronization process in dense collision environments.

1. Introduction

The experiments at HERA showed us that the parton density at small x grows rapidly $\sim 1/x^{1.3}$, as predicted by the perturbative BFKL pomeron. For pp collisions this implies a very large probability for gluon–gluon subcollisions, which implies that unitarity constraints are very important. These constraints lead to saturation of the gluon density, and suppression of partons with $k_\perp < Q_s^2$, which may explain why models based on multiple perturbative partonic subcollisions (such as Pythia[113,487]) are very successful at high energies. One can then ask: if perturbative physics dominates, is it then possible to calculate the result from basic principles, without experimentally determined parton densities as input?

Unitarity constraints and saturation are most easily described in impact parameter space. Rescattering, which is represented by a convolution in $\mathbf{k}_\perp$-space, simplifies to a product in $\mathbf{b}$-space, and the optical theorem reads

$$\mathrm{Im}A(b) = \frac{1}{2}\left\{|A(b)|^2 + \sum_n |A_n(b)|^2\right\}. \tag{1}$$

Here $A(b)$ is the elastic scattering amplitude and $A_n(b)$ describes the inelastic transition to state n.

The small size of Re $A_{\rm el}^{pp}$ indicates that the pp interaction is driven by absorption into inelastic channels. If the absorption probability into state n in the Born approximation equals $2F_n$, then in the eikonal approximation rescattering effects exponentiate, giving a total absorption probability

$$P_{\rm abs} = \sum_n |A_n(b)|^2 = 1 - e^{-2F(b)}, \quad \text{where } 2F(b) \equiv \sum_n 2F_n. \tag{2}$$

If the real part of the elastic amplitude can be neglected, then Eq. (1) gives

$$\text{Im } A(b) = 1 - e^{-F(b)}. \tag{3}$$

(The factor 2 in the definition of F in Eq. (2) is a convention which simplifies this result.) With the definition $T(b) \equiv \text{Im } A(b)$, we then get the following cross-sections:

$$d\sigma_{el}/d^2b = T^2 = (1 - e^{-F})^2, \tag{4}$$

$$d\sigma_{tot}/d^2b = 2T = 2(1 - e^{-F}), \tag{5}$$

$$d\sigma_{inel}/d^2b = 2T - T^2 = 1 - e^{-2F}. \tag{6}$$

In this chapter we will discuss an attempt to describe high energy interactions with a minimum of non-perturbative input, with the expressed aim to gain improved insight into the underlying dynamics rather than highest precision. The model is based on the perturbative BFKL pomeron in impact parameter space, with the inclusion of unitarity constraints and important non-leading effects. It is implemented in a Monte Carlo (MC) called DIPSY, applicable to collisions between photons, protons, and nuclei. Although the model cannot compete in precision with conventional MCs when it comes to rare events due to hard subcollisions, the DIPSY MC can give predictions for effects due to fluctuations and correlations based on fundamental principles. Here we will discuss diffractive excitation described in the Good–Walker formalism, and correlation effects like double (or multiple) parton distributions, and effects of high energy density. The latter may cause the formation of "ropes" and give transverse flow, which we will also discuss in this chapter.

2. Dipole Cascade Evolution

2.1. *Mueller's dipole model*

Mueller's dipole model[820–822] is a formulation of LL BFKL evolution in transverse coordinate space. Gluon radiation from the color charge in a parent quark or gluon is screened by the accompanying anticharge in the color dipole. This suppresses emissions at large transverse separation, which corresponds to the suppression of small $k_\perp$ in BFKL. For a dipole between transverse positions $\boldsymbol{x}$ and $\boldsymbol{y}$, the probability per unit rapidity (Y) for emission of a gluon at transverse position $\boldsymbol{z}$ is given by

$$\frac{d\mathscr{P}}{dY} = \frac{\bar{\alpha}}{2\pi} d^2\boldsymbol{z} \frac{(\boldsymbol{x}-\boldsymbol{y})^2}{(\boldsymbol{x}-\boldsymbol{z})^2(\boldsymbol{z}-\boldsymbol{y})^2}, \quad \text{with } \bar{\alpha} = \frac{3\alpha_s}{\pi}. \tag{7}$$

This emission implies that the dipole is split into two dipoles, which (in the large N_c limit) emit new gluons independently, as illustrated in Fig. 1. The result reproduces the BFKL evolution, with the number of dipoles growing exponentially with Y.

In a high energy collision, the dipole cascades in the projectile and the target are evolved from their rest frames to the rapidities they will have in the specific Lorentz frame chosen for the analysis. Two colliding dipoles interact via gluon exchange, which implies a color connection between the projectile and target remnants, as indicated in Fig. 2. In the Born approximation, the interaction probability between one dipole with coordinates $(\boldsymbol{x}_i, \boldsymbol{y}_i)$ in the projectile, and one with coordinates $(\boldsymbol{x}_j, \boldsymbol{y}_j)$ in the target, is given by

$$2f_{ij} = 2f(\boldsymbol{x}_i, \boldsymbol{y}_i | \boldsymbol{x}_j, \boldsymbol{y}_j) = \frac{\alpha_s^2}{4}\left[\log\left(\frac{(\boldsymbol{x}_i-\boldsymbol{y}_j)^2(\boldsymbol{y}_i-\boldsymbol{x}_j)^2}{(\boldsymbol{x}_i-\boldsymbol{x}_j)^2(\boldsymbol{y}_i-\boldsymbol{y}_j)^2}\right)\right]^2. \tag{8}$$

Assuming that the subcollisions are uncorrelated, multiple collisions are taken into account in the eikonal approximation, where the probability for an inelastic interaction is given by

$$P_{\text{abs}} = 1 - e^{-2F}, \quad \text{with } F = \sum f_{ij}. \tag{9}$$

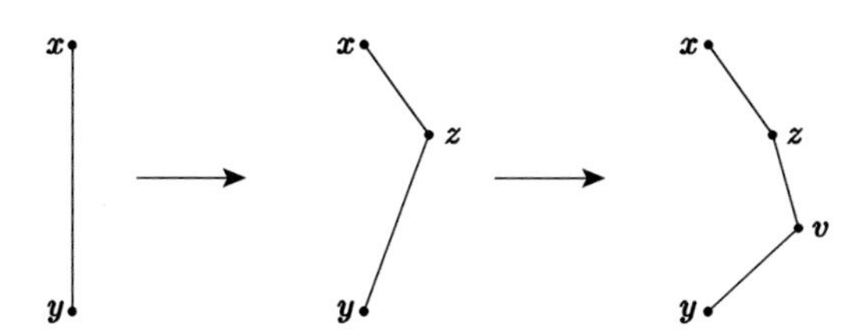

Fig. 1 The evolution of the dipole cascade in transverse coordinate space. In each step, a dipole can split into two new dipoles with decay probability given by Eq. (7).

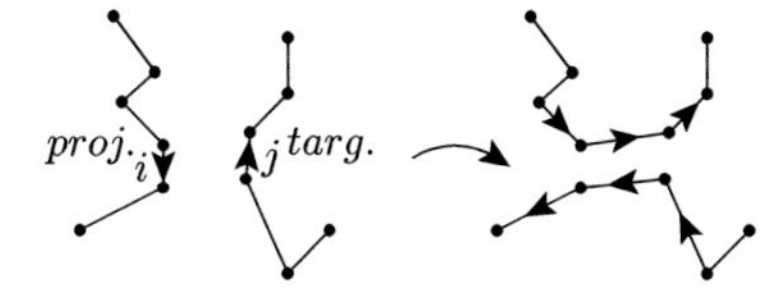

Fig. 2 A dipole–dipole interaction implies exchange of color and reconnection of the dipole chains in the colliding cascades. The arrows indicate the direction of the dipole going from color charge to anticharge.

The multiple interactions produce loops of dipole chains, corresponding to the pomeron loops in the reggeon formalism. As discussed in the introduction, assuming that the elastic scattering amplitude, T, is driven by absorption into inelastic states, we find via the optical theorem

$$T = 1 - e^{-F}. \tag{10}$$

To account for fluctuations in the cascade evolution, we should here take the average over all possible cascades, as discussed in Section 5 below. The differental cross-sections are then given by Eq. (6) with $e^{-F} \to \langle e^{-F} \rangle$.

2.2. *The Lund dipole cascade model DIPSY*

The Lund dipole cascade introduces a number of corrections to Mueller's original formulation, which are all beyond the leading logarithmic approximation. The corrections are described in greater detail in Refs. 416, 823 and 824, and we present here a short summary.

2.2.1. *Beyond LL BFKL evolution*

The NLL corrections to BFKL evolution have three major sources:[825]

Energy–momentum conservation: The non-singular terms in the splitting function suppress large z-values in the individual parton branchings, and correspond approximately to a veto for $\ln 1/z < 11/12$. This prevents the child from being faster than its recoiling parent. Most of this effect is taken care of by including energy–momentum conservation in the evolution, and ordering in the lightcone momentum p_+. This constraint is effectively taken into account by associating a dipole with transverse size r with a transverse momentum $k_\perp = 1/r$, and demanding conservation of p_+ in every step in the evolution. In addition this gives an effective cutoff for small dipoles, which eliminates the numerical problems encountered in the MC implementation by Mueller and Salam.[826]

Projectile-target symmetry: A parton chain should look the same if generated from the target end as from the projectile end. The corresponding corrections are also called energy scale terms, and are essentially equivalent to the so-called consistency constraint.[827] This effect is taken into account by conservation of both positive and negative lightcone momentum components, p_+ and p_-.

Running coupling: This is relatively easily included in an MC simulation process. The scale in the running coupling is chosen as the largest transverse momentum in the vertex, in accordance with the results in Ref. 828.

Resummations: It is well known that the NLL corrections are not sufficient to give a realistic result for small x evolution. As discussed in Refs. 829–831, it is important to resum subleading logs to all orders. In our MC simulation this is handled automatically in the Sudakov-veto algorithm.

2.2.2. *Nonlinear effects and subleading N_c corrections. The swing*

As mentioned above, multiple interactions produce loops of dipole chains corresponding to pomeron loops. Mueller's model includes all loops cut in the particular Lorentz frame used for the analysis, but not loops contained within the evolution of the individual cascades. A consistent model has to be independent of the frame used for the calculations, and thus color loops must also be possible within the evolution. As for dipole scattering the probability for such loops should be given by α_s^2, while a pair of dipole splittings is proportional to $\bar{\alpha}^2$, with $\bar{\alpha} = N_c\alpha_s/\pi$. Dipole loops are therefore formally color suppressed compared dipole splitting, and therefore related to the probability that two dipoles have the same color. Two dipoles with the same color form a quadrupole field. Such a field may be better approximated by two dipoles formed by the closest color–anticolor charges. This corresponds to a recoupling of the color dipole chains, and we call this process a dipole "swing".

Dipoles with the same color are allowed to swing back and forth, which results in an equilibrium, where the smaller dipoles have a larger weight. Although this scheme does not give an exactly frame independent result, the MC simulations show that it is a fairly good approximation.

In this formulation the number of dipoles is not reduced, and the saturation effect is obtained because the smaller dipoles have smaller cross-sections. When such a small dipole does not interact, it has to be regarded as virtual, be reabsorbed and thus not active in the scattering. The process can then also be interpreted as a $2 \to 1$ process, as, e.g., in the BK equation.[631,632]

2.2.3. *Confinement effects*

Confinement effects are included via an effective gluon mass, which gives an exponential suppression for very large dipoles.[824] This prevents the proton from growing too fast in transverse size, and is also essential to satisfy t-channel unitarity and Froisart's bound at high energies.[832]

2.2.4. *Initial dipole configurations*

In DIS an initial photon is split into a $q\bar{q}$ pair, and for larger Q^2 the wavefunction for a virtual photon can be determined perturbatively. For an initial proton we also need a starting configuration in the proton rest frame, even if the full evolution is calculated within the model. In DIPSY the starting configuration is represented by an equilateral triangle formed by three dipoles, and with a radius of 3 GeV^{-1} $\approx$ 0.6 fm. The model should be used at low x, and when the system is evolved over a large rapidity range, the observable results depend only weakly on the exact configuration of the initial dipoles.

3. Inclusive Observables

There are a number of parameters in the DIPSY model that needs to be tuned to data. The most influential are

- $R_{\max}$: Non-perturbative regularization scale, this corresponds to the maximum dipole size, above which emissions and interactions are exponentially suppressed.[824]
- R_p: The average size of the proton at rest.
- w_p: The width of the Gaussian fluctuations in proton size around R_p.
- Λ_{QCD}: This is the scale parameter of α_s, the running coupling constant of QCD.
- λ_r: This parameter controls the strength of swing effect.

In tuning these parameters to (semi-)inclusive cross-sections one typically ends up with quite reasonable values of $R_{\max} \approx R_p \approx 3$ GeV^{-1}. Λ_{QCD} needs to be somewhat above 200 MeV which may seem very high compared to current N^{n}LO fits, but is quite compatible with leading logarithmic fits of parton showers to LEP data.

The elastic and diffractive cross-sections are very sensitive to the fluctuations in the size of the initial proton state, and fits tend to favor small values of w_p, typically around 0.1 GeV^{-1}.[833]

Finally the strength of the swing mechanism should be taken as large as possible so that same-colored dipoles are allowed to swing back and forth and reach some kind of equilibrium. However, a large λ_r will slow down the simulations drastically. We have found[636] that setting $\lambda_r = 1$ is typically enough to reach this equilibrium, and increasing the value further does not give noticeable effects.

In Fig. 3 we show a very reasonable fit to the total and elastic pp cross-section. We are also able to obtain good fits to DIS data, both total and quasi-elastic (exclusive vector meson production and DVCS) $\gamma^\star$p cross-sections for a large range of Q^2, as well as some diffractive data.[833,835] Also many measurements of diffractive cross-sections in both DIS and pp are well described.[824]

4. Exclusive Final States

BFKL evolution properly reproduces inclusive observables in hadronic collisions. However, for exclusive final states it is necessary to take into account color coherence and angular ordering, as well as soft radiation, related to the $z=1$ singularity in the gluon splitting function. These effects are taken into account in the CCFM formalism,[390,391] which also reproduces the BFKL result for inclusive cross-sections.

An essential point is here the fact that the softer emissions in the CCFM formalism can be resummed, and the total cross-section, as well as the structure of the final states, are fully determined by the "$k_\perp$-changing" gluons. We denote the real emitted gluons in a ladder q_i and the virtual links k_i, and momentum

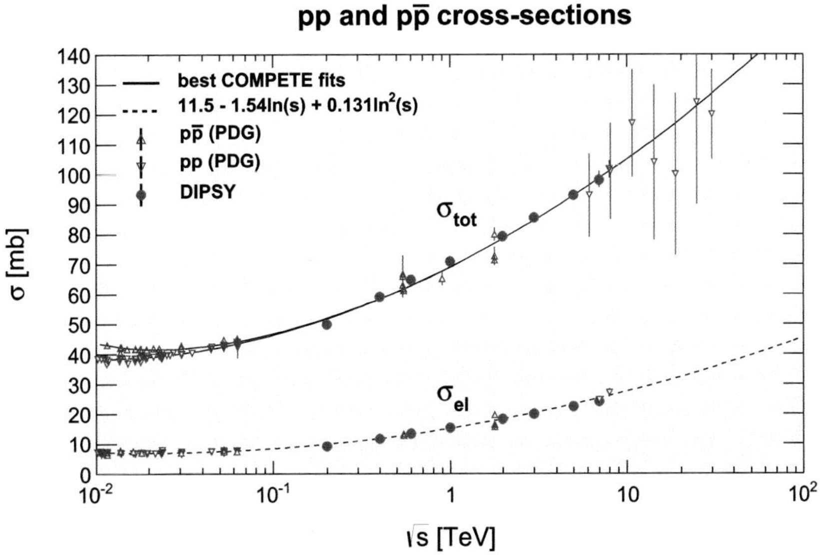

Fig. 3 Total and elastic cross-sections in pp and p$\overline{\text{p}}$ collisions plotted together with tuned DIPSY simulation results (red filled circles). The total cross-section data (open triangles) are described by the best fit of the COMPETE collaboration[834] and the inelastic cross-sections are also fitted by a phenomenological formula as shown in the figure.

conservation then implies $\mathbf{k}_{\perp i-1} = \mathbf{k}_{\perp i} + \mathbf{q}_{\perp i}$. For $k_\perp$-changing emissions $k_{\perp i}$ is either much larger or much smaller than $k_{\perp i-1}$. This also means that $q_{\perp i} \approx \max(k_{\perp i}, k_{\perp i-1})$. These emissions are called "primary gluons" in Ref. 395, and "backbone gluons" in Ref. 836, and the weight for such a backbone chain is given by

$$\text{weight} = \prod \bar{\alpha} \frac{dq^2_{\perp i}}{q^2_{\perp i}} dy_i. \tag{11}$$

As discussed in Ref. 837, this feature also gives a dynamical cutoff for small $q_\perp$, which grows slowly with energy, and gives a dynamical description of the cutoff for hard subcollisions needed in event generators like Pythia 8[192] or Herwig 7.[484]

The procedure for generating inelastic (non-diffractive) final states in DIPSY is therefore done in a number of steps as described in detail in Ref. 636:

- Generate two dipole cascades for the projectile and the target respectively, together with an impact parameter **b**.
- Determine which dipoles in the projectile and target become color connected via gluon exchange. This is simply given by the inelastic non-diffraction interaction probability for a given pair of projectile and target dipole, $1 - e^{-2f_{ij}}$ (see Eq. (9)).

- Any parton that has not taken part in a dipole–dipole interaction (either directly or indirectly via a daughter parton) is considered virtual and is reabsorbed.
- Any gluon that is not "$k_\perp$-changing", as discussed above, is considered to belong to final-state emissions and is therefore reabsorbed.
- All remaining partons now belong to "backbone" chains and are put on shell.
- Add softer emissions as final state radiation using ARIADNE [838] with appropriate Sudakov form factors, vetoing any emission that would correspond to a "$k_\perp$-changing" gluon.
- Hadronize the resulting parton chains.

The end result is a fair description of pp minimum bias data.[636] Although the model includes some non-leading and non-perturbative effects, it only deals with gluons in the initial state and we do not expect to reproduce data as well as e.g., Pythia 8. In addition, although inclusive cross-sections are rather insensitive to which frame is chosen for the interactions, final state observables change significantly as is seen in Fig. 4

Although the final-state model basically does not introduce additional parameters, the important ones described in Section 3 are only partly constrained by (semi-)inclusive cross-sections, therefore there is some freedom to tune also to final state observables. However, it should be noted that so far it has proven to be very difficult to tune away the much too hard transverse momentum spectrum in Fig. 4.

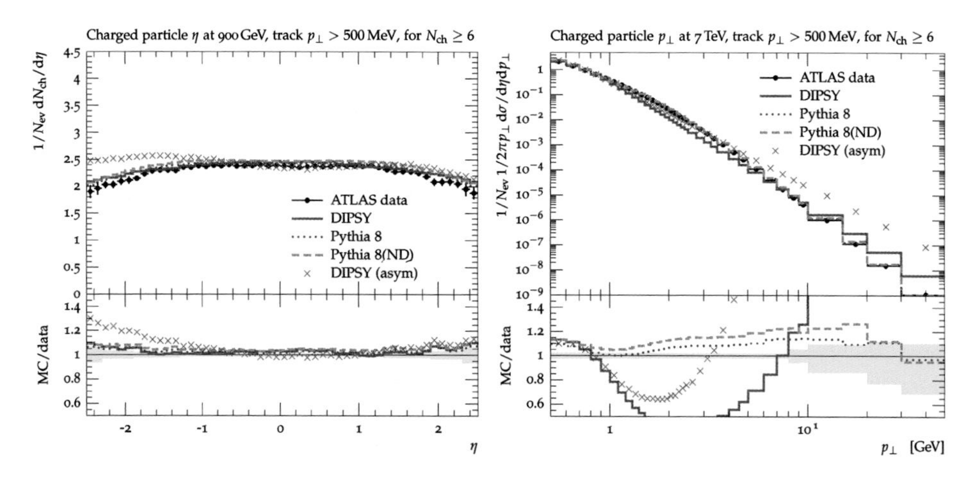

Fig. 4 The pseudo-rapidity (left) and transverse momentum (right) distribution of charged particles at 0.9 TeV (left) 7 TeV (right) compared to data from ATLAS[446] including only tracks with $p_\perp > 500$ MeV in events with more than six charged tracks. The full lines are the DIPSY results, the dotted lines are from Pythia (tune 4C) with diffractive and non-diffractive events, the dashed lines are Pythia with only non-diffractive events, and the crosses are from a DIPSY simulation in an asymmetric frame.

5. Fluctuations and Diffractive Excitation

Diffractive excitation contributes with a significant fraction of the pp cross-section, also at high energies (see, e.g., Refs. 839–841). This is often described within the Mueller–Regge formalism,[604] where high-mass diffraction is given by a triple-Pomeron diagram. It has also been described by the Good–Walker formalism,[603] as the result of fluctuations in the internal substructure of the colliding particles.[607,824,842] The stochastic nature of the BFKL Pomeron actually implies that the Good–Walker and the Mueller–Regge formalisms in fact describe the same dynamics, only seen from different sides.[843] At high energies the Good–Walker formulation has the advantage that saturation effects are easily accounted for within the dipole formalism, and the fluctuations in the cascade evolution are easily studied in the DIPSY MC. The saturation effects correspond to diagrams with Pomeron loops, which in the Mueller–Regge formalism lead to more complicated resummation schemes (see, e.g., Refs. 411, 844 and 845).

Good–Walker formalism: A projectile with a substructure may be diffractively excited to a different mass eigenstate. Assume that the projectile is a linear combination of diffractive eigenstates, Φ_n, with definite eigenvalues T_n. The elastic amplitude is then given by $\langle\Psi_{\rm in}|T(b)|\Psi_{\rm in}\rangle_{p,t}$, where the average is taken over both projectile and target states. The differential total and elastic cross-sections are then given by

$$d\sigma_{\rm tot}/d^2b = 2\langle T\rangle, \quad d\sigma_{\rm el}/d^2b = \langle T\rangle^2. \tag{12}$$

The total diffractive cross-section, including elastic scattering, is given by $\langle T^2\rangle$. Diffractive excitation is obtained subtracting the elastic, and thus given by the fluctuations $\langle T^2\rangle - \langle T\rangle^2$. For a fluctuating projectile scattering against a fluctuating target, the single and double excitations are obtained by separate averages over projectile and target states, below denoted by subscripts p and t respectively. Thus the cross-sections for single excitation of the projectile and of the target, and for double excitation are given by:

$$d\sigma_{SD,p}/d^2b = \langle\, \langle T\rangle_t^2\,\rangle_p - \langle T\rangle_{p,t}^2; \tag{13}$$

$$d\sigma_{SD,t}/d^2b = \langle\, \langle T\rangle_p^2\,\rangle_t - \langle T\rangle_{p,t}^2; \tag{14}$$

$$d\sigma_{DD}/d^2b = \langle T^2\,\rangle_{p,t} - \langle\, \langle T\rangle_t^2\,\rangle_p - \langle\, \langle T\rangle_p^2\,\rangle_t + \langle T\rangle_{p,t}^2. \tag{15}$$

Diffractive eigenstates: The parton cascades discussed in Section 2 can come on shell through interaction with a target. The BFKL evolution has a stochastic nature with large fluctuations, and following Hatta *et al.*[842] we assume that these cascades represent the diffractive eigenstates in high energy collisions. (A similar idea was early presented by Miettinen and Pumplin.[607])

Results: If the expression in Eq. 13 is calculated in a frame, where the projectile is evolved a rapidity range y_p, then the result gives the single diffractive cross-section,

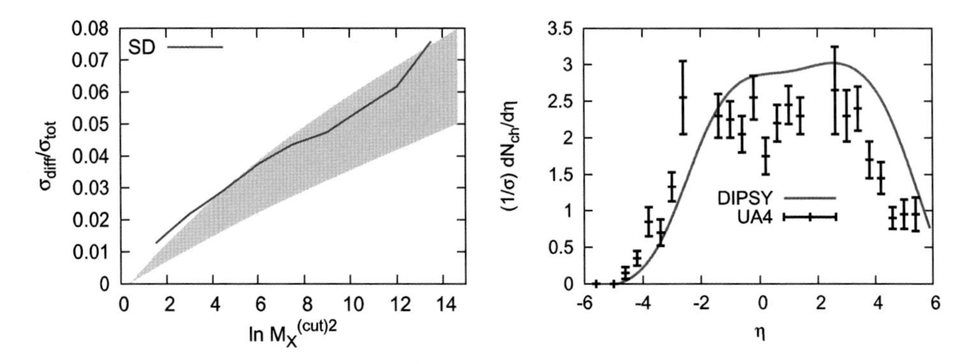

Fig. 5 Left: $\int dM_X^2\, d\sigma_{SD}/dM_X^2$ for $M_X < M_X^{(cut)}$. The shaded area is an estimate of CDF results. Figure taken from Ref. 964. Right: $dn_{ch}/d\eta$ in $\mathrm{p\bar{p}}$ collisions at 546 GeV and $\langle M_X \rangle = 140$ GeV. Data from UA4.[848]

including all excited projectiles confined within this range. This corresponds approximately to all diffractive masses in the range $M_X^2 < \exp(y_p)$ GeV^2. Varying the frame it is thus possible to calculate $d\sigma_{\rm SD}/dM_X^2$. Figure 5 (left) shows a comparison between DIPSY and CDF results for single diffraction *vs* max M_X^2.[846]

It is also possible to calculate diffractive final states, as described in Ref. 847. Due to interference between many terms, the calculations are, however, very slow. Figure 5 (right) shows the result for $dn_{\rm ch}/d\eta$ in $\mathrm{p\bar{p}}$ collisions. Note in particular that the MC is here only tuned to $\sigma_{\rm tot}$ and $\sigma_{\rm el}$, with no new parameter. (The MC result is too large for high η due to the lack of valence quarks in the model.) In Ref. 847 we also presented results for DIS in reasonable agreement with HERA data.

6. Correlations

6.1. *Multiparton interactions*

Evidence for multiparton interactions has most clearly been observed in events with four jets and with three jets + γ/W, first at the CERN ISR[162] and later at the Tevatron[97,148,153] and the LHC.[71,102,103] These results also show that the hard subcollisions are correlated, and double interactions occur with a larger probability than expected for uncorrelated hard interactions.

A measure of the correlation is conventionally given by the quantity $\sigma_{\rm eff}$, defined by the relation

$$\sigma^D_{(A,B)} = \frac{1}{(1+\delta_{AB})} \frac{\sigma^S_A \sigma^S_B}{\sigma_{\rm eff}}. \tag{16}$$

Here $\sigma^D_{(A,B)}$ is the cross-section for the two hard processes A and B, σ^S_A and σ^S_B the corresponding single inclusive cross-sections, and $(1+\delta_{AB})^{-1}$ is a symmetry factor equal to 1/2 if $A = B$. If the hard interactions were uncorrelated, σ_{eff} would be equal to the total non-diffractive cross-section. The TEVATRON and LHC measurements give instead $\sigma_{\text{eff}} \sim 15$ mb (with fairly large errors), which thus is significantly smaller.

Following Ref. 18, we define in Ref. 115 the "double parton distribution" $\Gamma_{ij}(x_1, x_2, b; Q_1^2, Q_2^2)$, which describes the inclusive density distribution for finding a parton of type i with energy fraction x_1 at scale Q_1^2, together with a parton of type j with energy fraction x_2 at scale Q_2^2, and with the two partons separated by a transverse distance b. The distributions Γ_{ij} are via a Fourier transformation related to the "two-parton generalized parton distributions" in transverse momentum space, $D(x_1, x_2, Q_1^2, Q_2^2, \mathbf{\Delta})$, studied by Blok *et al.*[82]

Assuming factorization of two hard subprocesses A and B, the cross-section for double scattering is given by

$$\sigma^D_{(A,B)} = \frac{1}{1+\delta_{AB}} \sum_{i,j,k,l} \int \Gamma_{ij}(x_1, x_2, b; Q_1^2, Q_2^2) \hat{\sigma}^A_{ik}(x_1, x_1') \hat{\sigma}^B_{jl}(x_2, x_2') \times \Gamma_{kl}(x_1', x_2', b; Q_1^2, Q_2^2) dx_1 dx_2 dx_1' dx_2' d^2b. \tag{17}$$

Here $\hat{\sigma}$ is the cross-section for a parton-level subprocess.

We define the distribution $F(b; x_1, x_2, Q_1^2, Q_2^2)$ by the relation

$$\Gamma(x_1, x_2, b; Q_1^2, Q_2^2) = D(x_1, Q_1^2) D(x_2, Q_2^2) F(b; x_1, x_2, Q_1^2, Q_2^2), \tag{18}$$

where D denotes the single parton distribution. Thus F is a density in transverse coordinate space $\mathbf{b}$, which may depend on all four variables x_1, x_2, Q_1^2, and Q_2^2, and which contains all information about correlations between the two partons. In case, e.g., some kind of "hot spots" develop for small x and/or large Q^2, this will show up as an increase in F for small b-values. We note here that the approximation used in the event generators Pythia and Herwig is to assume that F is a function of b and s only.

In DIPSY we determine F by colliding a proton with two virtual photons, separated by a distance b and with virtualities Q_1^2 and Q_2^2 respectively. The result is shown in Fig. 6. We note here that spikes or hotspots develop for small b-values, corresponding to a tail for large momentum imbalance Δ.

The effective cross-section σ_{eff} depends on both $\sqrt{s}$ and Q^2, as illustrated in Table 1. We see here that the variation with $\sqrt{s}$ for fixed Q^2 is much weaker than the variation of Q^2 for fixed $\sqrt{s}$. It is also shown in Ref. 115 that the variation with rapidity away from $y = 0$ is rather weak. Finally we note that part of the correlations is due to fluctuations. No fluctuations would imply $\int d^2b\, F(b) = 1$, while the MC gives $\int d^2b\, F(b) \sim 1.1$.

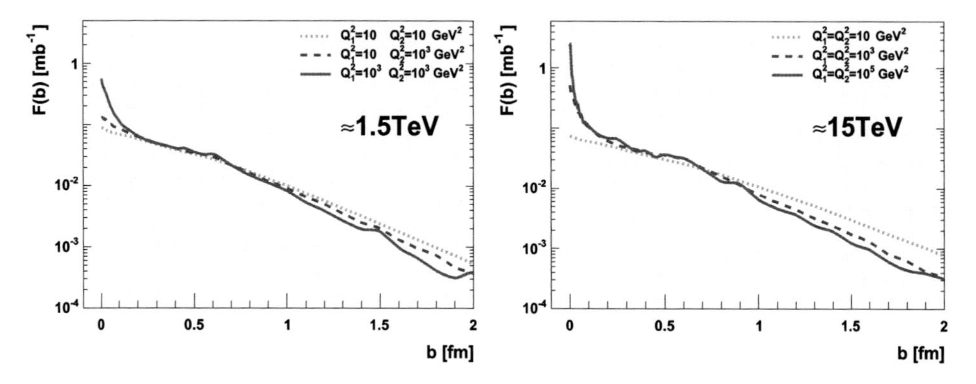

Fig. 6 Correlation function $F(b)$. Left: Central subcollisions at $\sqrt{s} \approx 1.5$ TeV, and three combinations of Q_1^2 and Q_2^2. The x-values correspond to $Q^2 = 10$ and 1000 GeV^2 are 10^{-3} and 10^{-2} respectively. Right: Corresponding result for $\sqrt{s} \approx 15$ TeV and $Q_1^2 = Q_2^2 = 10$, 10^3, and 10^5 GeV^2. The corresponding x-values are 10^{-4}, 10^{-3}, and 10^{-2}.

Table 1. Summary of results for σ_{eff} and corresponding integrals of the double distribution functions.

Q_1^2, Q_2^2[GeV^2], x_1, x_2				σ_{eff} [mb]	$\int F$
1.5 TeV, mid-rapidity					
10	10	0.001	0.001	35.3	1.09
10^3	10^3	0.01	0.01	23.1	1.06
15 TeV, mid-rapidity					
10	10	0.0001	0.0001	40.4	1.11
10^3	10^3	0.001	0.001	26.3	1.07
10^5	10^5	0.01	0.01	19.6	1.03

6.2. *Effects of high energy density*

6.2.1. *Ropes*

In most models for high energy collisions the hadronization mechanism is described via strings or cluster chains. The strings are often treated as independent, but in connection with nucleus collisions it was early suggested that the many strings produced within a limited space may interact and form "color ropes".[849,850] Such ropes have subsequently been studied by many authors with applications to high energy nucleus collisions.[851–857] The stronger field in a rope is expected to give larger rates for strangeness and baryon production.

Also in pp collisions the density of strings is here very high for high energies. This is illustrated in Fig. 7, which shows the extension of strings in $(\mathbf{r}_\perp, y)$-space for a pp collision at 7 TeV. The radius is set to 0.1 fm for a more clear picture, and

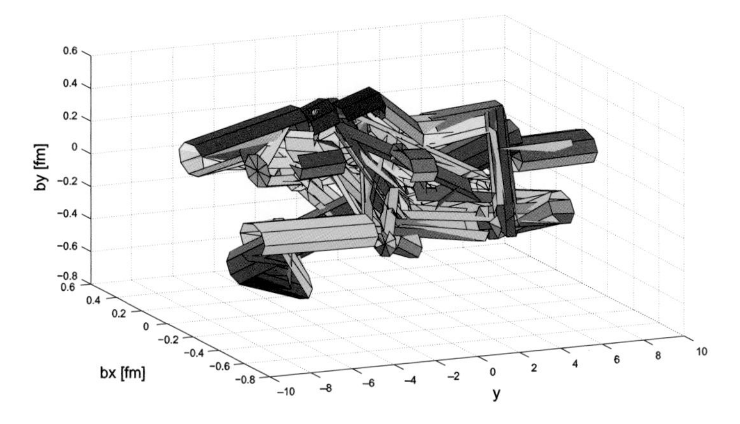

Fig. 7 Illustration of strings from a pp event at $\sqrt{s} = 7$ TeV in $(\mathbf{b}_\perp, Y)$-space before hadronization. Notice that the string radius is set at 0.1 fm — an order of magnitude less than in the calculation — in order to improve readability of the figure.

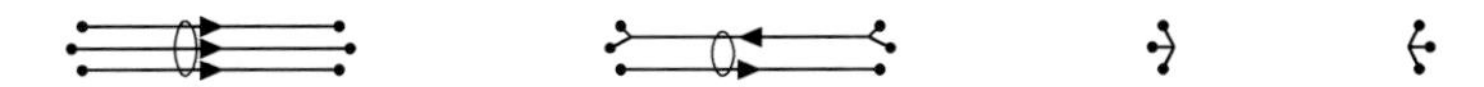

Fig. 8 Three uncorrelated triplet strings can form a decuplet, an octet, or a singlet. The decuplet and the octet can break in three or two steps respectively.

for a string radius of order 0.5 fm, there will clearly be a large amount of overlap. It may then not be surprising, that the rate of strangeness and baryons is somewhat higher in pp collisions than in e^+e^- collisions.

The Lund string hadronization model[858,859] is based on the assumption that a confined color field between a quark and an antiquark is compressed to a linear flux tube, similar to a vortex line in a superconductor. When the string is stretched between separated color charges, it can break by the production of $q\bar{q}$ pairs.[860–863] This process can be interpreted as the effect of a quantum-mechanical tunneling process,[864] similar to the production of e^+e^- pairs in a homogeneous electric field.[865] The production probability will then be suppressed for higher quark masses, μ, and transverse momenta, $k_\perp$:

$$P_{\text{tunnel}} \propto \exp\left(-\frac{\pi}{\kappa}(\mu^2 + k_\perp^2)\right). \tag{19}$$

As an example we can look at three uncorrelated parallel strings acting coherently as a rope. The three triplet charges at one end can form a color decuplet, an octet, or a singlet, as illustrated in Fig. 8.

We assume that the rope breaks by production of $q\bar{q}$ pairs. A decuplet can then break in three steps, first to a sextet and then to a triplet. Lattice calculations show that the tension in a rope is proportional to the quadratic Casimir operator. For a decuplet this gives a factor 9/2 and for a sextet 5/2 times the tension, κ_0, in the standard triplet string. When a quark pair is produced the *difference* in rope tension can be used to produce the new pair, and thus acts as an "effective string tension" κ_{eff}. For the first break in a decuplet rope this gives $\kappa_{\text{eff}} = 2\,\kappa_0$ and the second $\kappa_{\text{eff}} = 1.5\,\kappa_0$, before the final breakup of the remaining triplet string with $\kappa_{\text{eff}} = \kappa_0$. The enhanced effective string tension gives correspondingly larger ratios of strange quarks and of baryons, in accordance with Eq. (19). An octet rope has similarly a tension $9/4\,\kappa_0$ and breaks in two steps, while in case of a singlet no color field is stretched.

Some results from DIPSY are shown in Fig. 9. In general we see a clear improvement when adding ropes, with enhanced fractions of strange particles and baryons. We note that the enhancement is increasing with higher particle density, and that the baryon to meson ratio is increasing with $p_\perp$ for moderate $p_\perp$ in a way similar to a flow. We see, however, that the description is not good for higher $p_\perp$,

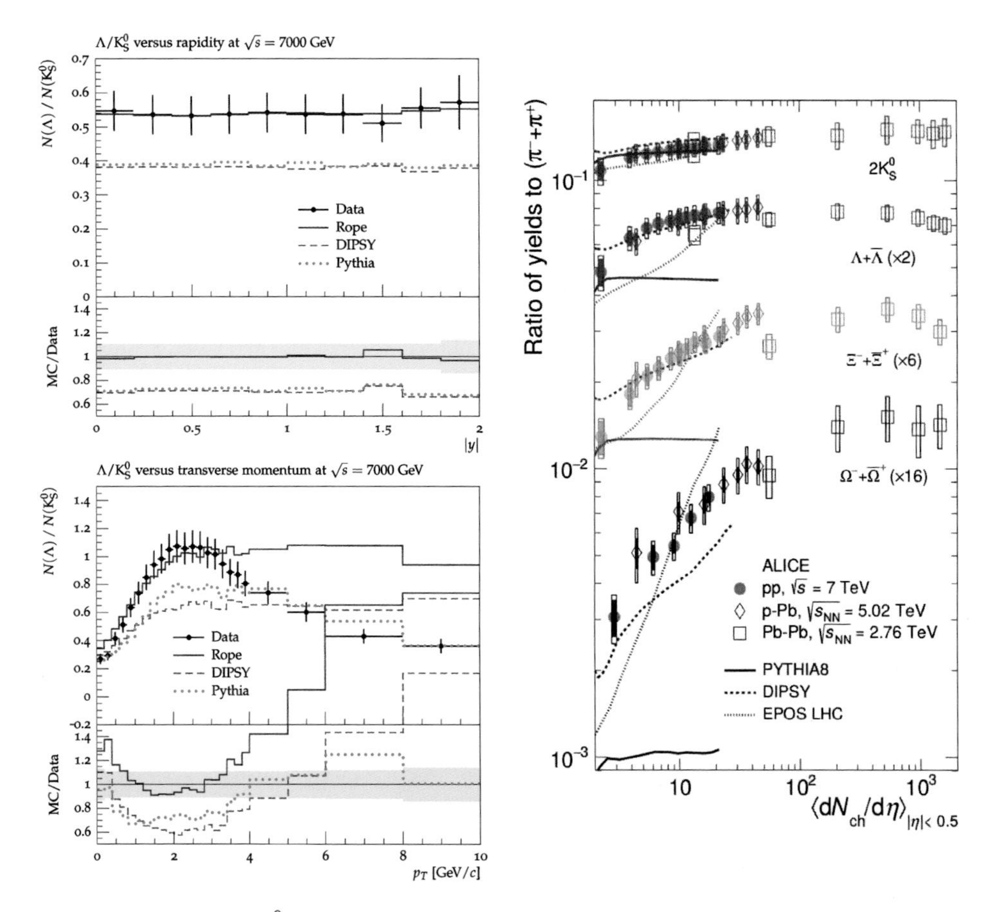

Fig. 9 Ratios of Λ/K_s^0 as a function of rapidity and transverse momentum from DIPSY with and without the rope model and default Pythia 8 compared to data from CMS[450] (left). Ratios of different particle species with respect to pions as a function of multiplicity for DIPSY with ropes, default Pythia 8 and EPOS[409] compared to data from ALICE.[866]

which may be related to the transverse momentum dependence in general being too hard in DIPSY, but further studies are needed.

6.2.2. *Transverse expansion*

The high density discussed above is also expected to give an outward pressure. This ought to give an outward flow similar to the one obtained from a high temperature quark–gluon plasma, also without thermal equilibrium. The flow due to overlapping parallel fluxtubes, with no gluonic excitations, has been studied in Refs. 867 and 868. A preliminary study of the effect expected in the DIPSY model is presented in Ref. 452. This problem will be further studied in a future publication.

7. Nuclei

Although developed for ep and pp collisions, the DIPSY model can be directly generalized to simulate collisions with nuclei. Since we are working in impact-parameter space, we just have to generate random positions for all the nucleons within a target nucleus, and let them collide with a projectile proton, a virtual photon, or a similarly generated projectile nucleus.

High energy nuclear collisions are usually analyzed within the Glauber formalism.[655,869] Here it is assumed that the projectile nucleon(s) travel along straight lines and undergo multiple diffractive subcollisions. Inspite of its pure geometric approach, it has been quite successful in describing many characteristics of reactions with nuclei, and has been widely used in experiments at RHIC and LHC, e.g., to estimate the number of binary nucleon–nucleon collisions and the number of participant nucleons as a function of centrality.

In DIPSY, the default procedure when generating the positions of the nucleons in the nuclei is based on the method in the GLISSANDO MC.[870,871] The nucleon positions are generated according to a Wood–Saxon distribution for the nucleons, with the following form:

$$\rho(r) = \frac{\rho_0(1 + wr^2/R^2)}{1 + \exp((r - R)/a)}. \tag{20}$$

Here R is the nuclear radius, a is "skin width", and ρ_0 is the central density. The parameter w describes a possible non-constant density, but is zero in the fits to nuclei used in this paper.

The nucleon centres are randomly generated in such a way that the charge distribution determined in Ref. 872 is recovered, when the result is convoluted with the charge distribution within nucleons. The nucleons are also generated with a hard-core, which thus introduces short-range correlations among the nucleons. As shown by Rybczynski and Broniowski,[873] the correct two-particle correlation can be obtained if the nucleons are generated with a minimum distance equal to $2r_{\text{core}}$.

We note that if the nucleons are generated within a specific volume, the resulting distribution will (for a finite nucleus) be confined within a smaller volume, and its centre will be shifted. According to Ref. 871, the correct charge distribution is, for mass numbers $A > 16$, obtained using randomly generated nucleon centres described by the Wood–Saxon form in Eq. (20) with the following parameters:

$$R_{\mathrm{NC}} = (1.1A^{1/3} - 0.656A^{-1/3})\ \mathrm{fm}, \quad a = 0.459\ \mathrm{fm}, \quad w = 0, \tag{21}$$

together with a hard core with radius $r_{\mathrm{core}} = 0.45$ fm.

Currently the DIPSY MC includes parametrizations for He, O, Cu, Au, and Pb, but other nuclei can easily be added by the user.

Concerning fluctuations we note that there are two different origins for fluctuations in collisions with nuclei. The first is due to fluctuations in the position of the nucleons in a nucleus, while the second is related to diffractive excitation of the wounded nucleons. As described in Section 5, diffractive excitation of a proton is determined by fluctuations in the internal proton substructure, as given by the Good–Walker formalism.

A very interesting question is also to what extent color charges in one nucleon can screen charges in other nucleons. In DIPSY this interference effect is described by the swing mechanism, and in a nucleus we also allow dipoles in different nucleons to swing, resulting in a kind of color reconnection between nucleons.

It should be noted that assuming the nucleon distribution is fixed from measurements of charge distributions, DIPSY introduces no new parameters to the generation of collisions with heavy ions as compared to pp collisions. This applies to inclusive as well as exclusive observables.

In Ref. 874 we presented a series of predictions for inclusive and semi-inclusive observables in pA and $\gamma^\star$A collisions compared to standard Glauber calculations. It is noteworthy, although maybe not surprising, that the DIPSY results for the total p–Pb inelastic cross-section at $\sqrt{s_{(\mathrm{NN})}} = 5$ TeV reproduces CMS[875] data very well. Here it was also shown that the effects of inter-nucleon swings are fairly limited in pA collisions, where the nucleus acts as an almost black absorber, whereas for $\gamma^\star$A, where the photon has a larger chance to penetrate the target, the corrections can reach well above 10% for heavy nuclei.

What makes DIPSY so interesting when it comes to collisions with nuclei is the detailed simulation of fluctuations and correlations in the initial state. In Ref. 874 we showed that predictions for semi-inclusive pA cross-sections related to diffraction are quite different in DIPSY as compared to standard Glauber calculations. This is because the latter only considers fluctuations and correlations in the distribution of nucleons, while DIPSY also includes such effects in the individual nucleons.

In a more recent study[876] we compared the DIPSY model for pA collisions with a more advanced Glauber model developed by Strikman and collaborators,[624,877] were also fluctuations in the nucleon–nucleon cross-section is taken into account using the parameterization

$$P_{\text{tot}}(\sigma) = \rho \frac{\sigma}{\sigma + \sigma_0} \exp\left\{ -\frac{(\sigma/\sigma_0 - 1)^2}{\Omega^2} \right\}, \tag{22}$$

$$\sigma_{\text{tot}}^{\text{pp}} = \int d\sigma\, \sigma\, P_{\text{tot}}(\sigma). \tag{23}$$

Here $\sigma_{\text{tot}}^{\text{pp}}$ is the observable average of the fluctuating pp cross-section σ. This formalism, often called the Glauber–Gribov model, is used in several experimental analyses of results from the LHC (e.g., Refs. 707 and 878).

One of the results of our comparisons is that the fluctuations in DIPSY have a longer tail for large cross-sections, than what can be achieved with Eq. (22). A reasonable approximation of DIPSY would rather be given by a log-normal distribution,

$$P_{\text{tot}}(\ln \sigma) \propto \exp\left(-\frac{\ln^2(\sigma/\sigma_0)}{2\Omega^2} \right). \tag{24}$$

A second difference is that intrinsic fluctuations in the nucleons in the target nucleus are included in the DIPSY results, but not fully accounted for in the Glauber–Gribov formalism. These fluctuations cause diffractive excitation of target nucleons, and thus strongly influence the final states. Both effects can greatly influence the modeling of the number of participant nucleons in pA collisions, which is important for understanding the centrality dependence of final-state properties.

8. Conclusion

The DIPSY model described in this chapter is in its present form not intended to give quantitatively precise predictions, neither for inclusive nor exclusive observables. Although in many cases it is able to fairly well reproduce a wide range of data, it is far from the accuracy achievable in programs like Pythia 8 and Herwig 7. The main advantage of the model is rather its detailed description of the initial parton evolution in hadronic collisions. The fact that the model is formulated in impact parameter space makes it possible to account for unitarity constraints and saturation including color-suppressed effects, thus providing a very realistic modelling of fluctuations and correlations. This gives us a unique possibility to study, e.g., the connection between fluctuations and diffraction, and effects of correlations in the final state.

The formulation in impact parameter space also gives us a unique starting point for studying the hadronization process in dense collision systems. We have seen that with our new rope model we can describe many features in pp collisions that are similar to effects in AA collisions, normally attributed flow effects in an expanded medium. Therefore, the natural next step is to apply DIPSY and the rope model also to pA, and then maybe even to AA, which could give important insights into the microscopic details of such very complex collisions.

Chapter 18

High Multiplicity Events in pp Collisions from the Color Glass Condensate and Lund String Fragmentation

Prithwish Tribedy

Physics Department, Brookhaven National Laboratory, Upton, NY 11973, USA
Variable Energy Cyclotron Centre, HBNI, 1/AF Bidhan Nagar, Kolkata 700064, India

The Color Glass Condensate (CGC) effective theory of high energy QCD provides an *ab initio* framework of multi-particle production in pp collisions. In the CGC framework, the origin of high multiplicity events is a natural consequence of the rare high color charge density configurations of the wave function of the colliding protons. With such wave functions, constrained by the HERA ep data, the CGC framework provides powerful tools for the phenomenology of high multiplicity pp collisions. When combined with a realistic scheme of hadronization such as the Lund string fragmentation in Pythia, one obtains an excellent description of a wide range measurements in pp collisions at the LHC including the long-range ridge-like correlations observed in the high multiplicity events.

1. Introduction

Recently, there has been a growing interest in studying high multiplicity events in small collision systems such as pp and pA at the LHC. These high multiplicity events are rare processes and can provide direct access to the underlying dynamics of multi-particle production in QCD at high energies. However, there are two major challenges in the theoretical description of particle production in such processes. First of all, it turns out that the average transverse momentum $\langle p_T \rangle$ of inclusive charged particles in a typical high multiplicity event is less than 1 GeV.[522] The bulk of the particles in such an event, therefore, are produced through soft partonic processes of QCD. An *ab initio* theoretical treatment of such soft processes from the conventional perturbative QCD approach is very challenging. The second challenge arises due to the lack of a first principle QCD based implementation of hadronization to treat the fragmentation of partons with low momentum.

Several effective theoretical approaches have been developed over the years to address the highly non-perturbative nature of the multi-particle production in QCD. In this chapter, we discuss one such approach which combines the *ab initio* Color Glass Condensate (CGC) effective theory of multi-particle production[879] with a state-of-the-art hadronization model based on the Lund-string fragmentation.[323] This newly developed CGC + Lund model[750] has been quite successful in describing several qualitative and quantitative features of the data in high multiplicity pp collisions. Although the CGC framework can be generalized for pA and AA collisions, in the following sections, we will restrict our discussion to the most simplistic case of pp collisions.

2. Multi-particle Production in the CGC

Before going into the details of particle production in pp (A) collisions at the LHC, it is better to take a step back and understand the nature of the colliding protons (nuclei) at high energies. In the high energy regime (the small-x regime), nonlinear processes of QCD lead to gluon saturation.[235,366] As a consequence, the wave functions of the colliding protons (nuclei) are dominated by gluons with momentum below a dynamically generated scale called the saturation scale Q_S. In high energy, collisions Q_S, being the only scale in the problem, can become large enough making $\alpha_S(Q_S^2) \ll 1$, therefore extending the applicability of perturbative approaches to perform first principles calculation of particle production. The CGC is an effective theory of high energy QCD developed to describe hadrons, nuclei, and their collisions in the regime of gluon saturation.[a]

In this framework, the two colliding protons (or nuclei) appear to each other as collections of color charges. Due to saturation, the occupation number of such color charges becomes sufficiently high so that they can be treated as classical sources. The classical gluon fields ($F^{\mu\nu}$) generated by such color sources are given by the solutions of the classical Yang–Mills equations $[D_\mu, F^{\mu\nu}] = J^\nu$. The current J^ν is generated due to the distribution of color charges $\rho_{A(B)}$ inside the target (A) and the projectile (B) protons.[b] The statistical properties of $\rho_{A(B)}$, according to the McLerran–Venugopalan (MV) model,[628,880] are described by a Gaussian distribution $\mathscr{W}[\rho]$ with a width related to the saturation scale Q_S of the colliding

[a]For an extensive discussion on the framework of CGC which employs various aspects of saturation physics, we refer the readers to the reviews.[635,879]
[b]D_μ denotes the covariant derivative.

protons.[c] The distribution $\mathscr{W}[\rho]$ or the saturation scale is an external input to the CGC framework and has to be constrained by experimental data. For this, one performs an independent analysis to fit the experimental ep DIS data and obtains a parameterization of Q_s and therefore the width of the $\mathscr{W}[\rho]$ in the MV model.[d] With such a constraint, one can obtain the gauge fields $\mathscr{A}_{A(B)}$ generated by individual colliding protons. The fields after the collision, referred to as the Glasma gluon fields,[884] can be expressed in terms of $\mathscr{A}_{A(B)}$.[885,886]

In a perturbative computation, one tries to obtain an analytical solution of the Yang–Mills equations by performing order-by-order expansion in the powers of $\rho_{A(B)}$. In such an approach one eventually obtains an expression of gluon production in terms of the product of unintegrated gluon distributions that are functions of x and k_T, also known as the k_T-factorization formalism[887e]. Alternatively, one can numerically solve the Yang–Mills equations with different input distributions of $\rho_{A(B)}$.[888–894]

The perturbative approaches in the CGC framework have been studied extensively for single, double and three-gluon productions and also been generalized for n-gluon productions.[895–897] The n-gluon production in CGC leads to a negative binomial probability distribution (NBD).[897] The mean and the variance of such distribution are proportional to $Q_S^2 S_\perp$, which represents the number of Glasma flux tubes (of size $\sim 1/Q_S^2$) in the transverse overlap area $S_\perp$ of the two colliding protons. Due to fluctuations in the collision geometry, the number of flux-tubes ($Q_S^2 S_\perp$) fluctuates from event-to-event, therefore, the final n-gluon distribution $P(n)$ becomes a convolution of many NBDs.[898,899] The high multiplicity events that populate the tail of such distribution originate as a consequence of the rare color charge configurations of the colliding protons (nuclei). One therefore finds a natural explanation of the origin of high multiplicity events in the CGC framework. We will revisit this point later on in this chapter.

The gluons produced in these high multiplicity events also show intrinsic momentum space (azimuthal) anisotropy of the distribution of the produced gluons.[895] In the CGC picture, one can estimate the two-gluon correlation under several approximation schemes.[804–807,895,900–914] In one such perturbative approach, known as the glasma-graph ansatz,[807,895,900,901] one can demonstrate that due to gluon saturation the wave functions of the colliding proton are highly populated (occupancy $\sim\mathscr{O}(1/g^2)$) with gluons of momentum close to Q_S which leads to enhancement of the probability of collimated two-gluons production peaked at relative azimuthal angle $\Delta\phi \sim 0, \pi$. The boost-invariant (rapidity-independent)

[c] Any final observable, computed this approach, has to be averaged over many such configurations of $\rho_{A(B)}$.

[d] MV model does not include rapidity or x dependence. The energy or rapidity dependence of the classical gluon fields are obtained by solving renormalization group (RG) evolution equations such as BK/JIMWLK.[631,632,634,881–883]

[e] k_T represents the transverse momentum of the produced gluon(s).

nature of the proton wave functions leads to the long-range (wide in $\Delta\eta$) ridge-like structure of such correlations. Along with such glasma-graph contribution, one can, in the same framework, estimate the back-to-back di-jet contribution (with a single peak structure at $\Delta\phi \sim \pi$) to the two-gluon correlation function that comprise a significant fraction of the observed anisotropy of produced gluons.[806,907] A consistent treatment of the multiparticle azimuthal correlations can also be obtained in this approach which is essential towards explaining the qualitative features of the of m-particle elliptic anisotropy coefficients $v_2\{m\}$ measured in small collision systems.[915] By far the perturbative diagrammatic approach of CGC explains several qualitative and quantitative features of the experimental data. However the most outstanding puzzle in this approach is that it cannot explain the origin of positive odd harmonics for gluon production.[804] Beyond the glasma-graph approximation, recent studies based on the numerical solutions of Yang–Mills equations have provided several additional insights about the multi-particle production in small systems including the origin of odd harmonics, mass ordering of azimuthal anisotropy etc.[805,893,894,916] which we discuss in the following section.

2.1. *The IP-Glasma model*

Numerical solutions of the classical Yang–Mills equations have also been extensively studied for the phenomenology of multiparticle production in small collision systems.[805,893,894,916] A very successful numerical implementation of such an approach is the IP-Glasma model. It combines the Impact-Parameter-dependent Saturation (IP-Sat) model[f] of HERA Deep-Inelastic Scattering (DIS),[917,919] to constrain the color charge densities $\rho_{A(B)}$ inside the colliding protons, with the Glasma gluon description[884] of the solutions of Yang–Mills equations. To provide a realistic description of the multi-particle production, the IP-Glasma model implements several sources of initial state fluctuations by including (1) the geometry of collisions, (2) the intrinsic fluctuations of the saturation scale of protons, (3) distribution of the color charge density inside the protons.[916,920] The first ingredient, the geometric fluctuations, is naturally included due to fluctuations of the impact parameter of the collisions. The fluctuations of the intrinsic saturation scale originate due to non-perturbative large-x effects.[921–926] Such effects are incorporated in the IP-Glasma model by fluctuating the saturation scale of the protons, Q_S, around its mean value $\langle Q_S \rangle$ according to a log-normal distribution.[920] The width of the log-normal distribution σ has been constrained by several independent analyses to be around $\sim$0.5.[918,920,927] The final and the most important ingredient in this context is

[f]The IP-Sat model parameterizes the differential cross-section in the DIS by incorporating the physics of saturation with a realistic model of the impact parameter dependence of the gluon distribution inside a proton. The most recent fits to the precision HERA data using the IP-Sat model can be found in Refs. 917 and 918.

the sub-nucleonic scale fluctuations of the color charge density inside a proton. The geometric profile of such densities is assumed to be a Gaussian, the width of which is constrained by the IP-Sat parameterization obtained from the fits to the HERA ep DIS data.[917,928] Recently, a modification to such parameterization has been performed by replacing the single Gaussian profile with three Gaussian hot-spots,[918,929] motivated by the constituent quark model. The parameters of such distributions are again constrained by incoherent diffractive HERA data.[918] More specifically, one obtains a parameterization of the cross-section of the diffractive vector meson production in DIS processes. With such a parameterization of the DIS cross-section, it is straightforward to obtain the distribution of the saturation scale inside the protons. The saturation scale inside the proton fixes the width of event-by-event distribution of the color charge densities $\rho_{A(B)}$ which is an input to the Yang–Mills equations.

With the constraints mentioned above for different sources of initial state fluctuations, the solutions of Yang–Mills equations lead to event-by-event fluctuations of the number of gluons produced in the collisions. The IP-Glasma calculations indicate that the final n-gluon production probability distribution $P(n)$ is a convolution of many NBDs.[894,916] Selecting the high multiplicity events in the tail of such distributions, one can estimate the two-gluon correlations and obtain a natural explanation of the origin of long-range ridge-like azimuthal correlations that show sizable elliptic v_2 and triangular v_3 anisotropies as seen in the experimental data.[750,805] More discussions on the origin of ridge-like azimuthal angular correlations can be found in Refs. 694, 914 and 930.

3. Hadronization

A major challenge in the phenomenology of multi-particle correlations is the unavailability of a first principles approach to convert the correlations from the partonic level to the hadronic level. In the conventional methods, one performs a convolution of fragmentation functions with the partonic cross-section to estimate hadronic observables for high p_T processes.[931] However, such approaches cannot be applied to bulk multi-particle production that is dominated by soft processes. Over the years several independent schemes have been developed to handle this issue. The most commonly used model for fragmentation, which describes the bulk particle production in event generators like Pythia is the Monte Carlo implementation of the Lund string fragmentation functions.[323,932] The inputs to such fragmentation functions are particular arrangements of partons in momentum space (see, e.g., Fig. 1), referred to as the Pythia strings.[932] Given such a collection of strings, Pythia performs hadronization of the partons using its Monte Carlo string fragmentation algorithm. The multi-particle correlations developed at the partonic level (e.g., in the gluon distributions from the CGC) are expected to appear in various hadron species after the hadronization.

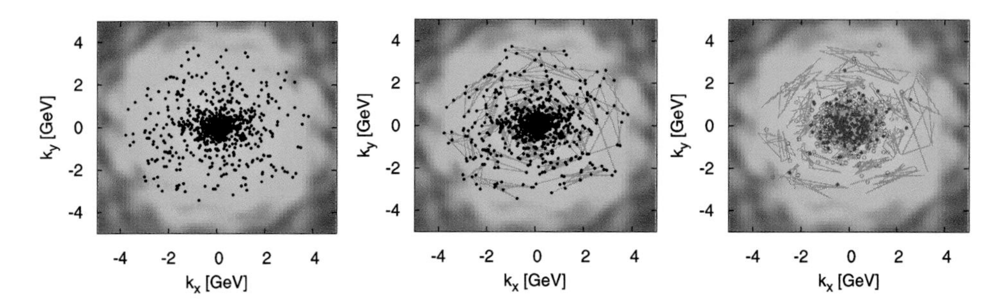

Fig. 1 Distribution of initial gluons and hadrons in momentum space in pp collisions obtained from the CGC + Lund (Pythia) model. (Left) The single event distribution of gluons in momentum space $dN_g/dk_\perp dy$ obtained from the IP-Glasma model shown by the color density plot, the black dots represent the sampled gluons in momentum space. (Middle) The gray lines represent strings connecting the sampled gluons. (Right) The momentum space distribution of different hadron species pions (red), kaons (yellow) and protons (purple) shown by color dots.

3.1. *The CGC+ Lund (CGC-Pythia) model*

The CGC + Lund (Pythia) model combines the IP-Glasma model of the CGC[893,894] and the Lund-string fragmentation algorithm[323,932] of the Pythia event generator.[750] In the IP-Glasma framework, one estimates the distribution of fields after collisions by solving the classical Yang–Mills equations. From such fields one can compute the event-by-event multiplicity density of the gluons in momentum space $dN_g/dk_\perp dy$. Such a distribution, as shown in Fig. 1 (left), already contains all information of multi-gluon correlations. One can then sample N_g gluons with momentum distributed according to such distribution, where one obtains N_g by integrating $dN_g/dk_\perp dy$ over a range of transverse momentum $0 < k_\perp < k_\perp^{\max}$ and rapidity $0 < y < y^{\max}$. In general, one has $y^{\max} \leq y^{\text{beam}} = \ln(\sqrt{s}/m_p)$ which is about 8.9 at the collision energy of $\sqrt{s} = 7$ TeV, with $m_p = 0.938$ GeV as the proton mass. A single configuration of sampled gluons from the distribution $dN_g/dk_\perp dy$ is also shown (by black dots) in Fig. 1 (left). The next step in the CGC + Lund model is to input the information of these sampled gluons to Pythia. For this one needs to first assign a color index to each sampled gluon and group a fixed number of them into Pythia strings. The number of sampled gluons N_{gs} that are grouped into a single string is a parameter in this approach. One guidance to fix the parameter comes from the Glasma flux-tube picture in which a natural choice of N_{gs} is the number of gluons per flux tubes, defined as $N_{gs} = N_g/\langle Q_S^2 S_\perp \rangle$[g]. The grouping of the gluons into Pythia-strings is not unique; one has the freedom of re-connecting the gluons into different possible configurations. In the CGC + Lund

[g]For pp collisions at 7 TeV, N_{gs} is about 18.

model a simple strategy is followed. One first arranges the gluons in rapidity, transverse momentum, and azimuthal angle and starts connecting any two adjacent ones until N_{gs} of them are connected.[h] In the end, one obtains strings that are stretched in rapidity and clustered in transverse momentum space as shown in Fig. 1 (middle). Finally, one needs to add a quark and an antiquark at the two ends of each string to guarantee color neutrality. The momentum of the quark and the antiquark are taken to be equal to the gluons attached to them. One finally inputs all the information of strings into the "hadron standalone mode" of Pythia to perform the fragmentation.[i] In the CGC + Lund model, Pythia is therefore only used for performing the hadronization. In such a mode of operation, Pythia provides an option to arrange the configuration of strings further though a scheme known as the "color-reconnection".[933] It was however seen that enabling color-reconnection has minimal effect on the final observables[750] in hadron standalone mode in contrast to the default mode of Pythia.[522,700,706,933]

4. Comparison with the Experimental Data

4.1. *Multiplicity distributions*

The multiplicity distribution in pp collisions at 7 TeV obtained from the CGC + Lund model[933] is compared to the experimental data from the CMS collaboration in Fig. 2 (left). On the same plot, the distribution for the gluons before fragmentation is also shown for a comparison. The quantity plotted is the probability distribution of the scaled charged hadron multiplicity $\langle N_{\text{ch}} \rangle$ which facilitates data-model comparison by removing several ambiguities related to absolute normalization of multiplicity.[j] It is evident from this plot that hadronization does not modify the probability distribution of multiplicity in a significant way. The underlying distribution which is a convolution of many NBDs is already generated at the gluon-level. On the same figure (left panel) the leading order Feynman diagrams referred to as the "Glasma graphs", that leads to NBDs are shown.[897] These diagrams are characterized by vertices connecting two large-x (valance) partons in the colliding

[h] The construction of strings is done simply by adding the color and momentum information of the gluons into Pythia's particle list.

[i] One also generates a copy of each configuration of strings by randomizing the azimuthal angle of every sampled gluon to use as baseline and to eliminate biases due to modeling of string topology.

[j] The absolute normalization of the initial gluon density can be shown to be sensitive to the choice of running coupling and the lattice parameters involved in the numerical solution of Yang–Mills equations.[934] It is therefore convenient to restrict the data-model comparison to observables that are by definition insensitive to absolute normalization of multiplicity density.

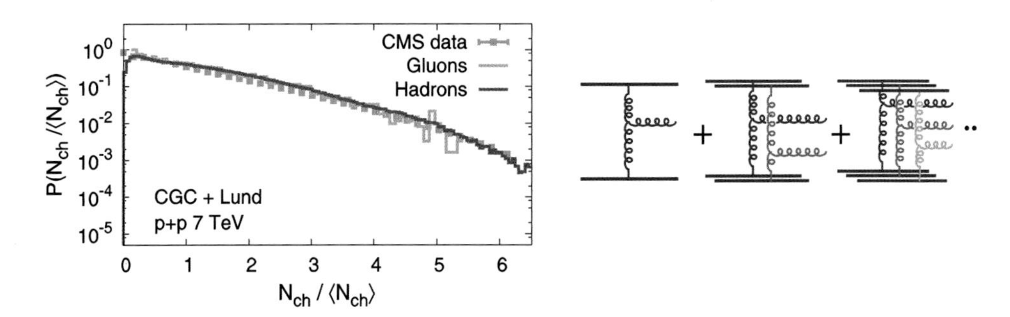

Fig. 2 (left) The probability distribution of inclusive charged hadrons in pp collisions measured over the pseudo-rapidity window of $\eta < 0.5$. (Figure from Ref. 750.) The distribution for the gluons from the IP-Glasma model is shown by the yellow histogram and the distribution after fragmentation using the CGC + Lund model is shown by the red histogram. (right) A series of Feynman diagrams (glasma-graphs with parton ladders and gluon emissions equal to one, two, three and so on) responsible for multi-particle correlations that leads to the negative binomial probability distribution of multiplicity. The blue horizontal lines denote large-x partons in the colliding proton. Figure reprinted with permission from Ref. 750. Copyright 2016 by the American Physical Society.

protons and ladder-like emission of gluons. In general these diagrams are power suppressed (by coupling g for each vertex), except in the regime of saturation when the colliding protons have occupancy of partons of order $\mathcal{O}(1/g^2)$. This changes the power counting for the vertices connecting a ladder and the large-x parton in the proton from $g \rightarrow 1/g$, therefore the amplitude for all the diagrams are enhanced. In a conventional perturbative approach of the CGC one computes such diagrams to estimate multi-particle correlations.[896,897] For example, the simplest diagram involving single ladder corresponds to single inclusive production.[887] The diagram including two parton ladders corresponds to two gluon emission leading to a symmetric structure in the relative azimuthal angle $\Delta\phi$ around $\pi/2$ that gives rise to the ridge phenomenon.[895] To estimate the probability distribution $P(n)$, in principle, one needs to compute all graphs corresponding to n-gluon emissions. A systematic treatment of which was first performed in Ref. 897 to demonstrate that the n-gluon emission probability in the CGC is indeed NBD; such observation was also confirmed in a non-perturbative approach using the IP-Glasma model.[893,894,934] Therefore as expected, in the CGC + Lund model, the underlying distribution of charged hadrons (see Fig. 2) in pp collisions is a convolution, due to various sources of initial state fluctuations, of many NBDs that are generated at the gluon-level. The tails of such distributions as shown in Fig. 2 are dominated by high multiplicity events (a few in every thousand events) in which the typical multiplicity goes above 5–6 times the average multiplicity. Such events are of prime experimental interest as they provide direct access to the rare configuration of color charges from the initial stages of colliding protons.

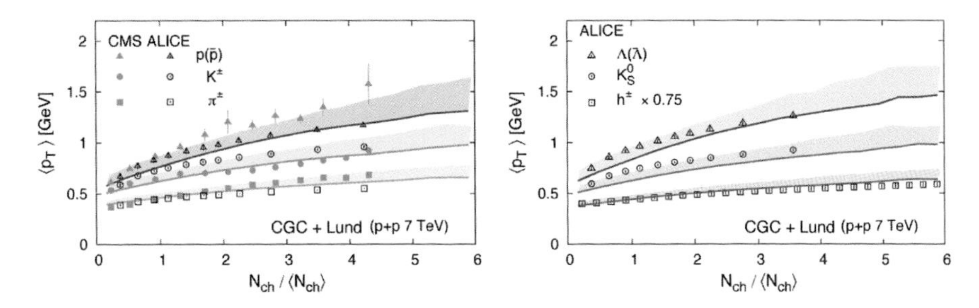

Fig. 3 Figures from Ref. 750 showing variation of average transverse momentum with event activity characterized by scaled charged hadron multiplicity $N_{\rm ch}/\langle N_{\rm ch}\rangle$. Experimental data (shown by points) in pp collisions at 7 TeV[522,525,532,683,935] for identified and inclusive hadrons are compared with CGC + Lund model calculations (shown by curves); the bands represent the uncertainties in CGC calculations due to running coupling. A strong growth of $\langle p_T\rangle$ with event activity and a clear signature of mass ordering is observed. Figure reprinted with permission from Ref. 750. Copyright 2016 by the American Physical Society.

4.2. *Average transverse momentum*

One widely studied bulk observables in pp collisions is the transverse momentum spectra of the produced particles. Such spectra can be characterized through the measurements of average transverse momentum $\langle p_T\rangle$. Figure 3 shows the variation of $\langle p_T\rangle$ with scaled event multiplicity $N_{\rm ch}/\langle N_{\rm ch}\rangle$ for identified particles $\pi^{\pm}, K^{\pm}, p(\bar{p}), \Lambda(\bar{\Lambda}), K_S^0$ and inclusive charged hadrons $h^{\pm}$ measured by the ALICE and the CMS collaboration in pp collisions at 7 TeV.[522,525,532,683,935] The kinematic window of these measurements are $0.15 < p_T < 10$ GeV, $|\eta| < 0.3$ for inclusive hadrons and $|y| < 0.5$ with no cuts on p_T for identified hadrons. Two important observations from these measurements are the strong growth of $\langle p_T\rangle$ with $N_{\rm ch}/\langle N_{\rm ch}\rangle$ and the mass ordering of different hadron species, i.e., $\langle p_T\rangle_p > \langle p_T\rangle_K > \langle p_T\rangle_\pi$ and $\langle p_T\rangle_\Lambda > \langle p_T\rangle_{K_S^0} > \langle p_T\rangle_h$ over the entire range of $N_{\rm ch}/\langle N_{\rm ch}\rangle$. Such observations are often attributed to signatures of radial flow due to the collective expansion of the system produced in pp collisions.[710,748,936,937] As shown in Fig. 3, the CGC + Lund model provides good explanation to both the features of the data purely based on initial state correlations.

An explanation for the strong growth of $\langle p_T\rangle$ with $N_{\rm ch}/\langle N_{\rm ch}\rangle$ has been previously obtained within the color-reconnection scheme of Pythia.[35,933] In such an implementation, different independent parton showers (MPIs) are correlated by reconnection of strings that enables the exchange of momentum and color information. With different MPIs being able to exchange information, any p_T kick generated due to a (semi-)hard interaction can be shared among all the produced hadrons. Color-reconnection, therefore, leads to a correlation of $\langle p_T\rangle$ with the number of independent MPIs leading to its growth with $N_{\rm ch}$ which is also proportional to the number of independent MPIs.

It must be noted that such color-reconnection does not play any significant role in the CGC + Lund model for describing the growth of $\langle p_T \rangle$ with $N_{\rm ch}/\langle N_{\rm ch}\rangle$. Aforementioned, the CGC + Lund model uses Pythia in hadron standalone mode in which color-reconnection scheme plays a minimal role in re-arranging strings and exchanging information.[750] This is because, unlike the default mode of Pythia, in this case the sampled gluons are not associated with separate MPIs and are already assigned to strings.[750] The strong growth of transverse momentum with multiplicity is already generated at the gluonic level, i.e., in the CGC initial state before hadronization. This is because, in some sense, the concept of parton showers, MPIs, and color-reconnection is already built in the framework of CGC. In the flux-tube picture, different independent ladders, as shown in Fig. 2 (right), that produce gluons, are correlated over a length scale of $1/Q_S^2$. One finds that the typical number of produced gluons to be $N_g \propto Q_S^2 S_\perp$, i.e., proportional to the number of flux tubes. Also since the saturation scale is the only scale in the CGC, one finds the typical momentum of produced gluons to be $\langle p_T\rangle_g \propto \langle Q_S\rangle$, leading to $\langle p_T\rangle_g \propto \sqrt{N_g/S_\perp}$. One naturally expects a strong growth of average transverse momentum with multiplicity in CGC. Such a dependence is already incorporated in the IP-Glasma model that initializes the CGC + Lund model and get propagated to the level of hadrons. The effect of mass ordering comes purely from the Lund string fragmentation.

4.3. *Long-range ridge-like correlations*

The experimental two-dimensional di-hadron correlation function in $\Delta\eta - \Delta\phi$ is shown in Fig. 4 for pp, pA and AA collisions. One of the most striking observations in high multiplicity pp (and also pA) collisions in recent times has been the appearance of near side ($\Delta\phi \sim 0$) ridge-like structure in such correlation functions that spread over a long range in pseudorapidity[282,283,728–730,938] as

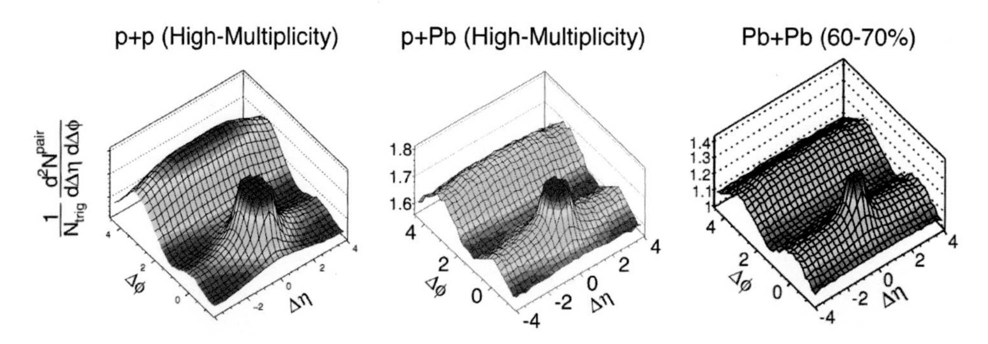

Fig. 4 The long-range ridge like correlations in different collision systems. Figures are obtained from Refs. 728, 736 and 938.

shown in the left and the middle panel of Fig. 4. The interesting feature of this data is that the structure of such correlations looks very similar to what has been observed in heavy-ion collisions[736,939–941] as shown on the same plot. Like heavy-ion collisions, the ridge-like component of the correlation function in pp and pA can be Fourier decomposed into both odd and even harmonic components. In heavy-ion collisions the appearance of such correlations is attributed to the hydrodynamic response of the system to a nearly boost invariant initial state spatial anisotropy. Whether similar explanation[711,748,749,936,942–949] applies for high multiplicity pp (pA) collisions is still a matter of debate.[718] It is, therefore, important to explore if the underlying multiparton dynamics of high energy QCD also generates intrinsic initial state momentum anisotropy and provides a natural explanation for such observation without invoking the formation of a medium describable by hydrodynamics. This is important also because there are regimes of kinematics, such as the large transverse momentum, where hydrodynamics is not applicable.[717,943,949] One, therefore, naturally resorts to an alternative explanation of the observed azimuthal anisotropy that might be driven by initial state partonic processes of QCD involving a (semi-)hard momentum scale.

The two-dimensional correlation function also includes a component which is due to mini-jet (di-jet) production and therefore kinematically constrained to be back-to-back in $\Delta\phi$. Such mini-jets give rise to a long-range structure in the away-side ($\Delta\phi \sim \pi$) but no contribution in the near-side except the peak-like structure at ($\Delta\phi, \Delta\eta \sim 0$). It must be noted that although MPIs in the default version of the event generators like Pythia can explain the mini-jet component of the di-hadron correlations, they do not explain the origin of the near side ridge[282]; even with the color-reconnection schemes.[950]

The CGC framework provides a natural explanation of the appearance of ridge in high multiplicity collisions as a consequence of gluon saturation.[804–807,902–914] The phenomenologically successful approach based on the glasma-graph approximation[k] and BFKL-dynamics has described both qualitative and quantitative features[806] of the long-range structure of the correlation function in pp and pA shown in Fig. 4.[l]

The IP-Glasma model that incorporates all features of the glasma-graph and beyond, also naturally produces the ridge-like structure of the two-gluon correlations. The CGC + Lund model calculation demonstrates that such structure of correlation can survive the hadronization and lead to di-hadron correlations that

[k]Approximation schemes of CGC, besides glasma-graph approach, that also describe several systematics of the ridge include Bose-enhancement of partons, spatial density variations of gluons, formation of color domains in the target, multiple interactions of the glasma gluon fields, etc.[914]

[l]One important characteristic of such correlations is that, unlike hydrodynamic flow, the azimuthal anisotropy that leads to ridge-like structure driven by initial state momentum space correlations are not correlated to the axis of global spatial anisotropy.[805]

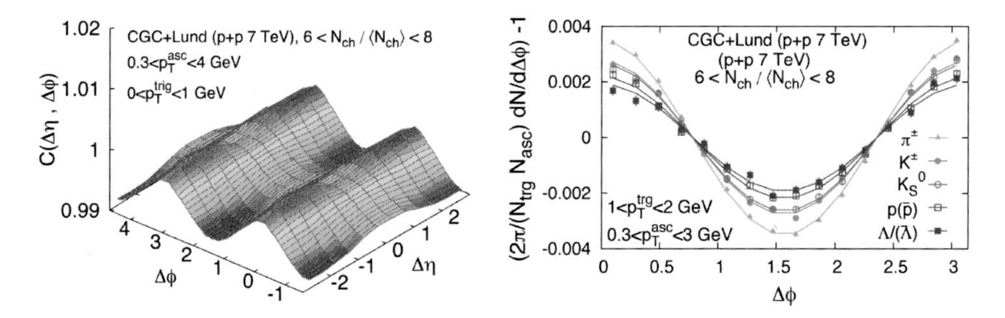

Fig. 5 (left) Two-dimensional ($\Delta\eta - \Delta\phi$) di-hadron correlation function with inclusive charged hadrons as both trigger and associated particles from CGC + Lund model in high multiplicity pp collisions, a figure obtained from Ref. 750. Reprinted with permission. Copyright 2016 by the American Physical Society. (right) The normalized two-particle azimuthal correlation for identified hadrons from CGC + Lund model shown by points. The curves show fit with $2V_{2\Delta}\cos(2\Delta\phi)$ modulation. The distributions show mass ordering with lighter hadron having higher value of $V_{2\Delta}$.

look very similar to the data as shown in Fig. 5 (left). It must be noted that CGC + Lund model, does not include the mini-jet production. Therefore, Fig. 5 (left) does not show (1) the di-jet peak at ($\Delta\phi, \Delta\eta \sim 0$) and (2) a much stronger away side, compared to the near side structure of the correlation functions as observed in the data.

An experimental quantity of further interest is the $\Delta\eta$ integrated azimuthal correlation function per trigger ($N_{\rm trig}$) and associated particles ($N_{\rm asco}$) defined as $2\pi/(N_{\rm trig}N_{\rm asco})dN^{\rm pair}/d\Delta\phi$. An interesting observation is that such distributions from the CGC + Lund model show a clear mass-ordering as demonstrated in Fig. 5 (right). The individual distributions in Fig. 5 (right) can be very well described by a fit function of $1 + 2V_{2\Delta}\cos(2\Delta\phi)$ indicating the presence of a strong dominant second-order Fourier anisotropy coefficient V_2. Therefore, Fig. 5 (right) indicates that lighter hadrons have higher values of $V_{2\Delta}$, an observation qualitatively consistent to what has been seen by the CMS collaboration in high multiplicity pp collisions.[451,951]

The observation of mass ordering of Fourier harmonic coefficients of azimuthal correlation in pp collisions is reminiscent of flow in heavy-ion collisions and often attributed as a signature of hydrodynamic flow. The CGC + Lund is the only model which provides an alternative explanation to such phenomenon purely based on initial state dynamics. It is therefore clear that observations like mass ordering of the azimuthal correlations and the average transverse momentum cannot be considered as decisive signatures of collectivity driven by hydrodynamics. More discussions on this topic, i.e., the interplay between initial state and final state effects in small collision systems, such as pp, can be found in many recent reviews.[694,930]

5. Summary

Many interesting recent observations in the high multiplicity events of pp collisions demand *ab initio* QCD based explanations. However, several theoretical challenges arise due to highly non-perturbative nature of the underlying processes leading to such observations. In this chapter, we have discussed one possible approach to such a problem that combines the CGC effective theory of high energy QCD and the Lund string fragmentation algorithm implemented in Pythia. In such a framework one naturally explains the origin of high multiplicity events and describe multiparticle production involving soft processes that comprise the bulk of the particle production in pp collisions. The main ingredient of this framework, the IP-Glasma model, is a very successful numerical implementation of the CGC effective theory. In the IP-Glasma model, the multiparticle production is computed from the numerical solutions of the classical Yang–Mills equations. This model includes a realistic description of initial state fluctuations and multiple interactions of the strong color fields produced in the collisions. It provides a natural explanation for the origin of high multiplicity events in pp collisions as an outcome of the rare color charge configurations of the wave functions of the colliding protons. Interestingly, IP-Glasma model also predicts the exact nature (negative-binomial) of the underlying probability distribution of multiplicity. Apart from explaining the origin of high multiplicity events, this model also provides explanations to (1) rapid growth of average transverse momentum of produced particles with multiplicity and (2) the origin of the long-range ridge like correlations in high multiplicity pp collisions. One limitation of the IP-Glasma model is that it describes multiparticle production only at the level of gluons. For successful phenomenological purposes, it becomes essential to combine such an approach with a realistic framework of hadronization such as the Lund string fragmentation algorithm implemented in Pythia. The development of the new Monte Carlo model combining two such approaches, known as the CGC + Lund model (also referred to as CGC-PYTHIA), turns out to be a very successful attempt in this context. The CGC + Lund model describes several features of the experimental data in pp collisions, e.g., the mass ordering of average transverse momentum, the origin of the long-range di-hadron correlations, and the mass ordering of the elliptic anisotropy coefficient extracted from the di-hadron correlation functions. Many such observations in the high multiplicity pp collisions have often been attributed to signatures of collectivity driven by hydrodynamic flow. It turns out that the CGC + Lund model can provide alternative explanations to such phenomenon, purely from initial state dynamics driven by the physics of gluon saturation. The CGC + Lund is a fully operational Monte Carlo event generator which will enable us to explore several observables in high multiplicity pp collisions; in principle it can also be extended to pA and AA collisions.

Acknowledgments

This work was supported under Department of Energy Contract No. DE-SC0012704. I would like to thanks my collaborators Kevin Dusling, Larry McLerran, Bjoern Schenke, Soeren Schlichting and Raju Venugopalan for important discussions.

Chapter 19

Multiple Scattering in EPOS

K. Werner*, B. Guiot*,†, Iu. Karpenko*,‡, A. G. Knospe§, C. Markert¶, T. Pierog‖, G. Sophys* and M. Stefaniak*,**

*SUBATECH, University of Nantes — IN2P3/CNRS — IMT Atlantique, Nantes, France

†Universidad Técnica Federico Santa Maria, Valparaiso, Chile

‡Bogolyubov Institute for Theoretical Physics, Kiev 143, 03680, Ukraine

§Department of Physics, University of Houston, Houston, TX 77204, USA

¶Physics Department, The University of Texas at Austin, Austin, TX, USA

‖Institute für Kernphysik, Karlsruhe Institute of Technology, KIT, Campus North, Germany

**Warsaw University of Technology, Warsaw, Poland

Crucial for any quantitative understanding of high energy proton–proton scattering and its relation to proton–nucleus and nucleus–nucleus collisions is the implementation of multiple parton scattering in models and event generators. We discuss the way multiple scattering is realized in the EPOS approach.

1. Introduction

Multiple parton interactions seem to be a "recent discovery" from studying proton–proton (pp) scattering at the LHC, but the idea of multiple scattering in pp collisions is actually quite old. Following the work of Regge[952] and Gribov,[953] the so-called Gribov–Regge theory (GRT) has been developed, based on the exchange of multiple pomerons, where the latter ones were hypothetical objects with properties obtained from Regge theory (Regge poles). In the 1980s, several quite successful models based on GRT emerged, like the Dual Parton Model,[345] the Quark Gluon String model,[352] and the VENUS model.[954] At the time, first attempts were undertaken to consider soft (based on Regge poles) and hard interactions (based on perturbative QCD).

In 2001, we presented "Parton Based Gribov Regge Theory" (PBGRT)[955] with a rigorous treatment of energy sharing in the GRT multiple scattering framework, where we consider soft and hard pomerons, the latter ones being parton ladders according to DGLAP parton evolution.[392–394] This approach (PBGRT) is the theoretical basis of the EPOS event generator, or more precisely of the "primary interactions" happening (at high energies) instantaneously at $t = 0$. We also consider "secondary interactions", which amounts to a hydrodynamical expansion of a core part of matter (determined from the primary scatterings). The EPOS approach uses precisely the same concepts for proton–proton (pp), proton–nucleus (pA) and nucleus–nucleus (AA) scattering.

All EPOS versions, also the most recent ones, are composed of primary and secondary interactions, also referred to as initial state and final state scatterings. The former ones are based on PBGRT,[955] almost unchanged over the years. The only issue which evolved significantly is the way of treating so-called "high density effects", referred nowadays as saturation effects. We will discuss this topic in detail. Also common to all EPOS versions is a core–corona separation mechanism,[956] which defines the initial conditions of the secondary interactions. This mechanism allows to identify a core part which expands collectively and a corona part of particles escaping from the dense core region. The core part will eventually decay statistically, whereas the corona particles have been produced from string decay.

The first public EPOS version was EPOS 1.99, available in 2009. Concerning secondary interactions, we simply introduced collective effects — meant to mimic a hydrodynamical expansion phase — via a parameterization of collective flow obtained directly from the initial conditions, i.e., the distribution of string segments in space at some given initial time τ_0. The corresponding parameters were tuned comparing to Tevatron, SPS, and RHIC data.

With the start of the LHC era many new experimental data became available. Although EPOS 1.99 could be used to make predictions for the LHC,[312] there was some need to retune the parameters, taking into account the new LHC pp scattering data. The resulting code was named EPOS LHC,[409] the code is freely available and heavily used to simulate LHC collisions, mainly for pp.

In parallel to the EPOS LHC work, we developed EPOS 2,[957] where the main new ingredient was an explicit treatment of the hydrodynamical evolution of the core part according to the equations of ideal relativistic hydrodynamics. The code was mainly applied to analyze RHIC heavy-ion data. We then switched to viscous hydrodynamics in EPOS 3,[693] and focused on analyzing small systems at the LHC (pp and pA) using EPOS 3.0xx, EPOS 3.1xx and EPOS 3.2xx (the main difference between these versions concerns parton saturation).

Currently we work on a "reunification" of EPOS LHC and EPOS 3, which should equally apply to small (pp, pA) and big systems (AA), and which should equally well describe leading particle production in pp and central rapidity particle production in AA collisions. The version numbers will be 3.3xx and higher. An important aspect will be the treatment of parton saturation, with preliminary results already available in EPOS 3.2xx, to be discussed later. All data from LHC run 1 (including diffraction, underlying event, flow, etc.) will be considered.

A common feature of all EPOS versions is a collective evolution of matter in the secondary scattering stage. And this collective behavior is present (more or less dominant) in all reactions, from pp to AA. This picture is supported by many experimental LHC results, showing flow-like behavior also for small systems. How can one actually detect flow and equilibration? One may measure for example particular properties of particle production from the flowing medium, and particle ratios. In Fig. 1, we show the time evolution of the energy density distribution in the transverse plane, for an EPOS simulation of pp scattering at 7 TeV. One can clearly see that there is a fast radial expansion (radial flow), and in addition we see that the initial anisotropy of the distribution leads to an anisotropic expansion. It is clear that particle production from such an anisotropically expanding medium will lead to azimuthal anisotropies in particle production. In Fig. 2, we show transverse momentum spectra from EPOS of pions, kaons, protons, and lambdas. We show results from string decay (dotted) and from a transversely expanding plasma (solid). Whereas the different particle spectra from string decay have similar shapes, the situation for particle production from an expanding plasma is very different: the heavier the particle, the more the low p_t particles are shifted towards higher p_t values, pushed by the collective flow (momentum is mass times velocity). So here we see a clear flow "mass effect".

The possible anisotropic expansion as seen in Fig. 1 leads to azimuthal anisotropies in particle production, which can be seen in dihadron correlation functions, $R = \frac{1}{N_{\rm trigg}} \frac{dn}{d\Delta\phi\Delta\eta}$, where $N_{\rm trigg}$ is the number of triggers, and $\Delta\phi$ and $\Delta\eta$ the differences in azimuthal angle and pseudorapidity between the two hadrons. In Fig. 3, we show a dihadron correlation function in p–Pb scattering at 5 TeV from EPOS simulations. One clearly sees the near side jet peak at $\Delta\phi = 0$ and $\Delta\eta = 0$. The away side jet peak at $\Delta\phi = \pi$ is expected to be quite broad and not so easy to isolate, but the near-side ridge at $\Delta\phi = 0$ away from the jet peak is clearly a "non-jet contribution" and in our simulation due to the anisotropic flow.

All this discussion about flow in small systems is very interesting, but the main requirement of having a flowing medium is a sufficiently high density of strings after the primary scattering stage, and here multiple scattering plays a crucial role. We will therefore in Section 2 discuss in detail the multiple scattering approach of primary interactions in EPOS.

2. Multiple Scattering Approach of Primary Interactions in EPOS

Before discussing multiple scattering, let us consider a single scattering (single pomeron). Here we consider what is well known in the parton model approach for pp scattering, namely successive parton emissions from both protons, with finally an elementary parton–parton (hard) scattering in the middle, see Fig. 4(a). The emissions are assumed to follow the DGLAP equations. Subsequent parton emissions occur towards smaller x-values (light cone momentum fractions) and

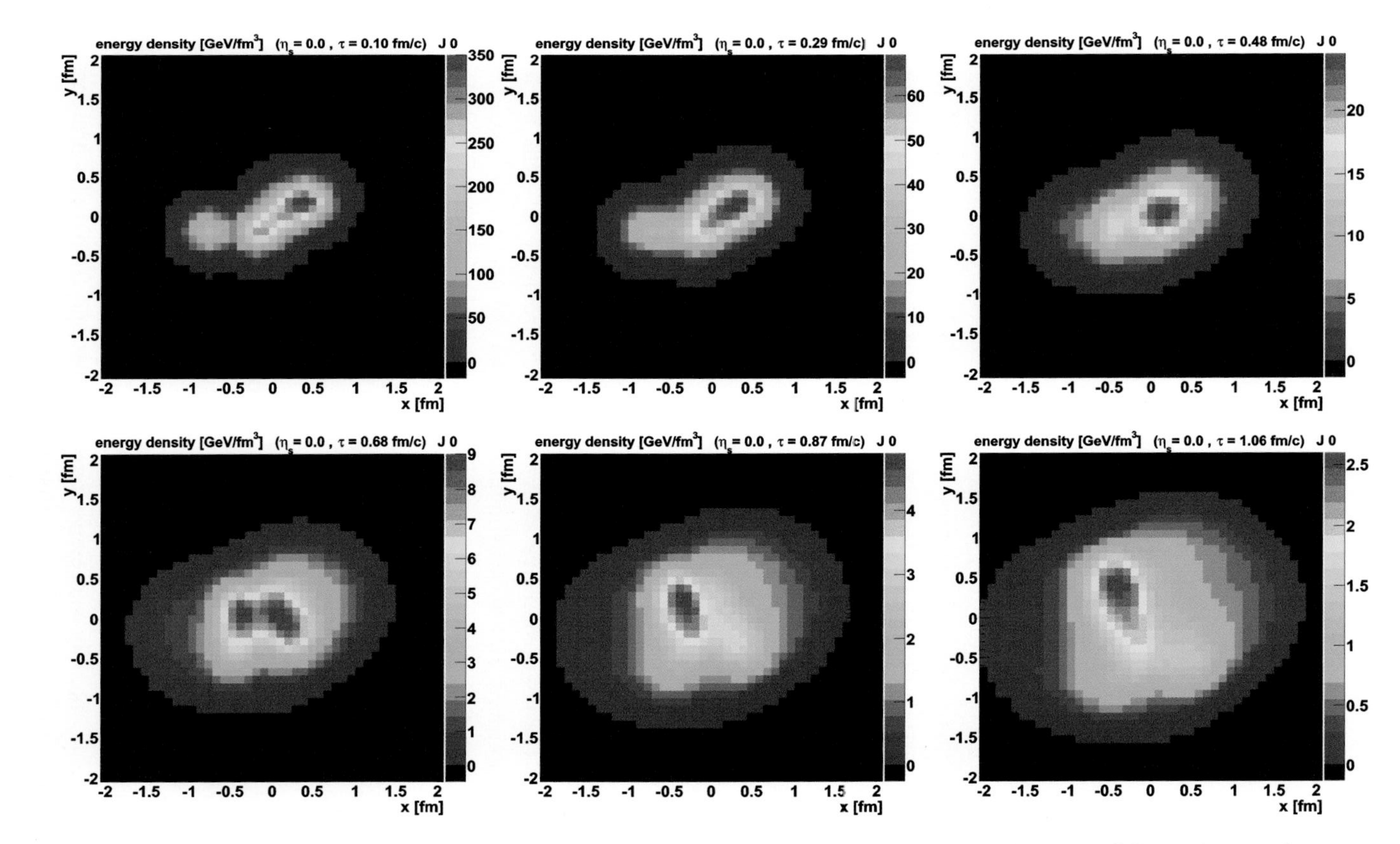

Fig. 1 Distribution of the energy density in the transverse plane for $z = 0$, at different times, for an EPOS simulation of pp scattering at 7 TeV.

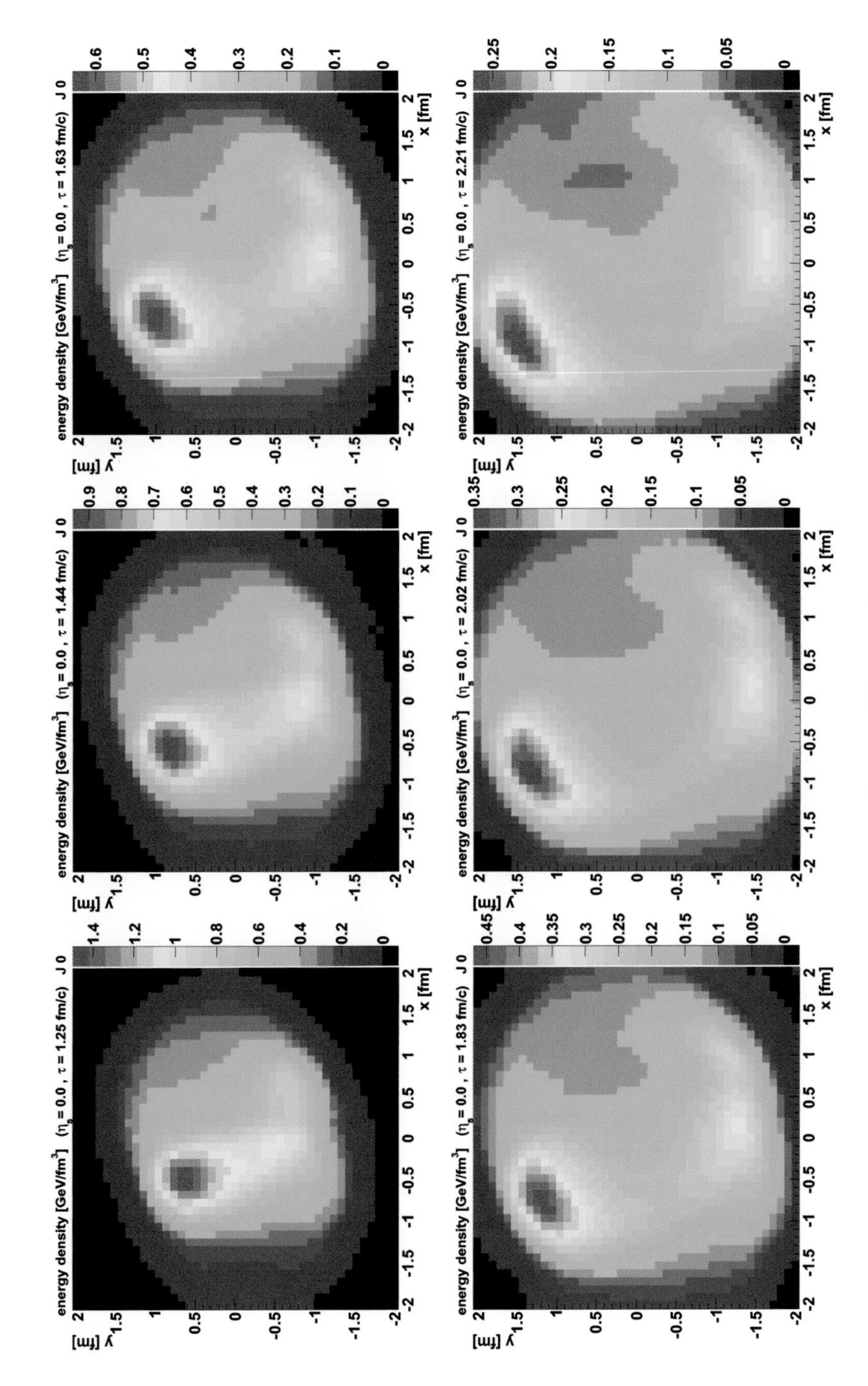

Fig. 1 *(Continued)*

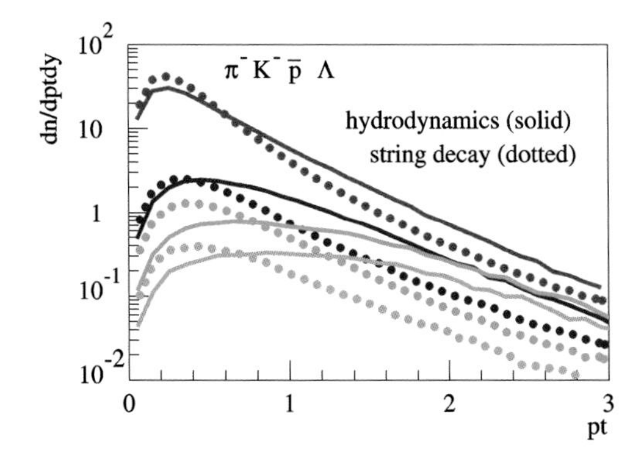

Fig. 2 Transverse momentum spectra from EPOS of (from top to bottom) pions, kaons, protons, and lambdas. We show results from string decay (dotted) and from a transversely expanding plasma (solid lines).

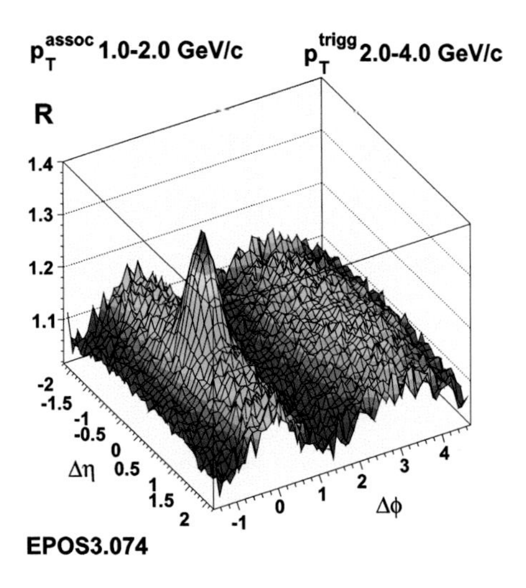

Fig. 3 Dihadron correlation function in p–Pb scattering at 5 TeV, from EPOS simulations. Figure reprinted with permission from Ref. 748. Copyright 2014 by the American Physical Society.

larger virtualities (from both sides). We therefore have rapidity ordering. The time interval between two successive emissions is (in particular for the first emissions) very large, due to the big gamma factors, so the whole interaction takes a very long time. This explains why multiple scattering must occur in parallel (to be discussed in detail later).

For $t > 0$, such a parton ladder represents actually a (mainly) longitudinal color field, see Fig. 4(b), where the ladder rungs (mainly gluons) represent small transverse momentum components. Such a picture has been used in the Lund string model approach,[323] first used to understand particle production in electron–positron annihilation, then generalized to pp scattering. The picture is also related

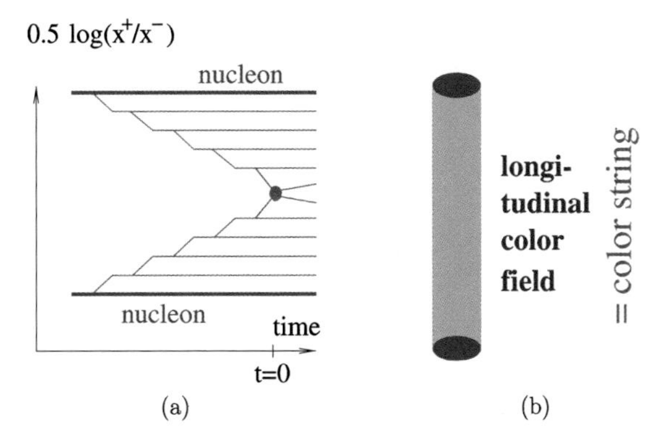

Fig. 4 (a) A single scattering: successive parton emissions from both protons, with finally an elementary parton–parton scattering in the middle (Born process). (b) The parton ladder represents a (mainly) longitudinal color field (color string).

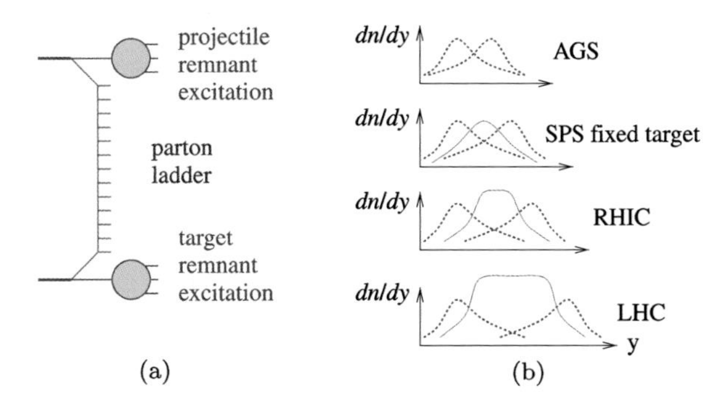

Fig. 5 (a) Complete picture with remnants. (b) Particle production from remnants and strings.

to the gluon fields emerging in the CGC approach.[628,880,958] The string picture in EPOS is described in detail in Ref. 955. The fields decay via pair production (Schwinger mechanism). The one-dimensional character of the fields allow to treat their evolution and decay via the classical string theory (which does not use much more than some general symmetries). One does a one-to-one mapping of partons to kinks, then treats (rigorously) the space-time evolution of the kinky string, and finally realizes (rigorously) the string decay via the area law.

The complete picture includes projectile and target remnant excitations, see Fig. 5(a). The remnants are important sources of particle production. There is the "inner contribution", from the parton ladder or in other words from string decay (full lines), and in addition there are the "outer contributions", from the remnant decays (dashed lines). In Fig. 5(b), we show the rapidity distribution of hadrons from remnants and strings at different energies (artists view). At least at LHC, even in minimum bias pp collisions, single scattering is not enough, so the picture will be

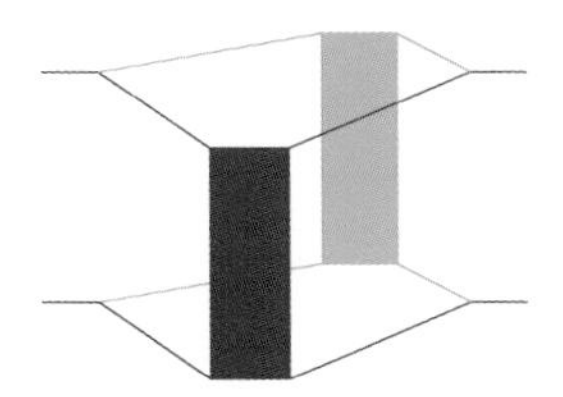

Fig. 6 Two pomeron exchange.

modified in the central region, a large number of parton ladders may contribute, not only a single one.

We will now discuss how to generalize the picture discussed above, to account for multiple scattering, which means to consider multiple parton ladders. All details of this PBGRT approach can be found in Ref. 955. Let T be the elastic (pp, pA, or AA) scattering T-matrix, which means that the total cross-section is given as

$$2s\,\sigma_{\mathrm{tot}} = \frac{1}{\mathrm{i}}\mathrm{disc}\,T, \tag{1}$$

where the discontinuity of T is defined as disc $T = T(s+i\epsilon,t) - T(s-i\epsilon)$. The quantities s and t are the Mandelstam variables. The basic assumption of PBGRT is the hypothesis that the T-matrix can be expressed as a sum of products of elementary objects called pomerons, in the case of pp scattering (AA is slightly more complicated)

$$T = \sum_k \frac{1}{k!}\{T_{\mathrm{Pom}} \times \cdots \times T_{\mathrm{Pom}}\}. \tag{2}$$

As discussed earlier from space-time considerations, the multiple pomeron structure must be parallel, as shown in Fig. 6 for the case of two pomerons, so energy–momentum sharing is an important issue, and Eq. (2) is meant to be symbolic: in reality it contains multidimensional integrations over momentum fractions. For the moment, the pomerons are considered to be black boxes (actually the blue and green boxes in the figure). The next step is the evaluation of the discontinuities ("cuts"),

$$\frac{1}{\mathrm{i}}\mathrm{disc}\{T_{\mathrm{Pom}} \times \cdots \times T_{\mathrm{Pom}}\}, \tag{3}$$

which is done using "cutting rules": a "cut" multi-pomeron diagram amounts to the sum of all possible cuts, as shown in Fig. 7 for the example of two pomerons. Based on these cutting rules, one may express the total cross-section in terms of cut and uncut pomerons, as sketched in Fig. 8. The great advantage of this approach: doing partial summations, one obtains expressions for partial cross-sections $d\sigma_{\mathrm{exclusive}}$, for particular multiple scattering configurations, based on which the Monte Carlo generation of configurations can be done. No additional approximations are needed. The above multiple scattering picture is used for pp, pA, and AA.

So far the multiple scattering formalism is general, we did not specify the nature of the pomeron yet, which we are going to do in the following. The pomeron

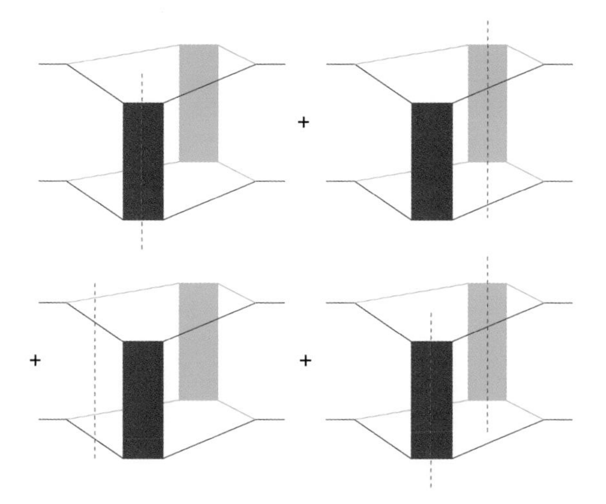

Fig. 7 Cutting the two pomeron diagram.

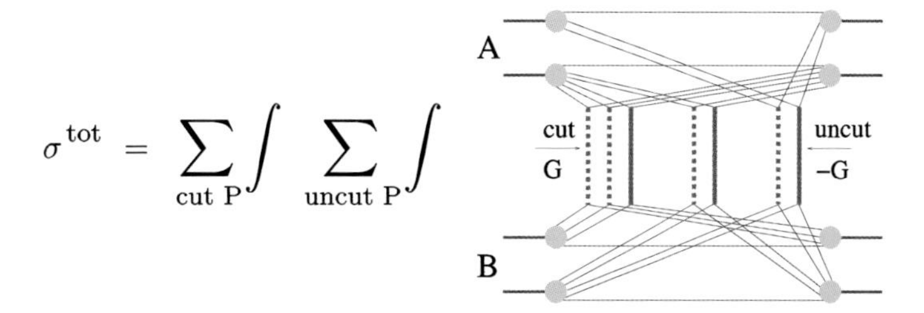

Fig. 8 PBGRT formalism: The total cross-section expressed in terms of cut (dashed lines) and uncut (solid lines) pomerons, for nucleus–nucleus, proton–nucleus, and proton–proton collisions. Partial summations allow to obtain exclusive cross-sections, the mathematical formulas can be found in Ref. 955 or in a somewhat simplified form in Ref. 693.

is — as already discussed earlier for the case of "single scattering" — a parton ladder following DGLAP parton evolution from both ends, with an elementary hard parton–parton scattering in the middle.

It is clear that DGLAP evolution is not enough, in particular when it comes to nuclear collisions. In our approach, they will be accommodated via a saturation scale $\boldsymbol{Q_s}$. This is the major improvement of our approach over the past years, so we will discuss this topic later in detail.

3. Secondary Interactions in EPOS

In heavy-ion collisions and also in high multiplicity events in proton–proton and proton–nucleus scattering at very high energies, the density of strings will be so high that the strings cannot decay independently as described above. Here we have to modify the procedure as discussed in the following. The starting point is still the flux tubes (kinky strings) discussed earlier. Some of these flux tubes will constitute bulk matter which thermalizes and expands collectively — this is the so-called "core". Other segments, being close to the surface or having a large transverse momentum, will leave the "bulk matter" and show up as hadrons (including jet-hadrons); this is the so-called "corona".

In principle the core–corona separation is a dynamical process. However, the knowledge of the initial transverse momenta p_t of string segments and their density $\rho(x, y)$ allows already an estimate about the fate of these string segments. By "initial" we mean some early proper time τ_0, which is a parameter of the model. String segments constitute bulk matter or escape, depending on their transverse momenta p_t and the local string density ρ. Also low p_t segments corresponding to a very high p_t jet may escape.

Our core–corona separation procedure is based on "jet cones". We identify for each hard process (in other words for each semihard pomeron) the primary produced partons, and then the string segments corresponding to the same process and being within a cone with respect to the primary parton axis, referred to as "jet cone", see Fig. 9. The jet-cone is defined as

$$(\Delta\eta)^2 + (\Delta\phi)^2 < R^2 \,, \tag{4}$$

with $\Delta\eta$ and $\Delta\phi$ being respectively the difference in pseudorapidity and azimuthal angle, with respect to the primary partons, and R is a parameter. Segments inside and outside of the cone are treated differently. At the moment, we use the same procedure with different parameters, always at initial time τ_0. In the future, one may imagine a more sophisticated treatment for the "inside-cone" part, considering the time evolution of the partons in the medium.

We compute for each string segment

$$p_t^{\rm new} = p_t - f_{\rm Eloss} \int_\gamma \rho \, dL, \tag{5}$$

where γ is the trajectory of the segment. If a segment has a positive $p_t^{\rm new}$, it is allowed to escape — it is a corona particle. Otherwise, the segment contributes to the core.

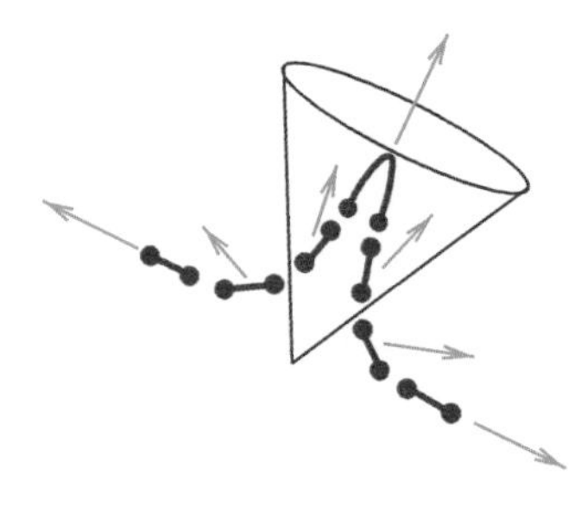

Fig. 9 String segments inside and outside the "jet cone".

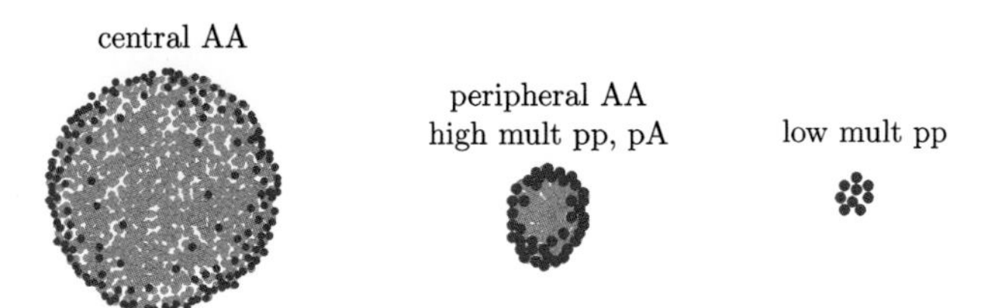

Fig. 10 Schematic view of core–corona separation in different systems. Red dots are core segments, blue ones corona segments.

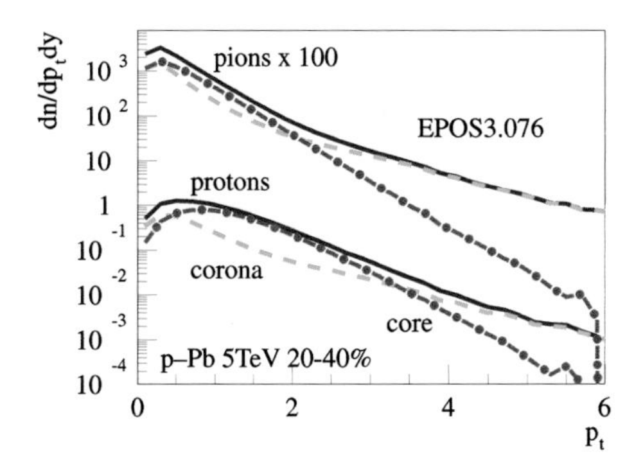

Fig. 11 Core and corona contributions. Reprinted from Ref. 693, Copyright 2014, with permission from Elsevier.

As sketched in Fig. 10, we have a non-zero core contribution not only in central heavy-ion collisions, but even in pp. String segments contributing to the core are shown as red dots, while the blue ones represent the corona. The latter ones will show up as hadrons, whereas the core provides the initial condition of a hydrodynamical evolution, where the particles will be produced later at "freeze-out" from the flowing medium, which occurs at some "hadronization temperature" T_H. After this "hadronization", the hadrons still interact among each other, realized via a hadronic cascade procedure. For details about hydroevolution and hadronic cascade see Refs. 693 and 959–961.

In Fig. 11, we show how core (red dashed-dotted lines) and corona (green dashed lines) contribute to the production of pions (upper curves, multiplied by 100) and protons, in semi-peripheral p–Pb collisions at 5 TeV. The blue solid lines are the sum of core and corona. The calculations are done based on the hydrodynamical evolution as described in the next chapter, without employing a hadronic cascade. The corona contributions dominate completely the high p_t regions. The core becomes important for both pions and protons at intermediate p_t, but the core over corona fraction is much bigger for protons, and the crossing (core = corona) happens at larger p_t. The fact that the core is much more visible in protons compared to pions is a consequence of radial flow: when particles are produced in a radially flowing medium, the heavier particles acquire more transverse momentum than the light ones. It is a mass effect (lambdas look similar to protons, kaons are in between pions and protons).

4. Recent Developments Concerning Saturation

Our multiple scattering approach PBGRT allows to compute partial cross-sections for particular multiple scattering configurations (after summing all uncut pomerons), where a multiple scattering configuration is defined by the number of pomerons and their properties, exchanged between any possible pair of a projectile and a target nucleon. For each cut pomeron, we have an expression $G = \frac{1}{2s\,i}\mathrm{disc}\,T_{\mathrm{Pom}}$, where T_{Pom} represents a parton ladder, computed using the DGLAP equations, using some soft cutoff Q_0.

The functions G can be computed in an iterative fashion using numerical integration. Fortunately, their dependence on the light cone momentum fractions x^+ and x^- can be perfectly fitted as

$$G = \alpha\,(x^+x^-)^\beta, \tag{6}$$

with coefficients α and β which depend on s and the impact parameter b, and of course on the cutoff Q_0. This form is extremely important, because it allows analytical computation of the multidimensional integrals to obtain the cross-sections. To indicate explicitly the Q_0 dependence (omitting for simplicity the s and b dependence) we write

$$G = G(Q_0;\ x^+, x^-). \tag{7}$$

We know that DGLAP evolution is not enough, see the detailed discussion in Ref. 955. There must be nonlinear effects, like the absorption of an emitted gluon by a ladder gluon (gluon fusion), see Fig. 12. To mimic the effects of gluon fusion, our fits are modified (for pp) by adding an exponent ε, which means instead of G we use

$$G_{\mathrm{eff}}(Q_0,\ x^+, x^-) = \alpha\,(x^+x^-)^{\beta+\varepsilon}, \tag{8}$$

with α and β still being the above-mentioned coefficient used to fit G. The exponent $\varepsilon = \varepsilon(s)$ is chosen to reproduce the energy dependence of cross-sections. This is the procedure employed in EPOS LHC, which has proven to quite successfully describe LHC data.

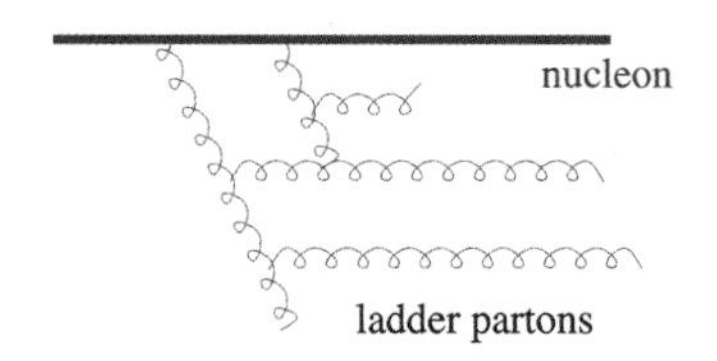

Fig. 12 Nonlinear contribution: Gluon fusion.

Nevertheless, the problem remains that adding an exponent ε must be accompanied by a corresponding modification of the internal structure of the pomeron, otherwise the whole approach is inconsistent! It is therefore a crucial issue. It is important since we use G not only for computing cross-sections, but we use the formula to obtain G also to generate explicitly the partons. So cross-section calculations and parton generation must be done in a consistent fashion.

We regain perfect consistency by defining a saturation scale Q_s via

$$G_{\text{eff}}(Q_0\,;\,x^+,x^-) = C \times G(Q_s\,;\,x^+,x^-), \tag{9}$$

(with some coefficient C) and then considering the parton ladder with the cutoff Q_s, and thus changing the internal structure of the pomeron, important for the parton generation. The saturation scale depends on x^+ and x^-, and we find $Q_s \propto (x^+x^-)^{0.30}$, which is equal to the CGC result when we replace x^+x^- by $1/x$.

We could simply use a constant coefficient C in Eq. (9), if the only issue would be internal consistency. But there is a second problem. We know that the saturation scale in pA scattering (for a nucleus with atomic number A) should be proportional to the number of participating nucleons N_{part}, which is some measure of the "event activity". We expect a similar effect in pp scattering, with the corresponding event activity being the number N_{Pom} of (cut) pomerons. So we set

$$C = A\,(N_{\text{Pom}})^B, \tag{10}$$

where B should be close to unity.

As mentioned above, the main reason to introduce a saturation scale via Eq. (9) is the need to generate partons in a way which is consistent with the formulas we use for cross-section calculations. Changing Q_s will change parton transverse momentum distributions, as sketched in Fig. 13. For large values of Q_s, small p_t partons will be suppressed, the partons get harder, which is the same as saying that the pomerons are getting harder. This also means that the corresponding strings get harder (the kinks carry higher p_t) after the mapping of partons to kinks. Considering the usual case of two strings corresponding to a pomeron, we have the situation

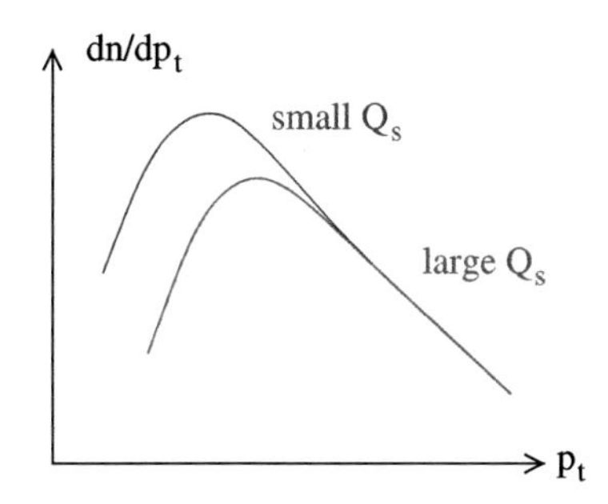

Fig. 13 Sketch of parton transverse momentum distributions for small and large Q_s.

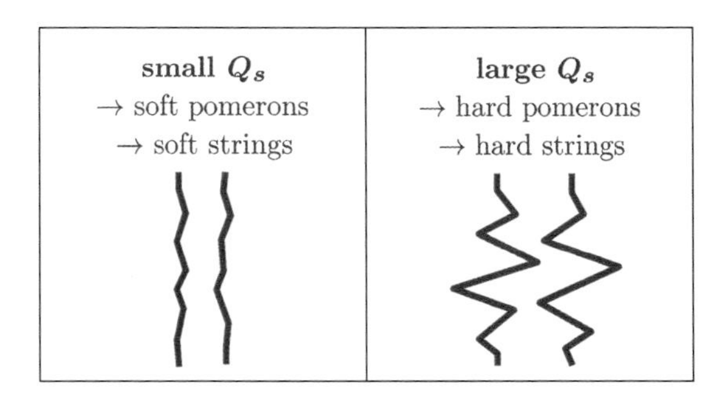

Fig. 14 Comparing small and large Q_s.

as sketched in Fig. 14. Whereas small Q_s corresponds to soft strings, large Q_s corresponds to hard strings (with high p_t kinks).

How can we test all this? Our fundamental equation

$$G_{\text{eff}}(Q_0) = A\,(N_{\text{Pom}})^B G(Q_s) \tag{11}$$

means that large N_{Pom} leads to large Q_s, in other words Q_s increases monotonically with N_{Pom}. On the other hand we have a very strong correlation between N_{Pom} and the multiplicity, for example $dn/d\eta(0)$, which increases monotonically with N_{Pom}. So increasing $dn/d\eta(0)$ will lead to increasing Q_s, hard strings, and finally large p_t hadrons. So we should study the multiplicity dependence of the average p_t of hadrons.

However, these saturation effects concern only the corona! So we have to also investigate the multiplicity dependence of the core–corona separation — by studying particle yields and ratios.

5. Multiplicity Dependence of Particle Yields and Mean Transverse Momenta

We compare simulations (mainly) to ALICE data[449,519,662,669,678,679,710,713,962] concerning particle ratios to pions versus multiplicity (more precisely $\langle dn_{\text{ch}}/d\eta(0)\rangle$) and mean transverse momentum versus multiplicity, for different particle species, for minimum bias pp scattering at 7 TeV as well as p–Pb at 5 TeV and Pb–Pb at 2.76 TeV for different multiplicity bins. The results, collected in Fig. 15 are either presented in the above-mentioned ALICE references or extracted from data presented. Whereas the ratios show an essentially continuous behavior when going from pp and p–Pb to Pb–Pb results, the mean transverse momentum results show a much stronger increase in case of p–Pb compared to Pb–Pb.

The discussion of these results in the EPOS framework is based very much on the concept of core–corona separation, as discussed earlier. We therefore show in Fig. 16 the pion production rate at central rapidity as a function of the multiplicity

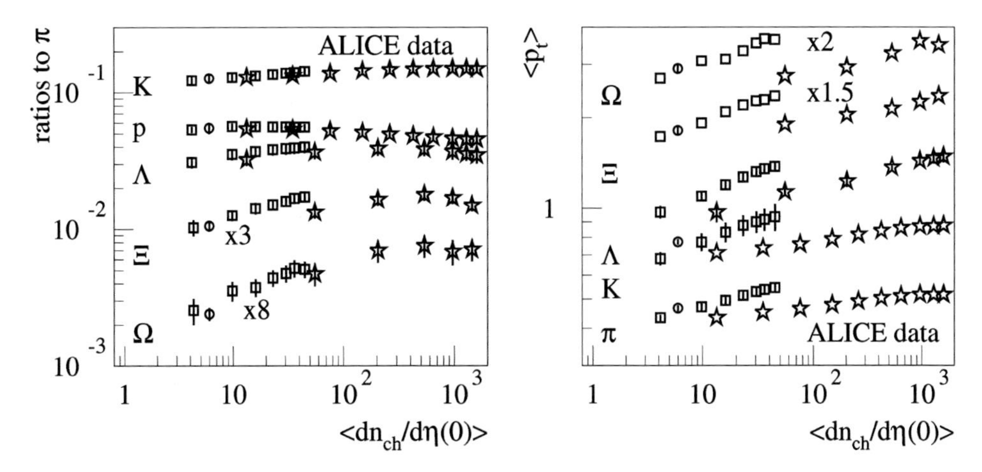

Fig. 15 Particle ratios to pions (left) and average transverse momenta (right) as a function of multiplicity, for different particle species, for minimum bias pp scattering (circles) as well as p–Pb (squares) and Pb–Pb (stars).

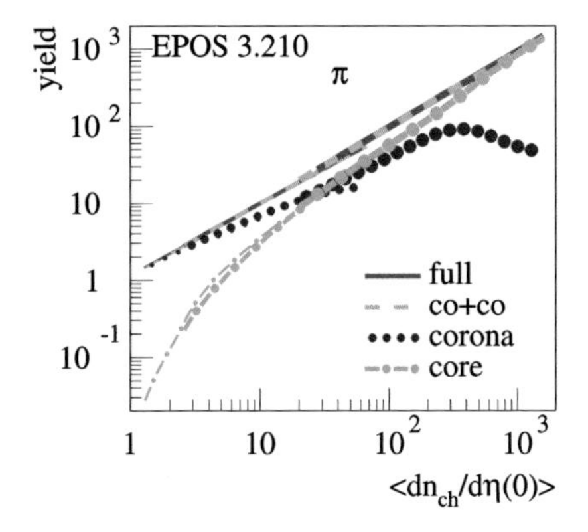

Fig. 16 The pion production rate as a function of the multiplicity, for pp scattering (thin lines), p–Pb (normal lines) and Pb–Pb (thick lines), for different contributions. Without the hadronic cascade: from core only (dashed-dotted), from corona only (dotted), and the sum of core and corona (co+co) shown as dashed line. The complete simulation, including hadronic cascade (full) is plotted as full line.

$\langle \frac{dn_{\mathrm{ch}}}{d\eta}(0) \rangle$, for pp scattering (thin lines) as well as p–Pb (normal lines) and Pb–Pb (thick lines), for different contributions. We consider first results without hadronic cascade: From core only (dashed-dotted), from corona only (dotted), and the sum of core and corona "co+co" shown as dashed line. The complete simulation, including hadronic cascade referred to as "full" is plotted as full line. The curves corresponding to the three colliding systems (pp, p–Pb, Pb–Pb) have considerably overlapping multiplicity ranges. Amazingly, we observe for any of the contributions (full, core, ...) essentially unique curves, so the yields do not depend on the system,

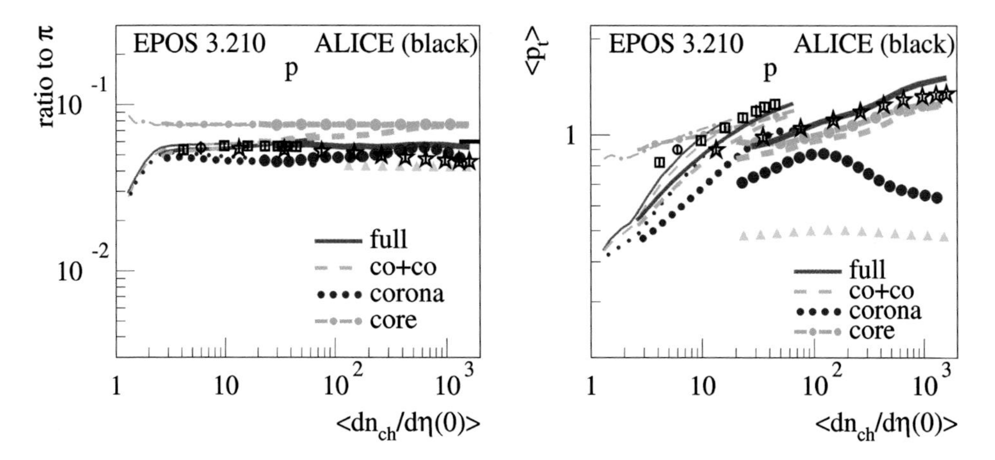

Fig. 17 Particle ratios to pions (left) and average transverse momenta (right) of protons and antiprotons as a function of multiplicity, as obtained from EPOS simulations, for pp scattering (thin lines), p–Pb (normal lines) and Pb–Pb (thick lines), for different contributions (as explained in the figure caption of Fig. 16). We also show the result from a "pure" EPOS simulation, without hydro and hadronic cascade (triangles). We compare with ALICE data for minimum bias pp scattering (circles) as well as p–Pb (squares) and Pb–Pb (stars).

but rather on the multiplicity. The yields for pp, p–Pb, and Pb–Pb agree, as long as the multiplicity is the same. Considering then these unique continuous curves, which extend over the whole multiplicity range, we see that the relative core rate (the core relative to core+corona(co-co) shows a smooth transition from 0% to 100%. Low multiplicity pp is pure corona, high multiplicity Pb–Pb is pure core.

We will now study ratios to pions versus multiplicity and mean transverse momenta versus multiplicity, as obtained from EPOS simulations. In Fig. 17 (left), we plot the ratios of protons and antiprotons to pions as a function of multiplicity, as obtained from EPOS simulations, for pp scattering (thin lines), p–Pb (normal lines) and Pb–Pb (thick lines), for different contributions. In addition to the corona contribution, we also show the result from a "pure" EPOS simulation, without hydro and hadronic cascade (triangles). Whereas the origin of both is simply kinky string fragmentation, they are not identical, because the corona particles may suffer energy loss (of the underlying partons), and the core–corona separation introduces biases. Despite this complication, both corona and core contributions are universal curves (pp = p–Pb = Pb–Pb in the overlap regions), and in addition these two universal curves are flat. Since we know from Fig. 16 that the relative core contribution increases with multiplicity, we understand that the core+corona curve (co+co, dashed) simply interpolates between the corona level at small multiplicity towards the core level at high multiplicity. The "full" results shows some reduction with respect to the "co+co" case, with increasing multiplicity, due to baryon–antibaryon annihilation, so at the end the "full" contribution is almost a constant curve.

Also shown on the figure as short black horizontal line on the right-hand side is the result from a statistical model calculation.[963]

In Fig. 17 (right), we plot the mean transverse momentum of protons and antiprotons versus multiplicity, again for pp, p–Pb, and Pb–Pb, for the different contributions. Here we do not observe universal curves. The "core" curves (dashed-dotted green lines) have increasing values in the overlap region, from AA via p–Pb to pp, which we can attribute to increasing radial flow for the smaller systems. For the "corona" curves we see a very steep increase for pp and p–Pb, slightly steeper for pp compared to p–Pb. Concerning Pb–Pb, we first observe a flat curve at low level for the "pure" case (no hydro, no cascade). The "corona" curve is a non-monotonic curve above the "pure" one. The difference between "pure" and "corona" is first of all due to the fact that low p_t particles are more likely to end up in the core, and secondly there is also an effect of energy loss in the medium. From the individual "core" and "corona" results and the fact that the relative core rate increases with increasing multiplicity, we understand that the core+corona contributions (co+co, yellow dashed lines) do not show a single universal curve for pp, p–Pb, and Pb–Pb. Considering the "full" result, including hadronic cascade, we see in a substantial increase, due to the annihilation of mainly low p_t baryon–antibaryon pairs.

In Fig. 18 (left) we plot particle ratios to pions of Omegas and Antiomegas as a function of multiplicity, as obtained from EPOS simulations (same conventions as Fig. 17). Here again, both corona and core contributions are universal and flat curves (pp = p–Pb = Pb–Pb in the overlap regions). Again, due to the fact that the relative core contribution increases with multiplicity, we understand that the core+corona curve (co+co, dashed) interpolates between the corona level at small multiplicity towards the core level at high multiplicity. The "full" results show some reduction with respect to the "co+co" case. The main difference to the proton case of

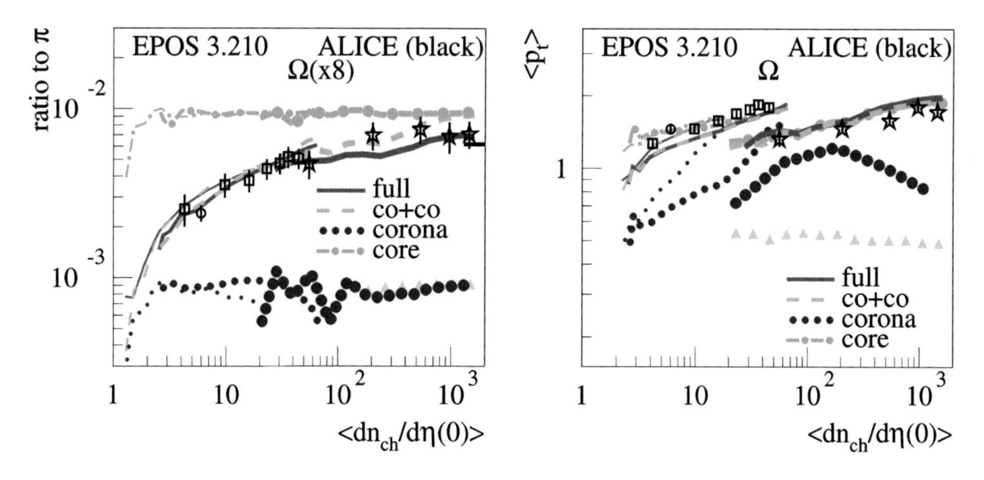

Fig. 18 Same as Fig. 17, but for Omegas and Antiomegas.

Fig. 17 is the big difference between the corona and the core level (Omega production from string decay is very rare), and therefore we get a strong enhancement from low towards high multiplicity. The average p_t results in Fig. 18 (right) are qualitatively similar to the ones for protons, just the values for Omegas are somewhat bigger compared to protons.

Apart of protons and Omegas, as discussed above, we also studied other hadrons; the results will be summarized in the following. Concerning the ratios h/π for $h = p, K, \Lambda, \Xi, \Omega$ versus multiplicity, the core and corona contributions separately are roughly constant, with the difference (core–corona) increasing for $p \to K \to \Lambda \to \Xi \to \Omega$. Since the relative core rate increases with increasing multiplicity, we get monotonically increasing curves for the total contributions, with increasing slopes for $p \to K \to \Lambda \to \Xi \to \Omega$. Concerning the average transverse momenta, we have a moderate increase of core contributions, the same for pp and p–Pb, similar to Pb–Pb, but with higher values for the smaller systems for given multiplicity. We have a strong increase of the corona contributions in pp and p–Pb, even stronger for pp compared to p–Pb. The Pb–Pb corona curves are non-monotonic, but lower compared to pp and p–Pb, at given multiplicity. Taking all together, we get as "core + corona" and "full" results discontinuous curves for pp $\to$ p–Pb $\to$ Pb–Pb, due to the discontinuities of both core and corona contributions.

6. Conclusion

EPOS is a universal approach, with all reactions (pp, pA, AA) described according to the same procedure. The primary interactions consist of a Gribov–Regge multiple scattering approach, where the elementary exchange object called a pomeron is a parton ladder according to DGLAP. High parton density effects are taken care of via a parton saturation scale. Secondary interactions are based on a core–corona approach to separate fluid and jet contributions. The core (fluid part) expands according to viscous hydrodynamics, and then decays at some given temperature according to statistical hadronization. The approach reproduces many flow-like features in pp, pA, and AA. The most recent developments concern a more sophisticated treatment of parton saturation, in particular a saturation scale which increases with the number of pomerons and not the number of participating nucleons, as done usually. As a result we get a very strong increase of the mean transverse momentum versus multiplicity in pp and pA, which (partly) explains the non-continuous increase of the mean p_t with multiplicity when considering pp, p–Pb, and Pb–Pb collisions.

Acknowledgments

This research was carried out within the scope of the GDRE (European Research Group) "Heavy ions at ultrarelativistic energies". The authors acknowledge the Texas Advanced Computing Center (TACC) at the University of Texas at Austin for providing computing resources that have contributed to the research results reported within this paper. URL: http://www.tacc.utexas.edu. This work was supported by U.S. Department of Energy Office of Science under contract number DE-SC0013391.

References

1. J. Collins, *Foundations of Perturbative QCD* (Cambridge University Press, 2013).
2. T. C. Rogers and P. J. Mulders, No generalized TMD-factorization in hadroproduction of high transverse momentum hadrons, *Phys. Rev. D* **81**, 094006, (2010). doi: 10.1103/PhysRevD.81.094006.
3. M. Diehl, D. Ostermeier and A. Schäfer, Elements of a theory for multiparton interactions in QCD, *J. High Energy Phys.* **03**, 089, (2012). doi: 10.1007/JHEP03(2012)089.
4. G. Calucci and D. Treleani, Incoherence and multiple parton interactions, *Phys. Rev. D* **80**, 054025, (2009). doi: 10.1103/PhysRevD.80.054025.
5. M. Diehl, J. R. Gaunt and K. Schönwald, Double hard scattering without double counting, *J. High Energy Phys.* **06**, 083, (2017). doi: 10.1007/JHEP06(2017)083.
6. N. Paver and D. Treleani, Multiple parton processes in the TeV region, *Z. Phys. C* **28**, 187, (1985). doi: 10.1007/BF01575722.
7. D. Treleani, AGK cutting rules and perturbative QCD, *Int. J. Mod. Phys. A* **11**, 613–654, (1996). doi: 10.1142/S0217751X96000286.
8. M. Cacciari, G. P. Salam and S. Sapeta, On the characterisation of the underlying event, *J. High Energy Phys.* **04**, 065, (2010). doi: 10.1007/JHEP04(2010)065.
9. B. Blok, Yu. Dokshitser, L. Frankfurt and M. Strikman, pQCD physics of multiparton interactions, *Eur. Phys. J. C* **72**, 1963, (2012). doi: 10.1140/epjc/s10052-012-1963-8.
10. B. Blok, Yu. Dokshitzer, L. Frankfurt and M. Strikman, Perturbative QCD correlations in multi-parton collisions, *Eur. Phys. J. C* **74**, 2926, (2014). doi: 10.1140/epjc/s10052-014-2926-z.
11. M. G. Ryskin and A. M. Snigirev, A fresh look at double parton scattering, *Phys. Rev. D* **83**, 114047, (2011). doi: 10.1103/PhysRevD.83.114047.
12. M. G. Ryskin and A. M. Snigirev, Double parton scattering in double logarithm approximation of perturbative QCD, *Phys. Rev. D* **86**, 014018, (2012). doi: 10.1103/PhysRevD.86.014018.
13. J. R. Gaunt, Single perturbative splitting diagrams in double parton scattering, *J. High Energy Phys.* **01**, 042, (2013). doi: 10.1007/JHEP01(2013)042.
14. A. V. Manohar and W. J. Waalewijn, What is double parton scattering? *Phys. Lett. B* **713**, 196–201, (2012). doi: 10.1016/j.physletb.2012.05.044.
15. R. Kirschner, Generalized Lipatov–Altarelli–Parisi equations and jet calculus rules, *Phys. Lett. B* **84**, 266–270, (1979). doi: 10.1016/0370-2693(79)90300-9.
16. V. P. Shelest, A. M. Snigirev and G. M. Zinovev, The multiparton distribution equations in QCD, *Phys. Lett. B* **113**, 325, (1982). doi: 10.1016/0370-2693(82)90049-1.

17. A. M. Snigirev, Double parton distributions in the leading logarithm approximation of perturbative QCD, *Phys. Rev. D* **68**, 114012, (2003). doi: 10.1103/PhysRevD.68.114012.
18. J. R. Gaunt and W. J. Stirling, Double parton distributions incorporating perturbative QCD evolution and momentum and quark number sum rules, *J. High Energy Phys.* **03**, 005, (2010). doi: 10.1007/JHEP03(2010)005.
19. F. A. Ceccopieri, An update on the evolution of double parton distributions, *Phys. Lett. B* **697**, 482–487, (2011). doi: 10.1016/j.physletb.2011.02.047.
20. P. Plößl. DPD sum rules in QCD. (2017). URL https://inspirehep.net/record/1520880.
21. M. Diehl, J. R. Gaunt, D. Ostermeier, P. Plößl and A. Schäfer, Cancellation of Glauber gluon exchange in the double Drell–Yan process, *J. High Energy Phys.* **01**, 076, (2016). doi: 10.1007/JHEP01(2016)076.
22. M. G. A. Buffing, M. Diehl and T. Kasemets, Transverse momentum in double parton scattering: factorisation, evolution and matching, *J. High Energy Phys.* **01**, 044, (2018). doi: 10.1007/JHEP01(2018)044.
23. A. V. Manohar and W. J. Waalewijn, A QCD analysis of double parton scattering: Color correlations, interference effects and evolution, *Phys. Rev. D* **85**, 114009, (2012). doi: 10.1103/PhysRevD.85.114009.
24. A. Vladimirov, Soft factors for double parton scattering at NNLO, *J. High Energy Phys.* **12**, 038, (2016). doi: 10.1007/JHEP12(2016)038.
25. M. Mekhfi and X. Artru, Sudakov suppression of color correlations in multiparton scattering, *Phys. Rev. D* **37**, 2618–2622, (1988). doi: 10.1103/PhysRevD.37.2618.
26. J. R. Gaunt, Glauber gluons and multiple parton interactions, *J. High Energy Phys.* **07**, 110, (2014). doi: 10.1007/JHEP07(2014)110.
27. J. C. Collins, D. E. Soper and G. F. Sterman, Soft gluons and factorization, *Nucl. Phys. B* **308**, 833–856, (1988). doi: 10.1016/0550-3213(88)90130-7.
28. A. Vladimirov, Structure of rapidity divergences in multi-parton scattering soft factors, *J. High Energy Phys.* **04**, 045, (2008). doi: 10.1007/JHEP04(2018)045.
29. D. Binosi, J. Collins, C. Kaufhold and L. Theussl, JaxoDraw: A graphical user interface for drawing Feynman diagrams. Version 2.0 release notes, *Comput. Phys. Commun.* **180**, 1709–1715, (2009). doi: 10.1016/j.cpc.2009.02.020.
30. N. Paver and D. Treleani, Multi-quark scattering and large p_T jet production in hadronic collisions, *Nuovo Cim. A* **70**, 215, (1982). doi: 10.1007/BF02814035.
31. B. Humpert, Are there multi-quark interactions? *Phys. Lett. B* **131**, 461–467, (1983). doi: 10.1016/0370-2693(83)90540-3.
32. M. Mekhfi, Multiparton processes: An application to double Drell–Yan, *Phys. Rev. D* **32**, 2371, (1985). doi: 10.1103/PhysRevD.32.2371.
33. M. Mekhfi, Correlations in color and spin in multiparton processes, *Phys. Rev. D* **32**, 2380, (1985). doi: 10.1103/PhysRevD.32.2380.
34. L. Ametller and D. Treleani, Shadowing in semihard interactions, *Int. J. Mod. Phys. A* **3**, 521–530, (1988). doi: 10.1142/S0217751X88000217.
35. T. Sjöstrand and M. van Zijl, A multiple interaction model for the event structure in hadron collisions, *Phys. Rev. D* **36**, 2019, (1987). doi: 10.1103/PhysRevD.36.2019.
36. J. M. Butterworth, J. R. Forshaw and M. H. Seymour, Multiparton interactions in photoproduction at HERA, *Z. Phys. C* **72**, 637–646, (1996). doi: 10.1007/BF02909195,10.1007/s002880050286.
37. M. Bahr, S. Gieseke and M. H. Seymour, Simulation of multiple partonic interactions in Herwig++, *J. High Energy Phys.* **07**, 076, (2008). doi: 10.1088/1126-6708/2008/07/076.

38. V. A. Abramovsky, V. N. Gribov and O. V. Kancheli, Character of inclusive spectra and fluctuations produced in inelastic processes by multi-pomeron exchange, *Yad. Fiz.* **18**, 595–616, (1973). [*Sov. J. Nucl. Phys.* **18**, 308, (1974)].
39. G. Calucci and D. Treleani, A functional formalism for multiparton interactions in high-energy collisions, *Int. J. Mod. Phys. A* **6**, 4375–4393, (1991). doi: 10.1142/S0217751X91002112.
40. A. Bialas, M. Bleszynski and W. Czyz, Multiplicity distributions in nucleus–nucleus collisions at high-energies, *Nucl. Phys. B* **111**, 461–476, (1976). doi: 10.1016/0550-3213(76)90329-1.
41. G. Calucci and D. Treleani, Multi-parton correlations and 'exclusive' cross sections, *Phys. Rev. D* **79**, 074013, (2009). doi: 10.1103/PhysRevD.79.074013.
42. M. H. Seymour and A. Siodmok, Extracting $\sigma_{\text{effective}}$ from the LHCb double-charm measurement. (2013). arXiv:1308.6749.
43. D. d'Enterria and A. M. Snigirev, Triple parton scatterings in high-energy proton–proton collisions, *Phys. Rev. Lett.* **118** (12), 122001, (2017). doi: 10.1103/PhysRevLett.118.122001.
44. G. Calucci and D. Treleani, Disentangling correlations in multiple parton interactions, *Phys. Rev. D* **83**, 016012, (2011). doi: 10.1103/PhysRevD.83.016012.
45. G. Calucci and D. Treleani, Proton structure in transverse space and the effective cross section, *Phys. Rev. D* **60**, 054023, (1999). doi: 10.1103/PhysRevD.60.054023.
46. M. Diehl, Introduction to GPDs and TMDs, *Eur. Phys. J. A* **52** (6), 149, (2016). doi: 10.1140/epja/i2016-16149-3.
47. S. Scopetta and V. Vento, Generalized parton distributions in constituent quark models, *Eur. Phys. J. A* **16**, 527–535, (2003). doi: 10.1140/epja/i2002-10120-y.
48. S. Boffi, B. Pasquini and M. Traini, Linking generalized parton distributions to constituent quark models, *Nucl. Phys. B* **649**, 243–262, (2003). doi: 10.1016/S0550-3213(02)01016-7.
49. R. Dupre, M. Guidal and M. Vanderhaeghen, Tomographic image of the proton, *Phys. Rev. D* **95** (1), 011501, (2017). doi: 10.1103/PhysRevD.95.011501.
50. J. Gaunt, *Double Parton Scattering in Proton–Proton Collisions* (Cambridge University Thesis, 2012).
51. F. A. Ceccopieri, A second update on double parton distributions, *Phys. Lett. B* **734**, 79–85, (2014). doi: 10.1016/j.physletb.2014.05.015.
52. H.-M. Chang, A. V. Manohar and W. J. Waalewijn, Double parton correlations in the bag model, *Phys. Rev. D* **87** (3), 034009, (2013). doi: 10.1103/PhysRevD.87.034009.
53. M. Rinaldi, S. Scopetta and V. Vento, Double parton correlations in constituent quark models, *Phys. Rev. D* **87**, 114021, (2013). doi: 10.1103/PhysRevD.87.114021.
54. M. Rinaldi, S. Scopetta, M. Traini and V. Vento, Double parton correlations and constituent quark models: a Light Front approach to the valence sector, *J. High Energy Phys.* **12**, 028, (2014). doi: 10.1007/JHEP12(2014)028.
55. W. Broniowski, E. Ruiz Arriola and K. Golec-Biernat, Generalized valon model for double parton distributions, *Few Body Syst.* **57** (6), 405–410, (2016). doi: 10.1007/s00601-016-1087-z.
56. T. Kasemets and A. Mukherjee, Quark–gluon double parton distributions in the light-front dressed quark model, *Phys. Rev. D* **94** (7), 074029, (2016). doi: 10.1103/PhysRevD.94.074029.
57. G. Parisi and R. Petronzio, On the breaking of Bjorken scaling, *Phys. Lett. B* **62**, 331–334, (1976). doi: 10.1016/0370-2693(76)90088-5.
58. R. L. Jaffe and G. G. Ross, Normalizing the renormalization group analysis of deep inelastic leptoproduction, *Phys. Lett. B* **93**, 313–317, (1980). doi: 10.1016/0370-2693(80)90521-3.

59. M. Diehl, T. Kasemets and S. Keane, Correlations in double parton distributions: effects of evolution, *J. High Energy Phys.* **05**, 118, (2014). doi: 10.1007/JHEP05(2014)118.
60. M. Rinaldi, S. Scopetta, M. C. Traini and V. Vento, Correlations in double parton distributions: Perturbative and non-perturbative effects, *J. High Energy Phys.* **10**, 063, (2016). doi: 10.1007/JHEP10(2016)063.
61. S. Chekanov *et al.*, Exclusive photoproduction of J/psi mesons at HERA, *Eur. Phys. J. C* **24**, 345–360, (2002). doi: 10.1007/s10052-002-0953-7.
62. A. Aktas *et al.*, Elastic J/psi production at HERA, *Eur. Phys. J. C* **46**, 585–603, (2006). doi: 10.1140/epjc/s2006-02519-5.
63. M. Diehl, From form factors to generalized parton distributions, *PoS* **DIS2013**, 224, (2013).
64. P. Hagler, Hadron structure from lattice quantum chromodynamics, *Phys. Rept.* **490**, 49–175, (2010). doi: 10.1016/j.physrep.2009.12.008.
65. L. Frankfurt, M. Strikman and C. Weiss, Dijet production as a centrality trigger for *pp* collisions at CERN LHC, *Phys. Rev. D* **69**, 114010, (2004). doi: 10.1103/PhysRevD.69.114010.
66. R. Corke and T. Sjöstrand, Multiparton interactions with an x-dependent proton size, *J. High Energy Phys.* **05**, 009, (2011). doi: 10.1007/JHEP05(2011)009.
67. M. Diehl and A. Schäfer, Theoretical considerations on multiparton interactions in QCD, *Phys. Lett. B* **698**, 389–402, (2011). doi: 10.1016/j.physletb.2011.03.024.
68. T. Kasemets and M. Diehl, Angular correlations in the double Drell–Yan process, *J. High Energy Phys.* **01**, 121, (2013). doi: 10.1007/JHEP01(2013)121.
69. T. Kasemets and P. J. Mulders, Constraining double parton correlations and interferences, *Phys. Rev. D* **91**, 014015, (2015). doi: 10.1103/PhysRevD.91.014015.
70. M. Diehl and T. Kasemets, Positivity bounds on double parton distributions, *J. High Energy Phys.* **05**, 150, (2013). doi: 10.1007/JHEP05(2013)150.
71. M. Aaboud *et al.*, Study of hard double-parton scattering in four-jet events in pp collisions at $\sqrt{s} = 7$ TeV with the ATLAS experiment, *J. High Energy Phys.* **11**, 110, (2016). doi: 10.1007/JHEP11(2016)110.
72. M. Rinaldi and F. A. Ceccopieri, Relativistic effects in model calculations of double parton distribution function, *Phys. Rev. D* **95** (3), 034040, (2017). doi: 10.1103/PhysRevD.95.034040.
73. M. Rinaldi, S. Scopetta, M. Traini and V. Vento, Double parton scattering: a study of the effective cross section within a Light–Front quark model, *Phys. Lett. B* **752**, 40–45, (2016). doi: 10.1016/j.physletb.2015.11.031.
74. M. Traini, M. Rinaldi, S. Scopetta and V. Vento, The effective cross section for double parton scattering within a holographic AdS/QCD approach, *Phys. Lett. B* **768**, 270–273, (2017). doi: 10.1016/j.physletb.2017.02.061.
75. F. A. Ceccopieri, M. Rinaldi and S. Scopetta, Parton correlations in same-sign W pair production via double parton scattering at the LHC, *Phys. Rev. D* **95** (11), 114030, (2017). doi: 10.1103/PhysRevD.95.114030.
76. CMS Collaboration, Measurement of double parton scattering in same-sign WW production in p–p collisions at $\sqrt{s} = 13$ TeV with the CMS experiment, (2017). http://cds.cern.ch/record/2257583.
77. A. Accardi *et al.*, Electron ion collider: the next QCD frontier, *Eur. Phys. J. A* **52** (9), 268, (2016). doi: 10.1140/epja/i2016-16268-9.
78. R. Aaij *et al.*, Observation of double charm production involving open charm in pp collisions at $\sqrt{s} = 7$ TeV, *J. High Energy Phys.* **06**, 141, (2012). doi: 10.1007/JHEP03(2014)108,10.1007/JHEP06(2012)141. [Addendum: *J. High Energy Phys.* **03**, 108, (2014)].

79. M. G. Echevarria, T. Kasemets, P. J. Mulders and C. Pisano, Polarization effects in double open-charm production at LHCb, *J. High Energy Phys.* **04**, 034, (2015). doi: 10.1007/JHEP04(2015)034.
80. M. Strikman and D. Treleani, Measuring double parton distributions in nucleons at proton–nucleus colliders, *Phys. Rev. Lett.* **88**, 031801, (2002). doi: 10.1103/PhysRevLett.88.031801.
81. A. Grebenyuk, F. Hautmann, H. Jung, P. Katsas and A. Knutsson, Jet production and the inelastic pp cross section at the LHC, *Phys.Rev. D* **86**, 117501, (2012). doi: 10.1103/PhysRevD.86.117501.
82. B. Blok, Yu. Dokshitzer, L. Frankfurt and M. Strikman, The four jet production at LHC and Tevatron in QCD, *Phys. Rev. D* **83**, 071501, (2011). doi: 10.1103/PhysRevD.83.071501.
83. M. Diehl, Multiple interactions and generalized parton distributions, *PoS* **DIS2010**, 223, (2010).
84. J. R. Gaunt and W. J. Stirling, Double parton scattering singularity in one-loop integrals, *J. High Energy Phys.* **06**, 048, (2011). doi: 10.1007/JHEP06(2011)048.
85. B. Blok, Yu. Dokshitzer, L. Frankfurt and M. Strikman, Origins of parton correlations in nucleon and multi-parton collisions, *arXiv:1206.5594*. (2012).
86. J. R. Gaunt, R. Maciuła and A. Szczurek, Conventional versus single-ladder-splitting contributions to double parton scattering production of two quarkonia, two Higgs bosons and $c\bar{c}c\bar{c}$, *Phys. Rev. D* **90** (5), 054017, (2014). doi: 10.1103/PhysRevD.90.054017.
87. K. Golec-Biernat and E. Lewandowska, Electroweak boson production in double parton scattering, *Phys. Rev. D* **90** (9), 094032, (2014). doi: 10.1103/PhysRevD.90.094032.
88. R. Astalos *et al.*, *Proceedings of the Sixth International Workshop on Multiple Partonic Interactions at the Large Hadron Collider.* (2015). URL https://inspirehep.net/record/1377199/files/arXiv:1506.05829.pdf.
89. H. Jung, D. Treleani, M. Strikman and N. van Buuren (eds.), *Proceedings, 7th International Workshop on Multiple Partonic Interactions at the LHC (MPI@LHC 2015)*, (2016). URL https://bib-pubdb1.desy.de/record/297386.
90. B. Blok and P. Gunnellini, Dynamical approach to MPI four-jet production in Pythia, *Eur. Phys. J. C* **75** (6), 282, (2015). doi: 10.1140/epjc/s10052-015-3520-8.
91. B. Blok and P. Gunnellini, Dynamical approach to MPI in W+dijet and Z+dijet production within the Pythia event generator, *Eur. Phys. J. C* **76** (4), 202, (2016). doi: 10.1140/epjc/s10052-016-4035-7.
92. B. Blok and M. Strikman, Shadowing in multiparton proton–deuteron collisions, *Eur. Phys. J. C* **74**, 3038, (2014). doi: 10.1140/epjc/s10052-014-3038-5.
93. B. Blok, M. Strikman and U. A. Wiedemann, Hard four-jet production in pA collisions, *Eur. Phys. J. C* **73** (6), 2433, (2013). doi: 10.1140/epjc/s10052-013-2433-7.
94. B. Blok and M. Strikman, Double parton interactions in $\gamma p, \gamma A$ collisions in the direct photon kinematics, *Eur. Phys. J. C* **74** (12), 3214, (2014). doi: 10.1140/epjc/s10052-014-3214-7.
95. B. Blok and M. Strikman, Open charm production in double parton scattering processes in the forward kinematics, *Eur. Phys. J. C* **76** (12), 694, (2016). doi: 10.1140/epjc/s10052-016-4551-5.
96. B. Blok and M. Strikman, Interplay of soft and perturbative correlations in multiparton interactions at central rapidities, *Phys. Lett. B* **772**, 219–224, (2017). doi: 10.1016/j.physletb.2017.06.049.
97. F. Abe *et al.*, Double parton scattering in $\bar{p}p$ collisions at $\sqrt{s} = 1.8$TeV, *Phys. Rev. D* **56**, 3811–3832, (1997). doi: 10.1103/PhysRevD.56.3811.

98. V. M. Abazov *et al.*, Double parton interactions in γ+3 jet events in p$\bar{\text{p}}$ collisions $\sqrt{s} = 1.96$ TeV *Phys. Rev. D* **81**, 052012, (2010). doi: 10.1103/PhysRevD.81.052012.
99. V. M. Abazov *et al.*, Azimuthal decorrelations and multiple parton interactions in γ+2 jet and γ+3 jet events in $p\bar{p}$ collisions at $\sqrt{s} = 1.96$ TeV, *Phys. Rev. D* **83**, 052008, (2011). doi: 10.1103/PhysRevD.83.052008.
100. S. Chatrchyan *et al.*, Measurement of four-jet production in proton–proton collisions at $\sqrt{s} = 7$ TeV, *Phys. Rev. D* **89**, 092010, (2014). doi: 10.1103/PhysRevD.89.092010.
101. G. Aad *et al.*, Measurement of underlying event characteristics using charged particles in pp collisions at $\sqrt{s} = 900$ GeV and 7 TeV with the ATLAS detector, *Phys. Rev. D* **83**, 112001, (2011). doi: 10.1103/PhysRevD.83.112001.
102. G. Aad *et al.*, Measurement of hard double-parton interactions in $W(\rightarrow l\nu) + 2$ jet events at $\sqrt{s} = 7$ TeV with the ATLAS detector, *New J. Phys.* **15**, 033038, (2013). doi: 10.1088/1367-2630/15/3/033038.
103. S. Chatrchyan *et al.*, Study of double parton scattering using W + 2-jet events in proton–proton collisions at $\sqrt{s} = 7$ TeV, *J. High Energy Phys.* **03**, 032, (2014). doi: 10.1007/JHEP03(2014)032.
104. I. M. Belyaev, Talk at MPI-2015 conference.
105. R. Aaij *et al.*, Prompt charm production in pp collisions at $\sqrt{s} = 7$ TeV, *Nucl. Phys. B* **871**, 1–20, (2013). doi: 10.1016/j.nuclphysb.2013.02.010.
106. R. Aaij *et al.*, Production of associated Υ and open charm hadrons in pp collisions at $\sqrt{s} = 7$ and 8 TeV via double parton scattering, *J. High Energy Phys.* **07**, 052, (2016). doi: 10.1007/JHEP07(2016)052.
107. T. Sjöstrand and P. Z. Skands, Transverse-momentum-ordered showers and interleaved multiple interactions, *Eur. Phys. J. C* **39**, 129–154, (2005). doi: 10.1140/epjc/s2004-02084-y.
108. S. Gieseke, C. Röhr and A. Siódmok, Multiple partonic interaction developments in Herwig++. (2011). arXiv:1110.2675.
109. S. Gieseke *et al.*, Herwig++ 2.5 Release Note. URL https://arxiv.org/abs/1102.1672. arXiv:1102.1672 [hep-ph], (2011).
110. M. H. Seymour and A. Siodmok, Constraining MPI models using σ_{eff} and recent Tevatron and LHC underlying event data, *J. High Energy Phys.* **10**, 113, (2013). doi: 10.1007/JHEP10(2013)113.
111. S. Gieseke, C. Röhr and A. Siodmok, Colour reconnections in Herwig++, *Eur. Phys. J. C* **72**, 2225, (2012). doi: 10.1140/epjc/s10052-012-2225-5.
112. M. Bahr, M. Myska, M. H. Seymour and A. Siodmok, Extracting $\sigma_{\text{effective}}$ from the CDF $\gamma + 3$ jets measurement, *J. High Energy Phys.* **03**, 129, (2013). doi: 10.1007/JHEP03(2013)129.
113. T. Sjöstrand, S. Ask, J. R. Christiansen, R. Corke, N. Desai, P. Ilten, S. Mrenna, S. Prestel, C. O. Rasmussen and P. Z. Skands, An Introduction to Pythia 8.2, *Comput. Phys. Commun.* **191**, 159–177, (2015). doi: 10.1016/j.cpc.2015.01.024.
114. R. Corke and T. Sjöstrand, Interleaved parton showers and tuning prospects, *J. High Energy Phys.* **03**, 032, (2011). doi: 10.1007/JHEP03(2011)032.
115. C. Flensburg, G. Gustafson, L. Lönnblad and A. Ster, Correlations in double parton distributions at small x, *J. High Energy Phys.* **06**, 066, (2011). doi: 10.1007/JHEP06(2011)066.
116. L. Frankfurt, M. Strikman and C. Weiss, Small-x physics: From HERA to LHC and beyond, *Ann. Rev. Nucl. Part. Sci.* **55**, 403–465, (2005). doi: 10.1146/annurev.nucl.53.041002.110615.
117. L. Frankfurt, M. Strikman and C. Weiss, Transverse nucleon structure and diagnostics of hard parton–parton processes at LHC, (2010). *Phys. Rev. D* **83** (2011) 054012. doi: 10.1103/PhysRevD.83.054012.

118. M. Strikman, Transverse nucleon structure and multiparton interactions, *Acta Phys. Polon. B* **42**, 2607–2630, (2011). doi: 10.5506/APhysPolB.42.2607.
119. Y. L. Dokshitzer, D. Diakonov and S. I. Troian, Hard processes in quantum chromodynamics, *Phys. Rept.* **58**, 269–395, (1980). doi: 10.1016/0370-1573(80)90043-5.
120. M. Glück, E. Reya and A. Vogt, Parton distributions for high-energy collisions, *Z. Phys. C* **53**, 127–134, (1992). doi: 10.1007/BF01483880.
121. V. Gribov, Y. Dokshitzer and J. Nyiri, *Strong Interactions of Hadrons at High Energies: Gribov Lectures on Theoretical Physics.* Cambridge Monographs on Partic, (Cambridge University Press, 2009). URL https://books.google.fr/books?id=idIi5ryJ_AYC.
122. C. Adloff *et al.*, Diffraction dissociation in photoproduction at HERA, *Z. Phys. C* **74**, 221–236, (1997). doi: 10.1007/s002880050385.
123. E. G. S. Luna, V. A. Khoze, A. D. Martin and M. G. Ryskin, Diffractive dissociation re-visited for predictions at the LHC, *Eur. Phys. J. C* **59**, 1–12, (2009). doi: 10.1140/epjc/s10052-008-0793-1.
124. M. Glück, E. Reya and A. Vogt, Dynamical parton distributions revisited, *Eur. Phys. J. C* **5**, 461–470, (1998). doi: 10.1007/s100529800978,10.1007/s100520050289.
125. F. Hautmann, Z. Kunszt and D. E. Soper, Hard scattering factorization and light cone Hamiltonian approach to diffractive processes, *Nucl. Phys. B* **563**, 153–199, (1999). doi: 10.1016/S0550-3213(99)00568-4.
126. F. D. Aaron *et al.*, Diffractive electroproduction of rho and phi Mesons at HERA, *J. High Energy Phys.* **05**, 032, (2010). doi: 10.1007/JHEP05(2010)032.
127. S. Salvini, D. Treleani and G. Calucci, Double parton scatterings in high-energy proton–nucleus collisions and partonic correlations, *Phys. Rev. D* **89** (1), 016020, (2014). doi: 10.1103/PhysRevD.89.016020.
128. T. Affolder *et al.*, Charged jet evolution and the underlying event in $p\bar{p}$ collisions at 1.8 TeV, *Phys. Rev. D* **65**, 092002, (2002). doi: 10.1103/PhysRevD.65.092002.
129. S. Chatrchyan *et al.*, Measurement of the underlying event activity at the LHC with $\sqrt{s} = 7$ TeV and comparison with $\sqrt{s} = 0.9$ TeV, *J. High Energy Phys.* **09**, 109, (2011). doi: 10.1007/JHEP09(2011)109.
130. ATLAS Collaboration, Measurements of underlying-event properties using neutral and charged particles in pp collisions at 900 GeV and 7 TeV with the ATLAS detector at the LHC, *Eur. Phys. J. C* **71**, 1636, (2011). doi: 10.1140/epjc/s10052-011-1636-z.
131. M. Yu. Azarkin, I. M. Dremin and M. Strikman, Jets in multiparticle production in and beyond geometry of proton–proton collisions at the LHC, *Phys. Lett. B* **735**, 244–249, (2014). doi: 10.1016/j.physletb.2014.06.040.
132. Y. L. Dokshitzer, V. A. Khoze, A. H. Mueller and S. I. Troian, *Basics of Perturbative QCD.* 1991.
133. M. Strikman, Comments on the observation of high multiplicity events at the LHC, *Phys. Rev. D* **84**, 011501, (2011). doi: 10.1103/PhysRevD.84.011501.
134. T. C. Rogers, A. M. Stasto and M. I. Strikman, Unitarity constraints on semi-hard jet production in impact parameter space, *Phys. Rev. D* **77**, 114009, (2008). doi: 10.1103/PhysRevD.77.114009.
135. T. C. Rogers and M. Strikman, Multiple hard partonic collisions with correlations in proton–proton scattering, *Phys. Rev. D* **81**, 016013, (2010). doi: 10.1103/PhysRevD.81.016013.
136. V. Khachatryan *et al.*, Measurement of prompt J/ψ pair production in pp collisions at $\sqrt{s} = 7$ Tev, *J. High Energy Phys.* **09**, 094, (2014). doi: 10.1007/JHEP09(2014)094.

137. J.-P. Lansberg and H.-S. Shao, Double-quarkonium production at a fixed-target experiment at the LHC (After@LHC), *Nucl. Phys. B* **900**, 273–294, (2015). doi: 10.1016/j.nuclphysb.2015.09.005.
138. J.-P. Lansberg and H.-S. Shao, Associated production of a quarkonium and a Z boson at one loop in a quark–hadron-duality approach, *J. High Energy Phys.* **10**, 153, (2016). doi: 10.1007/JHEP10(2016)153.
139. J.-P. Lansberg, H.-S. Shao and N. Yamanaka, Indication for double parton scatterings in W+ prompt J/Ψ production at the LHC, *Phys. Lett. B* **781**, 485–491, (2018). doi: 10.1016/j.physletb.2018.04.020.
140. P. Schweitzer, M. Strikman and C. Weiss, Intrinsic transverse momentum and parton correlations from dynamical chiral symmetry breaking, *J. High Energy Phys.* **01**, 163, (2013). doi: 10.1007/JHEP01(2013)163.
141. M. Strikman and W. Vogelsang, Multiple parton interactions and forward double pion production in pp and *dA* scattering, *Phys. Rev. D* **83**, 034029, (2011). doi: 10.1103/PhysRevD.83.034029.
142. J. C. Collins, L. Frankfurt and M. Strikman, Factorization for hard exclusive electroproduction of mesons in QCD, *Phys. Rev. D* **56**, 2982–3006, (1997). doi: 10.1103/PhysRevD.56.2982.
143. L. Frankfurt, C. E. Hyde-Wright, M. Strikman and C. Weiss, Generalized parton distributions and rapidity gap survival in exclusive diffractive pp scattering, *Phys. Rev. D* **75**, 054009, (2007). doi: 10.1103/PhysRevD.75.054009.
144. ATLAS collaboration, Double parton scattering in pp collisions at $\sqrt{s} = 63\,\text{GeV}$, *Z. Phys. C* **34**, 163, (1987).
145. UA2 collaboration and J. Alilti *et al.*, *A Study of multi-jet events at the CERN $\bar{p}p$ collider and a search for double parton scattering*, *Phys. Lett. B* **268**, 145–154, (1991). doi: 10.1016/0370-2693(91)90937-L.
146. DØ Collaboration, Double parton interactions in photon+3 jet events in $p\bar{p}$ collisions at $\sqrt{s} = 1.96\,\text{TeV}$, *Phys. Rev. D* **82**, (2010).
147. V. Khachatryan *et al.*, Studies of inclusive four-jet production with two *b*-tagged jets in proton–proton collisions at 7 TeV, *Phys. Rev. D* **94** (11), 112005, (2016). doi: 10.1103/PhysRevD.94.112005.
148. V. M. Abazov *et al.*, Double parton interactions in $\gamma + 3$ jet and $\gamma + b/cjet + 2$ jet events in $p\bar{p}$ collisions at $\sqrt{s} = 1.96\,\text{TeV}$, *Phys. Rev. D* **89** (7), 072006, (2014). doi: 10.1103/PhysRevD.89.072006.
149. P. Gunnellini. *Study of double parton scattering using four-jet scenarios in proton–proton collisions at $\sqrt{s} = 7\,TeV$ with the CMS experiment at the Large Hadron Collider*, PhD thesis, University of Hamburg, (2014).
150. O. Gueta. *Study of hard double parton scattering in four-jet events in pp collisions at $\sqrt{s} = 7\,TeV$ with the ATLAS experiment at the LHC*, PhD thesis, Tel Aviv University, (2015-09-30). URL http://inspirehep.net/record/1504136/files/CERN-THESIS-2015-359.pdf.
151. K. Kutak, R. Maciula, M. Serino, A. Szczurek and A. van Hameren, Four-jet production in single- and double-parton scattering within high-energy factorization, *J. High Energy Phys.* **04**, 175, (2016). doi: 10.1007/JHEP04(2016)175.
152. V. Khachatryan *et al.*, Event generator tunes obtained from underlying event and multiparton scattering measurements, *Eur. Phys. J. C* **76** (3), 155, (2016). doi: 10.1140/epjc/s10052-016-3988-x.
153. F. Abe *et al.*, Study of four jet events and evidence for double parton interactions in $p\bar{p}$ collisions at $\sqrt{s} = 1.8$ TeV, *Phys. Rev. D* **47**, 4857–4871, (1993). doi: 10.1103/PhysRevD.47.4857.

154. S. Chatrchyan *et al.*, Jet and underlying event properties as a function of charged-particle multiplicity in proton–proton collisions at $\sqrt{s} = 7$ TeV, *Eur. Phys. J. C* **73** (12), 2674, (2013). doi: 10.1140/epjc/s10052-013-2674-5.
155. CMS Collaboration, Study of double parton scattering in photon + 3 jets final state in proton–proton collisions at 7 TeV, (2015). https://cds.cern.ch/record/2007815
156. G. Corcella, I. G. Knowles, G. Marchesini, S. Moretti, K. Odagiri, P. Richardson, M. H. Seymour and B. R. Webber, Herwig 6: an event generator for hadron emission reactions with interfering gluons (including supersymmetric processes), *J. High Energy Phys.* **01**, 010, (2001). doi: 10.1088/1126-6708/2001/01/010.
157. T. Gleisberg, S. Hoeche, F. Krauss, M. Schonherr, S. Schumann, F. Siegert and J. Winter, Event generation with SHERPA 1.1, *J. High Energy Phys.* **02**, 007, (2009). doi: 10.1088/1126-6708/2009/02/007.
158. T. Sjöstrand, S. Mrenna and P. Z. Skands, Pythia 6.4 physics and manual, *J. High Energy Phys.* **05**, 026, (2006). doi: 10.1088/1126-6708/2006/05/026.
159. V. M. Abazov *et al.*, Observation and studies of double J/ψ production at the Tevatron, *Phys. Rev. D* **90** (11), 111101, (2014). doi: 10.1103/PhysRevD.90.111101.
160. D. Treleani, Double parton scattering, diffraction and effective cross section, *Phys. Rev. D* **76**, 076006, (2007). doi: 10.1103/PhysRevD.76.076006.
161. C. Goebel, F. Halzen and D. M. Scott, Double Drell–Yan annihilations in hadron collisions: novel tests of the constituent picture, *Phys. Rev. D* **22**, 2789, (1980). doi: 10.1103/PhysRevD.22.2789.
162. T. Akesson *et al.*, Double parton scattering in pp collisions at $\sqrt{s} = 63$-GeV, *Z. Phys. C* **34**, 163, (1987). doi: 10.1007/BF01566757.
163. F. Halzen, P. Hoyer and W. J. Stirling, Evidence for multiple parton interactions from the observation of multi-muon events in Drell–Yan experiments, *Phys. Lett. B* **188**, 375–378, (1987). doi: 10.1016/0370-2693(87)91400-6.
164. G. Arnison *et al.*, Experimental observation of isolated large transverse energy electrons with associated missing energy at $\sqrt{s} = 540$ GeV, *Phys. Lett. B* **122**, 103–116, (1983). doi: 10.1016/0370-2693(83)91177-2.
165. M. Banner *et al.*, Observation of single isolated electrons of high transverse momentum in events with missing transverse energy at the CERN $\bar{p}p$ collider, *Phys. Lett. B* **122**, 476–485, (1983). doi: 10.1016/0370-2693(83)91605-2.
166. G. Arnison *et al.*, Experimental observation of lepton pairs of invariant mass around 95-GeV/c**2 at the CERN SPS collider, *Phys. Lett. B* **126**, 398–410, (1983). doi: 10.1016/0370-2693(83)90188-0.
167. P. Bagnaia *et al.*, Evidence for $Z^0 \to e^+e^-$ at the CERN anti-*pp* collider, *Phys. Lett. B* **129**, 130–140, (1983). doi: 10.1016/0370-2693(83)90744-X.
168. B. Humpert, The production of gauge boson pairs by *p* anti-*p* colliders, *Phys. Lett. B* **135**, 179–186, (1984). doi: 10.1016/0370-2693(84)90479-9.
169. R. M. Godbole *et al.*, Double parton scattering contribution to W+jets, *Z. Phys. C* **47**, 69, (1990). doi: 10.1007/BF01551914.
170. O. J. P. Eboli, F. Halzen and J. K. Mizukoshi, The associated production of weak bosons and jets by multiple parton interactions, *Phys. Rev. D* **57**, 1730–1734, (1998). doi: 10.1103/PhysRevD.57.1730.
171. E. A. Kuraev, L. N. Lipatov and V. S. Fadin, The Pomeranchuk singularity in nonabelian gauge theories, *Sov. Phys. JETP.* **45**, 199–204, (1977). [*Zh. Eksp. Teor. Fiz.* **72**, 377, (1977)].
172. I. I. Balitsky and L. N. Lipatov, The Pomeranchuk singularity in quantum chromodynamics, *Sov. J. Nucl. Phys.* **28**, 822–829, (1978). [*Yad. Fiz.* **28**, 1597, (1978)].

173. M. Drees and T. Han, Signals for double parton scattering at the Fermilab Tevatron, *Phys. Rev. Lett.* **77**, 4142–4145, (1996). doi: 10.1103/PhysRevLett.77.4142.
174. V. M. Abazov *et al.*, Study of double parton interactions in diphoton + dijet events in $p\bar{p}$ collisions at $\sqrt{s} = 1.96$ TeV, *Phys. Rev. D* **93** (5), 052008, (2016). doi: 10.1103/PhysRevD.93.052008.
175. A. Kulesza and W. J. Stirling, Like sign W boson production at the LHC as a probe of double parton scattering, *Phys. Lett. B* **475**, 168–175, (2000). doi: 10.1016/S0370-2693(99)01512-9.
176. A. Del Fabbro and D. Treleani, A double parton scattering background to Higgs boson production at the LHC, *Phys. Rev. D* **61**, 077502, (2000). doi: 10.1103/PhysRevD.61.077502.
177. E. L. Berger, C. B. Jackson, S. Quackenbush and G. Shaughnessy, Calculation of $\mathrm{Wb\bar{b}}$ production via double parton scattering at the LHC, *Phys. Rev. D* **84**, 074021, (2011). doi: 10.1103/PhysRevD.84.074021.
178. E. Maina, Multiple parton interactions, top-antitop and $W + 4j$ production at the LHC, *J. High Energy Phys.* **04**, 098, (2009). doi: 10.1088/1126-6708/2009/04/098.
179. E. Maina, Multiple parton interactions in $Z + 4j$, $W^{\pm}W^{\pm} + 0/2j$ and $W^{+}W^{-} + 2j$ production at the LHC, *J. High Energy Phys.* **09**, 081, (2009). doi: 10.1088/1126-6708/2009/09/081.
180. E. Cattaruzza, A. Del Fabbro and D. Treleani, Fractional momentum correlations in multiple production of W bosons and of $\mathrm{b\bar{b}}$ pairs in high energy pp collisions, *Phys. Rev. D* **72**, 034022, (2005). doi: 10.1103/PhysRevD.72.034022.
181. D. Bandurin, G. Golovanov and N. Skachkov, Double parton interactions as a background to associated HW production at the Tevatron, *J. High Energy Phys.* **04**, 054, (2011). doi: 10.1007/JHEP04(2011)054.
182. G. Aad *et al.*, Electron performance measurements with the ATLAS detector using the 2010 LHC proton–proton collision data, *Eur. Phys. J. C* **72**, 1909, (2012). doi: 10.1140/epjc/s10052-012-1909-1.
183. G. Aad *et al.*, Study of jets produced in association with a W boson in pp collisions at $\sqrt{s} = 7$ TeV with the ATLAS detector, *Phys. Rev. D* **85**, 092002, (2012). doi: 10.1103/PhysRevD.85.092002.
184. G. Aad *et al.*, Muon reconstruction efficiency and momentum resolution of the ATLAS experiment in proton–proton collisions at $\sqrt{s} = 7\,\mathrm{TeV}$ in 2010, *Eur. Phys. J. C* **74** (9), 3034, (2014). doi: 10.1140/epjc/s10052-014-3034-9.
185. M. Cacciari, G. P. Salam and G. Soyez, The anti-k(t) jet clustering algorithm, *J. High Energy Phys.* **0804**, 063, (2008). doi: 10.1088/1126-6708/2008/04/063.
186. M. L. Mangano, M. Moretti, F. Piccinini, R. Pittau and A. D. Polosa, ALPGEN, a generator for hard multiparton processes in hadronic collisions, *J. High Energy Phys.* **07**, 001, (2003). doi: 10.1088/1126-6708/2003/07/001.
187. J. Pumplin, D. R. Stump, J. Huston, H. L. Lai, P. M. Nadolsky and W. K. Tung, New generation of parton distributions with uncertainties from global QCD analysis, *J. High Energy Phys.* **07**, 012, (2002). doi: 10.1088/1126-6708/2002/07/012.
188. ATLAS Collaboration. New ATLAS event generator tunes to 2010 data. Technical Report ATL-PHYS-PUB-2011-008, CERN, (2011). URL http://cds.cern.ch/record/1345343.
189. F. Maltoni and T. Stelzer, MadEvent: automatic event generation with MadGraph, *J. High Energy Phys.* **02**, 027, (2003). doi: 10.1088/1126-6708/2003/02/027.
190. J. Alwall, M. Herquet, F. Maltoni, O. Mattelaer and T. Stelzer, MadGraph 5: going beyond, *J. High Energy Phys.* **06**, 128, (2011). doi: 10.1007/JHEP06(2011)128.

191. R. Field. Early LHC underlying event data — findings and surprises, In *Hadron Collider Physics. Proceedings, 22nd Conference, HCP 2010*, Toronto, Canada, August 23–27, 2010, (2010). URL https://inspirehep.net/record/873443/files/arXiv:1010.3558.pdf.
192. T. Sjöstrand, S. Mrenna and P. Z. Skands, A Brief Introduction to Pythia 8.1, *Comput. Phys. Commun.* **178**, 852–867, (2008). doi: 10.1016/j.cpc.2008.01.036.
193. G. Aad *et al.*, Measurement of the production cross section of prompt J/ψ mesons in association with a $W^{\pm}$ boson in pp collisions at $\sqrt{s} = 7$ TeV with the ATLAS detector, *J. High Energy Phys.* **04**, 172, (2014). doi: 10.1007/JHEP04(2014)172.
194. G. Aad *et al.*, Measurement of the differential cross-sections of inclusive, prompt and non-prompt J/ψ production in proton–proton collisions at $\sqrt{s} = 7$ TeV, *Nucl. Phys. B* **850**, 387–444, (2011). doi: 10.1016/j.nuclphysb.2011.05.015.
195. G. Aad *et al.*, Observation and measurements of the production of prompt and non-prompt J/ψ mesons in association with a Z boson in *pp* collisions at $\sqrt{s} = 8$ TeV with the ATLAS detector, *Eur. Phys. J. C* **75** (5), 229, (2015). doi: 10.1140/epjc/s10052-015-3406-9.
196. CMS Collaboration, Double parton scattering cross section limit from same-sign W bosons pair production in di-muon final state at LHC, (2015). https://cds.cern.ch/record/2103756
197. A. Donnachie and P. V. Landshoff, Total cross-sections, *Phys. Lett. B* **296**, 227–232, (1992). doi: 10.1016/0370-2693(92)90832-O.
198. H. Abramowicz *et al.*, Summary of the workshop on multi-parton interactions (MPI@LHC 2012), (2013). arXiv:1306.5413.
199. S. Bansal *et al.*, Progress in double parton scattering studies. In *Workshop on Multi-parton Interactions at the LHC (MPI@LHC 2013) Antwerp, Belgium, December 2–6, 2013*, (2014). URL https://inspirehep.net/record/1323623/files/arXiv:1410.6664.pdf.
200. V. M. Abazov *et al.*, Evidence for simultaneous production of J/ψ and Υ mesons, *Phys. Rev. Lett.* **116** (8), 082002, (2016). doi: 10.1103/PhysRevLett.116.082002.
201. A. Szczurek, A short review of some double-parton scattering processes, *Acta Phys. Polon. Supp.* **8** (2), 483, (2015). doi: 10.5506/APhysPolBSupp.8.483.
202. G. Calucci and D. Treleani, Minijets and the two-body parton correlation, *Phys. Rev. D* **57**, 503–511, (1998). doi: 10.1103/PhysRevD.57.503.
203. A. Del Fabbro and D. Treleani, Scale factor in double parton collisions and parton densities in transverse space, *Phys. Rev. D* **63**, 057901, (2001). doi: 10.1103/PhysRevD.63.057901.
204. V. L. Korotkikh and A. M. Snigirev, Double parton correlations versus factorized distributions, *Phys. Lett. B* **594**, 171–176, (2004). doi: 10.1016/j.physletb.2004.05.012.
205. J. R. Gaunt, C.-H. Kom, A. Kulesza and W. J. Stirling, Same-sign W pair production as a probe of double parton scattering at the LHC, *Eur. Phys. J. C* **69**, 53–65, (2010). doi: 10.1140/epjc/s10052-010-1362-y.
206. M. W. Krasny and W. Płaczek, The LHC excess of four-lepton events interpreted as Higgs-boson signal: background from double Drell–Yan process? *Acta Phys. Polon. B* **45** (1), 71–87, (2014). doi: 10.5506/APhysPolB.45.71.
207. M. W. Krasny and W. Płaczek, On the contribution of the double Drell–Yan process to WW and ZZ production at the LHC, *Acta Phys. Polon. B* **47**, 1045, (2016).
208. M. W. Krasny and W. Płaczek, In search of elementary spin 0 particles, *Ann. Phys.* **352**, 11, (2015). doi: 10.1016/j.aop.2014.07.035.

209. R. Aaij *et al.*, Measurements of prompt charm production cross-sections in pp collisions at $\sqrt{s} = 5$ TeV, *J. High Energy Phys.* **06**, 147, (2017). doi: 10.1007/JHEP06(2017)147.
210. R. Aaij *et al.*, Measurements of prompt charm production cross-sections in pp collisions at $\sqrt{s} = 13$ TeV, *J. High Energy Phys.* **03**, 159, (2016). doi: 10.1007/JHEP03(2016)159.
211. G. Aad *et al.*, Measurement of $D^{*\pm}$, $D^{\pm}$ and $D_s^{\pm}$ meson production cross sections in pp collisions at $\sqrt{s} = 7$ TeV with the ATLAS detector, *Nucl. Phys. B* **907**, 717–763, (2016). doi: 10.1016/j.nuclphysb.2016.04.032.
212. B. Abelev *et al.*, Measurement of charm production at central rapidity in proton–proton collisions at $\sqrt{s} = 2.76$ TeV, *J. High Energy Phys.* **07**, 191, (2012). doi: 10.1007/JHEP07(2012)191.
213. B. Abelev *et al.*, Measurement of charm production at central rapidity in proton–proton collisions at $\sqrt{s} = 7$ TeV, *J. High Energy Phys.* **01**, 128, (2012). doi: 10.1007/JHEP01(2012)128.
214. B. Abelev *et al.*, D_s^+ meson production at central rapidity in proton–proton collisions at $\sqrt{s} = 7$ TeV, *Phys. Lett. B* **718**, 279, (2012). doi: 10.1016/j.physletb.2012.10.049.
215. D. Acosta *et al.*, Measurement of prompt charm meson production cross sections in $p\bar{p}$ collisions at $\sqrt{s} = 1.96$ TeV, *Phys. Rev. Lett.* **91**, 241804, (2003). doi: 10.1103/PhysRevLett.91.241804.
216. B. A. Kniehl, G. Kramer, I. Schienbein and H. Spiesberger, Inclusive charmed-meson production at the CERN LHC, *Eur. Phys. J. C* **72**, 2082, (2012). doi: 10.1140/epjc/s10052-012-2082-2.
217. B. A. Kniehl, G. Kramer, I. Schienbein and H. Spiesberger, Inclusive $D^{*\pm}$ production in $p\bar{p}$ collisions with massive charm quarks, *Phys. Rev. D* **71**, 014018, (2005). doi: 10.1103/PhysRevD.71.014018.
218. B. A. Kniehl, G. Kramer, I. Schienbein and H. Spiesberger, Reconciling open charm production at the Fermilab Tevatron with QCD, *Phys. Rev. Lett.* **96**, 012001, (2006). doi: 10.1103/PhysRevLett.96.012001.
219. T. Kneesch, B. A. Kniehl, G. Kramer and I. Schienbein, Charmed-meson fragmentation functions with finite-mass corrections, *Nucl. Phys. B* **799**, 34, (2008). doi: 10.1016/j.nuclphysb.2008.02.015.
220. B. A. Kniehl, G. Kramer, I. Schienbein and H. Spiesberger, Open charm hadroproduction and the charm content of the proton, *Phys. Rev. D* **79**, 094009, (2009). doi: 10.1103/PhysRevD.79.094009.
221. R. Gauld, J. Rojo, L. Rottoli and J. Talbert, Charm production in the forward region: constraints on the small-x gluon and backgrounds for neutrino astronomy, *J. High Energy Phys.* **11**, 009, (2015). doi: 10.1007/JHEP11(2015)009.
222. M. Cacciari, M. L. Mangano and P. Nason, Gluon PDF constraints from the ratio of forward heavy-quark production at the LHC at $\sqrt{s} = 7$ and 13 TeV, *Eur. Phys. J. C* **75** (12), 610, (2015). doi: 10.1140/epjc/s10052-015-3814-x.
223. M. Cacciari, M. Greco and P. Nason, The p_T spectrum in heavy flavor hadroproduction, *J. High Energy Phys.* **05**, 007, (1998). doi: 10.1088/1126-6708/1998/05/007.
224. M. Cacciari and P. Nason, Charm cross-sections for the Tevatron Run II, *J. High Energy Phys.* **09**, 006, (2003). doi: 10.1088/1126-6708/2003/09/006.
225. M. Cacciari, P. Nason and C. Oleari, A study of heavy flavored meson fragmentation functions in e^+e^- annihilation, *J. High Energy Phys.* **04**, 006, (2006). doi: 10.1088/1126-6708/2006/04/006.
226. R. Aaij *et al.*, Observation of J/ψ pair production in pp collisions at $\sqrt{s} = 7$ TeV, *Phys. Lett. B* **707**, 52–59, (2012). doi: 10.1016/j.physletb.2011.12.015.

227. R. Aaij *et al.*, Measurement of the J/ψ pair production cross-section in pp collisions at $\sqrt{s} = 13$ TeV, *J. High Energy Phys.* **06**, 047, (2017). doi: 10.1007/JHEP06(2017)047.
228. A. V. Berezhnoy, V. V. Kiselev, A. K. Likhoded and A. I. Onishchenko, Doubly charmed baryon production in hadronic experiments, *Phys. Rev. D* **57**, 4385–4392, (1998). doi: 10.1103/PhysRevD.57.4385.
229. S. P. Baranov, Topics in associated J/ψ $+ c + \bar{c}$ production at modern colliders, *Phys. Rev. D* **73**, 074021, (2006). doi: 10.1103/PhysRevD.73.074021.
230. J.-P. Lansberg, On the mechanisms of heavy-quarkonium hadroproduction, *Eur. Phys. J. C* **61**, 693–703, (2009). doi: 10.1140/epjc/s10052-008-0826-9.
231. M. Łuszczak, R. Maciuła and A. Szczurek, Production of two $c\bar{c}$ pairs in double-parton scattering, *Phys. Rev. D* **85**, 094034, (2012). doi: 10.1103/PhysRevD.85.094034.
232. R. Maciuła and A. Szczurek, Single and double charmed meson production at the LHC, *EPJ Web Conf.* **81**, 01007, (2014). doi: 10.1051/epjconf/20148101007.
233. R. Aaij *et al.*, Measurement of J/ψ production in pp collisions at $\sqrt{s} = 7$ TeV, *Eur. Phys. J. C* **71**, 1645, (2011). doi: 10.1140/epjc/s10052-011-1645-y.
234. A. V. Berezhnoy and A. K. Likhoded, Associated production of Υ and open charm at LHC, *Int. J. Mod. Phys. A* **30**, 1550125, (2015). doi: 10.1142/S0217751X15501250.
235. L. V. Gribov, E. M. Levin and M. G. Ryskin, Semihard processes in QCD, *Phys. Rept.* **100**, 1–150, (1983). doi: 10.1016/0370-1573(83)90022-4.
236. E. M. Levin and M. G. Ryskin, High-energy hadron collisions in QCD, *Phys. Rept.* **189**, 267, (1990). doi: 10.1016/0370-1573(90)90016-U.
237. S. P. Baranov, Highlights from the k_T factorization approach on the quarkonium production puzzles, *Phys. Rev. D* **66**, 114003, (2002). doi: 10.1103/PhysRevD.66.114003.
238. B. Andersson *et al.*, Small-x phenomenology: summary and status, *Eur. Phys. J. C* **25**, 77, (2002). doi: 10.1007/s10052-002-0998-7.
239. J. R. Andersen *et al.*, Small-x phenomenology: summary and status, *Eur. Phys. J. C* **35**, 67, (2004). doi: 10.1140/epjc/s2004-01775-7.
240. J. R. Andersen *et al.*, Small-x phenomenology: summary of the 3rd Lund small-x workshop in 2004, *Eur. Phys. J. C* **48**, 53, (2006). doi: 10.1140/epjc/s2006-02615-6.
241. S. P. Baranov, Associated $\Upsilon + b + \bar{b}$ production at the Fermilab Tevatron and CERN LHC, *Phys. Rev. D* **74**, 074002, (2006). doi: 10.1103/PhysRevD.74.074002.
242. S. P. Baranov, Prompt Υ(nS) production at the LHC in view of the k_T-approach, *Phys. Rev. D* **86**, 054015, (2012). doi: 10.1103/PhysRevD.86.054015.
243. R. Aaij *et al.*, Forward production of Υ mesons in pp collisions at $\sqrt{s} = 7$ and 8 TeV, *J. High Energy Phys.* **11**, 103, (2015). doi: 10.1007/JHEP11(2015)103.
244. J. Badier *et al.*, Evidence for $\psi\psi$ production in π^- interactions at 150 GeV/c and 280 GeV/c, *Phys. Lett. B* **114**, 457, (1982). doi: 10.1016/0370-2693(82)90091-0.
245. J. Badier *et al.*, ψψ production and limits on beauty meson production from 400 GeV/c protons, *Phys. Lett. B* **158**, 85, (1985). doi: 10.1016/0370-2693(85)90745-2.
246. M. Aaboud *et al.*, Measurement of the prompt J/ψ pair production cross-section in pp collisions at $\sqrt{s} = 7$ TeV with the ATLAS detector, *Eur. Phys. J. C* **77** (2), 76, (2017). doi: 10.1140/epjc/s10052-017-4644-9.
247. A. V. Berezhnoy, A. K. Likhoded, A. V. Luchinsky and A. A. Novoselov, Double J/ψ-meson production at LHC and 4c-tetraquark state, *Phys. Rev. D* **84**, 094023, (2011). doi: 10.1103/PhysRevD.84.094023.

248. S. P. Baranov, A. M. Snigirev and N. P. Zotov, Double heavy meson production through double parton scattering in hadronic collisions, *Phys. Lett. B* **705**, 116–119, (2011). doi: 10.1016/j.physletb.2011.09.106.
249. S. P. Baranov, A. M. Snigirev, N. P. Zotov, A. Szczurek and W. Schäfer, Interparticle correlations in the production of J/ψ pairs in proton–proton collisions, *Phys. Rev. D* **87** (3), 034035, (2013). doi: 10.1103/PhysRevD.87.034035.
250. C.-F. Qiao, L.-P. Sun and P. Sun, Testing charmonium production mechamism via polarized J/ψ pair production at the LHC, *J. Phys. G* **37**, 075019, (2010). doi: 10.1088/0954-3899/37/7/075019.
251. L.-P. Sun, H. Han and K.-T. Chao, Impact of J/ψ pair production at the LHC and predictions in nonrelativistic QCD, *Phys. Rev. D* **94** (7), 074033, (2016). doi: 10.1103/PhysRevD.94.074033.
252. C. H. Kom, A. Kulesza and W. J. Stirling, Pair production of J/Ψ as a probe of double parton scattering at LHCb, *Phys. Rev. Lett.* **107**, 082002, (2011). doi: 10. 1103/PhysRevLett.107.082002.
253. J.-P. Lansberg and H.-S. Shao, J/Ψ-pair production at large momenta: Indications for double parton scatterings and large α_s^5 contributions, *Phys. Lett. B* **751**, 479–486, (2015). doi: 10.1016/j.physletb.2015.10.083.
254. C. Borschensky and A. Kulesza, Double parton scattering in pair production of J/ψ mesons at the LHC revisited, *Phys. Rev. D* **95** (3), 034029, (2017). doi: 10.1103/PhysRevD.95.034029.
255. J.-P. Lansberg and H.-S. Shao, Production of $J/\psi + \eta_c$ versus $J/\psi + J/\psi$ at the LHC: importance of real α_s^5 corrections, *Phys. Rev. Lett.* **111**, 122001, (2013). doi: 10.1103/ PhysRevLett.111.122001.
256. A. K. Likhoded, A. V. Luchinsky and S. V. Poslavsky, Production of $J/\psi + \chi_c$ and $J/\psi + J/\psi$ with real gluon emission at LHC, *Phys. Rev. D* **94**, 054017, (2016). doi: 10.1103/PhysRevD.94.054017.
257. H.-S. Shao, HELAC-ONIA: an automatic matrix element generator for heavy quarkonium physics, *Comput. Phys. Commun.* **184**, 2562–2570, (2013). doi: 10.1016/ j.cpc.2013.05.023.
258. H.-S. Shao, HELAC-ONIA 2.0: an upgraded matrix-element and event generator for heavy quarkonium physics, *Comput. Phys. Commun.* **198**, 238–259, (2016). doi: 10.1016/j.cpc.2015.09.011.
259. R. Aaij *et al.*, Measurement of forward J/ψ production cross-sections in pp collisions at $\sqrt{s} = 13$ TeV, *J. High Energy Phys.* **10**, 172, (2015). doi: 10.1007/JHEP10(2015) 172.
260. S. P. Baranov, Pair production of J/ψ mesons in the k_t-factorization approach, *Phys. Rev. D.* **84**, 054012 (Sep, 2011). doi: 10.1103/PhysRevD.84.054012. URL https:// link.aps.org/doi/10.1103/PhysRevD.84.054012.
261. V. Khachatryan *et al.*, Observation of Υ(1S) pair production in proton–proton collisions at $\sqrt{s} = 8$ TeV, *J. High Energy Phys.* **05**, 013, (2017). doi: 10.1007/ JHEP05(2017)013.
262. A. Novoselov, Double parton scattering as a source of quarkonia pairs in LHCb, (2011). arXiv:1106.2184.
263. A. V. Berezhnoy, A. K. Likhoded and A. A. Novoselov, Υ-meson pair production at LHC, *Phys. Rev. D* **87** (5), 054023, (2013). doi: 10.1103/PhysRevD. 87.054023.
264. S. Chatrchyan *et al.*, Measurement of associated W + charm production in pp collisions at $\sqrt{s} = 7$ TeV, *J. High Energy Phys.* **02**, 013, (2014). doi: 10.1007/ JHEP02(2014)013.

265. G. Aad *et al.*, Measurement of the production of a W boson in association with a charm quark in pp collisions at $\sqrt{s} = 7$ TeV with the ATLAS detector, *J. High Energy Phys.* **05**, 068, (2014). doi: 10.1007/JHEP05(2014)068.
266. R. Aaij *et al.*, Observation of associated production of a Z boson with a D meson in the forward region, *J. High Energy Phys.* **04**, 091, (2014). doi: 10.1007/JHEP04(2014)091.
267. R. Maciula and A. Szczurek, Can the triple-parton scattering be observed in open charm meson production at the LHC? *Phys. Lett. B* **772**, 849–853, (2017). doi: 10.1016/j.physletb.2017.07.061.
268. P. Bartalini *et al.*, Multi-parton interactions at the LHC. (2011). URL https://inspirehep.net/record/944170/files/arXiv:1111.0469.pdf.
269. A. Del Fabbro and D. Treleani, b anti-b b anti-b production in proton–nucleus collisions at the LHC. (2003). arXiv:hep-ph/0301175.
270. L. Frankfurt, M. Strikman and C. Weiss, 3D parton imaging of the nucleon in high-energy pp and pA collisions, *Ann. Phys.* **13**, 665–672, (2004). doi: 10.1002/andp.200410108.
271. E. Cattaruzza, A. Del Fabbro and D. Treleani, Heavy-quark production in proton–nucleus collision at the CERN LHC, *Phys. Rev. D* **70**, 034022, (2004). doi: 10.1103/PhysRevD.70.034022.
272. A. Del Fabbro and D. Treleani, $b\bar{b}b\bar{b}$ production in proton–proton and proton–nucleus collisions at the CERN LHC, *Eur. Phys. J. A* **19S1**, 229–237, (2004). doi: 10.1140/epjad/s2004-03-038-4.
273. E. Cattaruzza, A. Del Fabbro and D. Treleani, Heavy-quark production in proton–nucleus collisions at the LHC, *Int. J. Mod. Phys. A* **20**, 4462–4468, (2005). doi: 10.1142/S0217751X05028077.
274. D. Treleani and G. Calucci, Collisions of protons with light nuclei shed new light on nucleon structure, *Phys. Rev. D* **86**, 036003, (2012). doi: 10.1103/PhysRevD.86.036003.
275. D. d'Enterria and A. M. Snigirev, Same-sign WW production in proton–nucleus collisions at the LHC as a signal for double parton scattering, *Phys. Lett. B* **718**, 1395–1400, (2013). doi: 10.1016/j.physletb.2012.12.032.
276. D. d'Enterria and A. M. Snigirev, Pair production of quarkonia and electroweak bosons from double-parton scatterings in nuclear collisions at the LHC, *Nucl. Phys. A* **931**, 303–308, (2014). doi: 10.1016/j.nuclphysa.2014.09.089.
277. D. d'Enterria and A. M. Snigirev, Double-parton scattering cross sections in proton–nucleus and nucleus–nucleus collisions at the LHC, *Nucl. Phys. A* **932**, 296–301, (2014). doi: 10.1016/j.nuclphysa.2014.07.005.
278. D. d'Enterria and A. M. Snigirev, Triple-parton scatterings in proton–nucleus collisions at high energies, *Eur. Phys. J.* **C78** (5) 359 (2018). doi: 10.1140/epjc/s10052-018-5687-2.
279. D. d'Enterria and A. M. Snigirev, Enhanced J/Ψ-pair production from double parton scatterings in nucleus–nucleus collisions at the large hadron collider, *Phys. Lett. B* **727**, 157–162, (2013). doi: 10.1016/j.physletb.2013.10.004.
280. T. Sjöstrand, The development of MPI modelling in Pythia, arXiv:1706.02166. (2017).
281. D. d'Enterria, G. K. Eyyubova, V. L. Korotkikh, I. P. Lokhtin, S. V. Petrushanko, L. I. Sarycheva and A. M. Snigirev, Estimates of hadron azimuthal anisotropy from multiparton interactions in proton–proton collisions at $\sqrt{s} = 14$ TeV, *Eur. Phys. J. C* **66**, 173–185, (2010). doi: 10.1140/epjc/s10052-009-1232-7.

282. V. Khachatryan *et al.*, Observation of long-range near-side angular correlations in proton–proton collisions at the LHC, *J. High Energy Phys.* **09**, 091, (2010). doi: 10.1007/JHEP09(2010)091.
283. G. Aad *et al.*, Observation of long-range elliptic azimuthal anisotropies in $\sqrt{s} = 13$ and 2.76 TeV pp collisions with the ATLAS detector, *Phys. Rev. Lett.* **116** (17), 172301, (2016). doi: 10.1103/PhysRevLett.116.172301.
284. F. Abe *et al.*, Measurement of double parton scattering in $\bar{p}p$ collisions at $\sqrt{s} = 1.8$ TeV, *Phys. Rev. Lett.* **79**, 584–589, (1997). doi: 10.1103/PhysRevLett.79.584.
285. M. L. Mangano *et al.*, Physics at a 100 TeV pp collider: standard model processes, *CERN Yellow Report.* (3), 1–254, (2017). doi: 10.23731/CYRM-2017-003.1.
286. A. Dainese *et al.*, Heavy ions at the future circular collider, *CERN Yellow Report.* (3), 635–692, (2017). doi: 10.23731/CYRM-2017-003.635.
287. J. C. Collins, D. E. Soper and G. F. Sterman, Factorization of hard processes in QCD, *Adv. Ser. Direct. High Energy Phys.* **5**, 1–91, (1989). doi: 10.1142/9789814503266_0001.
288. A. M. Snigirev, Triple parton scattering in collinear approximation of perturbative QCD, *Phys. Rev. D* **94** (3), 034026, (2016). doi: 10.1103/PhysRevD.94.034026.
289. A. V. Berezhnoy, A. K. Likhoded, A. V. Luchinsky and A. A. Novoselov, Double $c\bar{c}$ production at LHCb, *Phys. Rev. D* **86**, 034017, (2012). doi: 10.1103/PhysRevD.86.034017.
290. E. R. Cazaroto, V. P. Goncalves and F. S. Navarra, Heavy quark production and gluon saturation in double parton scattering at the LHC, *Phys. Rev. D* **88** (3), 034005, (2013). doi: 10.1103/PhysRevD.88.034005.
291. D. d'Enterria. Proceeds. Moriond-QCD (2017), to be submitted.
292. M. Czakon, P. Fiedler and A. Mitov, Total top-quark pair-production cross-section at hadron colliders through $O(\alpha_s^4)$, *Phys. Rev. Lett.* **110**, 252004, (2013). doi: 10.1103/PhysRevLett.110.252004.
293. S. Alekhin, J. Bluemlein, S.-O. Moch and R. Placakyte, The new ABMP16 PDF, *PoS* **DIS2016**, 016, (2016).
294. M. Cacciari, S. Frixione, N. Houdeau, M. L. Mangano, P. Nason and G. Ridolfi, Theoretical predictions for charm and bottom production at the LHC, *J. High Energy Phys.* **10**, 137, (2012). doi: 10.1007/JHEP10(2012)137.
295. M. L. Mangano, P. Nason and G. Ridolfi, Heavy quark correlations in hadron collisions at next-to-leading order, *Nucl. Phys. B* **373**, 295–345, (1992). doi: 10.1016/0550-3213(92)90435-E.
296. D. d'Enterria and T. Pierog, Global properties of proton–proton collisions at $\sqrt{s} = 100$ TeV, *J. High Energy Phys.* **08**, 170, (2016). doi: 10.1007/JHEP08(2016)170.
297. N. Armesto, Nuclear shadowing, *J. Phys. G* **32**, R367–R394, (2006). doi: 10.1088/0954-3899/32/11/R01.
298. D. G. d'Enterria, Hard scattering cross-sections at LHC in the Glauber approach: From pp to pA and AA collisions, (2003). arXiv:nucl-ex/0302016.
299. C. W. deJager, H. deVries and C. deVries, Atomic data and nuclear data tables. **14**, 485, (1974).
300. http://mcfm.fnal.gov.
301. J. M. Campbell, R. K. Ellis and C. Williams, Vector boson pair production at the LHC, *J. High Energy Phys.* **07**, 018, (2011). doi: 10.1007/JHEP07(2011)018.
302. H.-L. Lai *et al.*, New parton distributions for collider physics, *Phys. Rev. D* **82**, 074024, (2010). doi: 10.1103/PhysRevD.82.074024.

303. K. J. Eskola, H. Paukkunen and C. A. Salgado, EPS09: a new generation of NLO and LO nuclear parton distribution functions, *J. High Energy Phys.* **04**, 065, (2009). doi: 10.1088/1126-6708/2009/04/065.
304. http://www-itp.particle.uni-karlsruhe.de/vbfnlo.
305. K. Arnold *et al.*, Release Note — Vbfnlo-2.6.0. (2012). arXiv:1207.4975.
306. H. Paukkunen and C. A. Salgado, Constraints for the nuclear parton distributions from Z and W production at the LHC, *J. High Energy Phys.* **03**, 071, (2011). doi: 10.1007/JHEP03(2011)071.
307. T. Melia, K. Melnikov, R. Rontsch and G. Zanderighi, Next-to-leading order QCD predictions for W^+W^+jj production at the LHC, *J. High Energy Phys.* **12**, 053, (2010). doi: 10.1007/JHEP12(2010)053.
308. R. Vogt, R. E. Nelson and A. D. Frawley, Improving the J/ψ production baseline at RHIC and the LHC, *Nucl. Phys. A* **910–911**, 231–234, (2013). doi: 10.1016/j.nuclphysa.2012.12.106.
309. J. Adam *et al.*, D-meson production in p–Pb collisions at $\sqrt{s_{\rm NN}}$ =5.02 TeV and in pp collisions at $\sqrt{s}$ =7 TeV, *Phys. Rev. C* **94** (5), 054908, (2016). doi: 10.1103/PhysRevC.94.054908.
310. K. Greisen, End to the cosmic ray spectrum? *Phys. Rev. Lett.* **16**, 748–750, (1966). doi: 10.1103/PhysRevLett.16.748.
311. G. T. Zatsepin and V. A. Kuzmin, Upper limit of the spectrum of cosmic rays, *JETP Lett.* **4**, 78–80, (1966). [*Pisma Zh. Eksp. Teor. Fiz.* **4**, 114, (1966)].
312. D. d'Enterria, R. Engel, T. Pierog, S. Ostapchenko and K. Werner, Constraints from the first LHC data on hadronic event generators for ultra-high energy cosmic-ray physics, *Astropart. Phys.* **35**, 98–113, (2011). doi: 10.1016/j.astropartphys.2011.05.002.
313. I. P. Lokhtin and A. M. Snigirev, Nuclear geometry of jet quenching, *Eur. Phys. J. C* **16**, 527–536, (2000). doi: 10.1007/s100520000437.
314. T. Matsui and H. Satz, J/ψ Suppression by quark–gluon plasma formation, *Phys. Lett. B* **178**, 416–422, (1986). doi: 10.1016/0370-2693(86)91404-8.
315. J. P. Lansberg, A. Rakotozafindrabe, P. Artoisenet, D. Blaschke, J. Cugnon, D. d'Enterria, A. C. Kraan, F. Maltoni, D. Prorok and H. Satz, Perspectives on heavy-quarkonium production at the LHC, *AIP Conf. Proc.* **1038**, 15–44, (2008). doi: 10.1063/1.2987169.
316. D. Acosta *et al.*, Measurement of the J/ψ meson and b-hadron production cross sections in $p\bar{p}$ collisions at $\sqrt{s}$ = 1960 GeV, *Phys. Rev. D* **71**, 032001, (2005). doi: 10.1103/PhysRevD.71.032001.
317. B. Abelev *et al.*, Inclusive J/ψ production in pp collisions at $\sqrt{s}$ = 2.76 TeV, *Phys. Lett. B* **718**, 295–306, (2012). doi: 10.1016/j.physletb.2012.10.078,10.1016/j.physletb.2015.06.058. [Erratum: *Phys. Lett. B* **748**, 472, (2015)].
318. R. Aaij *et al.*, Measurement of J/ψ production in pp collisions at $\sqrt{s}$ = 2.76 TeV, *J. High Energy Phys.* **02**, 041, (2013). doi: 10.1007/JHEP02(2013)041.
319. B. Abelev *et al.*, Measurement of prompt J/ψ and beauty hadron production cross sections at mid-rapidity in pp collisions at $\sqrt{s}$ = 7 TeV, *J. High Energy Phys.* **11**, 065, (2012). doi: 10.1007/JHEP11(2012)065.
320. V. Khachatryan *et al.*, Prompt and non-prompt J/ψ production in pp collisions at $\sqrt{s}$ = 7 TeV, *Eur. Phys. J. C* **71**, 1575, (2011). doi: 10.1140/epjc/s10052-011-1575-8.
321. A. Andronic, P. Braun-Munzinger, K. Redlich and J. Stachel, Heavy quark(onium) at LHC: the statistical hadronization case, *J. Phys. G* **37**, 094014, (2010). doi: 10.1088/0954-3899/37/9/094014.

322. H. U. Bengtsson, The Lund Monte Carlo for high p_T physics, *Comput. Phys. Commun.* **31**, 323, (1984). doi: 10.1016/0010-4655(84)90018-3.
323. B. Andersson, G. Gustafson, G. Ingelman and T. Sjöstrand, Parton fragmentation and string dynamics, *Phys. Rept.* **97**, 31–145, (1983). doi: 10.1016/0370-1573(83) 90080-7.
324. B. Andersson, G. Gustafson and T. Sjöstrand, How to find the gluon jets in e^+ e^- Annihilation, *Phys. Lett. B* **94**, 211–215, (1980). doi: 10.1016/0370-2693(80)90861-8.
325. W. Bartel *et al.*, Experimental study of jets in Electron — positron annihilation, *Phys. Lett. B* **101**, 129–134, (1981). doi: 10.1016/0370-2693(81)90505-0.
326. T. Sjöstrand, The Lund Monte Carlo for jet fragmentation, *Comput. Phys. Commun.* **27**, 243, (1982). doi: 10.1016/0010-4655(82)90175-8.
327. T. Sjöstrand, A model for initial state parton showers, *Phys. Lett. B* **157**, 321–325, (1985). doi: 10.1016/0370-2693(85)90674-4.
328. T. Sjöstrand. *Multiple Parton-Parton Interactions in Hadronic Events.* Oregon Workshop on Super High Energy Physics, FERMILAB-PUB-85-119-T, (1985).
329. G. J. Alner *et al.*, Scaling violation favoring high multiplicity events at 540-GeV CMS energy, *Phys. Lett. B* **138**, 304–310, (1984). doi: 10.1016/0370-2693(84)91666-6.
330. R. E. Ansorge *et al.*, Charged particle multiplicity distributions at 200-GeV and 900-GeV center-of-mass energy, *Z. Phys. C* **43**, 357, (1989). doi: 10.1007/BF01506531.
331. Z. Koba, H. B. Nielsen and P. Olesen, Scaling of multiplicity distributions in high-energy hadron collisions, *Nucl. Phys. B* **40**, 317–334, (1972). doi: 10.1016/ 0550-3213(72)90551-2.
332. R. E. Ansorge *et al.*, Charged particle correlations in $\bar{p}p$ collisions at c.m. energies of 200-GeV, 546-GeV and 900-GeV, *Z. Phys. C* **37**, 191–213, (1988). doi: 10.1007/ BF01579906.
333. G. Arnison *et al.*, Transverse momentum spectra for charged particles at the CERN proton–anti-proton collider, *Phys. Lett. B* **118**, 167–172, (1982). doi: 10.1016/ 0370-2693(82)90623-2.
334. C. Albajar *et al.*, A study of the general characteristics of $p\bar{p}$ collisions at $\sqrt{s} = 0.2$-TeV to 0.9-TeV, *Nucl. Phys. B* **335**, 261–287, (1990). doi: 10.1016/ 0550-3213(90)90493-W.
335. A. Breakstone *et al.*, Multiplicity dependence of transverse momentum spectra at ISR energies, *Phys. Lett. B* **132**, 463–466, (1983). doi: 10.1016/0370-2693(83)90348-9.
336. F. Ceradini, Study of minimum bias trigger events at s**(1/2) = 0.2-TeV to 0.9-TeV with magnetic and calorimetric analysis at the CERN proton–anti-proton collider, *Conf. Proc. C* **850718**, 363, (1985).
337. C. Albajar *et al.*, Production of low transverse energy clusters in anti-p p collisions at s**(1/2) = 0.2-TeV to 0.9-TeV and their interpretation in terms of QCD jets, *Nucl. Phys. B* **309**, 405–425, (1988). doi: 10.1016/0550-3213(88)90450-6.
338. G. Arnison *et al.*, Hadronic jet production at the CERN proton–anti-proton collider, *Phys. Lett. B* **132**, 214, (1983). doi: 10.1016/0370-2693(83)90254-X.
339. P. V. Landshoff and J. C. Polkinghorne, Calorimeter triggers for hard collisions, *Phys. Rev. D* **18**, 3344, (1978). doi: 10.1103/PhysRevD.18.3344.
340. N. Paver and D. Treleani, Multiple parton interactions and multi-jet events at collider and Tevatron energies, *Phys. Lett. B* **146**, 252–256, (1984). doi: 10.1016/ 0370-2693(84)91029-3.
341. B. Humpert and R. Oderico, Multiparton scattering and QCD radiation as sources of four jet events, *Phys. Lett. B* **154**, 211, (1985). doi: 10.1016/0370-2693(85)90587-8.
342. V. N. Gribov, A reggeon diagram technique, *Sov. Phys. JETP* **26**, 414–422, (1968). [*Zh. Eksp. Teor. Fiz.* **53**, 654, (1967)].

343. G. Veneziano, Regge intercepts and unitarity in planar dual models, *Nucl. Phys. B* **74**, 365–377, (1974). doi: 10.1016/0550-3213(74)90203-X.
344. G. F. Chew and C. Rosenzweig, Dual topological unitarization: an ordered approach to hadron theory, *Phys. Rept.* **41**, 263–327, (1978). doi: 10.1016/0370-1573(78) 90194-1.
345. A. Capella, U. Sukhatme, C.-I. Tan and J. Tran Thanh Van, Jets in small p(T) hadronic collisions, universality of quark fragmentation and rising rapidity plateaus, *Phys. Lett. B* **81**, 68–74, (1979). doi: 10.1016/0370-2693(79)90718-4.
346. H. Minakata, Universal quark — jet fragmentation in soft hadronic reactions, *Phys. Rev. D* **20**, 1656, (1979). doi: 10.1103/PhysRevD.20.1656.
347. G. Cohen-Tannoudji, A. El Hassouni, J. Kalinowski, O. Napoly and R. B. Peschanski, Partons at low $P(T)$, *Phys. Rev. D* **21**, 2699, (1980). doi: 10.1103/PhysRevD.21.2699.
348. K. Fialkowski and A. Kotanski, Hadron multiplicity distributions in a dual model, *Phys. Lett. B* **107**, 132, (1981). doi: 10.1016/0370-2693(81)91165-5.
349. P. Aurenche and F. W. Bopp, Rapidity spectra in proton–proton and proton–antiproton scattering up to 540-GeV in a dual parton model, *Phys. Lett. B* **114**, 363–368, (1982). doi: 10.1016/0370-2693(82)90363-X.
350. A. Capella and J. Tran Thanh Van, Multiplicity distributions up to $\sqrt{s} = 540$-GeV in the dual parton model, *Phys. Lett. B* **114**, 450–456, (1982). doi: 10.1016/ 0370-2693(82)90090-9.
351. A. B. Kaidalov, The quark–gluon structure of the pomeron and the rise of inclusive spectra at high-energies, *Phys. Lett. B* **116**, 459–463, (1982). doi: 10.1016/ 0370-2693(82)90168-X.
352. A. B. Kaidalov and K. A. Ter-Martirosian, Pomeron as quark–gluon strings and multiple hadron production at SPS collider energies, *Phys. Lett. B* **117**, 247–251, (1982). doi: 10.1016/0370-2693(82)90556-1.
353. T. K. Gaisser and F. Halzen, Soft hard scattering in the TeV range, *Phys. Rev. Lett.* **54**, 1754, (1985). doi: 10.1103/PhysRevLett.54.1754.
354. G. Pancheri and Y. Srivastava, Jets in minimum bias physics, *Conf. Proc. C* **850313**, 28, (1985). doi: 10.1016/0370-2693(85)90121-2. [*Phys. Lett. B* **159**, 69 (1985)].
355. A. D. Martin and C. J. Maxwell, Isolation of hard scattering effects in minimum bias data, *Phys. Lett. B* **172**, 248, (1986). doi: 10.1016/0370-2693(86)90844-0.
356. F. E. Paige and S. D. Protopopescu, Isajet 5.02: a monte carlo event generator for pp and anti-pp interactions, *Conf. Proc. C* **850809**, 41, (1985).
357. R. Odorico, Cojets: a Monte Carlo program simulating QCD in hadronic production of jets and heavy flavors with inclusion of initial QCD bremsstrahlung, *Comput. Phys. Commun.* **32**, 139, (1984). doi: 10.1016/0010-4655(84)90067-5.
358. R. D. Field, High-energy multi-jets at the CERN $\bar{p}p$ collider and the SSC, *Nucl. Phys. B* **264**, 687–720, (1986). doi: 10.1016/0550-3213(86)90504-3.
359. P. Aurenche, F. W. Bopp and J. Ranft, Particle production in hadron hadron collisions at collider energies in an exclusive multistring fragmentation model, *Z. Phys. C* **23**, 67–76, (1984). doi: 10.1007/BF01558042.
360. G. J. Alner *et al.*, The UA5 high-energy $\bar{p}p$ simulation program, *Nucl. Phys. B* **291**, 445–502, (1987). doi: 10.1016/0550-3213(87)90481-0.
361. R. D. Ball, V. Bertone, S. Carrazza, L. Del Debbio, S. Forte, A. Guffanti, N. P. Hartland and J. Rojo, Parton distributions with QED corrections, *Nucl. Phys. B* **877**, 290–320, (2013). doi: 10.1016/j.nuclphysb.2013.10.010.
362. V. V. Sudakov, Vertex parts at very high-energies in quantum electrodynamics, *Sov. Phys. JETP* **3**, 65–71, (1956). [*Zh. Eksp. Teor. Fiz.* **30**, 87, (1956)].

363. A. Buckley *et al.*, General-purpose event generators for LHC physics, *Phys. Rept.* **504**, 145–233, (2011). doi: 10.1016/j.physrep.2011.03.005.
364. R. K. Ellis and J. C. Sexton, QCD radiative corrections to parton–parton scattering, *Nucl. Phys. B* **269**, 445–484, (1986). doi: 10.1016/0550-3213(86)90232-4.
365. T. Sjöstrand and P. Z. Skands, Multiple interactions and the structure of beam remnants, *J. High Energy Phys.* **03**, 053, (2004).
366. A. H. Mueller and J.-w. Qiu, Gluon recombination and shadowing at small values of x, *Nucl. Phys. B* **268**, 427–452, (1986). doi: 10.1016/0550-3213(86)90164-1.
367. R. J. Glauber. High-energy collision theory. In eds. W. E. Brittin and L. G. Dunham, *Lectures in Theoretical Physics*, Vol. I, pp. 315–414. Interscience, New York, (1959).
368. T. T. Chou and C.-N. Yang, Model of elastic high-energy scattering, *Phys. Rev.* **170**, 1591–1596, (1968). doi: 10.1103/PhysRev.170.1591.
369. C. Bourrely, J. Soffer and T. T. Wu, Impact picture expectations for very high-energy elastic pp and $p\bar{p}$ scattering, *Nucl. Phys. B* **247**, 15–28, (1984). doi: 10.1016/0550-3213(84)90369-9.
370. P. L'Heureux, B. Margolis and P. Valin, Quark–gluon model for diffraction at high-energies, *Phys. Rev. D* **32**, 1681–1691, (1985). doi: 10.1103/PhysRevD.32.1681.
371. G. 't Hooft, A planar diagram theory for strong interactions, *Nucl. Phys. B* **72**, 461, (1974). doi: 10.1016/0550-3213(74)90154-0.
372. B. Andersson, G. Gustafson and B. Söderberg, A probability measure on parton and string states, *Nucl. Phys. B* **264**, 29, (1986). doi: 10.1016/0550-3213(86)90471-2.
373. H. Fritzsch, Producing heavy quark flavors in hadronic collisions: a test of quantum chromodynamics, *Phys. Lett. B* **67**, 217–221, (1977). doi: 10.1016/0370-2693(77)90108-3.
374. A. Ali, J. G. Körner, G. Kramer and J. Willrodt, Nonleptonic weak decays of bottom mesons, *Z. Phys. C* **1**, 269, (1979). doi: 10.1007/BF01440227.
375. H. Fritzsch, How to discover the B mesons, *Phys. Lett. B* **86**, 343–346, (1979). doi: 10.1016/0370-2693(79)90853-0.
376. T. Alexopoulos *et al.*, The role of double parton collisions in soft hadron interactions, *Phys. Lett. B* **435**, 453–457, (1998). doi: 10.1016/S0370-2693(98)00921-6.
377. T. Alexopoulos *et al.*, Charged particle multiplicity correlations in $p\bar{p}$ collisions at $\sqrt{s}$ = 0.3-TeV to 1.8-TeV, *Phys. Lett. B* **353**, 155–160, (1995). doi: 10.1016/0370-2693(95)00554-X.
378. T. Alexopoulos *et al.*, Multiplicity dependence of transverse momentum spectra of centrally produced hadrons in anti-pp collisions at 0.3-TeV, 0.54-TeV, 0.9-TeV and 1.8-TeV center-of-mass energy, *Phys. Lett. B* **336**, 599–604, (1994). doi: 10.1016/0370-2693(94)90578-9.
379. D. Acosta *et al.*, Soft and hard interactions in $p\bar{p}$ collisions at $\sqrt{s}$ = 1800-GeV and 630-GeV, *Phys. Rev. D* **65**, 072005, (2002). doi: 10.1103/PhysRevD.65.072005.
380. R. D. Field, The underlying event in hard scattering processes, *eConf. C* **010630**, P501, (2001).
381. D. Acosta *et al.*, The underlying event in hard interactions at the Tevatron $\bar{p}p$ collider, *Phys. Rev. D* **70**, 072002, (2004). doi: 10.1103/PhysRevD.70.072002.
382. D. Kar. *Using Drell–Yan to probe the underlying event in Run II at Collider Detector at Fermilab (CDF)*. PhD thesis, Florida University, (2008). URL http://lss.fnal.gov/cgi-bin/find_paper.pl?thesis-2008-54.
383. R. Field and R. C. Group, Pythia tune A, Herwig and Jimmy in Run 2 at CDF, (2005). arXiv:hep-ph/0510198.

384. R. D. Field, Studying the underlying event at CDF, *Conf. Proc.* **C060726**, 581–584, (2006).
385. G. A. Schuler and T. Sjöstrand, Hadronic diffractive cross-sections and the rise of the total cross-section, *Phys. Rev. D* **49**, 2257–2267, (1994). doi: 10.1103/PhysRevD.49.2257.
386. J. Dischler and T. Sjöstrand, A Toy model of color screening in the proton, *Eur. Phys. J. direct.* **3** (1), 2, (2001). doi: 10.1007/s1010501c0002.
387. M. Glück, E. Reya and A. Vogt, Dynamical parton distributions of the proton and small x physics, *Z. Phys. C* **67**, 433–448, (1995). doi: 10.1007/BF01624586.
388. G. A. Schuler and T. Sjöstrand, Towards a complete description of high-energy photoproduction, *Nucl. Phys. B* **407**, 539–605, (1993). doi: 10.1016/0550-3213(93)90091-3.
389. G. A. Schuler and T. Sjöstrand, A scenario for high-energy gamma gamma interactions, *Z. Phys. C* **73**, 677–688, (1997). doi: 10.1007/s002880050359.
390. M. Ciafaloni, Coherence effects in initial jets at small Q^2/s, *Nucl. Phys. B* **296**, 49–74, (1988). doi: 10.1016/0550-3213(88)90380-X.
391. S. Catani, F. Fiorani and G. Marchesini, Small x behavior of initial state radiation in perturbative QCD, *Nucl. Phys. B* **336**, 18–85, (1990). doi: 10.1016/0550-3213(90)90342-B.
392. V. N. Gribov and L. N. Lipatov, Deep inelastic e p scattering in perturbation theory, *Sov. J. Nucl. Phys.* **15**, 438–450, (1972). [*Yad. Fiz.* **15**, 781, (1972)].
393. G. Altarelli and G. Parisi, Asymptotic freedom in parton language, *Nucl. Phys. B* **126**, 298–318, (1977). doi: 10.1016/0550-3213(77)90384-4.
394. Y. L. Dokshitzer, Calculation of the structure functions for deep inelastic scattering and $e^+ e^-$ annihilation by perturbation theory in quantum chromodynamics, *Sov. Phys. JETP* **46**, 641–653, (1977). [*Zh. Eksp. Teor. Fiz.* **73**, 1216, (1977)].
395. B. Andersson, G. Gustafson and J. Samuelsson, The linked dipole chain model for DIS, *Nucl. Phys. B* **467**, 443–478, (1996). doi: 10.1016/0550-3213(96)00114-9.
396. G. Marchesini and B. R. Webber, Monte Carlo simulation of general hard processes with coherent QCD radiation, *Nucl. Phys. B* **310**, 461–526, (1988). doi: 10.1016/0550-3213(88)90089-2.
397. I. Borozan and M. H. Seymour, An eikonal model for multiparticle production in hadron hadron interactions, *J. High Energy Phys.* **09**, 015, (2002). doi: 10.1088/1126-6708/2002/09/015.
398. J. Bellm *et al.*, Herwig 7.0/Herwig++ 3.0 release note, *Eur. Phys. J. C* **76** (4), 196, (2016). doi: 10.1140/epjc/s10052-016-4018-8.
399. S. Gieseke, F. Loshaj and P. Kirchgaesser, Soft and diffractive scattering with the cluster model in Herwig, *Eur. Phys. J. C* **77** (3), 156, (2017). doi: 10.1140/epjc/s10052-017-4727-7.
400. A. D. Martin, H. Hoeth, V. A. Khoze, F. Krauss, M. G. Ryskin and K. Zapp, Diffractive physics, *PoS* **QNP2012**, 017, (2012).
401. P. Aurenche, F. W. Bopp, A. Capella, J. Kwiecinski, M. Maire, J. Ranft and J. Tran Thanh Van, Multiparticle production in a two component dual parton model, *Phys. Rev. D* **45**, 92–105, (1992). doi: 10.1103/PhysRevD.45.92.
402. P. Aurenche *et al.*, DTUJET-93: Sampling inelastic proton–proton and anti-proton–proton collisions according to the two component dual parton model, *Comput. Phys. Commun.* **83**, 107–123, (1994). doi: 10.1016/0010-4655(94)90037-X.
403. R. Engel, Photoproduction within the two component dual parton model: amplitudes and cross-sections, *Z. Phys. C* **66**, 203–214, (1995). doi: 10.1007/BF01496594.

404. R. Engel and J. Ranft, Hadronic photon–photon interactions at high-energies, *Phys. Rev. D* **54**, 4244–4262, (1996). doi: 10.1103/PhysRevD.54.4244.
405. F. W. Bopp, J. Ranft, R. Engel and S. Roesler, Antiparticle to particle production ratios in hadron–hadron and d-Au collisions in the DPMJET-III Monte Carlo, *Phys. Rev. C* **77**, 014904, (2008). doi: 10.1103/PhysRevC.77.014904.
406. R. S. Fletcher, T. K. Gaisser, P. Lipari and T. Stanev, SIBYLL: an event generator for simulation of high-energy cosmic ray cascades, *Phys. Rev. D* **50**, 5710–5731, (1994). doi: 10.1103/PhysRevD.50.5710.
407. E.-J. Ahn, R. Engel, T. K. Gaisser, P. Lipari and T. Stanev, Cosmic ray interaction event generator SIBYLL 2.1, *Phys. Rev. D* **80**, 094003, (2009). doi: 10.1103/PhysRevD.80.094003.
408. K. Werner, F.-M. Liu and T. Pierog, Parton ladder splitting and the rapidity dependence of transverse momentum spectra in deuteron gold collisions at RHIC, *Phys. Rev. C* **74**, 044902, (2006). doi: 10.1103/PhysRevC.74.044902.
409. T. Pierog, I. Karpenko, J. M. Katzy, E. Yatsenko and K. Werner, EPOS LHC: test of collective hadronization with data measured at the CERN Large Hadron Collider, *Phys. Rev. C* **92** (3), 034906, (2015). doi: 10.1103/PhysRevC.92.034906.
410. S. Ostapchenko, Nonlinear screening effects in high energy hadronic interactions, *Phys. Rev. D* **74** (1), 014026, (2006). doi: 10.1103/PhysRevD.74.014026.
411. S. Ostapchenko, Monte Carlo treatment of hadronic interactions in enhanced Pomeron scheme: I. QGSJET-II model, *Phys. Rev. D* **83**, 014018, (2011). doi: 10.1103/PhysRevD.83.014018.
412. T. Sjöstrand and P. Z. Skands, Baryon number violation and string topologies, *Nucl. Phys. B* **659**, 243, (2003). doi: 10.1016/S0550-3213(03)00193-7.
413. R. Corke and T. Sjöstrand, Multiparton interactions and rescattering, *J. High Energy Phys.* **01**, 035, (2010). doi: 10.1007/JHEP01(2010)035.
414. K. Konishi, A. Ukawa and G. Veneziano, Jet calculus: a simple algorithm for resolving QCD jets, *Nucl. Phys. B* **157**, 45–107, (1979). doi: 10.1016/0550-3213(79)90053-1.
415. J. W. Cronin *et al.*, Production of hadrons with large transverse momentum at 200-GeV, 300-GeV and 400-GeV, *Phys. Rev. D* **11**, 3105, (1975). doi: 10.1103/PhysRevD.11.3105.
416. E. Avsar, G. Gustafson and L. Lönnblad, Small-x dipole evolution beyond the large-N(c) limit, *J. High Energy Phys.* **01**, 012, (2007). doi: 10.1088/1126-6708/2007/01/012.
417. C. Bierlich, G. Gustafson, L. Lönnblad and A. Tarasov, Effects of overlapping strings in pp collisions, *J. High Energy Phys.* **03**, 148, (2015). doi: 10.1007/JHEP03(2015)148.
418. G. Ingelman and P. E. Schlein, Jet structure in high mass diffractive scattering, *Phys. Lett. B* **152**, 256–260, (1985). doi: 10.1016/0370-2693(85)91181-5.
419. R. Bonino *et al.*, Evidence for transverse jets in high mass diffraction, *Phys. Lett. B* **211**, 239, (1988). doi: 10.1016/0370-2693(88)90840-4.
420. P. Bruni and G. Ingelman, Diffractive W and Z production at p anti-p colliders and the pomeron parton content, *Phys. Lett. B* **311**, 317–323, (1993). doi: 10.1016/0370-2693(93)90576-4.
421. S. Navin, Diffraction in Pythia, (2010). arXiv:1005.3894.
422. R. Ciesielski and K. Goulianos, MBR Monte Carlo simulation in PYTHIA8, *PoS* **ICHEP2012**, 301, (2013).
423. Y. L. Dokshitzer, V. A. Khoze and T. Sjöstrand, Rapidity gaps in Higgs production, *Phys. Lett. B* **274**, 116–121, (1992). doi: 10.1016/0370-2693(92)90312-R.

424. J. D. Bjorken, Rapidity gaps and jets as a new physics signature in very high-energy hadron hadron collisions, *Phys. Rev. D* **47**, 101–113, (1993). doi: 10.1103/PhysRevD.47.101.
425. T. Affolder *et al.*, Diffractive dijets with a leading antiproton in $\bar{p}p$ collisions at $\sqrt{s} = 1800$ GeV, *Phys. Rev. Lett.* **84**, 5043–5048, (2000). doi: 10.1103/PhysRevLett.84.5043.
426. C. O. Rasmussen and T. Sjöstrand, Hard diffraction with dynamic gap survival, *J. High Energy Phys.* **02**, 142, (2016). doi: 10.1007/JHEP02(2016)142.
427. J. R. Christiansen and T. Sjöstrand, Color reconnection at future e^+ e^- colliders, *Eur. Phys. J. C* **75** (9), 441, (2015). doi: 10.1140/epjc/s10052-015-3674-4.
428. T. Sjöstrand. Colour reconnections from LEP to future colliders. In *Proceedings, Parton Radiation and Fragmentation from LHC to FCC-ee: CERN, Geneva, Switzerland, November 22-23, 2016*, pp. 144–148, (2017). URL http://inspirehep.net/record/1513009/files/1512294_144-148.pdf.
429. T. Sjöstrand and V. A. Khoze, On color rearrangement in hadronic W^+W^- events, *Z. Phys. C* **62**, 281–310, (1994). doi: 10.1007/BF01560244.
430. S. Schael *et al.*, Electroweak measurements in electron–positron collisions at W-Boson–pair energies at LEP, *Phys.Rept.* **532**, 119–244, (2013). doi: 10.1016/j.physrep.2013.07.004.
431. J. R. Christiansen and P. Z. Skands, String formation beyond leading colour, *J. High Energy Phys.* **08**, 003, (2015). doi: 10.1007/JHEP08(2015)003.
432. P. Skands and D. Wicke, Non-perturbative QCD effects and the top mass at the Tevatron, *Eur. Phys. J. C* **52**, 133–140, (2007). doi: 10.1140/epjc/s10052-007-0352-1.
433. S. Argyropoulos and T. Sjöstrand, Effects of color reconnection on $t\bar{t}$ final states at the LHC, *J. High Energy Phys.* **1411**, 043, (2014). doi: 10.1007/JHEP11(2014)043.
434. P. Z. Skands, Tuning Monte Carlo generators: the Perugia tunes, *Phys. Rev. D* **82**, 074018, (2010). doi: 10.1103/PhysRevD.82.074018.
435. H. Schulz and P. Z. Skands, Energy scaling of minimum-bias tunes, *Eur. Phys. J. C* **71**, 1644, (2011). doi: 10.1140/epjc/s10052-011-1644-z.
436. P. Skands, S. Carrazza and J. Rojo, Tuning Pythia 8.1: the Monash 2013 tune, *Eur. Phys. J. C* **74** (8), 3024, (2014). doi: 10.1140/epjc/s10052-014-3024-y.
437. C. M. Buttar, D. Clements, I. Dawson and A. Moraes, Simulations of minimum bias events and the underlying event, MC tuning and predictions for the LHC, *Acta Phys. Polon. B* **35**, 433–441, (2004).
438. G. Aad *et al.*, Measurement of the Z/γ^* boson transverse momentum distribution in pp collisions at $\sqrt{s} = 7$ TeV with the ATLAS detector, *J. High Energy Phys.* **09**, 145, (2014). doi: 10.1007/JHEP09(2014)145.
439. A. Buckley, H. Hoeth, H. Lacker, H. Schulz and J. E. von Seggern, Systematic event generator tuning for the LHC, *Eur. Phys. J. C* **65**, 331–357, (2010). doi: 10.1140/epjc/s10052-009-1196-7.
440. A. Buckley, J. Butterworth, L. Lönnblad, D. Grellscheid, H. Hoeth, J. Monk, H. Schulz and F. Siegert, Rivet user manual, *Comput. Phys. Commun.* **184**, 2803–2819, (2013). doi: 10.1016/j.cpc.2013.05.021.
441. A. Karneyeu, L. Mijovic, S. Prestel and P. Z. Skands, MCPLOTS: a particle physics resource based on volunteer computing, *Eur. Phys. J. C* **74**, 2714, (2014). doi: 10.1140/epjc/s10052-014-2714-9.
442. N. Fischer and T. Sjöstrand, Thermodynamical string fragmentation, *J. High Energy Phys.* **01**, 140, (2017). doi: 10.1007/JHEP01(2017)140.
443. J. Adam *et al.*, Enhanced production of multi-strange hadrons in high-multiplicity proton–proton collisions, *Nature Phys.* **13**, 535–539, (2017). doi: 10.1038/nphys4111.

444. B. B. Abelev *et al.*, Production of $\Sigma(1385)^{\pm}$ and $\Xi(1530)^0$ in proton–proton collisions at $\sqrt{s} = 7\,\text{TeV}$, *Eur. Phys. J. C* **75** (1), 1, (2015). doi: 10.1140/epjc/s10052-014-3191-x.
445. B. I. Abelev *et al.*, Strange particle production in p+p collisions at s**(1/2) = 200-GeV, *Phys. Rev. C* **75**, 064901, (2007). doi: 10.1103/PhysRevC.75.064901.
446. G. Aad *et al.*, Charged-particle multiplicities in pp interactions measured with the ATLAS detector at the LHC, *New J. Phys.* **13**, 053033, (2011). doi: 10.1088/1367-2630/13/5/053033.
447. S. Chatrchyan *et al.*, Charged particle transverse momentum spectra in *pp* collisions at $\sqrt{s} = 0.9$ and 7 TeV, *J. High Energy Phys.* **08**, 086, (2011). doi: 10.1007/JHEP08(2011)086.
448. J. Adam *et al.*, Pseudorapidity and transverse-momentum distributions of charged particles in proton–proton collisions at $\sqrt{s} = 13$ TeV, *Phys. Lett. B* **753**, 319–329, (2016). doi: 10.1016/j.physletb.2015.12.030.
449. J. Adam *et al.*, Measurement of pion, kaon and proton production in proton–proton collisions at $\sqrt{s} = 7$ TeV, *Eur. Phys. J. C* **75** (5), 226, (2015). doi: 10.1140/epjc/s10052-015-3422-9.
450. V. Khachatryan *et al.*, Strange particle production in *pp* collisions at $\sqrt{s} = 0.9$ and 7 TeV, *J. High Energy Phys.* **05**, 064, (2011). doi: 10.1007/JHEP05(2011)064.
451. V. Khachatryan *et al.*, Evidence for collectivity in pp collisions at the LHC, *Phys. Lett. B* **765**, 193–220, (2017). doi: 10.1016/j.physletb.2016.12.009.
452. C. Bierlich, G. Gustafson and L. Lönnblad, A shoving model for collectivity in hadronic collisions. (2016). arXiv:1612.05132.
453. K. J. Eskola, K. Kajantie and J. Lindfors, Quark and gluon production in high-energy nucleus–nucleus collisions, *Nucl. Phys. B* **323**, 37–52, (1989). doi: 10.1016/0550-3213(89)90586-5.
454. A. Bialas and R. B. Peschanski, Moments of rapidity distributions as a measure of short range fluctuations in high-energy collisions, *Nucl. Phys. B* **273**, 703–718, (1986). doi: 10.1016/0550-3213(86)90386-X.
455. A. Bialas and R. B. Peschanski, Intermittency in multiparticle production at high-energy, *Nucl. Phys. B* **308**, 857–867, (1988). doi: 10.1016/0550-3213(88)90131-9.
456. C. Albajar *et al.*, Intermittency studies in $\bar{p}p$ collisions at $\sqrt{s} = 630\,\text{GeV}$, *Nucl. Phys. B* **345**, 1–21, (1990). doi: 10.1016/0550-3213(90)90606-E.
457. E. A. De Wolf, I. M. Dremin and W. Kittel, Scaling laws for density correlations and fluctuations in multiparticle dynamics, *Phys. Rept.* **270**, 1–141, (1996). doi: 10.1016/0370-1573(95)00069-0.
458. G. Arnison *et al.*, (UA1 Collaboration), Hadronic jet production at the CERN proton–antiproton collider, *Phys. Lett. B* **132** (1–3), 214 – 222, (1983). ISSN 0370-2693. doi: http://dx.doi.org/10.1016/0370-2693(83)90254-X.
459. T. Affolder and others (CDF Collaboration), Charged jet evolution and the underlying event in proton–antiproton collisions at 1.8 TeV, *Phys. Rev. D* **65**, 092002 (Apr, 2002). doi: 10.1103/PhysRevD.65.092002.
460. M. Cacciari and G. P. Salam, Pileup subtraction using jet areas, *Phys. Lett. B* **659**, 119–126, (2008). doi: 10.1016/j.physletb.2007.09.077.
461. ATLAS Collaboration. Measurement of the W-boson mass in pp collisions at $\sqrt{s} = 7$ TeV with the ATLAS detector. arXiv:1701.07240 [hep-ex], (2017).
462. A. Altheimer *et al.*, Jet substructure at the tevatron and LHC: new results, new tools, new benchmarks, *J. Phys. G* **39**, 063001, (2012). doi: 10.1088/0954-3899/39/6/063001.
463. S. Porteboeuf, T. Pierog and K. Werner, Producing hard processes regarding the complete event: the EPOS event generator. arXiv:1006.2967 [hep-ph], (2010).

464. S. Ostapchenko, QGSJET-II: towards reliable description of very high energy hadronic interactions, *Nucl. Phys. Proc. Suppl.* **151**, 143–146, (2006). doi: 10.1016/j.nuclphysbps.2005.07.026.
465. T. Sjöstrand *et al.*, High-energy physics event generation with Pythia 6.1, *Comput. Phys. Commun.* **135**, 238–259, (2001). doi: 10.1016/S0010-4655(00)00236-8.
466. T. Sjöstrand and M. van Zijl, A multiple-interaction model for the event structure in hadron collisions, *Phys. Rev. D* **36**, 2019–2041 (1987). doi: 10.1103/PhysRevD.36.2019.
467. R. Field, Min-bias and the underlying event at the LHC, *Acta Phys. Polon. B* **42**, 2631–2656, (2011). doi: 10.5506/APhysPolB.42.2631.
468. H. Lai *et al.*, Global QCD analysis of parton structure of the nucleon: CTEQ5 parton distributions, *Eur. Phys. J. C* **12**, 375–392, (2000). doi: 10.1007/s100529900196.
469. Pumplin, J. and others, New generation of parton distributions with uncertainties from global QCD analysis, *J. High Energy Phys.* **07**, 012, (2002).
470. A. Moraes. Modeling the underlying event: generating predictions for the LHC. ATL-PHYS-PROC-2009-045, (2009).
471. ATLAS Collaboration. ATLAS Monte Carlo tunes for MC09. URL http://cdsweb.cern.ch/record/1247375. ATL-PHYS-PUB-2010-002, (2010).
472. A. Sherstnev and R. S. Thorne, Parton distributions for LO generators, *Eur. Phys. J. C* **55**, 553–575, (2008). doi: 10.1140/epjc/s10052-008-0610-x.
473. P. Skands. The Perugia tunes. arXiv:0905.3418v1 [hep-ph], (2009).
474. G. Corcella *et al.*, Herwig 6.5 release note. arXiv:hep-ph/0210213 [hep-ph], (2002).
475. ATLAS Collaboration. Charged particle multiplicities in inelastic *pp* interactions with ATLAS and the ATLAS Minimum Bias Tune 1. URL http://cdsweb.cern.ch/record/1277665. ATLAS-CONF-2010-031, (2010).
476. ATLAS Collaboration. New ATLAS event generator tunes to 2010 data. URL http://cdsweb.cern.ch/record/1345343. ATL-PHYS-PUB-2011-008, (2011).
477. ATLAS Collaboration. ATLAS tunes of Pythia 6 and Pythia 8 for MC11. URL http://cdsweb.cern.ch/record/1363300. ATL-PHYS-PUB-2011-009, (2010).
478. ATLAS Collaboration. First tuning of Herwig/Jimmy to ATLAS data. URL https://cds.cern.ch/record/1303025. ATL-PHYS-PUB-2010-014, (2010).
479. ATLAS Collaboration. Summary of ATLAS Pythia 8 tunes. URL https://cds.cern.ch/record/1474107. ATL-PHYS-PUB-2012-003, (2012).
480. G. Watt and R. Thorne, Study of Monte Carlo approach to experimental uncertainty propagation with MSTW 2008 PDFs, *J. High Energy Phys.* **1208**, 052, (2012). doi: 10.1007/JHEP08(2012)052.
481. ATLAS Collaboration. ATLAS Pythia 8 tunes to 7 TeV data. URL https://cdsweb.cern.ch/record/1966419. ATL-PHYS-PUB-2014-021, (2014).
482. R. D. Ball, L. Del Debbio, S. Forte, A. Guffanti, J. I. Latorre, J. Rojo and M. Ubiali, A first unbiased global NLO determination of parton distributions and their uncertainties, *Nucl. Phys. B* **838**, 136–206, (2010). doi: 10.1016/j.nuclphysb.2010.05.008.
483. ATLAS Collaboration. The Pythia 8 A3 tune description of ATLAS minimum bias and inelastic measurements incorporating the Donnachie-Landshoff diffractive model. URL https://cdsweb.cern.ch/record/2206965. ATL-PHYS-PUB-2016-027, (2016).
484. M. Bahr *et al.*, Herwig++ physics and manual, *Eur. Phys. J. C* **58**, 639–707, (2008). doi: 10.1140/epjc/s10052-008-0798-9.
485. Herwig++ mpi model parameters. URL https://herwig.hepforge.org/tutorials/mpi/tunes.html, (2016 (accessed February 1, 2017)).

486. L. A. Harland-Lang, A. D. Martin, P. Motylinski and R. S. Thorne, Parton distributions in the LHC era: MMHT 2014 PDFs, *Eur. Phys. J. C* **75** (5), 204, (2015). doi: 10.1140/epjc/s10052-015-3397-6.
487. T. Sjöstrand and M. van Zijl, Multiple parton–parton interactions in an impact parameter picture, *Phys. Lett. B* **188**, 149–154, (1987). doi: 10.1016/0370-2693(87) 90722-2.
488. F. E. Paige, S. D. Protopopescu, H. Baer and X. Tata, ISAJET 7.37: a Monte Carlo event generator for pp, anti-pp, and e^+e^- reactions. (1998). arXiv:hep-ph/ 9804321.
489. T. Aaltonen *et al.*, Measurement of particle production and inclusive differential cross sections in $p\bar{p}$ collisions at $\sqrt{s} = 1.96$ TeV, *Phys. Rev. D* **79**, 112005 (2009). doi: 10.1103/PhysRevD.79.112005. URL http://link.aps.org/doi/10.1103/PhysRevD.79. 112005.
490. D. Acosta *et al.*, (CDF Collaboration), Underlying event in hard interactions at the fermilab tevatron $\bar{p}p$ collider, *Phys. Rev. D* **70**, 072002 (2004). doi: 10.1103/ PhysRevD.70.072002. URL http://link.aps.org/doi/10.1103/PhysRevD.70.072002.
491. D. Acosta *et al.*, (CDF Collaboration), Soft and hard interactions in $p\bar{p}$ collisions at $\sqrt{s} = 1800$ and 630 GeV, *Phys. Rev. D* **65**, 072005 (2002). doi: 10.1103/PhysRevD. 65.072005. URL http://link.aps.org/doi/10.1103/PhysRevD.65.072005.
492. T. Aaltonen *et al.*, Studying the underlying event in Drell–Yan and high transverse momentum jet production at the Tevatron, *Phys. Rev. D* **82**, 034001, (2010). doi: 10.1103/PhysRevD.82.034001.
493. G. Marchesini and B. R. Webber, Associated transverse energy in hadronic jet production, *Phys. Rev. D* **38**, 3419, (1988). doi: 10.1103/PhysRevD.38.3419.
494. J. Pumplin, Hard underlying event correction to inclusive jet cross sections, *Phys. Rev. D* **57** (9), 5787–5792, (1998). doi: 10.1103/PhysRevD.57.5787.
495. T. Aaltonen *et al.*, (CDF Collaboration), Measurement of event shapes in proton–antiproton collisions at center-of-mass energy 1.96 TeV, *Phys. Rev. D* **83**, 112007, (2011). doi: 10.1103/PhysRevD.83.112007.
496. T. Aaltonen *et al.*, (CDF Collaboration), A study of the energy dependence of the underlying event in proton–antiproton collisions, *Phys. Rev. D* **92** (9), 092009, (2015). doi: 10.1103/PhysRevD.92.092009.
497. B. Abelev *et al.*, Underlying Event measurements in pp collisions at $\sqrt{s} = 0.9$ and 7 TeV with the ALICE experiment at the LHC, *J. High Energy Phys.* **07**, 116, (2012). doi: 10.1007/JHEP07(2012)116.
498. ATLAS Collaboration, Measurement of charged-particle distributions sensitive to the underlying event in $\sqrt{s} = 13$ TeV proton–proton collisions with the ATLAS detector at the LHC, *J. High Energy Phys.* **03**, 157, (2017). doi: 10.1007/JHEP03(2017)157.
499. CMS Collaboration, Underlying event measurements with leading particles and jets in pp collisions at $\sqrt{s} = 13$ TeV, *CMS-PAS-FSQ-15-007*. (2015).
500. V. Khachatryan *et al.*, Measurement of the underlying event activity using charged-particle jets in proton–proton collisions at $\sqrt{(s)} = 2.76$ TeV, *J. High Energy Phys.* **09**, 137, (2015). doi: 10.1007/JHEP09(2015)137.
501. ATLAS Collaboration, Underlying event characteristics and their dependence on jet size of charged-particle jet events in pp collisions at $\sqrt{(s)} = 7\,\text{TeV}$ with the ATLAS detector, *Phys. Rev. D* **86**, 072004, (2012). doi: 10.1103/PhysRevD.86. 072004.
502. G. Aad *et al.*, Measurement of the underlying event in jet events from 7 TeV proton–proton collisions with the ATLAS detector, *Eur. Phys. J. C* **74** (8), 2965, (2014). doi: 10.1140/epjc/s10052-014-2965-5.

503. G. Aad *et al.*, Measurement of distributions sensitive to the underlying event in inclusive Z-boson production in pp collisions at $\sqrt{s} = 7$ TeV with the ATLAS detector, *Eur. Phys. J. C* **74** (12), 3195, (2014). doi: 10.1140/epjc/s10052-014-3195-6.
504. CMS Collaboration, Measurement of the underlying event in the Drell–Yan process in proton–proton collisions at $\sqrt{s} = 7$ TeV, *Eur. Phys. J. C* **72**, 2080, (2012). doi: 10.1140/epjc/s10052-012-2080-4.
505. Sirunyan, A. M. *et al.*, Measurement of the underlying event activity in inclusive Z boson production in proton-proton collisions at $\sqrt{s} = 13$ TeV, *JHEP*, **7**, 032 (2018). doi: 10.1007/JHEP07(2018)032.
506. CMS Collaboration, Study of the underlying event at forward rapidity in pp collisions at $\sqrt{s} = 0.9$, 2.76 and 7 TeV, *J. High Energy Phys.* **04**, 072, (2013). doi: 10.1007/JHEP04(2013)072.
507. CMS Collaboration, Measurement of the underlying event activity in pp collisions at $\sqrt{s} = 0.9$ and 7 TeV with the novel jet-area/median approach, *J. High Energy Phys.* **08**, 130, (2012). doi: 10.1007/JHEP08(2012)130.
508. CMS Collaboration, Underlying event measurement with $tt + X$ events with pp collision data at $\sqrt{(s)} = 13\,\text{TeV}$, CMS-PAS-TOP-15-017. (2015).
509. CMS Collaboration, Study of the underlying event, b-quark fragmentation and hadronization properties in ttbar events, *CMS-PAS-TOP-13-007.* (2013).
510. H. Caines. Jet & underlying event measurements in p-p collisions at RHIC, *Nucl. Phys. A* **855**, 376–379, (2011). doi: 10.1016/j.nuclphysa.2011.02.084.
511. Minimum-bias and underlying event WG. URL https://lpcc.web.cern.ch/lpcc/index.php?page=mb_ue_wg, (accessed April 1, 2017).
512. T. Martin, P. Skands and S. Farrington, Probing collective effects in hadronisation with the extremes of the underlying event, *Eur. Phys. J. C* **76** (5), 299, (2016). doi: 10.1140/epjc/s10052-016-4135-4.
513. R. Aaij *et al.*, Measurement of charged particle multiplicities in pp collisions at $\sqrt{s} = 7$TeV in the forward region, *Eur. Phys. J. C* **72**, 1947, (2012). doi: 10.1140/epjc/s10052-012-1947-8.
514. C. Patrignani *et al.*, Review of particle physics, *Chin. Phys. C* **40** (10), 100001, (2016). doi: 10.1088/1674-1137/40/10/100001.
515. K. Aamodt *et al.*, Charged-particle multiplicity measurement in proton–proton collisions at $\sqrt{s} = 0.9$ and 2.36 TeV with ALICE at LHC, *Eur. Phys. J. C* **68**, 89–108, (2010). doi: 10.1140/epjc/s10052-010-1339-x.
516. V. Khachatryan *et al.*, Transverse momentum and pseudorapidity distributions of charged hadrons in pp collisions at $\sqrt{s} = 0.9$ and 2.36 TeV, *J. High Energy Phys.* **02**, 041, (2010). doi: 10.1007/JHEP02(2010)041.
517. G. Aad *et al.*, Charged-particle distributions in $\sqrt{s} = 13\,\text{TeV}$ pp interactions measured with the ATLAS detector at the LHC, *Phys. Lett. B* **758**, 67–88, (2016). doi: 10.1016/j.physletb.2016.04.050.
518. K. Aamodt *et al.*, First proton–proton collisions at the LHC as observed with the ALICE detector: Measurement of the charged particle pseudorapidity density at $\sqrt{s} = 900\,\text{GeV}$, *Eur. Phys. J. C* **65**, 111–125, (2010). doi: 10.1140/epjc/s10052-009-1227-4.
519. K. Aamodt *et al.*, Charged-particle multiplicity measurement in proton–proton collisions at $\sqrt{s} = 7$ TeV with ALICE at LHC, *Eur. Phys. J. C* **68**, 345–354, (2010). doi: 10.1140/epjc/s10052-010-1350-2.
520. K. Aamodt *et al.*, Transverse momentum spectra of charged particles in proton–proton collisions at $\sqrt{s} = 900$ GeV with ALICE at the LHC, *Phys. Lett. B* **693**, 53–68, (2010). doi: 10.1016/j.physletb.2010.08.026.

521. B. B. Abelev *et al.*, Energy dependence of the transverse momentum distributions of charged particles in pp collisions measured by ALICE, *Eur. Phys. J. C* **73** (12), 2662, (2013). doi: 10.1140/epjc/s10052-013-2662-9.
522. B. B. Abelev *et al.*, Multiplicity dependence of the average transverse momentum in pp, p–Pb and Pb–Pb collisions at the LHC, *Phys. Lett. B* **727**, 371–380, (2013). doi: 10.1016/j.physletb.2013.10.054.
523. B. Abelev *et al.*, Multiplicity dependence of two-particle azimuthal correlations in pp collisions at the LHC, *J. High Energy Phys.* **09**, 049, (2013). doi: 10.1007/JHEP09(2013)049.
524. ALICE Collaboration, Charged-particle multiplicity measurement with reconstructed tracks in pp collisions at $\sqrt{s} = 0.9$ and 7 TeV with ALICE at the LHC. (2013). http://cds.cern.ch/record/1562873.
525. J. Adam *et al.*, Charged-particle multiplicities in proton–proton collisions at $\sqrt{s} = 0.9$ to 8 TeV, *Eur. Phys. J. C* **77** (1), 33, (2017). doi: 10.1140/epjc/s10052-016-4571-1.
526. G. Aad *et al.*, Charged-particle multiplicities in pp interactions at $\sqrt{s} = 900$ GeV measured with the ATLAS detector at the LHC, *Phys. Lett. B* **688**, 21–42, (2010). doi: 10.1016/j.physletb.2010.03.064.
527. ATLAS Collaboration, Central charged-particle multiplicities in pp interactions with $|\eta| < 0.8$ and pt > 0.5 and 1 GeV measured with the ATLAS detector at the LHC. (2010). http://cds.cern.ch/record/1317333.
528. G. Aad *et al.*, Measurement of charged-particle spectra in Pb+Pb collisions at $\sqrt{s_{\mathrm{NN}}} = 2.76$ TeV with the ATLAS detector at the LHC, *J. High Energy Phys.* **09**, 050, (2015). doi: 10.1007/JHEP09(2015)050.
529. G. Aad *et al.*, Charged-particle distributions in pp interactions at $\sqrt{s} = 8$ TeV measured with the ATLAS detector, *Eur. Phys. J. C* **76** (7), 403, (2016). doi: 10.1140/epjc/s10052-016-4203-9.
530. M. Aaboud *et al.*, Charged-particle distributions at low transverse momentum in $\sqrt{s} = 13$ TeV *pp* interactions measured with the ATLAS detector at the LHC, *Eur. Phys. J. C* **76** (9), 502, (2016). doi: 10.1140/epjc/s10052-016-4335-y.
531. V. Khachatryan *et al.*, Transverse-momentum and pseudorapidity distributions of charged hadrons in pp collisions at $\sqrt{s} = 7$ TeV, *Phys. Rev. Lett.* **105**, 022002, (2010). doi: 10.1103/PhysRevLett.105.022002.
532. V. Khachatryan *et al.*, Charged particle multiplicities in pp interactions at $\sqrt{s} = 0.9$, 2.36 and 7 TeV, *J. High Energy Phys.* **01**, 079, (2011). doi: 10.1007/JHEP01(2011)079.
533. CMS Collaboration, Pseudorapidity distributions of charged particles in pp collisions at $\sqrt{s} = 7$ TeV with at least one central charged particles, (2011). http://cds.cern.ch/record/1341853
534. S. Chatrchyan *et al.*, Measurement of pseudorapidity distributions of charged particles in proton–proton collisions at $\sqrt{s} = 8$ TeV by the CMS and TOTEM experiments, *Eur. Phys. J. C* **74** (10), 3053, (2014). doi: 10.1140/epjc/s10052-014-3053-6.
535. V. Khachatryan *et al.*, Pseudorapidity distribution of charged hadrons in proton–proton collisions at $\sqrt{s} = 13$ TeV, *Phys. Lett. B* **751**, 143–163, (2015).doi: 10.1016/j.physletb.2015.10.004.
536. CMS Collaboration, Measurement of pseudorapidity distributions of charged particles in proton–proton collisions at $\sqrt{s} = 13$ TeV by the CMS experiment, (2016). http://cds.cern.ch/record/2145373.
537. R. Aaij *et al.*, Measurement of charged particle multiplicities and densities in *pp* collisions at $\sqrt{s} = 7$ TeV in the forward region, *Eur. Phys. J. C* **74** (5), 2888, (2014). doi: 10.1140/epjc/s10052-014-2888-1.

538. V. Blobel. Unfolding methods in high-energy physics experiments. In *Proceedings, CERN School of Computing: Aiguablava, Spain, September 9-22 1984*, (1984). doi: 10.5170/CERN-1985-009.88.
539. V. Blobel. An unfolding method for high-energy physics experiments. In *Advanced Statistical Techniques in Particle Physics. Proceedings, Conference, Durham, UK, March 18–22, 2002*, pp. 258–267, (2002). URL http://www.ippp.dur.ac.uk/Workshops/02/statistics/proceedings//blobel2.pdf.
540. J. F. Grosse-Oetringhaus and K. Reygers, Charged-particle multiplicity in proton–proton collisions, *J. Phys. G* **37**, 083001, (2010). doi: 10.1088/0954-3899/37/8/083001.
541. G. Arnison *et al.*, Charged particle multiplicity distributions in proton–anti-proton collisions at 540-GeV center-of-mass energy, *Phys. Lett. B* **123**, 108–114, (1983). doi: 10.1016/0370-2693(83)90969-3.
542. G. J. Alner *et al.*, An investigation of multiplicity distributions in different pseudorapidity intervals in anti-p p reactions at a CMS energy of 540-GeV, *Phys. Lett. B* **160**, 193–198, (1985). doi: 10.1016/0370-2693(85)91491-1.
543. F. Abe *et al.*, Pseudorapidity distributions of charged particles produced in $\bar{p}p$ interactions at $\sqrt{s} = 630$ GeV and 1800 GeV, *Phys. Rev. D* **41**, 2330, (1990). doi: 10.1103/PhysRevD.41.2330.
544. R. Harr *et al.*, Pseudorapidity distribution of charged particles in $\bar{p}p$ collisions at $\sqrt{s} = 630$-GeV, *Phys. Lett. B* **401**, 176–180, (1997). doi: 10.1016/S0370-2693(97)00385-7.
545. C. Fuglesang. UA5 multiplicity distributions and fits of various functions. In *Multiparticle Dynamics: A Meeting Ground Between Particle and Statistical Physics. A Dialog Between Experiment and Theory. Festschrift for Leon van Hove. Proceedings, Conference, La Thuile, Italy, March 20–22, 1989*, pp. 193–210, (1989).
546. A. Giovannini and R. Ugoccioni, Possible scenarios for soft and semihard components structure in central hadron hadron collisions in the TeV region, *Phys. Rev. D* **59**, 094020, (1999). doi: 10.1103/PhysRevD.59.094020,10.1103/PhysRevD.69.059903. [Erratum: *Phys. Rev. D* **69**, 059903, (2004)].
547. A. Giovannini and R. Ugoccioni, On signals of new physics in global event properties in pp collisions in the TeV energy domain, *Phys. Rev. D* **68**, 034009, (2003). doi: 10.1103/PhysRevD.68.034009.
548. S. Kama. *Automatic Monte-Carlo Tuning for Minimum Bias Events at the LHC.* PhD thesis, DESY, (2010-03-24). URL https://inspirehep.net/record/1186251/files/CERN-THESIS-2010-259.pdf.
549. P. Abreu *et al.*, Tuning and test of fragmentation models based on identified particles and precision event shape data, *Z. Phys. C* **73**, 11–60, (1996). doi: 10.1007/s002880050295.
550. R. Barate *et al.*, Studies of quantum chromodynamics with the ALEPH detector, *Phys. Rept.* **294**, 1–165, (1998). doi: 10.1016/S0370-1573(97)00045-8.
551. P. Pfeifenschneider *et al.*, QCD analyses and determinations of alpha(s) in $e^+ e^-$ annihilation at energies between 35-GeV and 189-GeV, *Eur. Phys. J. C* **17**, 19–51, (2000). doi: 10.1007/s100520000432.
552. B. Abelev *et al.*, Measurement of inelastic, single- and double-diffraction cross sections in proton–proton collisions at the LHC with ALICE, *Eur. Phys. J. C* **73** (6), 2456, (2013). doi: 10.1140/epjc/s10052-013-2456-0.
553. T. Aaltonen *et al.*, Erratum: Measurement of particle production and inclusive differential cross sections in $p\bar{p}$ collisions at $\sqrt{s} = 1.96$ TeV, *Phys. Rev. D* **79**, 112005, (2009). *Phys. Rev. D* **82**, 119903, (2010).

554. A. Buckley, Sensitivities to PDFs in parton shower MC generator reweighting and tuning, (2016). arXiv:1601.04229
555. A. Donnachie and P. V. Landshoff, Small x: Two pomerons!, *Phys. Lett. B* **437**, 408–416, (1998). doi: 10.1016/S0370-2693(98)00899-5.
556. S. Frixione, P. Nason and C. Oleari, Matching NLO QCD computations with Parton Shower simulations: the POWHEG method, *J. High Energy Phys.* **11**, 070, (2007). doi: 10.1088/1126-6708/2007/11/070.
557. M. Althoff *et al.*, Determination of α_s in first and second order QCD from e^+e^- annihilation into hadrons, *Z. Phys. C* **26**, 157, (1984).
558. W. Braunschweig *et al.*, Jet fragmentation and QCD models in e^+e^- annihilation at c.m. energies between 12-GeV and 41.5-GeV, *Z. Phys. C* **41**, 359–373, (1988). doi: 10.1007/BF01585620.
559. D. Buskulic *et al.*, Properties of hadronic Z decays and test of QCD generators, *Z. Phys. C* **55**, 209–234, (1992). doi: 10.1007/BF01482583.
560. K. Hamacher and M. Weierstall, The next round of hadronic generator tuning heavily based on identified particle data, (1995). arXiv:hep-ex/9511011.
561. F. Feroz, M. P. Hobson, E. Cameron and A. N. Pettitt, Importance nested sampling and the MultiNest algorithm, (2013). arXiv:1306.2144.
562. R. Field. CDF Run II Monte-Carlo tunes. In *TeV4LHC 2006 Workshop 4th meeting Batavia, Illinois, October 20–22, 2006*, (2006). URL http://lss.fnal.gov/cgi-bin/find_paper.pl?pub-06-408.
563. T. Affolder *et al.*, The transverse momentum and total cross section of e^+e^- pairs in the Z boson region from $p\bar{p}$ collisions at $\sqrt{s} = 1.8$ TeV, *Phys. Rev. Lett.* **84**, 845–850, (2000). doi: 10.1103/PhysRevLett.84.845.
564. V. M. Abazov *et al.*, Measurement of dijet azimuthal decorrelations at central rapidities in $p\bar{p}$ collisions at $\sqrt{s} = 1.96$ TeV, *Phys. Rev. Lett.* **94**, 221801, (2005). doi: 10.1103/PhysRevLett.94.221801.
565. ATLAS Run 1 Pythia8 tunes. Technical Report ATL-PHYS-PUB-2014-021, CERN, Geneva (Nov, 2014). URL https://cds.cern.ch/record/1966419.
566. Example ATLAS tunes of Pythia8, Pythia6 and Powheg to an observable sensitive to Z boson transverse momentum. Technical Report ATL-PHYS-PUB-2013-017, CERN, Geneva (Nov, 2013). URL https://cds.cern.ch/record/1629317.
567. A study of the sensitivity to the Pythia8 parton shower parameters of $t\bar{t}$ production measurements in pp collisions at $\sqrt{s} = 7$ TeV with the ATLAS experiment at the LHC. Technical Report ATL-PHYS-PUB-2015-007, CERN, Geneva (Mar, 2015). URL https://cds.cern.ch/record/2004362.
568. P. Ilten, M. Williams and Y. Yang, Event generator tuning using Bayesian optimization, *J. Instrumentation* **12** (04), P04028, (2017). doi: 10.1088/1748-0221/12/04/P04028.
569. Ll. Ametller *et al.*, Possible signature of multiple parton interactions in collider four jet events, *Phys. Lett. B* **169**, 289, (1986). doi: 10.1016/0370-2693(86)90668-4.
570. A. Knutsson, Multiple interactions at HERA, *Nucl. Phys. Proc. Suppl.* **191**, 141–150, (2009). doi: 10.1016/j.nuclphysbps.2009.03.121.
571. D. d'Enterria and A. Snigirev, Double, triple and n-parton scatterings in high-energy proton and nuclear collisions, arXiv:1708.07519. (2017).
572. S. P. Baranov, A. V. Lipatov, M. A. Malyshev, A. M. Snigirev and N. P. Zotov, Associated $W^{\pm}D^{(*)}$ production at the LHC and prospects to observe double parton interactions, *Phys. Lett. B* **746**, 100–103, (2015). doi: 10.1016/j.physletb.2015.04.059.

573. E. Maina, Multiple parton interactions in Z+ jets production at the LHC. A comparison of factorized and non-factorized double parton distribution functions, *J. High Energy Phys.* **01**, 061, (2011). doi: 10.1007/JHEP01(2011)061.
574. K. Golec-Biernat, E. Lewandowska, M. Serino, Z. Snyder and A. M. Stasto, Constraining the double gluon distribution by the single gluon distribution, *Phys. Lett. B* **750**, 559–564, (2015). doi: 10.1016/j.physletb.2015.09.067.
575. F. Hautmann and O. Kepka. Multiparton interactions, small-x and diffraction session: introduction. In *Proceedings, 7th International Workshop on Multiple Partonic Interactions at the LHC (MPI@LHC 2015): Miramare, Trieste, Italy, November 23–27, 2015*, pp. 145–147, (2016). URL https://inspirehep.net/record/1456039/files/MPIatLHC2015_C15-11-23.145.pdf.
576. J. Bartels, M. Salvadore and G. P. Vacca, AGK cutting rules and multiple scattering in hadronic collisions, *Eur. Phys. J. C* **42**, 53–71, (2005). doi: 10.1140/epjc/s2005-02258-1.
577. M. Salvadore, J. Bartels and G. P. Vacca, Multiple interactions and AGK rules in pQCD, arXiv:0709.3062. (2007).
578. V. A. Abramovsky and A. V. Popov, The role of AGK theorem in QCD, arXiv:1112.1297. (2011).
579. L. N. Lipatov, The parton model and perturbation theory, *Sov. J. Nucl. Phys.* **20**, 94–102, (1975).
580. M. Bahr, S. Gieseke, M. A. Gigg, D. Grellscheid, K. Hamilton, O. Latunde-Dada, S. Platzer, P. Richardson, M. H. Seymour, A. Sherstnev, J. Tully and B. R. Webber, Herwig++ physics and manual (03. 2008). URL http://arxiv.org/abs/0803.0883.
581. F. Hautmann, H. Jung, A. Lelek, V. Radescu and R. Zlebcik, Soft-gluon resolution scale in QCD evolution equations, *Phys. Lett. B* **772**, 446–451, (2017). doi: 10.1016/j.physletb.2017.07.005.
582. A. Bassetto, M. Ciafaloni and G. Marchesini, Jet structure and infrared sensitive quantities in perturbative QCD, *Phys. Rept.* **100**, 201–272, (1983). doi: 10.1016/0370-1573(83)90083-2.
583. Y. L. Dokshitzer, V. A. Khoze, S. I. Troian and A. H. Mueller, QCD coherence in high-energy reactions, *Rev. Mod. Phys.* **60**, 373, (1988). doi: 10.1103/RevModPhys.60.373.
584. S. Catani, M. Ciafaloni and F. Hautmann, Gluon contributions to small x heavy flavour production, *Phys. Lett. B* **242**, 97–102, (1990). doi: 10.1016/0370-2693(90)91601-7.
585. S. Catani, M. Ciafaloni and F. Hautmann, High-energy factorization and small x heavy flavor production, *Nucl. Phys. B* **366**, 135–188, (1991). doi: 10.1016/0550-3213(91)90055-3.
586. J. C. Collins and R. K. Ellis, Heavy quark production in very high-energy hadron collisions, *Nucl. Phys. B* **360**, 3–30, (1991). doi: 10.1016/0550-3213(91)90288-9.
587. E. M. Levin, M. G. Ryskin, Y. M. Shabelski and A. G. Shuvaev, Heavy quark production in semihard nucleon interactions, *Sov. J. Nucl. Phys.* **53**, 657, (1991).
588. S. Catani and F. Hautmann, High-energy factorization and small x deep inelastic scattering beyond leading order, *Nucl. Phys. B* **427**, 475–524, (1994). doi: 10.1016/0550-3213(94)90636-X.
589. S. Catani, F. Fiorani and G. Marchesini, QCD coherence in initial state radiation, *Phys. Lett. B* **234**, 339, (1990).
590. G. Marchesini, QCD coherence in the structure function and associated distributions at small x, *Nucl. Phys. B* **445**, 49–80, (1995).

591. G. Marchesini and B. R. Webber, Final states in heavy quark leptoproduction at small x, *Nucl. Phys. B* **386**, 215–235, (1992). doi: 10.1016/0550-3213(92)90181-A.
592. F. Hautmann and H. Jung, Angular correlations in multi-jet final states from k-perpendicular-dependent parton showers, *J. High Energy Phys.* **10**, 113, (2008). doi: 10.1088/1126-6708/2008/10/113.
593. F. Hautmann and H. Jung, Jet correlations from unintegrated parton distributions, *AIP Conf. Proc.* **1056**, 79–86, (2008). doi: 10.1063/1.3013083.
594. B. Blok, Yu. Dokshitzer and M. Strikman, Rapidity distribution in DIS at small x, *Phys. Lett. B* **774** (2017) 26. doi: 10.1016/j.physletb.2017.09.038.
595. E. A. Kuraev, L. N. Lipatov and V. S. Fadin, Multi-reggeon processes in the Yang–Mills theory, *Sov. Phys. JETP* **44**, 443–450, (1976).
596. F. Hautmann and H. Jung, Transverse momentum dependent gluon density from DIS precision data, *Nucl. Phys. B* **883**, 1–19, (2014).
597. H. Jung, The CCFM Monte Carlo generator CASCADE, *Comput. Phys. Commun.* **143**, 100–111, (2002). doi: 10.1016/S0010-4655(01)00438-6.
598. H. Jung *et al.*, The CCFM Monte Carlo generator CASCADE version 2.2.03, *Eur. Phys. J. C* **70**, 1237–1249, (2010). doi: 10.1140/epjc/s10052-010-1507-z.
599. M. Deak, F. Hautmann, H. Jung and K. Kutak, Forward jets and energy flow in hadronic collisions, *Eur. Phys. J. C* **72**, 1982, (2012). doi: 10.1140/epjc/s10052-012-1982-5.
600. M. Deak, F. Hautmann, H. Jung and K. Kutak, Transverse energy flow with forward and central jets at the LHC, arXiv:1112.6386. (2011).
601. F. Hautmann, Energy flow observables in hadronic collisions, arXiv:1205.5411. (2012). doi: 10.3204/DESY-PROC-2012-03/52.
602. P. Kotko, A. M. Stasto and M. Strikman, Exploring minijets beyond the leading power, *Phys. Rev. D* **95** (5), 054009, (2017). doi: 10.1103/PhysRevD.95.054009.
603. M. L. Good and W. D. Walker, Diffraction disssociation of beam particles, *Phys. Rev.* **120**, 1857–1860, (1960). doi: 10.1103/PhysRev.120.1857.
604. A. H. Mueller, O(2,1) Analysis of single particle spectra at high-Energy, *Phys. Rev. D* **2**, 2963–2968, (1970). doi: 10.1103/PhysRevD.2.2963.
605. C. E. DeTar, C. E. Jones, F. E. Low, J. H. Weis, J. E. Young and C.-I. Tan, Helicity poles, triple-regge behavior and single-particle spectra in high-energy collisions, *Phys. Rev. Lett.* **26**, 675–676, (1971). doi: 10.1103/PhysRevLett.26.675.
606. A. B. Kaidalov and V. A. Khoze, Triple-Regge description of the inclusive hadron spectra, *Leningrad-73-62.* (1973).
607. H. I. Miettinen and J. Pumplin, Diffraction scattering and the parton structure of hadrons, *Phys. Rev. D* **18**, 1696, (1978). doi: 10.1103/PhysRevD.18.1696.
608. M. Grazzini, L. Trentadue and G. Veneziano, Fracture functions from cut vertices, *Nucl. Phys. B* **519**, 394–404, (1998). doi: 10.1016/S0550-3213(97)00840-7.
609. A. Berera and D. E. Soper, Behavior of diffractive parton distribution functions, *Phys. Rev. D* **53**, 6162–6179, (1996). doi: 10.1103/PhysRevD.53.6162.
610. J. C. Collins, Proof of factorization for diffractive hard scattering, *Phys. Rev. D* **57**, 3051–3056, (1998). doi: 10.1103/PhysRevD.61.019902,10.1103/PhysRevD.57.3051. [Erratum: Phys. Rev.D61,019902(2000)].
611. J. C. Collins, Factorization in hard diffraction, *J. Phys. G* **28**, 1069–1078, (2002). doi: 10.1088/0954-3899/28/5/327.
612. A. Aktas *et al.*, Measurement and QCD analysis of the diffractive deep-inelastic scattering cross-section at HERA, *Eur. Phys. J. C* **48**, 715–748, (2006). doi: 10. 1140/epjc/s10052-006-0035-3.

613. S. Chatrchyan *et al.*, Observation of a diffractive contribution to dijet production in proton–proton collisions at $\sqrt{s} = 7\,\text{TeV}$, *Phys. Rev. D* **87** (1), 012006, (2013). doi: 10.1103/PhysRevD.87.012006.
614. G. Aad *et al.*, Dijet production in $\sqrt{s} = 7\,\text{TeV}$ pp collisions with large rapidity gaps at the ATLAS experiment, *Phys. Lett. B* **754**, 214–234, (2016). doi: 10.1016/j.physletb.2016.01.028.
615. F. Hautmann and D. E. Soper, Color transparency in deeply inelastic diffraction, *Phys. Rev. D* **63**, 011501, (2001). doi: 10.1103/PhysRevD.63.011501.
616. F. Hautmann, Z. Kunszt and D. E. Soper, Diffractive deeply inelastic scattering of hadronic states with small transverse size, *Phys. Rev. Lett.* **81**, 3333–3336, (1998). doi: 10.1103/PhysRevLett.81.3333.
617. J. Bartels, J. R. Ellis, H. Kowalski and M. Wusthoff, An analysis of diffraction in deep inelastic scattering, *Eur. Phys. J. C* **7**, 443–458, (1999). doi: 10.1007/s100529801022.
618. W. Buchmüller, T. Gehrmann and A. Hebecker, Inclusive and diffractive structure functions at small x, *Nucl. Phys. B* **537**, 477–500, (1999). doi: 10.1016/S0550-3213(98)00682-8.
619. G. Ingelman, R. Pasechnik and D. Werder, Dynamic color screening in diffractive deep inelastic scattering, *Phys. Rev. D* **93** (9), 094016, (2016). doi: 10.1103/PhysRevD.93.094016.
620. F. Hautmann, Diffractive jet production in lepton hadron collisions, *J. High Energy Phys.* **10**, 025, (2002). doi: 10.1088/1126-6708/2002/10/025.
621. F. Hautmann, Q anti-Qg contribution to diffractive J/psi electroproduction, *J. High Energy Phys.* **04**, 036, (2002). doi: 10.1088/1126-6708/2002/04/036.
622. J. M. Grados Luyando. Recent minimum bias and UE measurements at CMS. MPI@LHC 2016, https://indico.nucleares.unam.mx/event/1100/session/1/contribution/32, (2016).
623. U. Amaldi and K. R. Schubert, Impact parameter interpretation of proton–proton scattering from a critical review of all ISR data, *Nucl. Phys. B* **166**, 301–320, (1980). doi: 10.1016/0550-3213(80)90229-1.
624. H. Heiselberg, G. Baym, B. Blaettel, L. L. Frankfurt and M. Strikman, Color transparency, color opacity and fluctuations in nuclear collisions, *Phys. Rev. Lett.* **67**, 2946–2949, (1991). doi: 10.1103/PhysRevLett.67.2946.
625. L. L. Frankfurt, G. A. Miller and M. Strikman, The geometrical color optics of coherent high-energy processes, *Ann. Rev. Nucl. Part. Sci.* **44**, 501–560, (1994). doi: 10.1146/annurev.ns.44.120194.002441.
626. G. Alberi and G. Goggi, Diffraction of subnuclear waves, *Phys. Rept.* **74**, 1–207, (1981). doi: 10.1016/0370-1573(81)90019-3.
627. A. B. Kaidalov, K. A. Ter-Martirosian and Yu. M. Shabelski, Inclusive spectra of secondary particles in proton–nucleus interactions in the quark–gluon string model, *Sov. J. Nucl. Phys.* **43**, 822–826, (1986) (in Russian). [*Yad. Fiz.* **43**, 1282, (1986)].
628. L. D. McLerran and R. Venugopalan, Gluon distribution functions for very large nuclei at small transverse momentum, *Phys. Rev. D* **49**, 3352–3355, (1994). doi: 10.1103/PhysRevD.49.3352.
629. K. J. Golec-Biernat and M. Wusthoff, Saturation effects in deep inelastic scattering at low Q^2 and its implications on diffraction, *Phys. Rev. D* **59**, 014017, (1998). doi: 10.1103/PhysRevD.59.014017.
630. A. H. Mueller, Parton saturation at small x and in large nuclei, *Nucl. Phys. B* **558**, 285–303, (1999). doi: 10.1016/S0550-3213(99)00394-6.

631. I. Balitsky, Operator expansion for high-energy scattering, *Nucl. Phys. B* **463**, 99–160, (1996). doi: 10.1016/0550-3213(95)00638-9.
632. Y. V. Kovchegov, Small x $F(2)$ structure function of a nucleus including multiple pomeron exchanges, *Phys. Rev. D* **60**, 034008, (1999). doi: 10.1103/PhysRevD.60.034008.
633. J. Jalilian-Marian, A. Kovner, A. Leonidov and H. Weigert, The Wilson renormalization group for low x physics: Towards the high density regime, *Phys. Rev. D* **59**, 014014, (1998). doi: 10.1103/PhysRevD.59.014014.
634. E. Iancu, A. Leonidov and L. D. McLerran, Nonlinear gluon evolution in the color glass condensate. 1, *Nucl. Phys. A* **692**, 583–645, (2001). doi: 10.1016/S0375-9474(01)00642-X.
635. F. Gelis, E. Iancu, J. Jalilian-Marian and R. Venugopalan, The color glass condensate, *Ann. Rev. Nucl. Part. Sci.* **60**, 463–489, (2010). doi: 10.1146/annurev.nucl.010909.083629.
636. C. Flensburg, G. Gustafson and L. Lönnblad, Inclusive and exclusive observables from dipoles in high energy collisions, *J. High Energy Phys.* **08**, 103, (2011). doi: 10.1007/JHEP08(2011)103.
637. F. Hautmann and D. E. Soper, Parton distribution function for quarks in an s-channel approach, *Phys. Rev. D* **75**, 074020, (2007). doi: 10.1103/PhysRevD.75.074020.
638. F. Hautmann, Power-like corrections and the determination of the gluon distribution, *Phys. Lett. B* **643**, 171–174, (2006). doi: 10.1016/j.physletb.2006.10.053.
639. V. Khachatryan *et al.*, Production of leading charged particles and leading charged-particle jets at small transverse momenta in pp collisions at $\sqrt{s} = 8\,\text{TeV}$, *Phys. Rev. D* **92** (11), 112001, (2015). doi: 10.1103/PhysRevD.92.112001.
640. K. Kutak and H. Jung, Saturation effects in final states due to CCFM with absorptive boundary, (2008). doi: 10.3204/DESY-PROC-2009-01/112.
641. A. Bacchetta, H. Jung, A. Knutsson, K. Kutak and F. Samson-Himmelstjerna, A method for tuning parameters of Monte Carlo generators and a its application to the determination of the unintegrated gluon density, *Eur. Phys. J. C* **70**, 503–511, (2010). doi: 10.1140/epjc/s10052-010-1464-6.
642. K. Kutak, Gluon saturation and entropy production in proton–proton collisions, *Phys. Lett. B* **705**, 217–221, (2011). doi: 10.1016/j.physletb.2011.09.113.
643. K. Kutak, K. Golec-Biernat, S. Jadach and M. Skrzypek, Nonlinear equation for coherent gluon emission, *J. High Energy Phys.* **02**, 117, (2012). doi: 10.1007/JHEP02(2012)117.
644. K. Kutak and S. Sapeta, Gluon saturation in dijet production in p–Pb collisions at Large Hadron Collider, *Phys. Rev. D* **86**, 094043, (2012). doi: 10.1103/PhysRevD.86.094043.
645. P. Kotko, K. Kutak, C. Marquet, E. Petreska, S. Sapeta and A. van Hameren, Improved TMD factorization for forward dijet production in dilute-dense hadronic collisions, *J. High Energy Phys.* **09**, 106, (2015). doi: 10.1007/JHEP09(2015)106.
646. A. H. Mueller and H. Navelet, An inclusive minijet cross-section and the bare pomeron in QCD, *Nucl. Phys. B* **282**, 727, (1987). doi: 10.1016/0550-3213(87)90705-X.
647. V. Khachatryan *et al.*, Azimuthal decorrelation of jets widely separated in rapidity in pp collisions at $\sqrt{s} = 7\,\text{TeV}$, *J. High Energy Phys.* **08**, 139, (2016). doi: 10.1007/JHEP08(2016)139.
648. S. Chatrchyan *et al.*, Ratios of dijet production cross sections as a function of the absolute difference in rapidity between jets in proton–proton collisions at $\sqrt{s} = 7$ TeV, *Eur. Phys. J. C* **72**, 2216, (2012). doi: 10.1140/epjc/s10052-012-2216-6.

649. G. Aad *et al.*, Measurements of jet vetoes and azimuthal decorrelations in dijet events produced in pp collisions at $\sqrt{s} = 7\,\text{TeV}$ using the ATLAS detector, *Eur. Phys. J. C* **74** (11), 3117, (2014). doi: 10.1140/epjc/s10052-014-3117-7.
650. G. Aad *et al.*, Measurement of dijet production with a veto on additional central jet activity in pp collisions at $\sqrt{s} = 7\,\text{TeV}$ using the ATLAS detector, *J. High Energy Phys.* **1109**, 053, (2011).
651. R. Maciula and A. Szczurek, Double-parton scattering contribution to production of jet pairs with large rapidity separation at the LHC, *Phys. Rev. D* **90** (1), 014022, (2014). doi: 10.1103/PhysRevD.90.014022.
652. B. Duclou, L. Szymanowski and S. Wallon, Evaluating the double parton scattering contribution to Mueller–Navelet jets production at the LHC, *Phys. Rev. D* **92** (7), 076002, (2015). doi: 10.1103/PhysRevD.92.076002.
653. F. Hautmann, H. Jung, A. Lelek, V. Radescu and R. Zlebcik, Collinear and TMD quark and gluon densities from parton branching solution of QCD evolution equations, arXiv:1708.03279. (2017).
654. B. B. Abelev *et al.*, Multiplicity dependence of jet-like two-particle correlation structures in pPb collisions at $\sqrt{s_{NN}} = 5.02\,\text{TeV}$, *Phys. Lett. B* **741**, 38–50, (2015). doi: 10.1016/j.physletb.2014.11.028.
655. M. L. Miller, K. Reygers, S. J. Sanders and P. Steinberg, Glauber modeling in high energy nuclear collisions, *Ann. Rev. Nucl. Part. Sci.* **57**, 205–243, (2007). doi: 10.1146/annurev.nucl.57.090506.123020.
656. U. W. Heinz, Concepts of heavy ion physics. In *2002 European School of High-Energy Physics, Pylos, Greece, 25 Aug–7 Sep 2002: Proceedings*, pp. 165–238, (2004). URL http://doc.cern.ch/yellowrep/CERN-2004-001.
657. C. Gale, S. Jeon and B. Schenke, Hydrodynamic modeling of heavy-ion collisions, *Int. J. Mod. Phys. A* **28**, 1340011, (2013). doi: 10.1142/S0217751X13400113.
658. J. Rafelski and B. Muller, Strangeness production in the quark–gluon plasma, *Phys. Rev. Lett.* **48**, 1066, (1982).
659. J. Rafelski and R. Hagedorn, From hadron gas to quark matter. 2. In *International Symposium on Statistical Mechanics of Quarks and Hadrons Bielefeld, Germany, August 24–31, 1980*, pp. 253–272, (1980). URL http://alice.cern.ch/format/showfull?sysnb=0036721.
660. F. Antinori *et al.*, Enhancement of hyperon production at central rapidity in 158-A-GeV/c Pb–Pb collisions, *J. Phys. G* **32**, 427–442, (2006). doi: 10.1088/0954-3899/32/4/003.
661. B. I. Abelev *et al.*, Enhanced strange baryon production in Au + Au collisions compared to p + p at s(NN)**(1/2) = 200-GeV, *Phys. Rev. C* **77**, 044908, (2008). doi: 10.1103/PhysRevC.77.044908.
662. B. B. Abelev *et al.*, Multi-strange baryon production at mid-rapidity in Pb–Pb collisions at $\sqrt{s_{NN}} = 2.76\,\text{TeV}$, *Phys. Lett. B* **728**, 216–227, (2014). doi: 10.1016/j.physletb.2014.05.052,10.1016/j.physletb.2013.11.048. [Erratum: *Phys. Lett. B* **734**, 409, (2014)].
663. F. Becattini and R. Fries, The QCD confinement transition: hadron formation, *Landolt-Börnstein* — Group I Elementary Particles, Nuclei and Atoms 23 (Relativistic Heavy Ion Physics) (2010). doi: 10.1007/978-3-642-01539-7_8.
664. M. Floris, Hadron yields and the phase diagram of strongly interacting matter, *Nucl. Phys. A* **931**, 103–112, (2014). doi: 10.1016/j.nuclphysa.2014.09.002.
665. F. Becattini, P. Castorina, A. Milov and H. Satz, A comparative analysis of statistical hadron production, *Eur. Phys. J. C* **66**, 377–386, (2010). doi: 10.1140/epjc/s10052-010-1265-y.

666. B. B. Abelev *et al.*, Performance of the ALICE Experiment at the CERN LHC, *Int. J. Mod. Phys. A* **29**, 1430044, (2014). doi: 10.1142/S0217751X14300440.
667. K. Aamodt *et al.*, The ALICE experiment at the CERN LHC, *JINST*. **3**, S08002, (2008). doi: 10.1088/1748-0221/3/08/S08002.
668. K. Aamodt *et al.*, Strange particle production in proton–proton collisions at $\sqrt(s) =$ 0.9 TeV with ALICE at the LHC, *Eur. Phys. J. C* **71**, 1594, (2011). doi: 10.1140/epjc/s10052-011-1594-5.
669. B. Abelev *et al.*, Multi-strange baryon production in pp collisions at $\sqrt{s} = 7$ TeV with ALICE, *Phys. Lett. B* **712**, 309–318, (2012). doi: 10.1016/j.physletb.2012.05.011.
670. B. Abelev *et al.*, Production of $K^*(892)^0$ and $\phi(1020)$ in pp collisions at $\sqrt{s} = 7$ TeV, *Eur. Phys. J. C* **72**, 2183, (2012). doi: 10.1140/epjc/s10052-012-2183-y.
671. B. B. Abelev *et al.*, Production of charged pions, kaons and protons at large transverse momenta in pp and Pb–Pb collisions at $\sqrt{s_{\mathrm{NN}}}$ =2.76 TeV, *Phys. Lett. B* **736**, 196–207, (2014). doi: 10.1016/j.physletb.2014.07.011.
672. J. Adam *et al.*, $\mathrm{K}^*(892)^0$ and $\phi(1020)$ meson production at high transverse momentum in pp and Pb–Pb collisions at $\sqrt{s_{\mathrm{NN}}} = 2.76$ TeV, *Phys. Rev. C* **95** (6), 064606, (2017). doi: 10.1103/PhysRevC.95.064606.
673. J. Adam *et al.*, Multiplicity dependence of charged pion, kaon and (anti)proton production at large transverse momentum in p–Pb collisions at $\sqrt{s_{\mathrm{NN}}} = 5.02$ TeV, *Phys. Lett. B* **760**, 720–735, (2016). doi: 10.1016/j.physletb.2016.07.050.
674. J. Adam *et al.*, Production of light nuclei and anti-nuclei in pp and Pb–Pb collisions at energies available at the CERN large hadron collider, *Phys. Rev. C* **93** (2), 024917, (2016). doi: 10.1103/PhysRevC.93.024917.
675. S. Acharya *et al.*, Production of deuterons, tritons, ^{3}He nuclei and their anti-nuclei in pp collisions at $\sqrt{s} = 0.9$, 2.76 and 7 TeV, *Phys. Rev. C* **97**, 024615, (2018). doi: 10.1103/PhysRevC.97.024615.
676. S. Acharya *et al.*, π^0 and η meson production in proton–proton collisions at $\sqrt{s} = 8$ TeV, *Eur. Phys. J. C* **78**, 263, (2018). doi: 10.1140/epjc/s10052-018-5612-8.
677. D. Adamova *et al.*, Production of $\Sigma(1385)^{\pm}$ and $\Xi(1530)^0$ in p–Pb collisions at $\sqrt{s_{\mathrm{NN}}} = 5.02$ TeV, *Eur. Phys. J. C* **77** (6), 389, (2017). doi: 10.1140/epjc/s10052-017-4943-1.
678. J. Adam *et al.*, Production of K^* $(892)^0$ and ϕ (1020) in p–Pb collisions at $\sqrt{s_{\mathrm{NN}}} =$ 5.02 TeV, *Eur. Phys. J. C* **76** (5), 245, (2016). doi: 10.1140/epjc/s10052-016-4088-7.
679. J. Adam *et al.*, Multi-strange baryon production in p–Pb collisions at $\sqrt{s_{\mathrm{NN}}} = 5.02$ TeV, *Phys. Lett. B* **758**, 389–401, (2016). doi: 10.1016/j.physletb.2016.05.027.
680. S. Acharya *et al.*, Production of π^0 and η mesons up to high transverse momentum in pp collisions at 2.76 TeV, *Eur. Phys. J. C* **77** (5), 339, (2017). doi: 10.1140/epjc/s10052-017-5144-7,10.1140/epjc/s10052-017-4890-x. [Erratum: *Eur. Phys. J. C* **77** (9), 586, (2017)].
681. B. B. Abelev *et al.*, Neutral pion production at midrapidity in pp and Pb–Pb collisions at $\sqrt{s_{\mathrm{NN}}} = 2.76$ TeV, *Eur. Phys. J. C* **74** (10), 3108, (2014). doi: 10.1140/epjc/s10052-014-3108-8.
682. B. Abelev *et al.*, Neutral pion and η meson production in proton–proton collisions at $\sqrt{s} = 0.9$ TeV and $\sqrt{s} = 7$ TeV, *Phys. Lett. B* **717**, 162–172, (2012). doi: 10.1016/j.physletb.2012.09.015.
683. S. Chatrchyan *et al.*, Study of the inclusive production of charged pions, kaons and protons in pp collisions at $\sqrt{s} = 0.9$, 2.76 and 7 TeV, *Eur. Phys. J. C* **72**, 2164, (2012). doi: 10.1140/epjc/s10052-012-2164-1.
684. V. Khachatryan *et al.*, Multiplicity and rapidity dependence of strange hadron production in pp, pPb and PbPb collisions at the LHC, *Phys. Lett. B* **768**, 103–129, (2017). ISSN 0370-2693. doi: http://dx.doi.org/10.1016/j.physletb.2017.01.075.

685. A. M. Sirunyan *et al.*, Measurement of charged pion, kaon and proton production in proton–proton collisions at $\sqrt{s} = 13\,\text{TeV}$, *Phys. Rev. D* **96**, 112003, (2017). doi: 10.1103/PhysRevD.96.112003.
686. R. Aaij *et al.*, Measurement of prompt hadron production ratios in pp collisions at $\sqrt{s} = 0.9$ and 7 TeV, *Eur. Phys. J. C* **72**, 2168, (2012). doi: 10.1140/epjc/s10052-012-2168-x.
687. R. Aaij *et al.*, Measurement of the inclusive ϕ cross-section in pp collisions at $\sqrt{s} = 7$ TeV, *Phys. Lett. B* **703**, 267–273, (2011). doi: 10.1016/j.physletb.2011.08.017.
688. R. Aaij *et al.*, Measurement of V^0 production ratios in pp collisions at $\sqrt{s} = 0.9$ and 7 TeV, *J. High Energy Phys.* **08**, 034, (2011). doi: 10.1007/JHEP08(2011)034.
689. R. Aaij *et al.*, Prompt K^0_s production in pp collisions at $\sqrt{s} = 0.9$ TeV, *Phys. Lett. B* **693**, 69–80, (2010). doi: 10.1016/j.physletb.2010.08.055.
690. G. Aad *et al.*, The differential production cross section of the ϕ (1020) meson in $\sqrt{s} = 7\,\text{TeV}$ pp collisions measured with the ATLAS detector, *Eur. Phys. J. C* **74** (7), 2895, (2014). doi: 10.1140/epjc/s10052-014-2895-2.
691. G. Aad *et al.*, Kshort and Λ production in pp interactions at $\sqrt{s} = 0.9$ and 7 TeV measured with the ATLAS detector at the LHC, *Phys. Rev. D* **85**, 012001, (2012). doi: 10.1103/PhysRevD.85.012001.
692. K. Aamodt *et al.*, Production of pions, kaons and protons in pp collisions at $\sqrt{s} = 900$ GeV with ALICE at the LHC, *Eur. Phys. J. C* **71**, 1655, (2011). doi: 10.1140/epjc/s10052-011-1655-9.
693. K. Werner, B. Guiot, I. Karpenko and T. Pierog, Analysing radial flow features in p–Pb and p-p collisions at several TeV by studying identified particle production in EPOS3, *Phys. Rev. C* **89** (6), 064903, (2014). doi: 10.1103/PhysRevC.89.064903.
694. K. Dusling, W. Li and B. Schenke, Novel collective phenomena in high-energy proton–proton and proton–nucleus collisions, *Int. J. Mod. Phys. E* **25**, 1630002, (2016). doi: 10.1142/S0218301316300022.
695. P. Romatschke, Relativistic fluid dynamics far from local equilibrium, *Phys. Rev. Lett.* **120**, 012301, (2018). doi: 10.1103/PhysRevLett.120.012301.
696. R. D. Weller and P. Romatschke, One fluid to rule them all: Viscous hydrodynamic description of event-by-event central p+p, p+Pb and Pb+Pb collisions at $\sqrt{s} =$ 5.02 TeV, *Phys. Lett. B* **774**, 351–356, (2017). doi: 10.1016/j.physletb.2017.09.077.
697. P. Castorina, S. Plumari and H. Satz, Universal strangeness production in hadronic and nuclear collisions, *Int. J. Mod. Phys. E* **25** (08), 1650058, (2016). doi: 10.1142/S0218301316500580.
698. E. Cuautle and G. Paić, The energy density representation of the strangeness enhancement from p+p to Pb+Pb. (2016). arXiv:1608.02101.
699. F. Becattini, J. Steinheimer, R. Stock and M. Bleicher, Hadronization conditions in relativistic nuclear collisions and the QCD pseudo-critical line, *Phys. Lett. B* **764**, 241–246, (2017). doi: 10.1016/j.physletb.2016.11.033.
700. C. Bierlich and J. R. Christiansen, Effects of color reconnection on hadron flavor observables, *Phys. Rev. D* **92** (9), 094010, (2015). doi: 10.1103/PhysRevD.92.094010.
701. S. Chatrchyan *et al.*, Study of the production of charged pions, kaons and protons in pPb collisions at $\sqrt{s_{NN}} = 5.02\,\text{TeV}$, *Eur. Phys. J. C* **74** (6), 2847, (2014). doi: 10.1140/epjc/s10052-014-2847-x.
702. M. M. Aggarwal *et al.*, K^{*0} production in Cu+Cu and Au+Au collisions at $\sqrt{s_N N} =$ 62.4 GeV and 200 GeV, *Phys. Rev. C* **84**, 034909, (2011). doi: 10.1103/PhysRevC.84.034909.
703. A. G. Knospe, Recent hadronic resonance measurements at ALICE, *J. Phys. Conf. Ser.* **779** (1), 012072, (2017). doi: 10.1088/1742-6596/779/1/012072.

704. B. B. Abelev *et al.*, $K^*(892)^0$ and $\phi(1020)$ production in Pb–Pb collisions at $\sqrt{sNN} = 2.76\,\text{TeV}$, *Phys. Rev. C* **91**, 024609, (2015). doi: 10.1103/PhysRevC.91.024609.
705. A. G. Knospe, C. Markert, K. Werner, J. Steinheimer and M. Bleicher, Hadronic resonance production and interaction in partonic and hadronic matter in the EPOS3 model with and without the hadronic afterburner UrQMD, *Phys. Rev. C* **93** (1), 014911, (2016). doi: 10.1103/PhysRevC.93.014911.
706. A. Ortiz, P. Christiansen, E. Cuautle, I. Maldonado and G. Paić, Color reconnection and flowlike patterns in pp collisions, *Phys. Rev. Lett.* **111** (4), 042001, (2013). doi: 10.1103/PhysRevLett.111.042001.
707. J. Adam *et al.*, Centrality dependence of particle production in p–Pb collisions at $\sqrt{s_{\rm NN}}$= 5.02 TeV, *Phys. Rev. C* **91** (6), 064905, (2015). doi: 10.1103/PhysRevC.91.064905.
708. D. V. Perepelitsa and P. A. Steinberg, Calculation of centrality bias factors in p+A collisions based on a positive correlation of hard process yields with underlying event activity. (2014). arXiv:1412.0976.
709. E. Schnedermann, J. Sollfrank and U. W. Heinz, Thermal phenomenology of hadrons from 200-A/GeV S+S collisions, *Phys. Rev. C* **48**, 2462–2475, (1993). doi: 10.1103/PhysRevC.48.2462.
710. B. B. Abelev *et al.*, Multiplicity dependence of pion, kaon, proton and lambda Production in p–Pb collisions at $\sqrt{s_{NN}} = 5.02\,\text{TeV}$, *Phys. Lett. B* **728**, 25–38, (2014). doi: 10.1016/j.physletb.2013.11.020.
711. E. Shuryak and I. Zahed, High-multiplicity pp and pA collisions: hydrodynamics at its edge, *Phys. Rev. C* **88** (4), 044915, (2013). doi: 10.1103/PhysRevC.88.044915.
712. N. Jacazio, Production of identified charged hadrons in Pb–Pb collisions at $\sqrt{s_{NN}}=5.02$ TeV, *Nucl. Phys. A* **967**, 421–424, (2017). doi: 10.1016/j.nuclphysa.2017.05.023.
713. B. Abelev *et al.*, Centrality dependence of π, K, p production in Pb–Pb collisions at $\sqrt{s_{NN}} = 2.76$ TeV, *Phys. Rev. C* **88**, 044910, (2013). doi: 10.1103/PhysRevC.88.044910.
714. R. J. Fries, V. Greco and P. Sorensen, Coalescence models for hadron formation from quark gluon plasma, *Ann. Rev. Nucl. Part. Sci.* **58**, 177–205, (2008). doi: 10.1146/annurev.nucl.58.110707.171134.
715. B. Schenke, Origins of collectivity in small systems, *Nucl. Phys. A* **967**, 105–112, (2017). doi: 10.1016/j.nuclphysa.2017.05.017.
716. P. Bok and W. Broniowski, Hydrodynamic approach to p–Pb, *Nucl. Phys. A* **926**, 16–23, (2014). doi: 10.1016/j.nuclphysa.2014.04.004.
717. H. Niemi and G. S. Denicol, How large is the Knudsen number reached in fluid dynamical simulations of ultrarelativistic heavy ion collisions? (2014). arXiv:1404.7327.
718. P. Romatschke, Do nuclear collisions create a locally equilibrated quark–gluon plasma? *Eur. Phys. J. C* **77** (1), 21, (2017). doi: 10.1140/epjc/s10052-016-4567-x.
719. B. Abelev *et al.*, J/ψ production as a function of charged particle multiplicity in pp collisions at $\sqrt{s} = 7$ TeV, *Phys. Lett. B* **712**, 165–175, (2012). doi: 10.1016/j.physletb.2012.04.052.
720. J. Adam *et al.*, Measurement of charm and beauty production at central rapidity versus charged-particle multiplicity in proton–proton collisions at $\sqrt{s} = 7\,\text{TeV}$, *J. High Energy Phys.* **09**, 148, (2015). doi: 10.1007/JHEP09(2015)148.
721. K. Eggert *et al.*, Angular correlations between the charged particles produced in pp collisions at ISR energies, *Nucl. Phys. B* **86**, 201, (1975). doi: 10.1016/0550-3213(75)90440-X.

722. B. Alver *et al.*, Cluster properties from two-particle angular correlations in p+p collisions at $\sqrt{s} = 200\,\text{GeV}$ and 410 GeV, *Phys. Rev. C.* **75**, 054913, (2007). doi: 10.1103/PhysRevC.75.054913.
723. I. Arsene *et al.*, Quark gluon plasma and color glass condensate at RHIC? The perspective from the BRAHMS experiment, *Nucl. Phys. A* **757**, 1, (2005). doi: 10.1016/j.nuclphysa.2005.02.130.
724. K. Adcox *et al.*, Formation of dense partonic matter in relativistic nucleus–nucleus collisions at RHIC: Experimental evaluation by the PHENIX collaboration, *Nucl. Phys. A* **757**, 184, (2005). doi: 10.1016/j.nuclphysa.2005.03.086.
725. B. B. Back *et al.*, The PHOBOS perspective on discoveries at RHIC, *Nucl. Phys. A* **757**, 28, (2005). doi: 10.1016/j.nuclphysa.2005.03.084.
726. J. Adams *et al.*, Experimental and theoretical challenges in the search for the quark gluon plasma: the STAR collaboration's critical assessment of the evidence from RHIC collisions, *Nucl. Phys. A* **757**, 102, (2005). doi: 10.1016/j.nuclphysa.2005.03.085.
727. W. Li, Observation of a 'Ridge' correlation structure in high multiplicity proton–proton collisions: a brief review, *Mod. Phys. Lett. A* **27**, 1230018, (2012). doi: 10.1142/S0217732312300182.
728. S. Chatrchyan *et al.*, Observation of long-range near-side angular correlations in proton-lead collisions at the LHC, *Phys. Lett. B* **718**, 795, (2013). doi: 10.1016/j.physletb.2012.11.025.
729. G. Aad *et al.*, Observation of associated near-side and away-side long-range correlations in $\sqrt{s_{NN}} = 5.02\,\text{TeV}$ proton–lead collisions with the ATLAS detector, *Phys. Rev. Lett.* **110** (18), 182302, (2013). doi: 10.1103/PhysRevLett.110.182302.
730. B. Abelev *et al.*, Long-range angular correlations on the near and away side in p–Pb collisions at $\sqrt{s_{NN}} = 5.02$ TeV, *Phys. Lett. B* **719**, 29–41, (2013). doi: 10.1016/j.physletb.2013.01.012.
731. S. Voloshin and Y. Zhang, Flow study in relativistic nuclear collisions by Fourier expansion of azimuthal particle distributions, *Z. Phys. C.* **70**, 665, (1996). doi: 10.1007/s002880050141.
732. B. H. Alver, C. Gombeaud, M. Luzum and J.-Y. Ollitrault, Triangular flow in hydrodynamics and transport theory, *Phys. Rev. C* **82**, 034913, (2010). doi: 10.1103/PhysRevC.82.034913.
733. B. Schenke, S. Jeon and C. Gale, Elliptic and triangular flow in event-by-event $D = 3 + 1$ viscous hydrodynamics, *Phys. Rev. Lett.* **106**, 042301, (2011). doi: 10.1103/PhysRevLett.106.042301.
734. Z. Qiu, C. Shen and U. Heinz, Hydrodynamic elliptic and triangular flow in Pb–Pb collisions at $\sqrt{s_{NN}} = 2.76\,\text{TeV}$, *Phys. Lett. B* **707**, 151, (2012). doi: 10.1016/j.physletb.2011.12.041.
735. S. Chatrchyan *et al.*, Long-range and short-range dihadron angular correlations in central PbPb collisions at a nucleon–nucleon center of mass energy of 2.76 TeV, *J. High Energy Phys.* **07**, 076, (2011). doi: 10.1007/JHEP07(2011)076.
736. S. Chatrchyan *et al.*, Centrality dependence of dihadron correlations and azimuthal anisotropy harmonics in Pb–Pb collisions at $\sqrt{s_{NN}} = 2.76$ TeV, *Eur. Phys. J. C* **72**, 2012, (2012). doi: 10.1140/epjc/s10052-012-2012-3.
737. S. Chatrchyan *et al.*, Measurement of the elliptic anisotropy of charged particles produced in Pb–Pb collisions at nucleon-nucleon center-of-mass energy = 2.76 TeV, *Phys. Rev. C* **87**, 014902, (2013). doi: 10.1103/PhysRevC.87.014902.
738. K. Aamodt *et al.*, Elliptic flow of charged particles in Pb–Pb collisions at 2.76 TeV, *Phys. Rev. Lett.* **105**, 252302, (2010). doi: 10.1103/PhysRevLett.105.252302.

739. K. Aamodt *et al.*, Higher harmonic anisotropic flow measurements of charged particles in Pb–Pb collisions at $\sqrt{s_{NN}} = 2.76$ TeV, *Phys. Rev. Lett.* **107**, 032301, (2011). doi: 10.1103/PhysRevLett.107.032301.
740. K. Aamodt *et al.*, Harmonic decomposition of two-particle angular correlations in Pb–Pb collisions at $\sqrt{s_{NN}} = 2.76$ TeV, *Phys. Lett. B* **708**, 249, (2012). doi: 10.1016/j.physletb.2012.01.060.
741. G. Aad *et al.*, Measurement of the pseudorapidity and transverse momentum dependence of the elliptic flow of charged particles in lead–lead collisions at $\sqrt{s_{NN}} = 2.76$ TeV with the ATLAS detector, *Phys. Lett. B* **707**, 330, (2012). doi: 10.1016/j.physletb.2011.12.056.
742. G. Aad *et al.*, Measurement of the azimuthal anisotropy for charged particle production in $\sqrt{s_{NN}} = 2.76$ TeV lead–lead collisions with the ATLAS detector, *Phys. Rev. C* **86**, 014907, (2012). doi: 10.1103/PhysRevC.86.014907.
743. M. Aaboud *et al.*, Measurements of long-range azimuthal anisotropies and associated Fourier coefficients for pp collisions at $\sqrt{s} = 5.02$ and 13 TeV and p+Pb collisions at $\sqrt{s_{\rm NN}} = 5.02$ TeV with the ATLAS detector, *Phys. Rev. C* **96**, 024908, (2017). doi: 10.1103/PhysRevC.96.024908.
744. V. Khachatryan *et al.*, Long-range two-particle correlations of strange hadrons with charged particles in pPb and Pb–Pb collisions at LHC energies, *Phys. Lett. B* **742**, 200, (2015). doi: 10.1016/j.physletb.2015.01.034.
745. B. Abelev *et al.*, Mass, quark-number and $\sqrt{s_{NN}}$ dependence of the second and fourth flow harmonics in ultra-relativistic nucleus-nucleus collisions, *Phys. Rev. C* **75**, 054906, (2007). doi: 10.1103/PhysRevC.75.054906.
746. A. Adare *et al.*, Deviation from quark-number scaling of the anisotropy parameter v_2 of pions, kaons and protons in Au+Au collisions at $\sqrt{s_{NN}} = 200$ GeV, *Phys. Rev. C* **85**, 064914, (2012). doi: 10.1103/PhysRevC.85.064914.
747. B. B. Abelev *et al.*, Long-range angular correlations of π, K and p in p–Pb collisions at $\sqrt{s_{\rm NN}} = 5.02\,$TeV, *Phys. Lett. B* **726**, 164–177, (2013). doi: 10.1016/j.physletb.2013.08.024.
748. K. Werner, M. Bleicher, B. Guiot, I. Karpenko and T. Pierog, Evidence for flow from hydrodynamic simulations of pPb collisions at 5.02 TeV from v_2 mass splitting, *Phys. Rev. Lett.* **112**, 232301, (2014). doi: 10.1103/PhysRevLett.112.232301.
749. P. Bozek, W. Broniowski and G. Torrieri, Mass hierarchy in identified particle distributions in proton–lead collisions, *Phys. Rev. Lett.* **111**, 172303, (2013). doi: 10.1103/PhysRevLett.111.172303.
750. B. Schenke, S. Schlichting, P. Tribedy and R. Venugopalan, Mass ordering of spectra from fragmentation of saturated gluon states in high multiplicity proton–proton collisions, *Phys. Rev. Lett.* **117** (16), 162301, (2016). doi: 10.1103/PhysRevLett.117.162301.
751. J. Adams *et al.*, Particle type dependence of azimuthal anisotropy and nuclear modification of particle production in Au+Au collisions at s(NN)**(1/2) = 200-GeV, *Phys. Rev. Lett.* **92**, 052302, (2004). doi: 10.1103/PhysRevLett.92.052302.
752. A. Adare *et al.*, Scaling properties of azimuthal anisotropy in Au+Au and Cu+Cu collisions at s(NN) ** (1/2) = 200-GeV, *Phys. Rev. Lett.* **98**, 162301, (2007). doi: 10.1103/PhysRevLett.98.162301.
753. D. Molnar and S. A. Voloshin, Elliptic flow at large transverse momenta from quark coalescence, *Phys. Rev. Lett.* **91**, 092301, (2003). doi: 10.1103/PhysRevLett.91.092301.
754. V. Greco, C. M. Ko and P. Levai, Parton coalescence and the anti-proton/pion anomaly at RHIC, *Phys. Rev. Lett.* **90**, 202302, (2003). doi: 10.1103/PhysRevLett.90.202302.

755. R. J. Fries, B. Muller, C. Nonaka and S. A. Bass, Hadronization in heavy ion collisions: Recombination and fragmentation of partons, *Phys. Rev. Lett.* **90**, 202303, (2003). doi: 10.1103/PhysRevLett.90.202303.
756. N. Borghini, P. M. Dinh and J.-Y. Ollitrault, Flow analysis from multiparticle azimuthal correlations, *Phys. Rev. C* **64**, 054901, (2001). doi: 10.1103/PhysRevC.64.054901.
757. R. Bhalerao, N. Borghini and J. Ollitrault, Genuine collective flow from Lee–Yang zeroes, *Phys. Lett. B* **580**, 157, (2004). doi: 10.1016/j.physletb.2003.11.056.
758. A. Bilandzic, R. Snellings and S. Voloshin, Flow analysis with cumulants: Direct calculations, *Phys. Rev. C* **83**, 044913, (2011). doi: 10.1103/PhysRevC.83.044913.
759. C. Alt *et al.*, Directed and elliptic flow of charged pions and protons in Pb+Pb collisions at 40-A-GeV and 158-A-GeV, *Phys. Rev. C* **68**, 034903, (2003). doi: 10.1103/PhysRevC.68.034903.
760. C. Adler *et al.*, Elliptic flow from two and four particle correlations in Au+Au collisions at $\sqrt{s_{NN}}$ = 130-GeV, *Phys. Rev. C* **66**, 034904, (2002). doi: 10.1103/PhysRevC.66.034904.
761. B. Abelev *et al.*, Anisotropic flow of charged hadrons, pions and (anti-)protons measured at high transverse momentum in Pb–Pb collisions at $\sqrt{s_{NN}} = 2.76$ TeV, *Phys. Lett. B* **719**, 18, (2013). doi: 10.1016/j.physletb.2012.12.066.
762. S. Chatrchyan *et al.*, Measurement of higher-order harmonic azimuthal anisotropy in PbPb collisions at $\sqrt{s_{NN}}$ = 2.76 TeV, *Phys. Rev. C.* **89**, 044906, (2014). doi: 10.1103/PhysRevC.89.044906.
763. G. Aad *et al.*, Measurement of flow harmonics with multi-particle cumulants in Pb+Pb collisions at $\sqrt{s_{\mathrm{NN}}}$ = 2.76 TeV with the ATLAS detector, *Eur. Phys. J. C* **74**, 3157, (2014). doi: 10.1140/epjc/s10052-014-3157-z.
764. G. Aad *et al.*, Measurement with the ATLAS detector of multi-particle azimuthal correlations in p+Pb collisions at $\sqrt{s_{NN}}$ = 5.02 TeV, *Phys. Lett. B* **725**, 60–78, (2013). doi: 10.1016/j.physletb.2013.06.057.
765. S. Chatrchyan *et al.*, Multiplicity and transverse-momentum dependence of two- and four-particle correlations in pPb and PbPb collisions, *Phys. Lett. B* **725**, 213, (2013). doi: 10.1016/j.physletb.2013.06.028.
766. B. B. Abelev *et al.*, Multiparticle azimuthal correlations in pPb and Pb–Pb collisions at the CERN Large Hadron Collider, *Phys. Rev. C* **90**, 054901, (2014). doi: 10.1103/PhysRevC.90.054901.
767. V. Khachatryan *et al.*, Evidence for collective multi-particle correlations in pPb collisions, *Phys. Rev. Lett.* **115**(1), 012301 (2015). doi: 10.1103/PhysRevLett.115.012301.
768. J. Jia, M. Zhou and A. Trzupek, Revealing long-range multi-particle collectivity in small collision systems via subevent cumulants, *Phys. Rev. C* **96**(3), 034906 (2017). doi: 10.1103/PhysRevC.96.034906.
769. P. Di Francesco, M. Guilbaud, M. Luzum and J.-Y. Ollitrault, Systematic procedure for analyzing cumulants at any order, *Phys. Rev. C* **95**, 044911, (2017). doi: 10.1103/PhysRevC.95.044911.
770. K. Gajdoov, Investigations of anisotropic collectivity using multi-particle correlations in pp, p–Pb and Pb–Pb collisions, *Nucl. Phys. A* **967**, 437, (2017). doi: 10.1016/j.nuclphysa.2017.04.033.
771. M. Aaboud *et al.*, Measurement of multi-particle azimuthal correlations in *pp*, *p*+Pb and low-multiplicity Pb+Pb collisions with the ATLAS detector, *Eur. Phys. J. C* **77**, 428, (2017). doi: 10.1140/epjc/s10052-017-4988-1.

772. A. Bilandzic, C. H. Christensen, K. Gulbrandsen, A. Hansen and Y. Zhou, Generic framework for anisotropic flow analyses with multiparticle azimuthal correlations, *Phys. Rev. C* **89**, 064904, (2014). doi: 10.1103/PhysRevC.89.064904.
773. A. M. Sirunyan *et al.*, Observation of correlated azimuthal anisotropy Fourier harmonics in pp and pPb collisions at the LHC, *Phys. Rev. Lett.* **120**, 092301, (2018). doi: 10.1103/PhysRevLett.120.092301.
774. M. Dasgupta and G. P. Salam, Event shapes in e^+e^- annihilation and deep inelastic scattering, *J. Phys. G: Nucl. Part. Phys.* **30** (5), R143, (2004). doi: 10.1088/0954-3899/30/5/R01.
775. P. A. Movilla Fernandez *et al.*, A Study of event shapes and determinations of alpha-s using data of e^+e^- annihilations at $\sqrt{s}$ = 22-GeV to 44-GeV, *Eur. Phys. J. C* **1**, 461–478, (1998). doi: 10.1007/s100520050096.
776. R. Ellis, D. Ross and A. Terrano, The perturbative calculation of jet structure in e+e annihilation, *Nucl. Phys. B* **178** (3), 421–456, (1981). ISSN 0550-3213. doi: http://dx.doi.org/10.1016/0550-3213(81)90165-6. URL http://www.sciencedirect.com/science/article/pii/0550321381901656.
777. S. Bethke, Qcd tests at e+e- colliders, *Nucl. Phys. B — Proc. Suppl.* **64** (1), 54–62, (1998). ISSN 0920-5632. doi: http://dx.doi.org/10.1016/S0920-5632(97)01036-0. URL http://www.sciencedirect.com/science/article/pii/S0920563297010360.
778. A. Banfi, G. P. Salam and G. Zanderighi, Semi-numerical resummation of event shapes, *J. High Energy Phys.* **01**, 018, (2002). doi: 10.1088/1126-6708/2002/01/018.
779. G. Hanson *et al.*, Evidence for jet structure in hadron production by e^+e^- annihilation, *Phys. Rev. Lett.* **35**, 1609–1612 (1975). doi: 10.1103/PhysRevLett.35.1609. URL http://link.aps.org/doi/10.1103/PhysRevLett.35.1609.
780. D. P. Barber *et al.*, Discovery of three-jet events and a test of quantum chromodynamics at petra, *Phys. Rev. Lett.* **43**, 830–833 (1979). doi: 10.1103/PhysRevLett.43.830. URL http://link.aps.org/doi/10.1103/PhysRevLett.43.830.
781. C. Berger *et al.*, Evidence for gluon bremsstrahlung in e^+e^- annihilations at high energies, *Phys. Lett. B* **86** (3), 418–425, (1979). doi: http://dx.doi.org/10.1016/0370-2693(79)90869-4. URL http://www.sciencedirect.com/science/article/pii/0370269379908694.
782. R. Brandelik *et al.*, Evidence for planar events in e^+e^- annihilation at high energies, *Phys. Lett. B* **86** (2), 243–249, (1979). ISSN 0370-2693. doi: http://dx.doi.org/10.1016/0370-2693(79)90830-X. URL http://www.sciencedirect.com/science/article/pii/037026937990830X.
783. S. Kluth, P. A. Movilla Fernandez, S. Bethke, C. Pahl and P. Pfeifenschneider, A Measurement of the QCD color factors using event shape distributions at $\sqrt{s}$ = 14-GeV to 189-GeV, *Eur. Phys. J. C* **21**, 199–210, (2001). doi: 10.1007/s100520100742.
784. P. Abreu *et al.* (DELPHI collaboration), Tuning and test of fragmentation models based on identified particles and precision event shape data, *Z. Phys. C* **73**(1), 11–59, (1997). ISSN 1431-5858. doi: 10.1007/s002880050295. URL http://dx.doi.org/10.1007/s002880050295.
785. A. Banfi, G. P. Salam and G. Zanderighi, Phenomenology of event shapes at hadron colliders, *J. High Energy Phys.* **06**, 038, (2010). doi: 10.1007/JHEP06(2010)038.
786. T. Becher, X. Garcia i Tormo and J. Piclum, Next-to-next-to-leading logarithmic resummation for transverse thrust, *Phys. Rev. D* **93**, 054038, (2016). doi: 10.1103/PhysRevD.93.054038. URL https://link.aps.org/doi/10.1103/PhysRevD.93.054038.

787. G. Aad *et al.*, Measurement of event shapes at large momentum transfer with the ATLAS detector in pp collisions at $\sqrt{s} = 7$ TeV, *Eur. Phys. J. C* **72**, 2211, (2012). doi: 10.1140/epjc/s10052-012-2211-y.
788. G. Aad *et al.*, Measurement of event-shape observables in $Z \to \ell^+\ell^-$ events in pp collisions at $\sqrt{s} = 7$ TeV with the ATLAS detector at the LHC, *Eur. Phys. J. C* **76**, 375, (2016). doi: 10.1140/epjc/s10052-016-4176-8.
789. S. Chatrchyan *et al.*, Event shapes and azimuthal correlations in events in pp collisions at, *Phys. Lett. B* **722** (4–5), 238–261, (2013). doi: http://dx.doi.org/10.1016/j.physletb.2013.04.025. URL http://www.sciencedirect.com/science/article/pii/S0370269313003043.
790. V. Khachatryan *et al.*, Study of hadronic event-shape variables in multijet final states in pp collisions at 7 Tev, *J. High Energy Phys.* **10**, 87, (2014). doi: 10.1007/JHEP10(2014)087. URL http://dx.doi.org/10.1007/JHEP10(2014)087.
791. R. M. Chatterjee, M. Guchait and D. Sengupta, Probing supersymmetry using event shape variables at 8 TeV LHC, *Phys. Rev. D* **86**, 075014, (2012). doi: 10.1103/PhysRevD.86.075014.
792. A. Roy and M. Cavaglia, Discriminating supersymmetry and black holes at the large hadron collider, *Phys. Rev. D* **77**, 064029, (2008). doi: 10.1103/PhysRevD.77.064029.
793. C. Quigg, Looking into particle production at the large hadron collider, *Nuovo Cim. C* **033N5**, 327–342, (2010). doi: 10.1393/ncc/i2011-10734-0.
794. C. Albajar *et al.*, Analysis of the highest transverse energy events seen in the UA1 detector at the S $p\bar{p}$ S collider, *Z. Phys. C* **36**, 33, (1987). doi: 10.1007/BF01556162.
795. C. Loizides, Experimental overview on small collision systems at the LHC, *Nucl. Phys. A* **956**, 200–207, (2016). doi: 10.1016/j.nuclphysa.2016.04.022.
796. E. Cuautle, R. Jimenez, I. Maldonado, A. Ortiz, G. Paić and E. Perez. Disentangling the soft and hard components of the pp collisions using the sphero(i)city approach, (2014). arXiv:1404.2372.
797. A. Ortiz, G. Paić and E. Cuautle, Mid-rapidity charged hadron transverse spherocity in pp collisions simulated with Pythia, *Nucl. Phys. A* **941**, 78–86, (2015). doi: 10.1016/j.nuclphysa.2015.05.010.
798. B. Abelev *et al.*, Transverse sphericity of primary charged particles in minimum bias proton–proton collisions at $\sqrt{s} = 0.9$, 2.76 and 7 TeV, *Eur. Phys. J. C* **72**, 2124, (2012). doi: 10.1140/epjc/s10052-012-2124-9.
799. V. Khachatryan *et al.*, First Measurement of Hadronic Event Shapes in pp Collisions at $\sqrt{s} = 7$ TeV, *Phys. Lett.* **B699**, 48–67, (2011). doi: 10.1016/j.physletb.2011.03.060.
800. J. Alwall, R. Frederix, S. Frixione, V. Hirschi, F. Maltoni, O. Mattelaer, H. S. Shao, T. Stelzer, P. Torrielli and M. Zaro, The automated computation of tree-level and next-to-leading order differential cross sections and their matching to parton shower simulations, *J. High Energy Phys.* **07**, 079, (2014). doi: 10.1007/JHEP07(2014)079.
801. S. Chatrchyan *et al.*, Event shapes and azimuthal correlations in $Z+$jets events in pp collisions at $\sqrt{s} = 7$ TeV, *Phys. Lett. B* **722**, 238–261, (2013). doi: 10.1016/j.physletb.2013.04.025.
802. R. Bala, I. Bautista, J. Bielcikova and A. Ortiz, Heavy-ion physics at the LHC: Review of Run I results, *Int. J. Mod. Phys. E* **25** (07), 1642006, (2016). doi: 10.1142/S0218301316420064.
803. J. Adam *et al.*, Measurement of charged jet production cross sections and nuclear modification in p–Pb collisions at $\sqrt{s_{\rm NN}} = 5.02$ TeV, *Phys. Lett. B* **749**, 68–81, (2015). doi: 10.1016/j.physletb.2015.07.054.
804. K. Dusling *et al.*, Energy dependence of the ridge in high multiplicity proton–proton collisions, *Phys. Rev. D* **93** (1), 014034, (2016). doi: 10.1103/PhysRevD.93.014034.

805. B. Schenke *et al.*, Azimuthal anisotropies in p+Pb collisions from classical Yang–Mills dynamics, *Phys. Lett. B* **747**, 76–82, (2015). doi: 10.1016/j.physletb.2015.05.051.
806. K. Dusling *et al.*, Evidence for BFKL and saturation dynamics from dihadron spectra at the LHC, *Phys. Rev. D* **87** (5), 051502, (2013). doi: 10.1103/PhysRevD.87.051502.
807. K. Dusling *et al.*, Azimuthal collimation of long range rapidity correlations by strong color fields in high multiplicity hadron–hadron collisions, *Phys. Rev. Lett.* **108**, 262001, (2012). doi: 10.1103/PhysRevLett.108.262001.
808. A. Ortiz, Mean pT scaling with m/n_q at the LHC: Absence of (hydro) flow in small systems? *Nucl. Phys. A* **943**, 9–17, (2015). doi: 10.1016/j.nuclphysa.2015.08.003.
809. B. G. Zakharov, Flavor dependence of jet quenching in pp collisions and its effect on R_{AA} for heavy mesons, *JETP Lett.* **103** (6), 363–368, (2016). doi: 10.1134/S0021364016060126.
810. A. Ortiz, G. Bencedi and H. Bello, Revealing the source of the radial flow patterns in proton–proton collisions using hard probes, *J. Phys. G: Nucl. Part. Phys.* **44** (6), 065001, (2017). doi: 10.1088/1361-6471/aa6594.
811. M. Veldhoen, p/π Ratio in di-hadron correlations, *Nucl. Phys. A* **910–911**, 306–309, (2013). doi: 10.1016/j.nuclphysa.2012.12.103.
812. X. Zhang, K^0_S and Λ production in charged particle jets in p–Pb collisions at $\sqrt{s_{\mathrm{NN}}} =$ 5.02 TeV with ALICE, *Nucl. Phys. A* **931**, 444–448, (2014). doi: 10.1016/j.nuclphysa.2014.08.102.
813. E. Cuautle, A. Ortiz and G. Paić, Effects produced by multi-parton interactions and color reconnection in small systems, *Nucl. Phys. A* **956**, 749–752, (2016). doi: 10.1016/j.nuclphysa.2016.02.031.
814. I. Bautista, J. G. Milhano, C. Pajares and J. Dias de Deus, Multiplicity in pp and AA collisions: the same power law from energy-momentum constraints in string production, *Phys. Lett. B* **715**, 230–233, (2012). doi: 10.1016/j.physletb.2012.07.029.
815. C. Andres, A. Moscoso and C. Pajares, Universal geometrical scaling for hadronic interactions, *Nucl. Phys. A* **901**, 14–21, (2013). doi: 10.1016/j.nuclphysa.2013.02.013.
816. Y.-L. Yan *et al.*, Simultaneously study for particle transverse sphericity and ellipticity in pp collisions at LHC energies, *Nucl. Phys. A* **930**, 187–194, (2014). doi: 10.1016/j.nuclphysa.2014.07.025.
817. G. Aad *et al.*, Measurement of charged-particle event shape variables in $\sqrt{s} = 7$ TeV proton–proton interactions with the ATLAS detector, *Phys. Rev. D* **88** (3), 032004, (2013). doi: 10.1103/PhysRevD.88.032004.
818. V. Khachatryan *et al.*, Dependence of the Υ(nS) production ratios on charged particle multiplicity in pp collisions at $\sqrt{s} = 7$ TeV. Technical Report CMS-PAS-BPH-14-009, CERN, Geneva, (2016). URL https://cds.cern.ch/record/2223879.
819. S. Chatrchyan *et al.*, Observation of sequential Upsilon suppression in PbPb collisions, *Phys. Rev. Lett.* **109**, 222301, (2012). doi: 10.1103/PhysRevLett.109.222301.
820. A. H. Mueller, Soft gluons in the infinite momentum wave function and the BFKL pomeron, *Nucl. Phys. B* **415**, 373–385, (1994). doi: 10.1016/0550-3213(94)90116-3.
821. A. H. Mueller and B. Patel, Single and double BFKL pomeron exchange and a dipole picture of high-energy hard processes, *Nucl. Phys. B* **425**, 471–488, (1994). doi: 10.1016/0550-3213(94)90284-4.
822. A. H. Mueller, Unitarity and the BFKL pomeron, *Nucl. Phys. B* **437**, 107–126, (1995). doi: 10.1016/0550-3213(94)00480-3.
823. E. Avsar, G. Gustafson and L. Lönnblad, Energy conservation and saturation in small-x evolution, *J. High Energy Phys.* **07**, 062, (2005). doi: 10.1088/1126-6708/2005/07/062.

824. E. Avsar, G. Gustafson and L. Lönnblad, Difractive excitation in DIS and pp collisions, *J. High Energy Phys.* **12**, 012, (2007). doi: 10.1088/1126-6708/2007/12/012.
825. G. P. Salam, An introduction to leading and next-to-leading BFKL, *Acta Phys. Polon. B* **30**, 3679–3705, (1999).
826. A. H. Mueller and G. P. Salam, Large multiplicity fluctuations and saturation effects in onium collisions, *Nucl. Phys. B* **475**, 293–320, (1996). doi: 10.1016/0550-3213(96)00336-7.
827. J. Kwiecinski, A. D. Martin and P. J. Sutton, Constraints on gluon evolution at small x, *Z. Phys. C* **71**, 585–594, (1996). doi: 10.1007/BF02907019,10.1007/s002880050206.
828. I. Balitsky and G. A. Chirilli, NLO evolution of color dipole, *Acta Phys. Polon. B* **39**, 2561–2566, (2008).
829. G. P. Salam, A resummation of large subleading corrections at small x, *J. High Energy Phys.* **07**, 019, (1998). doi: 10.1088/1126-6708/1998/07/019.
830. M. Ciafaloni and D. Colferai, The BFKL equation at next-to-leading level and beyond, *Phys. Lett. B* **452**, 372–378, (1999). doi: 10.1016/S0370-2693(99)00281-6.
831. M. Ciafaloni, D. Colferai and G. P. Salam, A collinear model for small x physics, *J. High Energy Phys.* **10**, 017, (1999). doi: 10.1088/1126-6708/1999/10/017.
832. E. Avsar, On the high energy behaviour of the total cross section in the QCD dipole model, *J. High Energy Phys.* **04**, 033, (2008). doi: 10.1088/1126-6708/2008/04/033.
833. C. Flensburg, G. Gustafson and L. Lönnblad, Elastic and quasi-elastic pp and $\gamma^* p$ scattering in the dipole model, *Eur. Phys. J. C* **60**, 233–247, (2009). doi: 10.1140/epjc/s10052-009-0868-7.
834. J. R. Cudell, V. V. Ezhela, P. Gauron, K. Kang, Yu. V. Kuyanov, S. B. Lugovsky, E. Martynov, B. Nicolescu, E. A. Razuvaev and N. P. Tkachenko, Benchmarks for the forward observables at RHIC, the Tevatron Run II and the LHC, *Phys. Rev. Lett.* **89**, 201801, (2002). doi: 10.1103/PhysRevLett.89.201801.
835. E. Avsar and G. Gustafson, Geometric scaling and QCD dynamics in DIS, *J. High Energy Phys.* **04**, 067, (2007). doi: 10.1088/1126-6708/2007/04/067.
836. G. P. Salam, Soft emissions and the equivalence of BFKL and CCFM final states, *J. High Energy Phys.* **03**, 009, (1999). doi: 10.1088/1126-6708/1999/03/009.
837. G. Gustafson and G. Miu, Minijets and transverse energy flow in high-energy collisions, *Phys. Rev. D* **63**, 034004, (2001). doi: 10.1103/PhysRevD.63.034004.
838. L. Lönnblad, ARIADNE version 4: a program for simulation of QCD cascades implementing the color dipole model, *Comput. Phys. Commun.* **71**, 15–31, (1992). doi: 10.1016/0010-4655(92)90068-A.
839. F. Abe *et al.*, Measurement of $\bar{p}p$ single diffraction dissociation at $\sqrt{s} = 546$ GeV and 1800 GeV, *Phys. Rev. D* **50**, 5535–5549, (1994). doi: 10.1103/PhysRevD.50.5535.
840. G. Aad *et al.*, Rapidity gap cross sections measured with the ATLAS detector in pp collisions at $\sqrt{s} = 7$ TeV, *Eur. Phys. J. C* **72**, 1926, (2012). doi: 10.1140/epjc/s10052-012-1926-0.
841. V. Khachatryan *et al.*, Measurement of diffraction dissociation cross sections in pp collisions at $\sqrt{s} = 7\,\mathrm{TeV}$, *Phys. Rev. D* **92** (1), 012003, (2015). doi: 10.1103/PhysRevD.92.012003.
842. Y. Hatta, E. Iancu, C. Marquet, G. Soyez and D. N. Triantafyllopoulos, Diffusive scaling and the high-energy limit of deep inelastic scattering in QCD at large N(c), *Nucl. Phys. A* **773**, 95–155, (2006). doi: 10.1016/j.nuclphysa.2006.04.003.
843. G. Gustafson, The relation between the Good–Walker and triple-regge formalisms for diffractive excitation, *Phys. Lett. B* **718**, 1054–1057, (2013). doi: 10.1016/j.physletb.2012.11.061.

844. E. Gotsman, E. Levin and U. Maor, A comprehensive model of soft interactions in the LHC era, *Int. J. Mod. Phys. A* **30** (08), 1542005, (2015). doi: 10.1142/S0217751X15420051.
845. V. A. Khoze, A. D. Martin and M. G. Ryskin, Elastic scattering and diffractive dissociation in the light of LHC data, *Int. J. Mod. Phys. A* **30** (08), 1542004, (2015). doi: 10.1142/S0217751X1542004X.
846. C. Flensburg and G. Gustafson, Fluctuations, saturation and diffractive excitation in high energy collisions, *J. High Energy Phys.* **10**, 014, (2010). doi: 10.1007/JHEP10(2010)014.
847. C. Flensburg, G. Gustafson and L. Lönnblad, Exclusive final states in diffractive excitation, *J. High Energy Phys.* **12**, 115, (2012). doi: 10.1007/JHEP12(2012)115.
848. D. Bernard *et al.*, Pseudorapidity distribution of charged particles in diffraction dissociation events at the CERN SPS collider, *Phys. Lett. B* **166**, 459–462, (1986). doi: 10.1016/0370-2693(86)91598-4.
849. T. S. Biro, H. B. Nielsen and J. Knoll, Color rope model for extreme relativistic heavy ion collisions, *Nucl. Phys. B* **245**, 449–468, (1984). doi: 10.1016/0550-3213(84)90441-3.
850. A. Bialas and W. Czyz, Chromoelectric flux tubes and the transverse momentum distribution in high-energy nucleus–nucleus collisions, *Phys. Rev. D* **31**, 198, (1985). doi: 10.1103/PhysRevD.31.198.
851. A. K. Kerman, T. Matsui and B. Svetitsky, Particle production in the central rapidity region of ultrarelativistic nuclear collisions, *Phys. Rev. Lett.* **56**, 219, (1986). doi: 10.1103/PhysRevLett.56.219.
852. M. Gyulassy and A. Iwazaki, Quark and gluon pair production in SU(n) covariant constant fields, *Phys. Lett. B* **165**, 157–161, (1985). doi: 10.1016/0370-2693(85)90711-7.
853. B. Andersson and P. A. Henning, On the dynamics of a color rope: the fragmentation of interacting strings and the longitudinal distributions, *Nucl. Phys. B* **355**, 82–105, (1991). doi: 10.1016/0550-3213(91)90303-F.
854. M. Braun and C. Pajares, A probabilistic model of interacting strings, *Nucl. Phys. B* **390**, 542–558, (1993). doi: 10.1016/0550-3213(93)90467-4.
855. M. Braun and C. Pajares, Cross-sections and multiplicities in hadron nucleus collisions with interacting color strings, *Phys. Rev. D* **47**, 114–122, (1993). doi: 10.1103/PhysRevD.47.114.
856. N. S. Amelin, M. A. Braun and C. Pajares, String fusion and particle production at high-energies: Monte Carlo string fusion model, *Z. Phys. C* **63**, 507–516, (1994). doi: 10.1007/BF01580331.
857. N. Armesto, M. A. Braun, E. G. Ferreiro and C. Pajares, Strangeness enhancement and string fusion in nucleus–nucleus collisions, *Phys. Lett. B* **344**, 301–307, (1995). doi: 10.1016/0370-2693(94)01511-A.
858. B. Andersson, G. Gustafson and B. Soderberg, A general model for jet fragmentation, *Z. Phys. C* **20**, 317, (1983). doi: 10.1007/BF01407824.
859. B. Andersson and G. Gustafson, Semiclassical models for gluon jets and leptoproduction based on the massless relativistic string, *Z. Phys. C* **3**, 223, (1980). doi: 10.1007/BF01577421.
860. A. Casher, H. Neuberger and S. Nussinov, Chromoelectric flux tube model of particle production, *Phys. Rev. D* **20**, 179–188, (1979). doi: 10.1103/PhysRevD.20.179.
861. B. Andersson, G. Gustafson and T. Sjöstrand, A three-dimensional model for quark and gluon jets, *Z. Phys. C* **6**, 235, (1980). doi: 10.1007/BF01557774.

862. E. G. Gurvich, The quark anti-quark pair production mechanism in a quark jet, *Phys. Lett. B* **87**, 386–388, (1979). doi: 10.1016/0370-2693(79)90560-4.
863. N. K. Glendenning and T. Matsui, Creation of anti-Q Q pair in a chromoelectric flux tube, *Phys. Rev. D* **28**, 2890–2891, (1983). doi: 10.1103/PhysRevD.28.2890.
864. E. Brezin and C. Itzykson, Pair production in vacuum by an alternating field, *Phys. Rev. D* **2**, 1191–1199, (1970). doi: 10.1103/PhysRevD.2.1191.
865. J. S. Schwinger, On gauge invariance and vacuum polarization, *Phys. Rev.* **82**, 664–679, (1951). doi: 10.1103/PhysRev.82.664.
866. J. Adam *et al.*, Enhanced production of multi-strange hadrons in high-multiplicity proton-proton collisions, *Nature Phys.* **13**, 535–539 (2017). doi: 10.1038/nphys4111.
867. V. A. Abramovsky, E. V. Gedalin, E. G. Gurvich and O. V. Kancheli, Long range azimuthal correlations in multiple production processes at high-energies, *JETP Lett.* **47**, 337–339, (1988). [*Pisma Zh. Eksp. Teor. Fiz.* **47**, 281, (1988)].
868. I. Altsybeev, Mean transverse momenta correlations in hadron-hadron collisions in MC toy model with repulsing strings, *AIP Conf. Proc.* **1701**, 100002, (2016). doi: 10.1063/1.4938711.
869. R. J. Glauber, Cross-sections in deuterium at high-energies, *Phys. Rev.* **100**, 242–248, (1955). doi: 10.1103/PhysRev.100.242.
870. W. Broniowski, M. Rybczynski and P. Bozek, GLISSANDO: Glauber initial-state simulation and more, *Comput. Phys. Commun.* **180**, 69–83, (2009). doi: 10.1016/j.cpc.2008.07.016.
871. M. Rybczynski, G. Stefanek, W. Broniowski and P. Bozek, GLISSANDO 2 : GLauber Initial-State Simulation AND more, ver.2, *Comput. Phys. Commun.* **185**, 1759–1772, (2014). doi: 10.1016/j.cpc.2014.02.016.
872. H. De Vries, C. W. De Jager and C. De Vries, Nuclear charge-density-distribution parameters from elastic electron scattering, *Atom. Data Nucl. Data Tabl.* **36**, 495–536, (1987).
873. M. Rybczynski and W. Broniowski, Two-body nucleon–nucleon correlations in Glauber-like models, *Phys. Part. Nucl. Lett.* **8**, 992–994, (2011). doi: 10.1134/S1547477111090299.
874. G. Gustafson, L. Lönnblad, A. Ster and T. Csörgő, Total, inelastic and (quasi-)elastic cross sections of high energy pA and $\gamma^\star$A reactions with the dipole formalism, *J. High Energy Phys.* **10**, 022, (2015). doi: 10.1007/JHEP10(2015)022.
875. S. Baur, H. Wöhrmann, C. Baus, I. Katkov, R. Ulrich and M. Akbiyik, Inelastic cross section measurement in p–Pb collisions at 5.02 TeV with the CMS experiment, *PoS* **ICRC2015**, 440, (2016).
876. C. Bierlich, G. Gustafson and L. Lönnblad, Diffractive and non-diffractive wounded nucleons and final states in pA collisions, *J. High Energy Phys.* **10**, 139, (2016). doi: 10.1007/JHEP10(2016)139.
877. B. Blaettel, G. Baym, L. L. Frankfurt, H. Heiselberg and M. Strikman, Hadronic cross-section fluctuations, *Phys. Rev. D* **47**, 2761–2772, (1993). doi: 10.1103/PhysRevD.47.2761.
878. G. Aad *et al.*, Measurement of the centrality dependence of the charged-particle pseudorapidity distribution in protonlead collisions at $\sqrt{s_{\rm NN}} = 5.02\,\text{TeV}$ with the ATLAS detector, *Eur. Phys. J. C* **76** (4), 199, (2016). doi: 10.1140/epjc/s10052-016-4002-3.
879. E. Iancu and R. Venugopalan. The color glass condensate and high-energy scattering in QCD, arXiv:hep-ph/0303204, (2003).
880. L. D. McLerran and R. Venugopalan, Computing quark and gluon distribution functions for very large nuclei, *Phys. Rev. D* **49**, 2233–2241, (1994). doi: 10.1103/PhysRevD.49.2233.

881. Y. V. Kovchegov, Unitarization of the BFKL pomeron on a nucleus, *Phys. Rev. D* **61**, 074018 (2000). doi: 10.1103/PhysRevD.61.074018. URL http://link.aps.org/doi/10.1103/PhysRevD.61.074018.
882. J. Jalilian-Marian, A. Kovner, L. D. McLerran and H. Weigert, The intrinsic glue distribution at very small x, *Phys. Rev. D* **55**, 5414–5428, (1997). doi: 10.1103/PhysRevD.55.5414.
883. E. Ferreiro, E. Iancu, A. Leonidov and L. McLerran, Nonlinear gluon evolution in the color glass condensate. II, *Nucl. Phys. A* **703**, 489–538, (2002). doi: 10.1016/S0375-9474(01)01329-X.
884. T. Lappi and L. McLerran, Some features of the glasma, *Nucl. Phys. A* **772**, 200–212, (2006). doi: 10.1016/j.nuclphysa.2006.04.001.
885. A. Kovner, L. D. McLerran and H. Weigert, Gluon production from nonAbelian Weizsacker-Williams fields in nucleus–nucleus collisions, *Phys. Rev. D* **52**, 6231–6237, (1995). doi: 10.1103/PhysRevD.52.6231.
886. A. Kovner, L. D. McLerran and H. Weigert, Gluon production at high transverse momentum in the McLerran–Venugopalan model of nuclear structure functions, *Phys. Rev. D* **52**, 3809–3814, (1995). doi: 10.1103/PhysRevD.52.3809.
887. J. P. Blaizot, F. Gelis and R. Venugopalan, High energy pA collisions in the color glass condensate approach. I: Gluon production and the Cronin effect, *Nucl. Phys. A* **743**, 13–56, (2004). doi: 10.1016/j.nuclphysa.2004.07.005.
888. A. Krasnitz and R. Venugopalan, Non-perturbative computation of gluon mini-jet production in nuclear collisions at very high energies, *Nucl. Phys. B* **557**, 237, (1999).
889. A. Krasnitz and R. Venugopalan, The initial energy density of gluons produced in very high energy nuclear collisions, *Phys. Rev. Lett.* **84**, 4309–4312, (2000).
890. A. Krasnitz and R. Venugopalan, The initial gluon multiplicity in heavy ion collisions, *Phys. Rev. Lett.* **86**, 1717–1720, (2001). doi: 10.1103/PhysRevLett.86.1717.
891. A. Krasnitz, Y. Nara and R. Venugopalan, Coherent gluon production in very high energy heavy ion collisions, *Phys. Rev. Lett.* **87**, 192302, (2001).
892. T. Lappi, Production of gluons in the classical field model for heavy ion collisions, *Phys. Rev. C* **67**, 054903, (2003). doi: 10.1103/PhysRevC.67.054903.
893. B. Schenke, P. Tribedy and R. Venugopalan, Fluctuating Glasma initial conditions and flow in heavy ion collisions, *Phys. Rev. Lett.* **108**, 252301, (2012). doi: 10.1103/PhysRevLett.108.252301.
894. B. Schenke, P. Tribedy and R. Venugopalan, Event-by-event gluon multiplicity, energy density and eccentricities in ultrarelativistic heavy-ion collisions, *Phys. Rev. C* **86**, 034908, (2012). doi: 10.1103/PhysRevC.86.034908.
895. A. Dumitru, F. Gelis, L. McLerran and R. Venugopalan, Glasma flux tubes and the near side ridge phenomenon at RHIC, *Nucl. Phys. A* **810**, 91–108, (2008). doi: 10.1016/j.nuclphysa.2008.06.012.
896. K. Dusling, D. Fernandez-Fraile and R. Venugopalan, Three-particle correlation from glasma flux tubes, *Nucl. Phys. A* **828**, 161–177, (2009). doi: 10.1016/j.nuclphysa.2009.06.017.
897. F. Gelis, T. Lappi and L. McLerran, Glittering Glasmas, *Nucl. Phys. A* **828**, 149–160, (2009). doi: 10.1016/j.nuclphysa.2009.07.004.
898. P. Tribedy and R. Venugopalan, Saturation models of HERA DIS data and inclusive hadron distributions in p+p collisions at the LHC, *Nucl. Phys. A* **850**, 136–156, (2011). doi: 10.1016/j.nuclphysa.2011.04.008,10.1016/j.nuclphysa.2010.12.006. [Erratum: *Nucl. Phys. A* **859**, 185, (2011)].
899. P. Tribedy and R. Venugopalan, QCD saturation at the LHC: comparisons of models to p+p and A+A data and predictions for p+Pb collisions, *Phys. Lett. B* **710**,

125–133, (2012). doi: 10.1016/j.physletb.2012.02.047,10.1016/j.physletb.2012.12.004. [Erratum: *Phys. Lett. B* **718**, 1154, (2013)].

900. K. Dusling, F. Gelis, T. Lappi and R. Venugopalan, Long range two-particle rapidity correlations in A+A collisions from high energy QCD evolution, *Nucl. Phys. A* **836**, 159–182, (2010). doi: 10.1016/j.nuclphysa.2009.12.044.
901. A. Dumitru *et al.*, The ridge in proton–proton collisions at the LHC, *Phys. Lett. B* **697**, 21–25, (2011). doi: 10.1016/j.physletb.2011.01.024.
902. A. Kovner and M. Lublinsky, Angular correlations in gluon production at high energy, *Phys. Rev. D* **83**, 034017, (2011). doi: 10.1103/PhysRevD.83.034017.
903. A. Kovner and M. Lublinsky, On angular correlations and high energy evolution, *Phys. Rev. D* **84**, 094011, (2011). doi: 10.1103/PhysRevD.84.094011.
904. E. Levin and A. H. Rezaeian, The ridge from the BFKL evolution and beyond, *Phys. Rev. D* **84**, 034031, (2011). doi: 10.1103/PhysRevD.84.034031.
905. K. Dusling and R. Venugopalan, Explanation of systematics of CMS p+Pb high multiplicity di-hadron data at $\sqrt{s_{\rm NN}} = 5.02$ TeV, *Phys. Rev. D* **87** (5), 054014, (2013). doi: 10.1103/PhysRevD.87.054014.
906. Y. V. Kovchegov and D. E. Wertepny, Long-range rapidity correlations in heavy-light ion collisions, *Nucl. Phys. A* **906**, 50–83, (2013). doi: 10.1016/j.nuclphysa.2013.03.006.
907. K. Dusling and R. Venugopalan, Comparison of the color glass condensate to dihadron correlations in proton–proton and proton–nucleus collisions, *Phys. Rev. D* **87** (9), 094034, (2013). doi: 10.1103/PhysRevD.87.094034.
908. Y. V. Kovchegov and D. E. Wertepny, Two-gluon correlations in heavy-light ion collisions: energy and geometry dependence, IR divergences and k_T-factorization, *Nucl. Phys. A* **925**, 254–295, (2014). doi: 10.1016/j.nuclphysa.2014.02.021.
909. A. Dumitru and A. V. Giannini, Initial state angular asymmetries in high energy p+A collisions: spontaneous breaking of rotational symmetry by a color electric field and C-odd fluctuations, *Nucl. Phys. A* **933** (2015) 212. doi: 10.1016/j.nuclphysa.2014.10.037.
910. A. Dumitru and V. Skokov, Anisotropy of the semiclassical gluon field of a large nucleus at high energy, *Phys. Rev. D* **91** (7), 074006, (2015). doi: 10.1103/PhysRevD.91.074006.
911. A. Dumitru, L. McLerran and V. Skokov, Azimuthal asymmetries and the emergence of "collectivity" from multi-particle correlations in high-energy pA collisions, *Phys. Lett. B* **743**, 134–137, (2015). doi: 10.1016/j.physletb.2015.02.046.
912. M. Gyulassy, P. Levai, I. Vitev and T. Biro, Non-Abelian bremsstrahlung and azimuthal asymmetries in high energy p+A reactions, *Phys. Rev. D* **90**, 054025, (2014). doi: 10.1103/PhysRevD.90.054025.
913. T. Lappi, Azimuthal harmonics of color fields in a high energy nucleus, *Phys. Lett. B* **744**, 315–319, (2015). doi: 10.1016/j.physletb.2015.04.015.
914. T. Lappi, B. Schenke, S. Schlichting and R. Venugopalan, Tracing the origin of azimuthal gluon correlations in the color glass condensate, *J. High Energy Phys.* **01**, 061, (2016). doi: 10.1007/JHEP01(2016)061.
915. K. Dusling, M. Mace and R. Venugopalan, Multiparticle collectivity from initial state correlations in high energy proton–nucleus collisions, *Phys. Rev. Lett.* **120**(4), 042002 (2018). doi: 10.1103/PhysRevLett.120.042002.
916. B. Schenke, P. Tribedy and R. Venugopalan, Initial-state geometry and fluctuations in Au+Au, Cu+Au and U+U collisions at energies available at the BNL Relativistic Heavy Ion Collider, *Phys. Rev. C* **89** (6), 064908, (2014). doi: 10.1103/PhysRevC.89.064908.

917. A. H. Rezaeian, M. Siddikov, M. Van de Klundert and R. Venugopalan, Analysis of combined HERA data in the Impact-Parameter dependent Saturation model, *Phys. Rev. D* **87** (3), 034002, (2013). doi: 10.1103/PhysRevD.87.034002.
918. H. Mantysaari and B. Schenke, Evidence of strong proton shape fluctuations from incoherent diffraction, *Phys. Rev. Lett.* **117**(5), 052301 (2016). doi: 10.1103/PhysRevLett.117.052301.
919. H. Kowalski and D. Teaney, An impact parameter dipole saturation model, *Phys. Rev. D* **68**, 114005, (2003). doi: 10.1103/PhysRevD.68.114005.
920. L. McLerran and P. Tribedy, Intrinsic fluctuations of the proton saturation momentum scale in high multiplicity p+p collisions, *Nucl. Phys. A* **945**, 216–225, (2016). doi: 10.1016/j.nuclphysa.2015.10.008.
921. E. Iancu, A. H. Mueller and S. Munier, Universal behavior of QCD amplitudes at high energy from general tools of statistical physics, *Phys. Lett. B* **606**, 342–350, (2005). doi: 10.1016/j.physletb.2004.12.009.
922. E. Iancu and D. N. Triantafyllopoulos, A Langevin equation for high energy evolution with pomeron loops, *Nucl. Phys. A* **756**, 419–467, (2005). doi: 10.1016/j.nuclphysa.2005.03.124.
923. S. Munier and R. B. Peschanski, Traveling wave fronts and the transition to saturation, *Phys. Rev. D* **69**, 034008, (2004). doi: 10.1103/PhysRevD.69.034008.
924. S. Munier and R. B. Peschanski, Geometric scaling as traveling waves, *Phys. Rev. Lett.* **91**, 232001, (2003). doi: 10.1103/PhysRevLett.91.232001.
925. A. H. Mueller and S. Munier, On parton number fluctuations at various stages of the rapidity evolution, *Phys. Lett. B* **737**, 303–310, (2014). doi: 10.1016/j.physletb.2014.08.058.
926. C. Marquet, G. Soyez and B.-W. Xiao, On the probability distribution of the stochastic saturation scale in QCD, *Phys. Lett. B* **639**, 635–641, (2006). doi: 10.1016/j.physletb.2006.07.022.
927. A. Bzdak and K. Dusling, Probing proton fluctuations with asymmetric rapidity correlations, *Phys. Rev. C* **93** (3), 031901, (2016). doi: 10.1103/PhysRevC.93.031901.
928. H. Kowalski, L. Motyka and G. Watt, Exclusive diffractive processes at HERA within the dipole picture, *Phys. Rev. D* **74**, 074016, (2006). doi: 10.1103/PhysRevD.74.074016.
929. S. Schlichting and B. Schenke, The shape of the proton at high energies, *Phys. Lett. B* **739**, 313–319, (2014). doi: 10.1016/j.physletb.2014.10.068.
930. S. Schlichting and P. Tribedy, Collectivity in small collision systems: an initial-state perspective, *Adv. High Energy Phys.* **2016**, 8460349, (2016). doi: 10.1155/2016/8460349.
931. R. D. Field and R. P. Feynman, A parametrization of the properties of quark jets, *Nucl. Phys. B* **136**, 1, (1978). doi: 10.1016/0550-3213(78)90015-9.
932. T. Sjöstrand, High-energy physics event generation with Pythia 5.7 and JETSET 7.4, *Comput. Phys. Commun.* **82**, 74–90, (1994). doi: 10.1016/0010-4655(94)90132-5.
933. T. Sjöstrand. Colour reconnection and its effects on precise measurements at the LHC. (2013). URL http://inspirehep.net/record/1262718/files/arXiv:1310.8073.pdf.
934. B. Schenke, P. Tribedy and R. Venugopalan, Multiplicity distributions in p+p, p+A and A+A collisions from Yang–Mills dynamics, *Phys. Rev. C* **89** (2), 024901, (2014). doi: 10.1103/PhysRevC.89.024901.
935. L. Bianchi, Strangeness production as a function of charged particle multiplicity in proton–proton collisions, *Nucl. Phys. A*, **956**, 777–780, (2016). doi: 10.1016/j.nuclphysa.2016.03.005.

936. T. Kalaydzhyan and E. Shuryak, Collective flow in high-multiplicity proton–proton collisions, *Phys. Rev. C* **91** (5), 054913, (2015). doi: 10.1103/PhysRevC.91.054913.
937. P. Levai and B. Muller, Transverse baryon flow as possible evidence for a quark-gluon plasma phase, *Phys. Rev. Lett.* **67**, 1519–1522, (1991). doi: 10.1103/PhysRevLett.67.1519.
938. V. Khachatryan *et al.*, Measurement of long-range near-side two-particle angular correlations in pp collisions at $\sqrt{s} = 13\,\text{TeV}$, *Phys. Rev. Lett.* **116** (17), 172302, (2016). doi: 10.1103/PhysRevLett.116.172302.
939. J. Adams *et al.*, Distributions of charged hadrons associated with high transverse momentum particles in pp and Au + Au collisions at $\sqrt{s_{NN}} = 200$-GeV, *Phys. Rev. Lett.* **95**, 152301, (2005). doi: 10.1103/PhysRevLett.95.152301.
940. A. Adare *et al.*, System size and energy dependence of jet-induced hadron pair correlation shapes in Cu+Cu and Au+Au collisions at $\sqrt{s_{NN}} = 200$ and 62.4-GeV, *Phys. Rev. Lett.* **98**, 232302, (2007). doi: 10.1103/PhysRevLett.98.232302.
941. B. Alver *et al.*, System size dependence of cluster properties from two-particle angular correlations in Cu+Cu and Au+Au collisions at $\sqrt{s_{NN}} = 200$-GeV, *Phys. Rev. C* **81**, 024904, (2010). doi: 10.1103/PhysRevC.81.024904.
942. P. Bozek and W. Broniowski, Collective dynamics in high-energy proton–nucleus collisions, *Phys. Rev. C* **88** (1), 014903, (2013). doi: 10.1103/PhysRevC.88.014903.
943. A. Bzdak, B. Schenke, P. Tribedy and R. Venugopalan, Initial state geometry and the role of hydrodynamics in proton–proton, proton–nucleus and deuteron–nucleus collisions, *Phys. Rev. C* **87** (6), 064906, (2013). doi: 10.1103/PhysRevC.87.064906.
944. G.-Y. Qin and B. Müller, Elliptic and triangular flow anisotropy in deuteron-gold collisions at $\sqrt{s_{NN}} = 200$ GeV at RHIC and in proton–lead collisions at $\sqrt{s_{NN}} = 5.02$ TeV at the LHC, *Phys. Rev. C* **89** (4), 044902, (2014). doi: 10.1103/PhysRevC.89.044902.
945. J. L. Nagle, A. Adare, S. Beckman, T. Koblesky, J. Orjuela Koop, D. McGlinchey, P. Romatschke, J. Carlson, J. E. Lynn and M. McCumber, Exploiting intrinsic triangular geometry in relativistic He3+Au collisions to disentangle medium properties, *Phys. Rev. Lett.* **113** (11), 112301, (2014). doi: 10.1103/PhysRevLett.113.112301.
946. B. Schenke and R. Venugopalan, Eccentric protons? Sensitivity of flow to system size and shape in p+p, p+Pb and Pb+Pb collisions, *Phys. Rev. Lett.* **113**, 102301, (2014). doi: 10.1103/PhysRevLett.113.102301.
947. A. Bzdak and G.-L. Ma, Elliptic and triangular flow in p+Pb and peripheral Pb+Pb collisions from parton scatterings, *Phys. Rev. Lett.* **113** (25), 252301, (2014). doi: 10.1103/PhysRevLett.113.252301.
948. P. Ghosh, S. Muhuri, J. K. Nayak and R. Varma, Indication of transverse radial flow in high-multiplicity proton–proton collisions at the large hadron collider, *J. Phys. G* **41**, 035106, (2014). doi: 10.1088/0954-3899/41/3/035106.
949. H. Mäntysaari, B. Schenke, C. Shen and P. Tribedy, Imprints of fluctuating proton shapes on flow in proton–lead collisions at the LHC, *Phys. Lett. B* **772**, 681–686, (2017). doi: 10.1016/j.physletb.2017.07.038.
950. S. Kar, S. Choudhury, S. Muhuri and P. Ghosh, Multiple parton interactions and production of charged particles up to the intermediate-p_T range in high-multiplicity *pp* events at the LHC, *Phys. Rev. D* **95** (1), 014016, (2017).
951. CMS Collaboration, Azimuthal anisotropy harmonics from long-range correlations in high multiplicity pp collisions at $\sqrt{s} = 7\,\text{TeV}$, (2015). http://cds.cern.ch/record/2056802.
952. T. Regge, Introduction to complex orbital momenta, *Nuovo Cim.* **14**, 951, (1959). doi: 10.1007/BF02728177.

953. V. N. Gribov, Partial waves with complex orbital angular momenta and the asymptotic behavior of the scattering amplitude, *Sov. Phys. JETP.* **14**, 1395, (1962). [*Zh. Eksp. Teor. Fiz.* **41**, 1962, (1961)].
954. K. Werner, An event generator for e^+e^- annihilation, neutrino p, anti-neutrino p, μp and soft pp scattering, *Phys. Lett. B* **197**, 225–231, (1987). doi: 10.1016/0370-2693(87)90372-8.
955. H. J. Drescher, M. Hladik, S. Ostapchenko, T. Pierog and K. Werner, Parton based Gribov–Regge theory, *Phys. Rept.* **350**, 93–289, (2001). doi: 10.1016/S0370-1573(00)00122-8.
956. K. Werner, Core-corona separation in ultra-relativistic heavy ion collisions, *Phys. Rev. Lett.* **98**, 152301, (2007). doi: 10.1103/PhysRevLett.98.152301.
957. K. Werner, I. Karpenko, T. Pierog, M. Bleicher and K. Mikhailov, Event-by-event simulation of the three-dimensional hydrodynamic evolution from flux tube initial conditions in ultrarelativistic heavy ion collisions, *Phys. Rev. C* **82**, 044904, (2010). doi: 10.1103/PhysRevC.82.044904.
958. L. D. McLerran and R. Venugopalan, Green's functions in the color field of a large nucleus, *Phys. Rev. D* **50**, 2225–2233, (1994). doi: 10.1103/PhysRevD.50.2225.
959. I. Karpenko, P. Huovinen and M. Bleicher, A 3+1 dimensional viscous hydrodynamic code for relativistic heavy ion collisions, *Comput. Phys. Commun.* **185**, 3016–3027, (2014). doi: 10.1016/j.cpc.2014.07.010.
960. M. Bleicher *et al.*, Relativistic hadron–hadron collisions in the ultrarelativistic quantum molecular dynamics model, *J. Phys. G* **25**, 1859–1896, (1999). doi: 10.1088/0954-3899/25/9/308.
961. H. Petersen, J. Steinheimer, G. Burau, M. Bleicher and H. Stocker, A fully integrated transport approach to heavy ion reactions with an intermediate hydrodynamic stage, *Phys. Rev. C* **78**, 044901, (2008). doi: 10.1103/PhysRevC.78.044901.
962. B. B. Abelev *et al.*, K^0_S and Λ production in Pb–Pb collisions at $\sqrt{s_{NN}} = 2.76$ TeV, *Phys. Rev. Lett.* **111**, 222301, (2013). doi: 10.1103/PhysRevLett.111.222301.
963. A. Andronic, P. Braun-Munzinger, K. Redlich and J. Stachel, Hadron yields, the chemical freeze-out and the QCD phase diagram, *J. Phys. Conf. Ser.* **779** (1), 012012, (2017). doi: 10.1088/1742-6596/779/1/012012.
964. G. Gustafson, The importance of fluctuations, *Acta Phys. Polon. B* **42**, 2571–2580, (2011). doi: 10.5506/APhysPolB.42.2571.